Lecture Notes in Mathematics

Edited by A. Dold and B. Eckmann

1179

Shi Jian-Yi

The Kazhdan-Lusztig Cells in Certain Affine Weyl Groups

Springer-Verlag
Berlin Heidelberg New York Tokyo

Author

SHI Jian-Yi
Department of Mathematics, East China Normal University
Shanghai, The People's Republic of China

Mathematics Subject Classification (1980): 20B27, 20H15

ISBN 3-540-16439-1 Springer-Verlag Berlin Heidelberg New York Tokyo
ISBN 0-387-16439-1 Springer-Verlag New York Heidelberg Berlin Tokyo

Printing and binding: Beltz Offsetdruck, Hemsbach/Bergstr.
2146/3140-543210

INTRODUCTION

Cells of a Coxeter group play an important role not only in the representation theory of the Coxeter group and its associated Hecke algebra, but also in the representation theory of algebraic groups, finite groups of Lie type, Lie algebras and enveloping algebras.

The concept of cells originally came from combinatorial theory. Robinson [R1] defined a map from the symmetric group S_n to the set of pairs (P,Q) of standard Young tableaux of the same shape and of size n. Then Schensted [Sch] proved that this map is bijective. Hence a left cell of S_n corresponds to the set of such pairs (P,Q) with P fixed. A two-sided cell of S_n corresponds to the set of such pairs (P,Q) with P,Q of fixed shape. So there is a 1-1 correspondence between the set of two-sided cells of S_n and the set of partitions of n. This is the prototype of cells which applies to any Coxeter group.

Then Joseph [Jo1] defined the concept of left cells in the Weyl group W in terms of primitive ideals in the enveloping algebra of a complex semisimple Lie algebra. For $w \in W$, let J_w be the annihilator of the irreducible module of the enveloping algebra with highest weight $-w\rho-\rho$, where ρ is half the sum of positive roots. Then w , w' are said to be in the same left cell precisely when $J_w = J_{w'}$. Joseph's definition of left cells and the corresponding Weyl group representations involves some deep results about the multiplicities of the composition factors of the Verma modules with highest weight $-w\rho-\rho$.

In 1979, Kazhdan and Lusztig [KL1] gave the definition of cells for an arbitrary Coxeter group. Their definition is elementary but it gives rise not only to representations of the Coxeter group, but also of the corresponding Hecke algebra. This makes possible applications of the results on cells to more general representation theory. On the other hand, the definition of Kazhdan-Lusztig cells coincides with that of Joseph, in the case of Weyl groups.

In this book, we adopt Kazhdan-Lusztig's definition of cells.

The first three chapters of this book are expository in nature and aim to give an account of the Kazhdan-Lusztig theory and its applications. The subsequent chapters form an original contribution to the theory.

In Chapter 1 we define Coxeter groups and Hecke algebras. We also define the decomposition of a Coxeter group into left, right or two-sided cells. We illustrate these definitions with a number of examples and explain how the cells give rise to representations of the Coxeter groups and the Hecke algebras.

The object of Chapter 2 is to discuss the further results about Coxeter groups and Hecke algebras as well as their representations, and to describe the most important applications of the Kazhdan-Lusztig polynomials.

We introduce in §2.1 left ϕ-cells of the Weyl group $W(B_n)$ of type B_n defined by Lusztig [L8], where ϕ is a map from $W(B_n)$ to the infinite cyclic group. Left ϕ-cells are a generalization of Kazhdan-Lusztig left cells. By proper choice of a map ϕ, we can get left ϕ-cells of $W(B_n)$ each of which affords an irreducible representation of $W(B_n)$.

Let H_K be the Hecke algebra over a ring K associated to a Weyl group W. Tits [Bou] showed that $H_{\mathbb{C}} \cong \mathbb{C}W$ but did not give any explicit isomorphism from $H_{\mathbb{C}}$ to $\mathbb{C}W$. Iwahori [I] conjectured that $H_{\mathbb{Q}} \cong \mathbb{Q}W$. But this conjecture is false in case W of type E_7 or E_8. We state with proof in §2.2 a result of Lusztig [L4] which shows that $H_{\mathbb{Q}[X^{\frac{1}{2}}]} \cong \mathbb{Q}[X^{\frac{1}{2}}]W$ and which gives an explicit isomorphism from $H_{\mathbb{Q}[X^{\frac{1}{2}}]}$ to $\mathbb{Q}[X^{\frac{1}{2}}]W$, where $\mathbb{Q}[X^{\frac{1}{2}}]$ is the ring of polynomials in an indeterminate $X^{\frac{1}{2}}$ with rational coefficients.

Let W be an indecomposable Weyl group and $\hat{W}$ the set of irreducible representations of W. Barbasch and Vogan [BV1], [BV2] divided $\hat{W}$ into families and showed that two irreducible representations of W lie in the same family if and only if they occur as components of the representation of W afforded by some two-sided cell of W. Lusztig [L10] also defined cell representations of W (In his book, Lusztig called them constructible representations) which are just the representations afforded by left cells of W. We state without proof all these results of Lusztig in §2.3.

Then we state the Kazhdan-Lusztig conjecture for composition factors of Verma modules in §2.4 [KL1] and the Lusztig conjecture for composition factors of Weyl modules in §2.6 [L2]. Both conjectures involve Kazhdan-Lusztig polynomials. The former was proved to be true by Brylinski and Kashiwara [BK] and also by Bernstein and Beilinson [BB] in 1981 independently. The latter still remains open.

In §2.5, we discuss without proof the classification of the primitive ideals in the enveloping algebra of a semisimple Lie algebra. The readers may find here the original definition of cells given by Joseph [Jo2]. Some results on the primitive ideals are very powerful for the decomposition of a Weyl group into cells.

We state without proof in §2.7 a formula, due to Lusztig [L5], giving a relation between a Kazhdan-Lusztig polynomial and a Foulkes polynomial in the case where W is an affine Weyl group of type $\tilde{A}$.

Then in Chapter 3 we give a brief description of a geometric interpretation of the Kazhdan-Lusztig polynomials of a Weyl group in terms of the intersection cohomology of a Schubert variety. We also give related geometric descriptions of the polynomials for their Coxeter groups. All of the results in Chapter 3 are stated without proof but with references.

The subsequent chapters are our own work. Here we concentrate on those infinite Coxeter groups which occur as affine Weyl groups of type $\tilde{A}$. We study these groups in detail and determine their decomposition into left, right and two-sided cells.

Let A_n be the affine Weyl group of type $\tilde{A}_{n-1}$. Then A_n can be described not only as a Coxeter group, but also as a subgroup of the permutation group on $\mathbb{Z}$ or a group of affine matrices of period n. The precise descriptions of the last two will be given in Chapter 4. As a set, A_n can also be regarded as the set of alcoves of V, where V is a subspace of the Euclidean space $\mathbb{R}^n$ of dimension n-1. We describe this situation in Chapter 6. The second and the third descriptions lead us to define a map σ from A_n to the set Λ_n of partitions of n by a procedure due to Lusztig [L9] explained in Chapter 5. The last description leads us to consider in Chapter 7 the geometrical structure of cells of A_n and to define a set S of equivalence classes of A_n called admissible sign types.

Our main results show that for any partition $\lambda = \{\lambda_1 \geq \cdots \geq \lambda_r\}$ of n the fibre $\sigma^{-1}(\lambda)$ is a two-sided cell. This was conjectured by Lusztig [L9] and is proved in Theorem 17.6. $\sigma^{-1}(\lambda)$ is also an RL-equivalence class of A_n as is shown in Theorem 15.1. $\sigma^{-1}(\lambda)$ consists of $\frac{n!}{\prod_{j=1}^{m} \mu_j!}$ left (resp. right) cells of A_n, where $\mu = \{\mu_1 \geq \cdots \geq \mu_m\}$ is the dual partition of λ. This was also conjectured by Lusztig [L9] and is proved in Theorem 14.4.5. We show in Theorem 18.2.1 that $\sigma^{-1}(\lambda)$ is also a connected set (for the definitions of a connected set as well as a maximal left connected component below, see §7.2). We also show in this theorem that each left cell in $\sigma^{-1}(\lambda)$ is a maximal left connected component of $\sigma^{-1}(\lambda)$ which, as shown in Propositions 9.3.7, 10.4 and 14.4.3, can be characterized by a combinatorial object called a λ-tabloid introduced in §4.4(xiii). Each left cell is also characterized by its generalized right τ-invariant in Theorem 16.1.2. Any element of S is in some left cell of A_n, as a consequence of Proposition 18.2.2. The cardinality of S is $(n+1)^{n-1}$ (Theorem 7.3.1). Any non-identity left cell of A_n is an infinite set of elements of A_n (Proposition 19.1.5). Let P be any proper standard parabolic subgroup of A_n isomorphic to the symmetric group S_n. Thus the intersection of P with any two-sided cell of A_n is non-empty and is just a two-sided cell of P (Theorem 16.2.8). The first part of this result was conjectured by Lusztig [L2]. The intersection of P with any left (resp. right) cell of A_n is either empty or a left (resp. right) cell of P (Theorem 16.2.9).

In Chapter 19 we define a map $\hat{T}$ from A_n to a set $\hat{C}$ of combinatorial objects called proper tabloids of rank n. We show that $\hat{T}$ induces a bijection between the set of left cells of A_n and the set $\hat{C}$.

Chapters 4 to 7 contain most of the basic definitions, terminology and some elementary properties of the affine Weyl group A_n for later use. Then in Chapters 8 to 18 we get most of our main results on the cells of A_n. Finally, in the remainder of this book, we discuss some further properties of cells of A_n. In particular, we explain how the map $\hat{T}: A_n \to \hat{C}$ referred to above generalizes the Robinson-Schensted map from the symmetric group to the affine Weyl group of type $\tilde{A}$.

In obtaining these results, we find it most convenient to express the elements of A_n as affine matrices of period n. This has the advantage of preserving all the symmetry inherent in the group A_n.

Most of our results are obtained by using combinatorial methods applied to such affine matrices, except at one point where we must show that any RL-equivalence class of A_n is just a two-sided cell of A_n (see Chapter 17). In order to settle this point we must use a recent result of Lusztig coming from the deep theory of ℓ-adic intersection cohomology.

This book is partly based on my Ph.D. thesis. The greater part of it was written during my period of study for a doctoral degree at the University of Warwick, except for the first three chapters which were written at East China Normal University in the Spring of 1985.

First and foremost, I wish to thank Professor R.W. Carter, who has been a constant source of encouragement and ideas, who has helped me over my difficulties not only in mathematics but also in English, and who made very helpful, detailed comments and suggestions on my manuscript. Further, I am indebted to Professor G. Lusztig who permitted me to use his recent unpublished results on the cells of the affine Weyl groups, by which the results of my book are made more satisfactory. I am very grateful to a number of referees for helpful comments and useful suggestions. I am also grateful to Terri Moss who has so patiently and cheerfully done the difficult job of typing.

I wish also to acknowledge the financial support of the Chinese government and the Committee of Vice-Chancellors and Principals of the universities of the United Kingdom.

Finally, I would like to thank my first supervisor, Professor S.H. Tsao at East China Normal University who brought me into the mathematical field in 1978.

East China Normal University

July 1985

J.Y. Shi

CONTENTS

CHAPTER 1 : COXETER GROUPS, HECKE ALGEBRAS AND THEIR REPRESENTATIONS

We start with some general discussion of Coxeter groups and Hecke algebras. Then we shall define left, right and two-sided cells of a Coxeter group. We then prove some results on cells due to Kazhdan and Lusztig [KL1]. Finally, a number of examples are given to illustrate these definitions and to explain how the cells give rise to representations of the Coxeter groups and the Hecke algebras.

§1.1 COXETER GROUPS AND WEYL GROUPS

A group W is called a Coxeter group if it has a presentation

$$W = (W,S) = \langle s \in S \mid s^2 = 1,\ (st)^{m_{st}} = 1 \text{ for all } s \neq t \text{ in } S\rangle$$

with m_{st} positive integers greater than 1, or $m_{st} = \infty$.

The Coxeter graph Γ associated to a given Coxeter group (W,S) is, by definition, a set of vertices labelled by elements of S with a set Y of edges (an edge is a subset of S consisting of two elements) such that for $s,t \in S$, $\{s,t\} \in Y$ if and only if $m_{st} \geq 3$. Each edge $\{s,t\} \in Y$ is labelled by the number m_{st} whenever $m_{st} > 3$.

A Coxeter group is entirely determined, up to isomorphism, by its associated Coxeter graph.

A Coxeter group (W,S) is defined to be indecomposable if its Coxeter graph is connected, i.e. for any $s,t \in S$, there exists a sequence $s_0 = s, s_1, \ldots, s_r = t$ in S for some $r \geq 0$ so that $\{s_{i-1}, s_i\}$ is an edge for every i, $1 \leq i \leq r$.

<u>Theorem 1.1.1</u> <u>If (W,S) is an indecomposable Coxeter group of finite order, then its Coxeter graph has one of the following forms.</u>

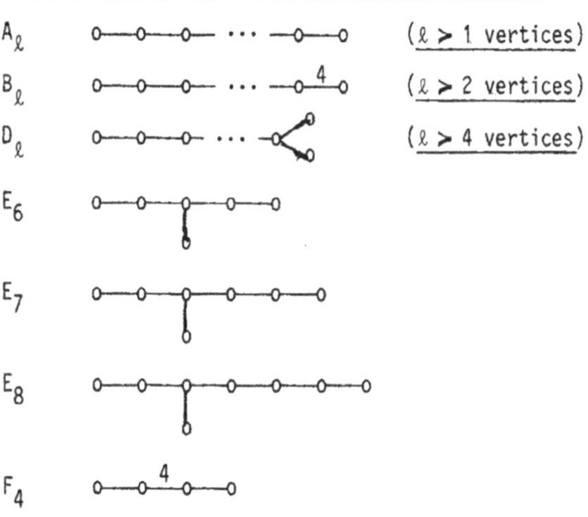

G_2 o—6—o

H_3 o—5—o——o

H_4 o—5—o——o——o

$I_2(p)$ o—p—o (p = 5 or p > 7)

These Coxeter groups are pairwise non-isomorphic. □

The above theorem gives a classification of all indecomposable Coxeter groups of finite order. A proof of this theorem can be found, for example, in [Bou].

Example 1.1.2 The symmetric group S_n, $n > 1$, is a Coxeter group. Let $s_i = (i,i+1)$ be the permutation which transposes i and $i+1$ for any i, $1 < i < n$. Then S_n has a presentation

$$S_n = \langle s_i \mid s_i^2 = 1,\ (s_is_j)^{m_{ij}} = 1 \text{ for } 1 < i\ ,\ j < n,\ i \neq j\rangle$$

with

$$m_{ij} = \begin{cases} 2 & \text{if } j \neq i \pm 1 \\ 3 & \text{if } j = i \pm 1 \end{cases}$$

The Coxeter graph associated to S_n is

s_1 o——o s_2 ——o s_3 —— ... ——o s_{n-2} ——o s_{n-1}

Assume that $W = \langle s_1,\ldots,s_\ell \mid s_i^2 = 1,\ (s_is_j)^{m_{ij}} = 1 \text{ for } 1 < i,\ j < \ell,\ i \neq j\rangle$ is a finite Coxeter group. Then W can be described as a group generated by reflections in a finite dimensional Euclidean space.

Let V be a vector space over the real field $\mathbb{R}$ with basis $\{\alpha_1,\alpha_2,\ldots,\alpha_\ell\}$. We define a bilinear form on V, $(v,v') \to \langle v,v'\rangle$ by $\langle\alpha_i,\alpha_j\rangle = -\cos \pi/m_{ij}$ and extended by linearity. This form is symmetric since $m_{ij} = m_{ji}$. In particular we have $\langle\alpha_i,\alpha_i\rangle = 1$ for each i. Let H_i be the subspace of V given by $H_i = \{v \in V; \langle\alpha_i, v\rangle = 0\}$. Then $\dim H_i = \ell-1$ and we have $V = \mathbb{R}\,\alpha_i \oplus H_i$.

We now define a linear map $\tau_i : V \to V$ by $\tau_i(v) = v - 2\langle\alpha_i,v\rangle\alpha_i$. Then $\tau_i(\alpha_i) = -\alpha_i$ and $\tau_i(v) = v$ whenever $v \in H_i$. Thus τ_i is the reflection in the hyperplane H_i. Thus $\tau_i^2 = 1$. It is also true that $(\tau_i\tau_j)^{m_{ij}} = 1$ if $i \neq j$. Thus there exists a homomorphism $\theta : W \to \langle\tau_1,\ldots,\tau_\ell\rangle$ from W into the group generated by the τ_i given by $\theta(s_i) = \tau_i$. Since $\langle\tau v,\tau v'\rangle = \langle v,v'\rangle$ for all $v,v' \in V$ this homomorphism θ gives a representation of W as a group of isometries of V.

The form $\langle v,v'\rangle$ on V can be shown to be non-singular and positive definite and so V may be regarded as a Euclidean space. It can also be shown that W acts faithfully on V [Bou]. Thus any finite Coxeter group can be described as a reflection group in a finite dimensional Euclidean space.

For any element w of a Coxeter group (W,S), we define $\ell(w)$ to be the smallest number r such that $w = s_1 s_2 \dots s_r$ with all s_i in S. $\ell(w)$ is called the length of w. The expression $w = s_1 s_2 \dots s_r$ is called a reduced form of w if $r = \ell(w)$ and $s_i \in S$, $1 \leq i \leq r$.

A Coxeter group is called crystallographic if, for all $i \neq j$, $m_{ij} \in \{2,3,4,6\}$. Thus the indecomposable Coxeter groups which are crystallographic are of type A_ℓ, B_ℓ, D_ℓ, E_6, E_7, E_8, F_4, G_2.

Let $\underline{g}$ be a simple Lie algebra of finite dimension over the complex field $\mathbb{C}$. Then there is a finite crystallographic Coxeter group W associated to $\underline{g}$, called the Weyl group of $\underline{g}$. For each such finite indecomposable crystallographic Coxeter group W there is just one simple Lie algebra which has W as its Weyl group except when W has type B_ℓ, $\ell \geq 3$, when there are two such Lie algebras, called B_ℓ and C_ℓ, [Hu1].

Any such simple Lie algebra has a Cartan decomposition $\underline{g} = \underline{h} \oplus \sum_{\alpha\in\Phi} \underline{g}_\alpha$ where $\underline{h}$ is a maximal commutative subalgebra called a Cartan subalgebra of $\underline{g}$, and each $\underline{g}_\alpha$ is a 1-dimensional $\underline{h}$-submodule of $\underline{g}$. The set Φ of 1-dimensional representations of $\underline{h}$ arising in this way is called the set of roots of $\underline{g}$. Φ has a subset Π, called a set of simple roots, such that each root in Φ is uniquely expressible as a linear combination of elements of Π with coefficients in $\mathbb{Z}$ which are either all non-negative or all non-positive. The set Φ decomposes as $\Phi = \Phi^+ \cup \Phi^-$ in this way where Φ^+, Φ^- are the positive and negative roots respectively. If $\alpha \in \Phi^+$ and $\alpha = n_1\alpha_1 + \dots + n_\ell\alpha_\ell$ where $\Pi = \{\alpha_1,\dots,\alpha_\ell\}$ then the positive integer $n_1 + \dots + n_\ell$ is called the height of α. There is exactly one root of maximal height, called the highest root in Φ.

The real subspace $V = h^*_{\mathbb{R}}$ of the space of linear functions h^* on h generated by the roots may be regarded as a Euclidean space with respect to the Killing form of $\underline{g}$. The Weyl group W of $\underline{g}$ is generated as a Coxeter group by the reflections $s_1,\dots,s_\ell$ in the hyperplanes of V orthogonal to the simple roots $\alpha_1,\dots,\alpha_\ell$.

The finite crystallographic Coxeter groups will from now on be called Weyl groups.

We conclude this section by introducing a family of infinite Coxeter groups called affine Weyl groups which we are particularly interested in. There is one affine Weyl group W_a for each simple Lie algebra over $\mathbb{C}$. It acts on the vector space $V = h^*_{\mathbb{R}}$ as a group generated by reflections $s_0,s_1,\dots,s_\ell$. $s_1,\dots,s_\ell$ act on V as above, as the reflections in the hyperplanes orthogonal to the simple roots. s_0 acts on V as the reflection in an affine hyperplane of V which does not pass through the origin. This affine hyperplane is parallel to the hyperplane orthogonal to the highest root. (Which particular parallel translate is taken does not affect the structure of W_a up to isomorphism). The reflections $s_0,s_1,\ldots,s_\ell$ generate W_a as a Coxeter group. If W is a Coxeter group of type A_ℓ then W_a is called a Coxeter group of type $\tilde{A}_\ell$, and similarly for the other types. The following result describes the Coxeter graphs of the affine Weyl groups.

<u>Theorem 1.1.3</u> <u>Let (W,S) be an affine Weyl group. Then its Coxeter graph has one of the following forms</u>

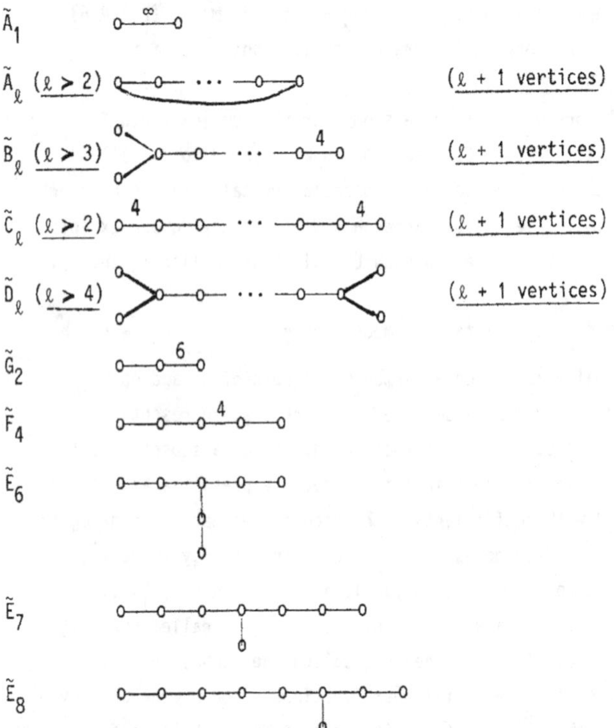

<u>These Coxeter groups are pairwise non-isomorphic.</u>

The reflecting hyperplanes fixed by the Coxeter generators $s_0,s_1,\ldots,s_\ell$ of W_a enclose a closed bounded region $\overline{A_1}$ which is a fundamental region for the action of W_a on V. Its interior A_1 is called the fundamental alcove. We define $A_w = (A_1)w$ for all $w \in W_a$. The subsets A_w are called the alcoves of W_a. The alcoves are the connected components of the complement in V of the union of the fixed hyperplanes of all reflections in W_a. The affine Weyl group W_a operates simply transitively on the set of alcoves. For a detailed description of an affine Weyl group in terms of alcoves, we refer to [L3], [Verm].

§1.2 HECKE ALGEBRAS

Hecke algebras arise naturally in the theory of Chevalley groups over finite fields. Let $\underline{g}$ be a simple Lie algebra of finite dimension over $\mathbb{C}$, with Cartan decomposition $\underline{g} = \underline{h} \oplus \sum_{\alpha\in\Phi} g_\alpha$. $\underline{g}$ has a special kind of basis, called a Chevalley basis, whose elements are $h_1,\dots,h_\ell$, e_α, $\alpha \in \Phi$, where $h_1,\dots,h_\ell \in \underline{h}$ and $e_\alpha \in \underline{g}_\alpha$. For each $x \in \underline{g}$, we define $\mathrm{ad}x:\underline{g} \to \underline{g}$ by $\mathrm{ad}x.y = [x,y]$. For each $\alpha \in \Phi$ and $\lambda \in \mathbb{C}$ the map $\mathrm{ad}(\lambda e_\alpha):\underline{g} \to \underline{g}$ is nilpotent. Thus we can form $\exp \mathrm{ad}(\lambda e_\alpha):\underline{g} \to \underline{g}$ which is an automorphism of $\underline{g}$. We write $x_\alpha(\lambda) = \exp \mathrm{ad}(\lambda e_\alpha)$. The Chevalley basis has the property that the matrix of each $x_\alpha(\lambda)$ with respect to this basis has entries which are polynomials in λ with coefficients in $\mathbb{Z}$. Moreover the Lie product of any two elements in the basis is a linear combination of basis elements with coefficients in $\mathbb{Z}$.

Now given any field k we can define a Lie algebra $\underline{g}_k$ over k by taking all k-combinations of elements in a Chevalley basis and taking Lie multiplication as before, interpreting the integers as elements of k. For each $\alpha \in \Phi$ and $\lambda \in k$ we can define an automorphism $x_\alpha(\lambda)$ of $\underline{g}_k$ by taking its matrix with respect to the Chevalley basis as before. The group of automorphisms of $\underline{g}_k$ generated by $x_\alpha(\lambda)$ for all $\alpha \in \Phi$, $\lambda \in k$ is called the adjoint Chevalley group of type $\underline{g}$ over k. In particular, if k is the finite field $\mathbb{F}_q$, we obtain a finite Chevalley group which will be denoted by $G(q)$ [Ca1].

For example, if $\underline{g}$ is the simple Lie algebra $s\ell_n(\mathbb{C})$ of all $n \times n$ matrices of trace o over $\mathbb{C}$, $G(q)$ will be the group $PSL_n(q)$ of all $n \times n$ matrices of determinant 1, factored by its centre.

The set Φ decomposes as $\Phi = \Phi^+ \cup \Phi^-$ where Φ^+, Φ^- are the positive and negative roots respectively. Let $U(q)$ be the subgroup of $G(q)$ generated by $x_\alpha(\lambda)$ for all $\alpha \in \Phi^+$ and all $\lambda \in \mathbb{F}_q$. Let $B(q) = N_{G(q)}(U(q))$. For example, when $G(q) = PSL_n(q)$, $U(q)$ is the subgroup of all upper unitriangular matrices and $B(q)$ is the subgroup of all upper triangular matrices.

Let $1_{B(q)}$ be the unit character of $B(q)$ and let $1_{B(q)}^{G(q)}$ be the induced character of $G(q)$. Let $H_{\mathbb{C}}(q)$ be the algebra of endomorphisms of the $\mathbb{C}G(q)$-module affording the character $1_{B(q)}^{G(q)}$. $H_{\mathbb{C}}(q)$ is called the Hecke algebra. Its dimension is given by $\dim H_{\mathbb{C}}(q) = |W|$, where W is the Weyl group of $\underline{g}$. It has a basis T_w, $w \in W$ such that, if $w = s_{i_1} \dots s_{i_k}$ is a reduced expression of w, then $T_w = T_{s_{i_1}} \dots T_{s_{i_k}}$. Moreover each basis element T_{s_i} satisfies the quadratic relation

$$T_{s_i}^s = q.1 + (q-1)T_{s_i} \quad (T_1 = 1 \text{ is the identity of } H_{\mathbb{C}}(q)).$$

Let W be defined as a Coxeter group by

$$W = \langle s_1,\ldots,s_\ell \mid s_i^2 = 1,\ \underbrace{s_is_j\ldots}_{m_{ij}} = \underbrace{s_js_i\ldots}_{m_{ij}}\rangle$$

Then $H_{\mathbb{C}}(q)$ is defined as an algebra by

$$H_{\mathbb{C}}(q) = \langle T_{s_1},\ldots,T_{s_\ell} \mid T_{s_i}^2 = q.1 + (q-1)T_{s_i},\ \underbrace{T_{s_i}T_{s_j}\ldots}_{m_{ij}} = \underbrace{T_{s_j}T_{s_i}\ldots}_{m_{ij}}\rangle$$

Now there is a bijective correspondence between the irreducible components of the induced representation $1_{B(q)}^{G(q)}$ and the irreducible representations of its endomorphism algebra $H_{\mathbb{C}}(q)$. However it was shown by Tits [Bou] that $H_{\mathbb{C}}(q)$ is isomorphic to the group algebra $\mathbb{C}W$ of the Weyl group. Thus irreducible components of $1_{B(q)}^{G(q)}$ correspond to irreducible characters of the Weyl group.

For example, when $G(q) = PSL_n(q)$ we have $W = S_n$ and so the irreducible components of $1_{B(q)}^{G(q)}$ in this case correspond to the irreducible characters of the symmetric group S_n.

Now the product of any two basis elements T_w of $H_{\mathbb{C}}(q)$ is a $\mathbb{Z}$-combination of basis elements. Thus we have a subring $H_{\mathbb{Z}}(q)$ of all $\mathbb{Z}$-combinations of the T_w. However in this subring the elements T_w are not invertible, since $T_s^{-1} = q^{-1}T_s + (q^{-1}-1)1$ for $s \in S$. We therefore extend the base ring to include q^{-1}. In addition there is in general no way of writing down an explicit isomorphism between $H_{\mathbb{C}}(q)$ and $\mathbb{C}W$ without introducing a square root of q. (c.f. Lusztig [L4]). For this reason we consider the Hecke algebra over the base ring $\mathbb{Z}[q^{\frac{1}{2}}, q^{-\frac{1}{2}}]$. More generally, we can define the generic Hecke algebra $H_A(X)$ for any Coxeter group (W,S) by generators and relations as before, where X is an indeterminate over $\mathbb{Z}$ and $A = \mathbb{Z}[X^{\frac{1}{2}}, X^{-\frac{1}{2}}]$. We shall denote this generic Hecke algebra over A by H.

§1.3 W-GRAPHS [KL1]

Assume that H is the generic Hecke algebra associated to a given Coxeter group (W,S). We shall construct a representation of H endowed with a special basis.

Definition 1.3.1 A W-graph is, by definition, a set of vertices Z, with a set Y of edges (each edge consists of two elements of Z) together with two additional data: for each vertex $x \in Z$, we are given a subset I_x of S and, for each ordered pair of vertices y, x such that $\{y,x\} \in Y$, we are given an integer $\mu(y,x) \neq 0$. These data are subject to the following requirements: Let E be the free A-module with basis Z. Then for any $s \in S$,

$$\tau_s(x) = \begin{cases} -x & \text{if } s \in I_x \\ Xx + X^{\frac{1}{2}} \sum\limits_{\substack{y \in Z \\ s \in I_y \\ \{y,x\} \in Y}} \mu(y,x)y & \text{if } s \notin I_x \end{cases} \qquad (1.3.1a)$$

defines an endomorphism of E (i.e. the sum over y is assumed to be always finite) and there is a unique representation $\phi: H \to \mathrm{End}(E)$ such that $\phi(T_s) = \tau_s$ for each $s \in S$.

Example 1.3.2 Let (W,S) be a Weyl group of type B_2 with $S = \{s_1,s_2\}$. Let Γ be the graph with vertex set $Z = \{x_1,x_2,x_3\}$ and edge set $Y = \{\{x_1,x_2\}, \{x_2,x_3\}\}$ such that $I_{x_1} = I_{x_3} = \{s_1\}$, $I_{x_2} = \{s_2\}$, and $\mu(x_1,x_2) = \mu(x_2,x_1) = \mu(x_2,x_3) = \mu(x_3,x_2) = 1$. Then Γ is a W-graph. In fact, let E be a free A-module with basis Z. Then there exists a unique representation $\phi: H \to \mathrm{End}(E)$ so that the matrices $\phi(T_{s_1})$, $\phi(T_{s_2})$ on the basis x_1, x_2, x_3 are $\begin{pmatrix} -1 & X^{\frac{1}{2}} & 0 \\ 0 & X & 0 \\ 0 & X^{\frac{1}{2}} & -1 \end{pmatrix}$ $\begin{pmatrix} X & 0 & 0 \\ X^{\frac{1}{2}} & -1 & X^{\frac{1}{2}} \\ 0 & 0 & X \end{pmatrix}$ respectively.

§1.4 KAZHDAN-LUSZTIG POLYNOMIALS [KL1]

In the remainder of this chapter, we shall construct some W-graphs for any Coxeter group. To do that, we shall first introduce Kazhdan-Lusztig polynomials.

Fix a Coxeter group (W,S). Let $a \to \bar{a}$ be the involution of the ring $A = \mathbb{Z}[X^{\frac{1}{2}}, X^{-\frac{1}{2}}]$ defined by $\overline{X^{\frac{1}{2}}} = X^{-\frac{1}{2}}$. This extends to an involution $h \to \bar{h}$ of the ring H, defined by $\overline{\Sigma a_w T_w} = \Sigma \bar{a}_w T^{-1}_{w^{-1}}$. (Note that T_w is an invertible element of H for any $w \in W$. For, if $s \in S$, we have $T_s^{-1} = X^{-1}T_s + (X^{-1}-1) \in H$. Assume that $w \in W$ has a reduced form $w = s_1s_2\ldots s_r$ with $s_i \in S$. Then $T_w^{-1} = T_{s_r}^{-1}T_{s_{r-1}}^{-1} \ldots T_{s_1}^{-1}$ is also in H). It can be shown that for h, $h_1,h_2 \in H$ and $\alpha \in A$ we have $\overline{h_1 + h_2} = \bar{h}_1 + \bar{h}_2$, $\overline{h_1h_2} = \bar{h}_1\bar{h}_2$ and $\overline{\alpha h} = \bar{\alpha}\bar{h}$. For any $w \in W$, we define $X_w = X^{\ell(w)}$, $\varepsilon_w = (-1)^{\ell(w)}$. Let $<$ be the Bruhat order relation on W. i.e. for $y,w \in W$, we say $y < w$ if there exist reduced forms $y = s_{i_1}s_{i_2}\ldots s_{i_t}$ and $w = s_1s_2\ldots s_r$ with all s lying in S such that $i_1,i_2,\ldots,i_t$ is a subsequence of $1,2,\ldots,r$. We can now state

Theorem 1.4.1 For any $w \in W$, there is a unique element $C_w \in H$ such that

$$\overline{C_w} = C_w \qquad (1.4.1a)$$

$$C_w = \sum_{y \leq w} \varepsilon_y \varepsilon_w X_w^{\frac{1}{2}} X_y^{-1} \overline{P_{y,w}}\, T_y \qquad (1.4.1b)$$

where $P_{y,w} \in A$ is a polynomial in X of degree less than or equal to $\frac{1}{2}(\ell(w)-\ell(y)-1)$ for $y < w$, and $P_{w,w} = 1$.

The polynomials $P_{y,w}$ in Theorem 1.4.1 are called Kazhdan-Lusztig polynomials.

To prove Theorem 1.4.1, we need to define for each $x,y \in W$ an element $R_{x,y} \in A$ which satisfies

$$T^{-1}_{y^{-1}} = \sum_x \overline{R_{x,y}}\, X_x^{-1}\, T_x \qquad (1.4.1c)$$

The following formulae provide an inductive procedure for computing $R_{x,y}$

$$R_{x,y} = \begin{cases} R_{sx,sy} & \text{if } sx < x \text{ and } sy < y \\ R_{xs,ys} & \text{if } xs < x \text{ and } ys < y \end{cases} \qquad (1.4.1d)$$

$$R_{x,y} = (X-1)R_{sx,y} + XR_{sx,sy} \quad \text{if } sx > x \text{ and } sy < y. \qquad (1.4.1e)$$

For, by the definition of the Hecke algebra and (1.4.1c), we have that $R_{x,y} \neq 0$ if and only if $x \leqslant y$. When $x \leqslant y$, we can compute $R_{x,y}$ by induction on $m = 2\ell(y) - \ell(x) \geqslant 0$. $m = 0$ implies $x = y = 1$. It is obvious that $R_{1,1} = 0$ by (1.4.1c). Now assume $m > 0$. Then $\ell(y) > 0$ and there exists some $s \in S$ satisfying $sy < y$. If $sx > x$ then $2\ell(sy) - \ell(sx)$ and $2\ell(y) - \ell(sx)$ are both less than m. So by inductive hypothesis we know $R_{sx,y}$ and $R_{sx,sy}$ and hence by (1.4.1e) we get $R_{x,y}$ as well. If $sx < x$ then $R_{sx,sy} = R_{x,y}$ by (1.4.1d). Since $2\ell(sy) - \ell(sx) < m$, we can get $R_{sx,sy}$ by inductive hypothesis and so we get $R_{x,y}$ also.

From (1.4.1d) and (1.4.1e), we see that when $x \leqslant y$, $R_{x,y}$ is a polynomial in X of degree $\ell(y) - \ell(x)$. The following lemma gives some further properties of $R_{x,y}$.

<u>Lemma 1.4.2</u> (i) $\overline{R_{x,y}} = \varepsilon_x\varepsilon_y X_x X_y^{-1} R_{x,y}$

(ii) $\sum_{x \leqslant t \leqslant y} \varepsilon_t\varepsilon_x R_{x,t} R_{t,y} = \delta_{x,y}$, for all $x \leqslant y$ in W.

(iii) $R_{x,y} = (X-1)^{\ell(y)-\ell(x)}$ for all $x \leqslant y$ such that $\ell(x) \geqslant \ell(y)-2$

(iv) If W is finite and w_0 is its longest element, we have $R_{w_0y,w_0x} = R_{x,y}$ for all $x,y \in W$.

<u>Proof</u> (i) follows from (1.4.1d), (1.4.1e) by applying induction on $2\ell(y) - \ell(x)$ (we may assume $x \leqslant y$ since otherwise the result is obvious). Apply the involution $h \to \bar{h}$ to (1.4.1c), we get

$$T_y = \sum_{x\leq y} R_{x,y} X_x T^{-1}_{x^{-1}} = \sum_{x\leq y} R_{x,y} X_x \sum_{z\leq x} \overline{R_{z,x}} X_z^{-1} T_z$$

$$= \sum_{z\leq x\leq y} R_{x,y} X_x X_z^{-1} (\varepsilon_z \varepsilon_x X_z X_x^{-1} R_{z,x}) T_z \qquad \text{by (i)}$$

$$= \sum_{z\leq x\leq y} \varepsilon_z \varepsilon_x R_{z,x} R_{x,y} T_z.$$

Since $\{T_w | w \in W\}$ forms a A-basis of H, we get (ii).

The formula (iii) is obvious for $x = y$. Now suppose $x < y$. If $\ell(x) = \ell(y) - 1$ then there is a reduced expression $y = s_1 \ldots s_i \ldots s_n$ such that $x = s_1 \ldots \hat{s}_i \ldots s_n$. By (1.4.1d), we may assume $1 = i = n$, but then (iii) is obvious. If $\ell(x) = \ell(y)-2$ then there is a reduced expression $y = s_1 \ldots s_i \ldots s_j \ldots s_n$ such that $x = s_1 \ldots \hat{s}_i \ldots \hat{s}_j \ldots s_n$. By (1.4.1d), we may assume $i = 1$, $j = n$. Applying (1.4.1e) with $s = s_i$, we have

$$R_{x,y} = (X-1) R_{s_1 \ldots \hat{s}_n, s_1 \ldots s_n} + X\, R_{s_1 \ldots \hat{s}_n, \hat{s}_1 \ldots s_n}$$

But $R_{s_1 \ldots \hat{s}_n, s_1 \ldots s_n} = X-1$ as we have shown, and $R_{s_1 \ldots \hat{s}_n, \hat{s}_1 \ldots s_n} = 0$ by the fact that $s_1 \ldots \hat{s}_n \not< \hat{s}_1 \ldots s_n$. So we get (iii).

Now we consider (iv). We know that $x < y$ if and only if $w_0 y < w_0 x$. We may assume $x < y$ since otherwise the result is obvious. Note that for any $x \in W$ and $s \in X$, we have $xs < x$ if and only if $w_0 xs > w_0 x$. Also note that there is the analogue of (1.4.1e):

$$R_{x,y} = (X-1) R_{xs,y} + X R_{xs,ys}, \text{ if } xs > x \text{ and } ys < y. \qquad (1.4.1f)$$

We apply induction on $m = 2\ell(y) - \ell(x)$ to prove (iv). When $m = 0$, we have $x = y = 1$ and $R_{1,1} = R_{w_0,w_0} = 1$ by (iii). Assume now $m > 0$. Then $\ell(y) > 0$ and there exists some $s \in S$ so that $ys < y$. If $xs < x$ then by (1.4.1d) and inductive hypothesis $R_{x,y} = R_{xs,ys} = R_{w_0 ys, w_0 xs} = R_{w_0 y, w_0 x}$. If $xs > x$ then we have

$$\begin{aligned} R_{x,y} &= (X-1) R_{xs,y} + X R_{xs,ys} && \text{by (1.4.1f)} \\ &= (X-1) R_{w_0 y, w_0 xs} + X R_{w_0 ys, w_0 xs} && \text{by inductive hypothesis} \\ &= (X-1) R_{w_0 ys, w_0 x} + X R_{w_0 ys, w_0 xs} && \text{by (1.4.1d)} \\ &= R_{w_0 y, w_0 x} && \text{by (1.4.1f)} \end{aligned}$$

So (iv) is proved. □

Definition 1.4.3 Given $y,w \in W$, we say that $y \prec w$ if the following conditions are satisfied: $y < w$, $\varepsilon_w = -\varepsilon_y$ and $P_{y,w}$ is a polynomial in X of degree exactly $\frac{1}{2}(\ell(w)-\ell(y)-1)$. In this case, the leading coefficient of $P_{y,w}$ is denoted by $\mu(y,w)$. It is a non-zero integer. If $w \prec y$, we set $\mu(w,y) = \mu(y,w)$.

Now we are ready to show Theorem 1.4.1.

1.4.4 Proof of Theorem 1.4.1

Uniqueness The polynomials $P_{y,w}$, $y < w$ in W, must satisfy the following system of equations.

$$\begin{cases} P_{w,w} = 1 & (1.4.4a) \\ X_w \overline{P_{x,w}} - X_x P_{x,w} = \sum_{x<y\prec w} \varepsilon_x \varepsilon_y X_y P_{y,w} \overline{R_{x,y}} \quad x < w & (1.4.4b) \end{cases}$$

It is enough to show that $P_{x,w}$ is uniquely determined for $x < w$ in W. Suppose that we have determined all $P_{y,w}$ with $x < y \prec w$. Then the right side of (1.4.4b) is determined. Since $P_{x,w}$ is a polynomial in X of degree less than or equal to $\frac{1}{2}(\ell(w)-\ell(x)-1)$, we see that the degree of any term of $X_w \overline{P_{x,w}}$ is different from those of $X_x P_{x,w}$. So by (1.4.4b), $P_{x,w}$ is uniquely determined.

Existence We have $C_1 = T_1$. Now assume that $w \neq 1$ and that we have constructed all required C_x with $\ell(x) < \ell(w)$. Thus we can define the relation $x \prec y$ and the integer $\mu(x,y)$ for $x,y \in W$ with $\ell(x)$, $\ell(y)$ both less than $\ell(w)$. There exists some $s \in S$ such that $w = sv$ and $\ell(w) = \ell(v)+1$. Let

$$C_w = (X^{-\frac{1}{2}}T_s - X^{\frac{1}{2}})C_v - \sum_{\substack{z \prec v \\ sz<z}} \mu(z,v)C_z \qquad (1.4.4c)$$

Then C_w satisfies (1.4.1a) since $\overline{X^{-\frac{1}{2}}T_s - X^{\frac{1}{2}}} = X^{-\frac{1}{2}}T_s - X^{\frac{1}{2}}$.

By substituting (1.4.1b) for C_v, C_z in (1.4.4c), we get

$$C_w = \sum_{y \preceq w} \varepsilon_y \varepsilon_w X_w^{\frac{1}{2}} X_y^{-1} \overline{P_{y,w}}\, T_y$$

with

$$P_{y,w} = X^{1-c}P_{sy,v} + X^c P_{y,v} - \sum_{\substack{y \preceq z \prec v \\ sz<z}} \mu(z,v) X_z^{-\frac{1}{2}} X_v^{\frac{1}{2}} X^{\frac{1}{2}} P_{y,z} \quad y \preceq w \qquad (1.4.4d)$$

where $c = 1$ if $sy < y$, $c = 0$ if $sy > y$. (assume $P_{x,v} = 0$ when $x \not\leq v$).

(1.4.4d) shows that $P_{y,w}$ is a polynomial in X of degree less than or equal to $\frac{1}{2}(\ell(w)-\ell(y)-1)$ if $y < w$ and that $P_{w,w} = 1$. So C_w satisfies (1.4.1b) and Theorem 1.4.1 is proved. □

Now we can show the following properties of Kazhdan-Lusztig polynomials.

Lemma 1.4.5 (i) For each $x < y$ in W, $P_{x,y}$ is a polynomial in X with constant term 1.

(ii) For each $y < w$ with $\ell(w) = \ell(y) + 1$, we have $P_{y,w} = 1$. In particular, we have $y \prec w$ and $\mu(y,w) = 1$.

(iii) For each $y < w$ with $\ell(w) = \ell(y) + 2$, we have $P_{y,w} = 1$.

(iv) For each $w \in W$, we have $X_w^{-1} \sum_{y \leq w} X_y P_{y,w} = \overline{\sum_{y \leq w} X_y P_{y,w}}$

(v) $P_{y,w} = P_{sy,sw}$ if $y \not\leq sw$ with $s \in S$ and $sw < w$.

(iv) $P_{y,w} = P_{sy,w}$ if $y < w$, $sw < w$ with $s \in S$.

(vii) If W is finite and w_0 is its longest element, then

(a) $P_{y,w_0} = 1$ for all $y \in W$.

(b) $\sum_{x \leq z \leq y} \varepsilon_x \varepsilon_z P_{x,z} P_{w_0 y, w_0 z} = \delta_{x,y}$ for all $x \leq y$ in W. (1.4.5a)

(viii) $P_{y,w} = P_{y^{-1},w^{-1}}$ for any $w,y \in W$.

Proof (i) and (v) follow immediately from the inductive formula (1.4.4d). (ii) follows from (1.4.4d) by induction on $\ell(w)$. Under the assumptions of (iii), we see from (1.4.4d) that

$$P_{y,w} = \begin{cases} P_{sy,v} & \text{if } sy < y \quad (1.4.5b) \\ P_{y,v} & \text{if } sy > y \quad (1.4.5c) \end{cases}$$

where $s \in S$ satisfies $w = sv$ and $\ell(w) = \ell(v)+1$. In case (1.4.5b), we can use induction on $\ell(w)$; in case (1.4.5c), we use (ii). In both cases we have $P_{y,w} = 1$, hence (iii). The identity (iv) is just the identity $\phi(C_w) = \phi(\bar{C}_w)$, where $\phi: H \to A$ is the algebra homomorphism defined by $\phi(T_y) = \varepsilon_y$ for all y. (vi) follows from (1.4.4d) by noting $\{z \in W | sz < z,\ y \leq z < v\} = \{z \in W | sz < z, sy \leq z \prec v\}$ and by induction on $\ell(w)$. (vii)(a) follows by applying repeatedly (vi). For (vii)(b), let $M_{x,y}$ be the left hand side of (1.4.5a). We may assume that $x < y$ and that $M_{t,s} = 0$ for all $t < s$ such that $\ell(s)-\ell(t) < \ell(y)-\ell(x)$. The following equation can be deduced easily from (1.4.4b).

$$P_{x,z} = \sum_{x<t<z} \varepsilon_x\varepsilon_t R_{x,t} \overline{P_{t,z}} \; X_t^{-1} X_z \quad (x < z \text{ in } W).$$

It follows that

$$M_{x,y} = \sum_{x<z<y} \varepsilon_x\varepsilon_z \sum_{\substack{x<t<z \\ z<s<y}} \varepsilon_x\varepsilon_t\varepsilon_y\varepsilon_s R_{x,t}\overline{P_{t,z}} R_{w_0y,w_0s} \overline{P_{w_0s,w_0z}} X_t^{-1}X_zX_z^{-1}X_s$$

$$= \sum_{\substack{t,s \\ x<t<s<y}} \varepsilon_y\varepsilon_s X_t^{-1} X_s R_{x,t} R_{w_0y,w_0s} \overline{M_{t,s}}$$

The only t,s which can contribute to this sum satisfy t = s or t = x, s = y. Thus

$$M_{x,y} = X_x^{-1}X_y \overline{M_{x,y}} + \sum_{x<t<y} \varepsilon_y\varepsilon_t R_{x,t} R_{w_0y,w_0t}.$$

Using Lemma 1.4.2(iv) and (ii), we see that the last sum (over t) is equal to $\sum_{x<t<y} \varepsilon_y\varepsilon_t R_{x,t}R_{t,y} = 0$.

Thus $M_{x,y} = X_x^{-1}X_y \overline{M_{x,y}}$ hence $X_x^{\frac{1}{2}} X_y^{-\frac{1}{2}} M_{x,y} = X_x^{-\frac{1}{2}} X_y^{\frac{1}{2}} \overline{M_{x,y}}$. The bounds on the degree of polynomials $P_{y,w}$ described in Theorem 1.4.1 imply that $X_x^{-\frac{1}{2}} X_y^{\frac{1}{2}} \overline{M_{x,y}}$ is a polynomial in $X^{\frac{1}{2}}$ without constant term. Hence it cannot be fixed by the involution $a \to \bar{a}$, unless it is zero. Thus $M_{x,y} = 0$, as required. This proves (vii)(b). Finally, (viii) follows directly from the definition of Kazhdan-Lusztig polynomials in Theorem 1.4.1. □

The following results are on the relation $\prec$.

Lemma 1.4.6. (i) Let $x,y \in W$, $s \in S$ be such that $x < y$, $sy < y$, $sx > x$. Then $x \prec y$ if and only if $y = sx$. Moreover, this implies that $\mu(x,y) = 1$.

(ii) Let $x,y \in W$, $s \in S$ be such that $x < y$, $ys < y$, $xs > x$. Then $x \prec y$ if and only if $y = xs$. Moreover, this implies that $\mu(x,y) = 1$.

(iii) Let $x < y$ be two elements of W (assumed to be finite). The following conditions are equivalent, $x \prec y$ and $w_0y \prec w_0x$. If these conditions are satisfied, we have $\mu(x,y) = \mu(w_0y,w_0x)$.

Proof If $sx \neq y$, it follows from Lemma 1.4.5(vi) that $\deg P_{x,y} = \deg P_{sx,y} < \frac{1}{2}(\ell(y) - \ell(x)-2) < \frac{1}{2}(\ell(y) - \ell(x)-1)$ hence the relation $x \prec y$ is not satisfied. If $sx = y$, it follows from Lemma 1.4.5(ii) that $P_{x,y} = 1$, hence $x \prec y$ and $\mu(x,y) = 1$. This proves (i). The proof of (ii) is entirely similar. For (iii), we can assume that $\varepsilon_x = -\varepsilon_y$. The difference

$P_{x,y} - P_{w_0y,w_0x}$ is equal to $\sum_{x<z<y} \varepsilon_x\varepsilon_z P_{x,z} P_{w_0y,w_0z}$ by Lemma 1.4.5(vii)(b) and one checks easily that the last expression is a polynomial in X of degree less than $\frac{1}{2}(\ell(y) - \ell(x)-1)$. Therefore, the $\frac{1}{2}(\ell(y)-\ell(x)-1)$-th power of X appears in $P_{x,y}$ with the same coefficients as in P_{w_0y,w_0x}. □

§1.5 CELLS OF A COXETER GROUP [KL1]

We keep all the notations in §1.4. We shall define, following Kazhdan and Lusztig [KL1], cells of any Coxeter group (W,S). Let L be the graph whose vertices are the elements of W and whose edges are the subsets of W of the form {y,w} with $y \prec w$. For each $w \in W$, let $I_w = \mathcal{L}(w) = \{s \in S | sw < w\}$.

The following result shows that we have constructed a W-graph.

Theorem 1.5.1 L, together with the assignment $w \to I_w$ and with the function μ defined in 1.4.3, is a W-graph.

Proof By (1.4.4c), we have

$$T_sC_v = XC_v + X^{\frac{1}{2}}C_{sv} + X^{\frac{1}{2}} \sum_{\substack{z\prec v\\ sz<z}} \mu(z,v)C_z, \text{ if } s \in S \text{ and } sv > v \qquad (1.5.1a)$$

We now show that

$$T_sC_v = -C_v \text{ if } s \in S \text{ and } sv < v \qquad (1.5.1b)$$

We may assume that (1.5.1b) has been shown for all v' with $sv' < v'$ and $\ell(v') < \ell(v)$. By replacing v by sv in (1.5.1a), we have

$$\begin{aligned} T_sC_v &= T_s(X^{-\frac{1}{2}}T_sC_{sv} - X^{\frac{1}{2}}C_{sv} - \sum_{\substack{z\prec sv\\ sz<z}} \mu(z,sv)C_z) \\ &= X^{-\frac{1}{2}}((X-1)T_s+X)C_{sv} - X^{\frac{1}{2}}T_sC_{sv} + \sum_{\substack{z\prec sv\\ sz<z}} \mu(z,sv)C_z \\ &= X^{\frac{1}{2}}C_{sv}-X^{-\frac{1}{2}}T_sC_{sv} + \sum_{\substack{z\ sv\\ sz<z}} \mu(z,sv)C_z \\ &= -C_v \end{aligned}$$

as required. Thus our proof is complete by Lemma 1.4.6(i), (ii). □

For $x,y \in W$, we denote $x - y$ if either $x \prec y$ or $y \prec x$ holds.

We define a preorder relation $w \underset{L}{\leq} w'$ on W if there exist elements

$w = x_1, x_2, \ldots, x_t = w'$ in W such that for each i we have $x_{i-1} — x_i$ and $\mathcal{L}(x_{i-1}) \not\subset \mathcal{L}(x_i)$. We may then define an equivalence relation $w \underset{L}{\sim} w'$ to mean $w \underset{L}{\leq} w' \underset{L}{\leq} w$. The equivalence classes with respect to the relation $\underset{L}{\sim}$ are called left cells. Each left cell, regarded as a full subgraph of L with the same sets I_x and the same function μ is itself a W-graph. One can similarly define right cells by replacing $\mathcal{L}(x_{i-1}) \not\subset \mathcal{L}(x_i)$ by $R(x_{i-1}) \not\subset R(x_i)$, where $R(w) = \{s \in S | ws < w\}$. One can also define two-sided cells of W by replacing $\mathcal{L}(x_{i-1}) \not\subset \mathcal{L}(x_i)$ by the condition that either $\mathcal{L}(x_{i-1}) \not\subset \mathcal{L}(x_i)$ or $R(x_{i-1}) \not\subset R(x_i)$. The notation $x \underset{R}{\sim} y$ (resp. $x \underset{\Gamma}{\sim} y$) means that x,y are in the same right (resp. two-sided) cell of W. $x \underset{R}{\leq} y$, $x \underset{\Gamma}{\leq} y$ denote the corresponding partial orders on W. A minimal non-empty set which is both a union of left cells and a union of right cells is called a RL-equivalence class, written $w \underset{RL}{\sim} y$ if w, y are in the same RL-equivalence class. Clearly, any two-sided cell is a union of some RL-equivalence classes. An RL-equivalence class is also called a Vogan two-sided cell. [L13] [Vo1].

Each left, right or two-sided cell of W is the vertex set of some W-graph, and so is an RL-equivalence class.

Let us give a result on the preorders $\underset{L}{\leq}$ and $\underset{R}{\leq}$ on W.

<u>Theorem 1.5.2</u>

(i) <u>If $x \underset{L}{\leq} y$, then $R(x) \supset R(y)$. Hence, if $x \underset{L}{\sim} y$, then $R(x) = R(y)$.</u>

(ii) <u>If $x \underset{R}{\leq} y$, then $\mathcal{L}(x) \supset \mathcal{L}(y)$. Hence, if $x \underset{R}{\sim} y$, then $\mathcal{L}(x) = \mathcal{L}(y)$.</u>

<u>Proof</u> (i) We may assume $x — y$ and $\mathcal{L}(x) \not\subset \mathcal{L}(y)$ without loss of generality. If $y \prec x$ then by Lemma 1.4.6(i) we have $x = sy$ for some $s \in \mathcal{L}(x) - \mathcal{L}(y)$. This implies $R(x) \supset R(y)$. Now assume $x \prec y$. Suppose $R(x) \not\supset R(y)$. We take $t \in R(y) - R(x)$. Then by Lemma 1.4.6(ii), we have $y = xt$. But then $\mathcal{L}(y) \supset \mathcal{L}(x)$, this contradicts our assumption. So we must have $R(x) \supset R(y)$. (i) is proved.

The proof of (ii) is entirely similar. □

Finally, we state without proof a recent result of Lusztig [L12] which comes from the deep theory of intersection cohomology [GM] [L11].

<u>Theorem 1.5.3</u> <u>Let W_a be any affine Weyl group. Let z, z' be two elements of W_a which satisfy the following conditions:</u>

(i) <u>$z — z'$</u>

(ii) <u>$R(z') \not\subset R(z)$ and $\mathcal{L}(z') \not\subset \mathcal{L}(z)$.</u>

Then z and z' are not in the same two-sided cell of W_a. □

§1.6 THE STAR OPERATIONS IN THE SETS $\mathcal{D}_L(s,t)$, $\mathcal{D}_R(s,t)$ [KL1]

The star operations we shall define in this section will be one of the main operations in our book.

Fix a Coxeter group (W,S). Recall that in the last section we defined $\mathcal{L}(w) = \{s \in S | sw < w\}$ and $\mathcal{R}(w) = \{s \in S | ws < w\}$ for $w \in (W,S)$.

Assume that $s,t \in S$ and st has order 3. We define

$$\mathcal{D}_L(s,t) = \{w \in W | \mathcal{L}(w) \cap \{s,t\} \text{ consists of one element}\}$$

$$\mathcal{D}_R(s,t) = \{w \in W | w^{-1} \in \mathcal{D}_L(s,t)\}$$

If $w \in \mathcal{D}_L(s,t)$ then $|\{sw,tw\} \cap \mathcal{D}_L(s,t)| = 1$. We denote $*w = \{sw,tw\} \cap \mathcal{D}_L(s,t)$. The map $w \to *w$ is an involution of $\mathcal{D}_L(s,t)$ and is called a left star operation on w. To see this, we may assume, without loss of generality, that w is in $\mathcal{D}_L(s,t)$ with $s \in \mathcal{L}(w)$ and $t \notin \mathcal{L}(w)$. Then we are in one of the following two situations:

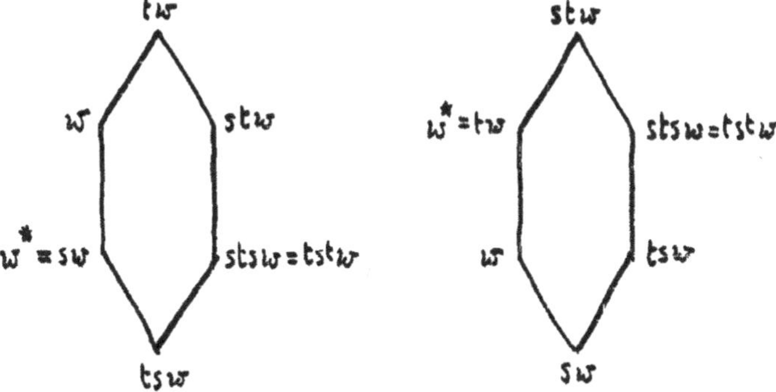

Similarly, we have an involution $w \to w^*$ of $\mathcal{D}_R(s,t)$ called a right star operation on w: $w^* = \{ws,wt\} \cap \mathcal{D}_R(s,t)$. Let $\langle s,t \rangle$ be the group generated by s,t. We have

<u>Theorem 1.6.1</u> <u>Assume $y,w \in \mathcal{D}_L(s,t)$</u>

(i) <u>Suppose $yw^{-1} \notin \langle s,t \rangle$. Then $y \prec w \iff *y \prec *w$. Also, $\mu(y,w) = \mu(*y,*w)$.</u>

(ii) <u>Suppose $yw^{-1} \in \langle s,t \rangle$. Then $y \prec w \iff *w \prec *y$. Also, $\mu(y,w) = \mu(*w,*y) = 1$</u>

<u>Assume $y,w \in \mathcal{D}_R(s,t)$</u>

(iii) <u>Suppose $y^{-1}w \notin \langle s,t \rangle$. Then $y \prec w \iff y^* \prec w^*$. Also, $\mu(y,w) = \mu(y^*,w^*)$.</u>

(iv) <u>Suppose $y^{-1}w \in \langle s,t \rangle$. Then $y \prec w \iff w^* \prec y^*$. Also, $\mu(y,w) = \mu(w^*,y^*) = 1$.</u>

<u>Proof</u> For any $x < x'$ in W with $\varepsilon_x = -\varepsilon_{x'}$, we denote $d(x,x') = \frac{1}{2}(\ell(x')-\ell(x)-1)$ and $\mu(x,x')$ the coefficient of $X^{d(x,x')}$ in $P_{x,x'}$. So $x \prec x' \Leftrightarrow \mu(x,x') \neq 0$. We say $P_{x,x'} \sim P'$ if $P_{x,x'} - P'$ is a polynomial in X of degree less than $d(x,x')$.

It is sufficient to show (i) and (ii). In case (ii), we see

$$y \prec w \Leftrightarrow y_* < w \quad \text{and} \quad \ell(w) = \ell(y)+1$$

by Lemmas 1.4.6(i) and 1.4.5(ii). We also have $\mu(y,w) = 1$. So (ii) follows immediately. Now assume that we are in case (i), i.e. $y,w \in \mathcal{D}_L(s,t)$ and $yw^{-1} \notin \langle s,t\rangle$. Since $\varepsilon_y = -\varepsilon_w$ is equivalent to $\varepsilon_{*y} = -\varepsilon_{*w}$, we may assume $\varepsilon_y = -\varepsilon_w$. Then two cases may occur.

<u>Case 1</u> $*y.y^{-1} = *w.w^{-1}$.

In this case, we may assume without loss of generality that $tsy < sy < y < ty$ and $tsw < sw < w < tw$ as in the following diagrams

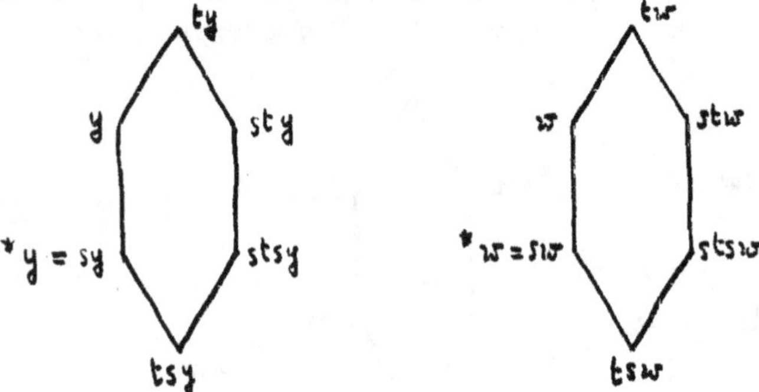

We have $*y = sy$ and $*w = sw$. We may assume that $y < w$ since the conditions $y < w$ and $sy < sw$ are equivalent. If $y \not\prec sw$ then we have $P_{y,w} = P_{sy,sw}$ by Lemma 1.4.5(v). Our result follows easily. Now assume that $y \prec sw$. By (1.4.4d), we have

$$P_{y,w} \sim P_{sy,sw} + XP_{y,sw} - \sum_{\substack{y \prec z \prec sw \\ sz<z}} \mu(y,z)\,\mu(z,sw)X^{d(y,w)} \tag{1.6.1a}$$

Since $t \in \mathcal{L}(sw)$, we also have $ty \prec sw$. We see from Lemma 1.4.6(i) that if there is z in the last sum satisfying $z \neq ty$ and $z \neq tsw$ then $t \in \mathcal{L}(sw)$ implies $t \in \mathcal{L}(z)$ and then implies $t \in \mathcal{L}(y)$. This contradicts our assumption. On the other hand, $s \in \mathcal{L}(ty)$ and $s \notin \mathcal{L}(tsw)$. So the sum over z has exactly one term: $z = ty$. Clearly, $\mu(y,ty) = 1$. Now (1.6.1a) becomes

$$P_{y,w} \sim P_{sy,sw} + XP_{y,sw} - \mu(ty,sw)X^{d(y,w)}$$

By Lemma 1.4.5(vi), we see $P_{y,sw} = P_{ty,sw}$ and so $\deg(XP_{y,sw}-\mu(ty\,,sw)X^{d(y,w)}) < d(y,w)$.

Therefore $P_{y,w} \sim P_{sy,sw}$ as required.

<u>Case 2</u> $\;{}^*y.y^{-1} \neq {}^*w.w^{-1}$.

We may assume without loss of generality that $tsy < sy < y < ty$ and $sw < w < tw < stw$ as follows.

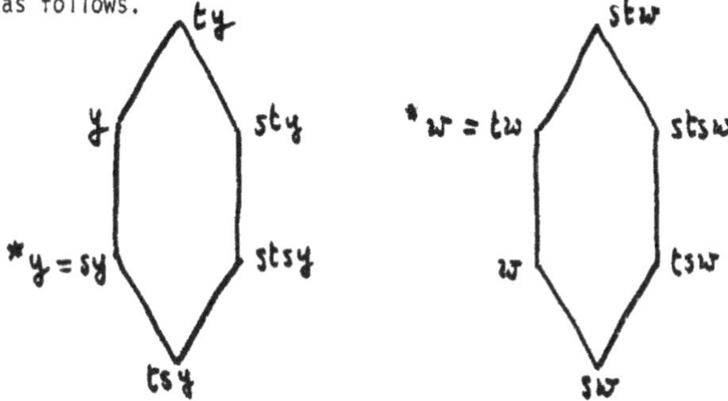

We have ${}^*y = sy$ and ${}^*w = tw$. We may assume that $sy < tw$ since otherwise we have both $y \not\prec w$ and ${}^*y \not\prec {}^*w$ and then there is nothing to do. So $tsy < w$ and $y < stw$. This follows from 1.4.5(v) that $P_{sy,tw} = P_{tsy,w}$ if $sy \not< w$. Since $s \in \mathcal{L}(w)$, $s \notin \mathcal{L}(tsy)$ and $w \neq stsy$, the relation $tsy \prec w$ cannot hold by Lemma 1.4.6(i). So, if $sy \nleq w$, we have $P_{sy,tw} \sim 0$ and so $sy \not\prec tw$. On the other hand, $sy \not< w$ implies $y \nleq w$ (since $s \in \mathcal{L}(w)$) and then $y \not\prec w$. So our conclusion is true when $sy \nleq w$. Now assume $sy < w$. By (1.4.4d), we have

$$P_{sy,tw} \sim P_{tsy,w} + XP_{sy,w} - \sum_{\substack{sy \prec z \prec w \\ tz < z}} \mu(sy,z)\mu(z,w)X^{d(sy,tw)} \qquad (1.6.1b)$$

We have shown that $tsy \not\prec w$. Thus (1.6.1b) becomes

$$P_{sy,tw} \sim XP_{sy,w} - \sum_{\substack{sy \prec z \prec w \\ tz < z}} \mu(sy,z)\,\mu(z,w)X^{d(sy,tw)} \qquad (1.6.1c)$$

We see from Lemma 1.4.6(i) that if there is z in the last sum satisfying $z \neq y$ and $z \neq sw$ then $s \in \mathcal{L}(w)$ implies $s \in \mathcal{L}(z)$ and then $s \in \mathcal{L}(sy)$. This is a contradiction. On the other hand, neither $z = y$ nor $z = sw$ satisfy $tz < z$. This implies that $P_{sy,tw} \sim XP_{sy,w}$. Since $sy < w$ and $s \in \mathcal{L}(w)$, we have $y < w$. So we have $P_{sy,w} = P_{y,w}$ by Lemma 1.4.5(vi). We get $P_{sy,tw} \sim XP_{y,w}$ and hence $\mu(sy,tw) = \mu(y,w)$, as required. □

Another result is for the relations $\underset{L}{\sim}$ and $\underset{R}{\sim}$ on W.

<u>Theorem 1.6.2</u> (i) <u>Let $y,w \in \mathcal{D}_L(s,t)$. Then $y \underset{R}{\sim} w$ if and only if ${}^*y \underset{R}{\sim} {}^*w$.</u>

(ii) <u>Let $y,w \in \mathcal{D}_R(s,t)$. Then $y \underset{L}{\sim} w$ if and only if $y^* \underset{L}{\sim} w^*$.</u>

Proof It is enough to show (i). For (i), it is enough to show that $y \underset{R}{\sim} w$ implies $*y \underset{R}{\sim} *w$. To show this, we need only to show that $y \underset{R}{\leq} w$ implies $*y \underset{R}{\leq} *w$. We may assume without loss of generality that $y — w$ and $R(y) \not\subset R(w)$. Then by Theorem 1.6.1,(ii), we have $*y — *w$. On the other hand, we have $y \underset{L}{\sim} *y$ and $w \underset{L}{\sim} *w$. So by Theorem 1.5.2(i), we have $R(y) = R(*y)$ and $R(w) = R(*w)$. This implies $R(*y) \not\subset R(*w)$ and hence $*y \underset{R}{\leq} *w$, as required. □

We now define an equivalence relation P_L on W generated by $w \underset{P_L}{\sim} *w$ in $\mathcal{D}_L(s,t)$ for some $s,t \in S$ with st having order 3. We call these equivalence classes the P_L-equivalence classes. Similarly, we can define an equivalence relation P_R on W by replacing "$w \underset{P_L}{\sim} *w$ in $\mathcal{D}_L(s,t)$" by "$w \underset{P_R}{\sim} w*$ in $\mathcal{D}_R(s,t)$", and also define a P-equivalence class on W to be a minimal non-empty set in W which is both a union of P_R-equivalence classes and a union of P_L-equivalence classes.

Theorem 1.6.3 For any $y,w \in W$,

(i) If $w \underset{P_L}{\sim} y$, then $w \underset{L}{\sim} y$,

(ii) If $w \underset{P_R}{\sim} y$, then $w \underset{R}{\sim} y$.

(iii) If $w \underset{P}{\sim} y$, then $w \underset{RL}{\sim} y$. In particular, $w \underset{\Gamma}{\sim} y$.

Proof (i) can be reduced to the case when $y = *w$ in $\mathcal{D}_L(s,t)$ for some $s,t \in S$ with st having order 3. Then $y — w$. Since the sets $\mathcal{L}(y) \cap \{s,t\}$ and $\mathcal{L}(w) \cap \{s,t\}$ both contain exactly one element and these elements are distinct, this implies $\mathcal{L}(y) \not\subset \mathcal{L}(w)$ and $\mathcal{L}(w) \not\subset \mathcal{L}(y)$. So $y \underset{L}{\sim} w$. Similarly for (ii), and (iii) follows from (i) and (ii). □

Finally we define, following Vogan [Vo1], the generalized right τ-invariant of $w \in W$ which will characterize left cells of the affine Weyl groups of type $\tilde{A}$ in Chapter 14.

Definition 1.6.4 We say $w,y \in W$ are equivalent to order zero, written $w \overset{R}{\underset{0}{\approx}} y$, if $R(w) = R(y)$. Inductively, we define equivalence to order n for $n \geq 1$. We say w,y are equivalent to order n, written $w \overset{R}{\underset{n}{\approx}} y$, if $w \overset{R}{\underset{n-1}{\approx}} y$ and for every $s,t \in S$ with st having order 3 and $y,w \in \mathcal{D}_R(s,t)$, we have $y' \overset{R}{\underset{n-1}{\approx}} w'$, where $y' = y*$ and $w' = w*$ in $\mathcal{D}_R(s,t)$. We say w,y have the same generalized right τ-invariant if $w \overset{R}{\underset{n}{\approx}} y$ for any $n \geq 0$.

§1.7 EXAMPLES OF CELLS

In this section, we shall give the explicit description of cells of some Coxeter groups (W,S). We shall draw the W-graphs associated to left cells. In all these graphs, the function μ is identically 1, hence it will be omitted. The vertices will be represented by circles, inside which we describe the corresponding subsets of S. We shall explain how the left cells give rise to representations of the Coxeter groups and the Hecke algebras.

(i) The dihedral group of order 2n, $n > 1$, is defined by

$$D_{2n} = \langle s_1, s_2 \mid s_1^2 = s_2^2 = (s_1 s_2)^n = 1 \rangle$$

For $y, w \in D_{2n}$, we have: $y < w \Leftrightarrow$ either $y = w$ or $\ell(w) > \ell(y)$. We see that if $\ell(w) - \ell(y) > 0$ is odd then either $\mathcal{L}(w) \not\subset \mathcal{L}(y)$ or $\mathcal{R}(w) \not\subset \mathcal{R}(y)$ holds. Thus by Lemma 1.4.6(i), (ii), we have: $y \prec w \Leftrightarrow$ either $w = s_i y$ or $w = y s_i$ for some i.

<u>Proposition 1.7.1</u> <u>$P_{y,w} = 1$ for any $y < w$ in D_{2n}.</u>

<u>Proof</u> We use induction on $\ell(w) \geq 0$. It is clear when $\ell(w) = 0$. Now assume $\ell(w) > 0$. We may assume $\ell(w) - \ell(y) > 2$ by Lemma 1.4.5. We may also assume $s_1 \in \mathcal{L}(w)$ without loss of generality.

By (1.4.4d) and the above remark, we have

$$P_{y,w} = \begin{cases} P_{s_1 y, s_1 w} + X P_{y, s_1 w} - X & \text{if } s_1 y < y \\ X P_{s_1 y, s_1 w} + P_{y, s_1 w} - X & \text{if } s_1 y > y. \end{cases}$$

In either case, we get $P_{y,w} = 1$ by inductive hypothesis. Our proof is complete. □

By Proposition 1.7.1, we see that D_{2n} has three two-sided cells: $\{1\}$, $\{w_0\}$ and $C = D_{2n} - \{1, w_0\}$, where w_0 is the longest element of D_{2n}. C contains two left cells C_1, C_2 and also contains two right cells C_1', C_2', where $C_i = \{w \in D_{2n} \mid \mathcal{R}(w) = s_i\}$ and $C_i' = \{w \in D_{2n} \mid \mathcal{L}(w) = s_i\}$, $i = 1,2$.

The W-graphs associated to the left cells of D_{2n} are (∅), (1,2), and

(1)—(2)—(1)—(2)- ⋯ -(1) and (2)—(1)—(2)—(1)- ⋯ -(2) if n is even.

(1)—(2)—(1)—(2)- ⋯ -(2) and (2)—(1)—(2)—(1)- ⋯ -(1) if n is odd.

The matrices representing T_{s_i}, $i = 1,2$, on the free A-modules with bases $\{1\}$, $\{w_0\}$, C_i, C_j, $j \neq i$, are

(X), (-1), $\begin{pmatrix} -1 & X^{1/2} & & & & \\ & X & & & & \\ & X^{1/2} & -1 & X^{1/2} & & \\ & & & X & & \\ & & & X^{1/2} & -1 & \\ & & & & & \ddots \end{pmatrix}_{(n-1)\times(n-1)}$, $\begin{pmatrix} X & & & & & \\ X^{1/2} & -1 & X^{1/2} & & & \\ & & X & & & \\ & & X^{1/2} & -1 & X^{1/2} & \\ & & & & X & \\ & & & & & \ddots \end{pmatrix}_{(n-1)\times(n-1)}$ respectively.

(ii) Let W be the Weyl group of type A_3 with $S = \{s_1,s_2,s_3\}$ such that $s_1s_3 = s_3s_1$. We can describe the Bruhat partial ordering on the elements of W as follows

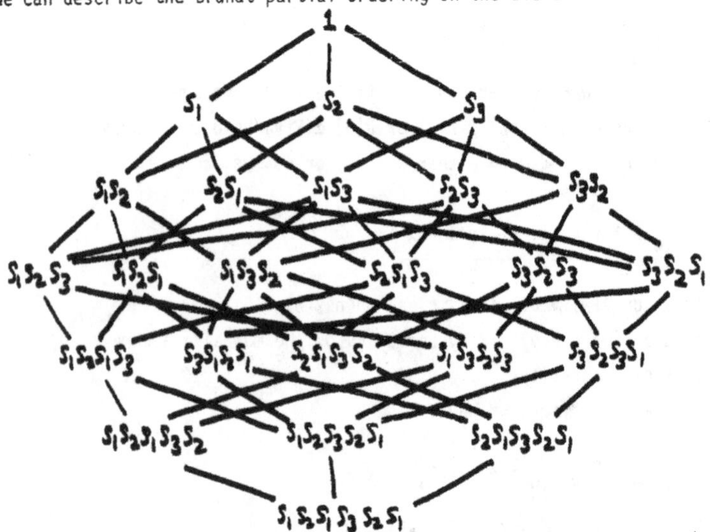

By the above diagram, we see that $y < w$ in W if and only if there exists a sequence $y_0 = y, y_1,\ldots,y_r = w$ in W for some $r > 0$ so that for each j, $1 < j < r$, y_{j-1} and y_j are joined by a bond with $\ell(y_j) > \ell(y_{j-1})$.

There are exactly two pairs of elements $y < w$ in W such that $y \not\prec w$, $\ell(w)-\ell(y) > 1$. They are $s_2 < s_2s_1s_3s_2$ and $s_1s_3 < s_1s_3s_2s_3s_1$. For both pairs, we have $P_{y,w} = X+1$.

There are ten left cells in W:

$A = \{s_1s_2s_1s_3s_2s_1\}$, $B_1 = \{s_1s_2s_1, s_3s_1s_2s_1, s_2s_3s_1s_2s_1\}$, $B_2 = \{s_2s_3s_2, s_1s_2s_3s_2, s_2s_1s_2s_3s_2\}$, $B_3 = \{s_1s_2s_1s_3, s_3s_1s_2s_1s_3, s_3s_2s_1s_3\}$, $C_1 = \{s_1s_3, s_2s_1s_3\}$, $C_2 = \{s_1s_3s_2, s_2s_1s_3s_2\}$, $D_1 = \{s_1, s_2s_1, s_3s_2s_1\}$, $D_2 = \{s_3, s_2s_3, s_1s_2s_3\}$, $D_3 = \{s_2, s_1s_2, s_3s_2\}$, $E = \{1\}$. There are five two-sided cells in W:

A, $B = B_1 \cup B_2 \cup B_3$, $C = C_1 \cup C_2$, $D = D_1 \cup D_2 \cup D_3$, E.

The following W-graphs are associated with the left cells of W:

(1,2,3) , (2,3)—(3,1)—(1,2) , (1,3)—(2) (1)—(2)—(3), (∅)

The matrices representing T_{s_1} on the free A-modules with bases A, B_i, C_j, D_ℓ, E, $1 < i, \ell < 3$, $1 < j < 2$, are

$$(-1), \quad \begin{pmatrix} -1 & 0 & 0 \\ 0 & -1 & X^{\frac{1}{2}} \\ 0 & 0 & X \end{pmatrix}, \quad \begin{pmatrix} -1 & X^{\frac{1}{2}} \\ 0 & X \end{pmatrix}, \quad \begin{pmatrix} -1 & X^{\frac{1}{2}} & 0 \\ 0 & X & 0 \\ 0 & 0 & X \end{pmatrix}, \quad (X), \text{ respectively.}$$

The matrices representing T_{s_3} on these bases are the same as those for T_{s_1} by reordering the elements in these bases. The matrices representing T_{s_2} on these bases are

$$(-1), \quad \begin{pmatrix} -1 & X^{\frac{1}{2}} & 0 \\ 0 & X & 0 \\ 0 & X^{\frac{1}{2}} & -1 \end{pmatrix}, \quad \begin{pmatrix} X & 0 \\ X^{\frac{1}{2}} & -1 \end{pmatrix}, \quad \begin{pmatrix} X & 0 & 0 \\ X^{\frac{1}{2}} & -1 & X^{\frac{1}{2}} \\ 0 & 0 & X \end{pmatrix}, \quad (X), \text{ respectively.}$$

(iii) Let W be the Weyl group of type B_3 with $S = \{s_1,s_2,s_3\}$ such that $s_1s_3 = s_3s_1$. We describe the Bruhat partial ordering on the elements of W by the following diagram:

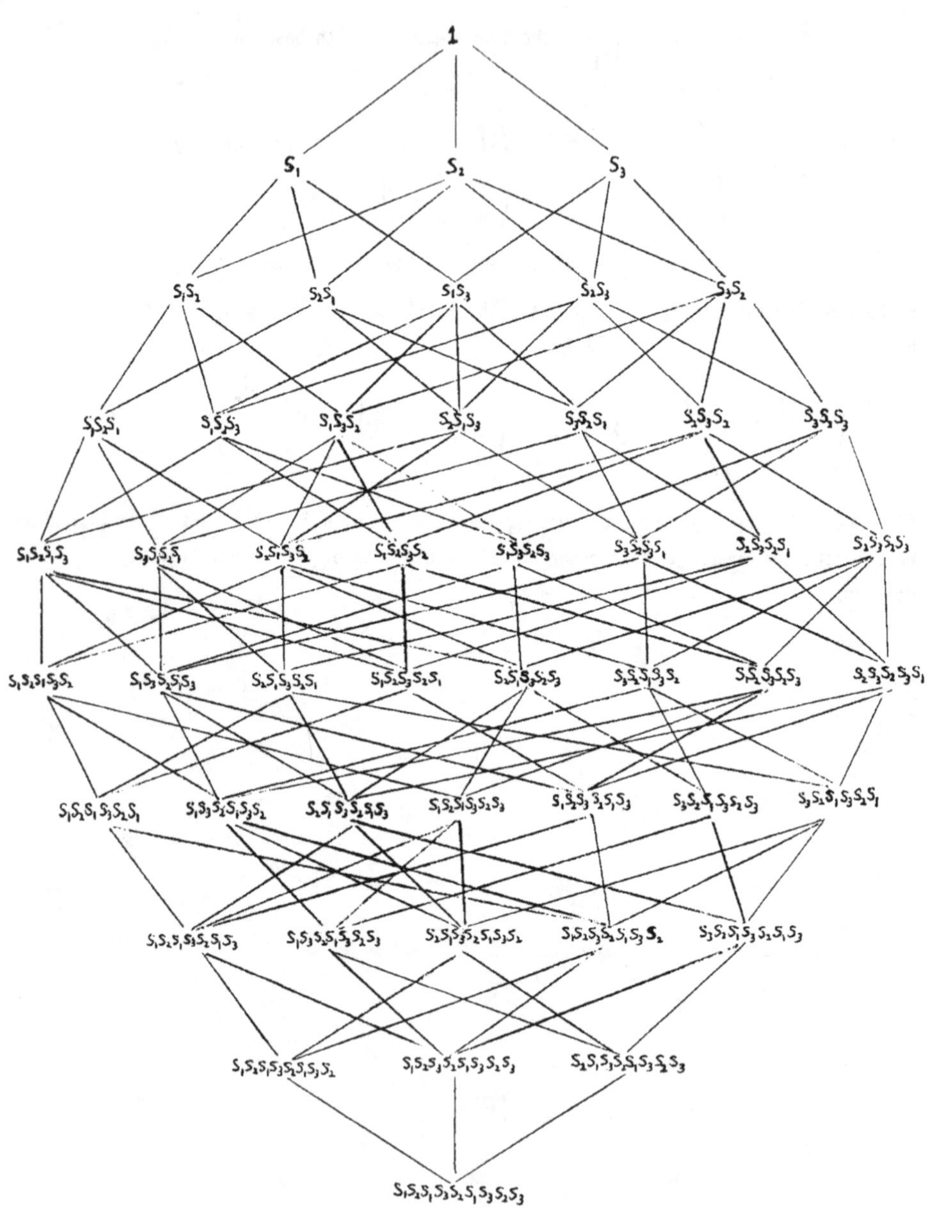
1
s_1 s_2 s_3
s_1s_2 s_2s_1 s_1s_3 s_2s_3 s_3s_2
$s_1s_2s_1$ $s_1s_2s_3$ $s_1s_3s_2$ $s_2s_1s_3$ $s_3s_2s_1$ $s_2s_3s_2$ $s_3s_2s_3$
$s_1s_2s_1s_3$ $s_3s_1s_2s_1$ $s_2s_1s_3s_2$ $s_1s_2s_3s_2$ $s_1s_3s_2s_3$ $s_3s_2s_3s_1$ $s_2s_3s_2s_1$ $s_2s_3s_2s_3$
$s_1s_2s_1s_3s_2$ $s_1s_3s_2s_1s_3$ $s_2s_1s_3s_2s_1$ $s_1s_2s_3s_2s_1$ $s_2s_1s_3s_2s_3$ $s_3s_2s_1s_3s_2$ $s_1s_2s_3s_2s_3$ $s_2s_3s_2s_3s_1$
$s_1s_2s_1s_3s_2s_1$ $s_1s_3s_2s_1s_3s_2$ $s_2s_1s_3s_2s_1s_3$ $s_1s_2s_1s_3s_2s_3$ $s_1s_2s_3s_2s_1s_3$ $s_3s_2s_1s_3s_2s_3$ $s_3s_2s_1s_3s_2s_1$
$s_1s_2s_1s_3s_2s_1s_3$ $s_1s_3s_2s_1s_3s_2s_3$ $s_2s_1s_3s_2s_1s_3s_2$ $s_1s_2s_3s_2s_1s_3s_2$ $s_3s_2s_1s_3s_2s_1s_3$
$s_1s_2s_1s_3s_2s_1s_3s_2$ $s_1s_2s_3s_2s_1s_3s_2s_3$ $s_2s_1s_3s_2s_1s_3s_2s_3$
$s_1s_2s_1s_3s_2s_1s_3s_2s_3$

The following pairs are the pairs $y < w$ in W such that $P_{y,w} \neq 1$, where we denote s_i by i, e.g. 121 denotes $s_1s_2s_1$. We write $y{:}x_1,x_2,\ldots,x_r$ which means that we consider all the polynomials P_{y,x_i}, $1 < i < r$. $P_{y,x_j} = X^2 + 1$ if we write $y{:}\ \ldots\ (x_j)\ldots$; $P_{y,x_j} = X^2 + X + 1$ if we write $y{:}\ \ldots\ [x_j]\ldots$; $P_{y,x_j} = X + 1$ otherwise.

1: 2132, 13213, 12321, 123213, 12321323, 21323, 32132, 132132, 213213, 1213213, 1232132, (321323), (12132132), [2132132].

1: 13213, 12321, 123213, 2132132, 12321323, 132132, 213213, 1213213, 1232132, (12132132).

2: 2132, 12321323, 21323, 32132, 132132, 213213, 1213213, 1232132, (12132132), [2132132].

3: 13213, 12321, 123213, 2132132, 12321323, 21323, 32132, 132132, 213213, 1213213, 1232132, (321323)

13: 13213, 12321, 123213, 132132, 213213, 1213213, 1232132.

12: 2132132, 12321323, 1213213, 1232132, 132132, (12132132).

21: 2132132, 12321323, 1213213, 1232132, 213213, (12132132).

23: 12321323, 21323, 213213, 1213213.

32: 12321323, 32132, 132132, 1232132.

213:12321323, 213213, 1213213.

312:12321323, 132132, 1232132.

121:2132132, 12321323, 1213213, 1232132, (12132132).

123:12321323, 1213213.

321:12321323, 1232132.

232:2132132.

323:12321323.

2132: 2132132.

1213: 12321323, 1213213.

1321: 12321323, 1232132.

1323: 12321323.

3213: 12321323.

2323: 2132132.

13213:12321323.

The following pairs are the only pairs $y < w$ in W such that $y \prec w$, $\ell(w)-\ell(y) > 1$:

$2 < 2132$, $3 < 321323$, $13 < 13213$, $13 < 12321$, $23 < 21323$, $32 < 32132$,

$213 < 213213$, $132 < 132132$, $121 < 12132132$, $2132 < 2132132$, $1321 < 1232132$,

$13213 < 12321323$, $2323 < 2132132$, $1213 < 1213213$,

where, $P_{y,w} = X + 1$ when $\ell(w) - \ell(y) = 3$, and $P_{y,w} = X^2+1$ when $\ell(w) - \ell(y) = 5$.

There are fourteen left cells in W:

$A = \{121321323\}$, $B_1 = \{2323, 12323, 212323, 3212323, 3212323\}$,

$B_2 = \{23231, 123231, 2123231, 32123231, 3213231\}$,

$B_3 = \{123121, 323121, 3123121, 23123121\}$, $C_1 = \{121, 3121, 23121\}$,

$C_2 = \{2123, 32123, 232123\}$, $C_3 = \{12312, 312312, 2312312\}$,

$D_1 = \{3231, 231, 31\}$, $D_2 = \{312, 2132, 32312\}$, $D_3 = \{3123, 23123, 323123\}$,

$E_1 = \{1, 21, 321, 2321, 12321\}$, $E_2 = \{12, 2, 32, 232, 1232\}$, $E_3 = \{3,23,123,323\}$,

$F = \{1\}$.

There are six two-sided cells in W:

A, $B = B_1 \cup B_2 \cup B_3$, $C = C_1 \cup C_2 \cup C_3$, $D = D_1 \cup D_2 \cup D_3$, $E = E_1 \cup E_2 \cup E_3$, F.

The W-graphs associated to the left cells of W are:

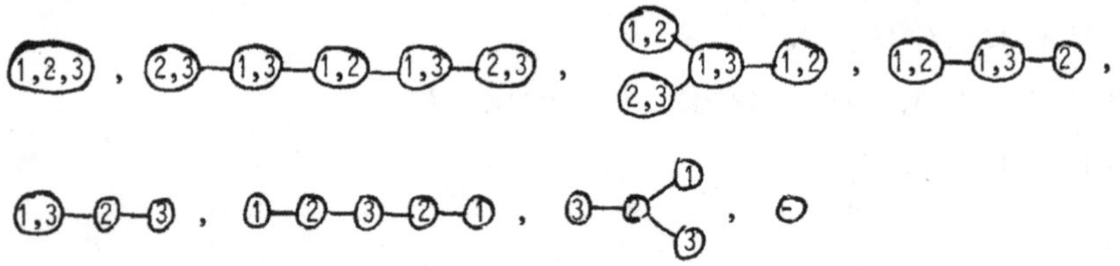

(1,2,3) , (2,3)—(1,3)—(1,2)—(1,3)—(2,3) , (1,2) and (2,3)—(1,3)—(1,2) , (1,2)—(1,3)—(2) ,

(1,3)—(2)—(3) , (1)—(2)—(3)—(2)—(1) , (3)—(2)—(1) and (3) , ∅

The matrices representing T_{s_1} on the free A-modules with bases A, B_i, B_3, C_j, D_ℓ, E_m, E_3, F, $1 \leq i,m \leq 2$, $1 \leq j, \ell \leq 3$, are

$$(-1)\quad \begin{pmatrix} X & 0 & 0 & 0 & 0 \\ \hline X^{\frac{1}{2}} & -1 & 0 & 0 & 0 \\ \hline 0 & 0 & -1 & 0 & 0 \\ \hline 0 & 0 & 0 & -1 & X^{\frac{1}{2}} \\ \hline 0 & 0 & 0 & 0 & X \end{pmatrix} \quad \begin{pmatrix} -1 & 0 & 0 & 0 \\ \hline 0 & X & 0 & 0 \\ \hline 0 & X^{\frac{1}{2}} & -1 & 0 \\ \hline 0 & 0 & 0 & -1 \end{pmatrix} \quad \begin{pmatrix} -1 & 0 & 0 \\ \hline 0 & -1 & X^{\frac{1}{2}} \\ \hline 0 & 0 & X \end{pmatrix} \quad \begin{pmatrix} -1 & X^{\frac{1}{2}} & 0 \\ \hline 0 & X & 0 \\ \hline 0 & X^{\frac{1}{2}} & X \end{pmatrix}$$

$$\begin{pmatrix} -1 & X^{\frac12} & 0 & 0 & 0 \\ 0 & X & 0 & 0 & 0 \\ 0 & 0 & X & 0 & 0 \\ 0 & 0 & 0 & X & 0 \\ 0 & 0 & 0 & X^{\frac12} & -1 \end{pmatrix}, \quad \begin{pmatrix} X & 0 & 0 & 0 \\ 0 & X & 0 & 0 \\ 0 & X^{\frac12} & -1 & 0 \\ 0 & 0 & 0 & X \end{pmatrix} \quad (X), \text{ respectively.}$$

The matrices representing T_{s_2} on these bases are

$$(-1) \quad \begin{pmatrix} -1 & X^{\frac12} & 0 & 0 & 0 \\ 0 & X & 0 & 0 & 0 \\ 0 & X^{\frac12} & -1 & X^{\frac12} & 0 \\ 0 & 0 & 0 & X & 0 \\ 0 & 0 & 0 & X^{\frac12} & -1 \end{pmatrix} \begin{pmatrix} -1 & 0 & X^{\frac12} & 0 \\ 0 & -1 & X^{\frac12} & 0 \\ 0 & 0 & X & 0 \\ 0 & 0 & X^{\frac12} & -1 \end{pmatrix} \begin{pmatrix} -1 & X^{\frac12} & 0 \\ 0 & X & 0 \\ 0 & X^{\frac12} & -1 \end{pmatrix} \begin{pmatrix} X & 0 & 0 \\ X^{\frac12} & -1 & X^{\frac12} \\ 0 & 0 & X \end{pmatrix}$$

$$\begin{pmatrix} X & 0 & 0 & 0 & 0 \\ X^{\frac12} & -1 & X^{\frac12} & 0 & 0 \\ 0 & 0 & X & 0 & 0 \\ 0 & 0 & X^{\frac12} & -1 & X^{\frac12} \\ 0 & 0 & 0 & 0 & X \end{pmatrix} \quad \begin{pmatrix} X & 0 & 0 & 0 \\ X^{\frac12} & -1 & X^{\frac12} & X^{\frac12} \\ 0 & 0 & X & 0 \\ 0 & 0 & 0 & X \end{pmatrix} \quad (X), \text{ respectively.}$$

The matrices representing T_{s_3} on these bases are

$$(-1) \quad \begin{pmatrix} -1 & 0 & 0 & 0 & 0 \\ 0 & -1 & X^{\frac12} & 0 & 0 \\ 0 & 0 & X & 0 & 0 \\ 0 & 0 & X^{\frac12} & -1 & 0 \\ 0 & 0 & 0 & 0 & -1 \end{pmatrix} \quad \begin{pmatrix} X & 0 & 0 & 0 \\ 0 & -1 & 0 & 0 \\ X^{\frac12} & 0 & -1 & X^{\frac12} \\ 0 & 0 & 0 & X \end{pmatrix} \quad \begin{pmatrix} X & 0 & 0 \\ X^{\frac12} & -1 & X^{\frac12} \\ 0 & 0 & X \end{pmatrix} \begin{pmatrix} -1 & X^{\frac12} & 0 \\ 0 & X & 0 \\ 0 & X^{\frac12} & -1 \end{pmatrix}$$

$$\begin{pmatrix} X & 0 & 0 & 0 & 0 \\ 0 & X & 0 & 0 & 0 \\ 0 & X^{\frac12} & -1 & X^{\frac12} & 0 \\ 0 & 0 & 0 & X & 0 \\ 0 & 0 & 0 & 0 & X \end{pmatrix} \quad \begin{pmatrix} -1 & X^{\frac12} & 0 & 0 \\ 0 & X & 0 & 0 \\ 0 & 0 & X & 0 \\ 0 & X^{\frac12} & 0 & -1 \end{pmatrix} \quad (X), \text{ respectively.}$$

(iv) Let W be the symmetric group S_n. Then the cells of W can be described by the Robinson-Schensted map which we shall now describe.

Any partition λ of n, $n \geq 1$, is visualized as a Young diagram F_λ with n cells. We shall adopt the convention that the parts of the partition λ give the lengths of the columns of the Young diagram. For example, the Young diagram of shape $\lambda = \{4 \geq 3 \geq 1\}$ is as follows.

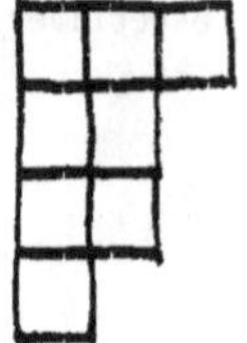

A Young tableau of shape λ is a labelling of the cells of F_λ with integers $1,2,\dots,n$. A Young tableau is standard if the numbers labelling its cells are increasing in the columns from top to bottom and are increasing in the row from left to right. The following is an example of a standard Young tableau of shape $\lambda = \{4 > 3 > 1\}$

1	3	7
2	5	
4	8	
6		

We know that any element $w \in W$ corresponds to a permutation, say $(j_1,j_2,\dots,j_n)$ of $1,2,\dots,n$. The Robinson-Schensted map $\theta : w \to (P(w),Q(w))$ gives a bijection from W to the pairs of standard Young tableaux on $[n]$ having the same shape, where $[n] = \{1,2,\dots,n\}$.

The Robinson-Schensted map for sending $w \in W$ to the standard Young tableau $P(w)$ is defined as follows. [Vie2].

Given $w = (j_1,j_2,\dots,j_n) \in W$. We define now recursively a process which inserts an integer j_t, $1 < t < n$, in a Young tableau T_{t-1}, where T_{t-1} has been determined by $(j_1,\dots,j_{t-1})$ with the convention that T_0 is an empty Young tableau. The result is a Young tableau denoted by T_t.

If j_t is greater than every element of the first row of T_{t-1} then T_t is the Young tableau obtained by adding j_t at the end of this first row.

If not, then let z be the smallest value of the first row of T_{t-1} greater than j_t. Let T'_{t-1} be the Young tableau obtained by deleting the first row of T_{t-1}. Then T_t is the tableau obtained from T_{t-1} by replacing z by j_t and by inserting (recursively) the value z in the tableau T'_{t-1}.

A lemma proves that the result is again a Young tableau. Note that the shape of T_t is obtained by adding a cell on the border of the shape of T_{t-1}. Let $P(w) = T_n$. An element of this "bumping process" is given in the following example.

Let w = (617258394). Then

$$T_1 = \begin{array}{|c|}\hline 6 \\ \hline\end{array}\,,\quad T_2 = \begin{array}{|c|}\hline 1 \\ \hline 6 \\ \hline\end{array}\,,\quad T_3 = \begin{array}{|c|c|}\hline 1 & 7 \\ \hline 6 \\ \cline{1-1}\end{array}\,,\quad T_4 = \begin{array}{|c|c|}\hline 1 & 2 \\ \hline 6 & 7 \\ \hline\end{array}\,,\quad T_5 = \begin{array}{|c|c|c|}\hline 1 & 2 & 5 \\ \hline 6 & 7 \\ \cline{1-2}\end{array}$$

$$T_6 = \begin{array}{|c|c|c|c|}\hline 1 & 2 & 5 & 8 \\ \hline 6 & 7 \\ \cline{1-2}\end{array}\,,\quad T_7 = \begin{array}{|c|c|c|c|}\hline 1 & 2 & 3 & 8 \\ \hline 5 & 7 \\ \cline{1-2} 6 \\ \cline{1-1}\end{array}\,,\quad T_8 = \begin{array}{|c|c|c|c|c|}\hline 1 & 2 & 3 & 8 & 9 \\ \hline 5 & 7 \\ \cline{1-2} 6 \\ \cline{1-1}\end{array}$$

$$T_9 = \begin{array}{|c|c|c|c|c|}\hline 1 & 2 & 3 & 4 & 9 \\ \hline 5 & 7 & 8 \\ \cline{1-3} 6 \\ \cline{1-1}\end{array}$$

The detailed process from T_6 to T_7 as follows.

$$T_6 = \begin{array}{|c|c|c|c|}\hline 1 & 2 & 5 & 8 \\ \hline 6 & 7 \\ \cline{1-2}\end{array} \leftarrow 3 = j_7 \qquad \begin{array}{|c|c|c|c|}\hline 1 & 2 & 3 & 8 \\ \hline 6 & 7 \\ \cline{1-2}\end{array} \leftarrow 5$$

$$\begin{array}{|c|c|c|c|}\hline 1 & 2 & 3 & 8 \\ \hline 5 & 7 \\ \cline{1-2}\end{array} \leftarrow 6 \qquad T_7 = \begin{array}{|c|c|c|c|}\hline 1 & 2 & 3 & 8 \\ \hline 5 & 7 \\ \cline{1-2} 6 \\ \cline{1-1}\end{array}$$

We define $Q(w) = P(w^{-1})$.

Now we can state the following result. [BV1]

<u>Theorem 1.7.2</u> <u>For $y, w \in S_n$, we have</u>

$$y \underset{L}{\sim} w \Leftrightarrow P(y) = P(w)$$

$$y \underset{R}{\sim} w \Leftrightarrow P(y^{-1}) = P(w^{-1})$$

$$y \underset{\Gamma}{\sim} w \Leftrightarrow P(y) \text{ and } P(w) \text{ have the same shape.} \qquad \square$$

This result will be generalized to the case of the affine Weyl groups of type $\tilde{A}$ later in the book.

Kazhdan and Lusztig [KL1] showed the following result on the representations of S_n afforded by left cells of S_n.

<u>Theorem 1.7.3</u> <u>Let X be a left cell of $W = S_n$, let Γ be the W-graph associated to X and let ρ be the representation of H over the quotient field of A corresponding</u>

to Γ. Then ρ is irreducible and the isomorphism class of the W-graph Γ depends only on the isomorphism class of ρ and not on X. □

For example, the five representations of H of type A_3 given in (ii) are precisely the five inequivalent irreducible representations of H.

Now we shall consider some infinite Coxeter groups.

(v) The infinite dihedral group is defined by $D_\infty = \langle s_1, s_2 | s_1^2 = s_2^2 = 1\rangle$ which is also the affine Weyl group of type $\tilde{A}_1$. By the same argument as that for Proposition 1.7.1, we can show

Proposition 1.7.4 $P_{y,w} = 1$ for any $y < w$ in D_∞. □

By this result, we see that D_∞ has two two-sided cells: $\{1\}$, $C = D_\infty - \{1\}$. C contains two left cells C_1, C_2 and also two right cells C_1', C_2', where $C_i = \{w \in D_\infty | R(w) = s_i\}$ and $C_i' = \{w \in D_\infty | \mathcal{L}(w) = s_i\}$, $i = 1,2$.

The W-graphs associated to the left cells of D_∞ are

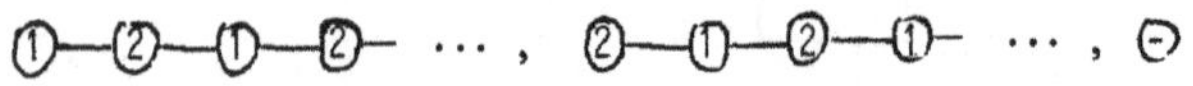

(1)—(2)—(1)—(2)— ... , (2)—(1)—(2)—(1)— ... , (∅)

The matrices representing T_{s_i}, $i = 1,2$, on the free A-modules with bases $\{1\}$, C_i, C_j, $j \neq i$, are

$$(X)\quad \begin{pmatrix} -1 & X^{\frac12} & 0 & 0 & 0 \\ 0 & X & 0 & 0 & 0 \\ 0 & X^{\frac12} & -1 & X^{\frac12} & 0 \\ 0 & 0 & 0 & X & 0 \\ 0 & 0 & 0 & X^{\frac12} & -1 \\ & & & & \ddots \end{pmatrix}_{\infty\times\infty} \qquad \begin{pmatrix} X & 0 & 0 & 0 & 0 \\ X^{\frac12} & -1 & X^{\frac12} & 0 & 0 \\ 0 & 0 & X & 0 & 0 \\ 0 & 0 & X^{\frac12} & -1 & X^{\frac12} \\ 0 & 0 & 0 & 0 & X \\ & & & & \ddots \end{pmatrix}_{\infty\times\infty}$$

respectively.

(vi) Let $W = (W,S)$ be an affine Weyl group of type $\tilde{A}_2$, $\tilde{B}_2$ or $\tilde{G}_2$. We denote the elements of S by s_1, s_2, s_3. In the case $\tilde{B}_2$, we assume $(s_1s_3)^4 = (s_2s_3)^4 = (s_1s_2)^2 = 1$. In the case $\tilde{G}_2$, we assume that $(s_1s_3)^3 = (s_2s_3)^6 = (s_1s_2)^2 = 1$. For any subset J of $\{1,2,3\}$, we denote by W^J the set of all $w \in W$ such that $R(w)$ consists of the s_j, $j \in J$.

We shall define following Lusztig [L12], a partition of W into finitely many subsets, as follows.

Type $\tilde{A}_2$: $A_{13} = W^{13}$, $A_{12} = W^{12}$, $A_{23} = W^{23}$, $A_2 = A_{13}s_2$, $A_3 = A_{12}s_3$, $A_1 = A_{23}s_1$,

$B_1 = W^1 - A_1$, $B_2 = W^2 - A_2$, $B_3 = W^3 - A_3$, $C_\emptyset = W^\emptyset$.

Type $\tilde{B}_2$: $A_{13} = W^{13}$, $A_{12} = A_{13}s_2$, $A_3 = A_{12}s_3$, $A_1 = A_3s_1$, $A_{23} = W^{23}$, $A'_{12} = A_{23}\, s_1$,
$A'_3 = A'_{12}s_3$, $A_2 = A'_3s_2$, $B_{12} = W^{12} - (A_{12} \cup A'_{12})$, $B_3 = B_{12}s_3$,
$B_1 = B_3s_1$, $B_2 = B_3s_2$,
$C_1 = W'-(A_1 \cup B_1)$, $C_2 = W^2-(A_2 \cup B_2)$, $C_3 = W^3 - (A_3 \cup A'_3 \cup B_3)$, $D_\emptyset = W^\emptyset$.

Type $\tilde{G}_2$: $A_{23} = W^{23}$, $A_{12} = A_{23}s_1$, $A_{13} = A_{12}s_3$, $A'_{12} = A_{13}s_2$, $A_3 = A'_{12}s_3$, $A_2 = A_3s_2$,
$A'_{13} = A_3s_1$, $A''_{12} = A'_{13}s_2$, $A'_3 = A''_{12}s_3$, $A'_2 = A'_3s_2$, $A''_3 = A'_2s_3$, $A_1 = A''_3s_1$,
$B_{13} = W^{13}-(A_{13} \cup A'_{13})$, $B_{12} = B_{13}s_2$, $B_3 = B_{12}s_3$, $B_2 = B_3s_2$, $B'_3 = B_2s_3$,
$B_1 = B_3s_1$,
$C_{12} = W^{12}-(A_{12} \cup A'_{12} \cup A''_{12} \cup B_{12})$, $C_3 = C_{12}s_3$, $C_2 = C_3s_2$, $C'_3 = C_2s_3$,
$C_1 = C'_3s_1$, $C'_2 = C'_3s_2$, $D_1 = W'-(A_1 \cup B_1 \cup C_1)$,
$D_2 = W^2-(A_2 \cup A'_2 \cup B_2 \cup C_2 \cup C'_2)$,
$D_3 = W^3 - (A_3 \cup A'_3 \cup A''_3 \cup B_3 \cup B'_3 \cup C_3 \cup C'_3)$, $E_\emptyset = W^\emptyset$.

Each of the subsets in the partition is contained in some W^J with J indicated as a subscript.

Lusztig [L12] showed that the partition of W just described coincides with the partition of W into left cells.

We now consider the union of all left cells in W whose name contains a fixed capital letter, we denote this union by that capital letter. (For example, for type $\tilde{G}_2$, we have $C = C_{12} \cup C_3 \cup C_2 \cup C'_3 \cup C'_2 \cup C_1$). Thus we have a partition into pieces: $W = A \cup B \cup C$ (for type $\tilde{A}_2$), $W = A \cup B \cup C \cup D$ (for type $\tilde{B}_2$), $W = A \cup B \cup C \cup D \cup E$ (for type $\tilde{G}_2$).

Lusztig [L12] showed that the pieces in this partition of W are just the two-sided cells of W.

Recall that at the end of §1.1 we got a one-to-one correspondence between the set of elements of an affine Weyl group of rank n and the set of alcoves in a Euclidean space V of dimension n. We now shall describe the left cells and the two-sided cells for the affine Weyl groups W of type $\tilde{A}_2$, $\tilde{B}_2$, $\tilde{G}_2$ in three figures: A two-sided cell is represented by the union of all closed alcoves shaded in the same way. If we remove from this union all faces of codimension greater than 1, the remaining set will have finitely many connected components; the closures of these components will be the subsets of V corresponding to the various left cells.

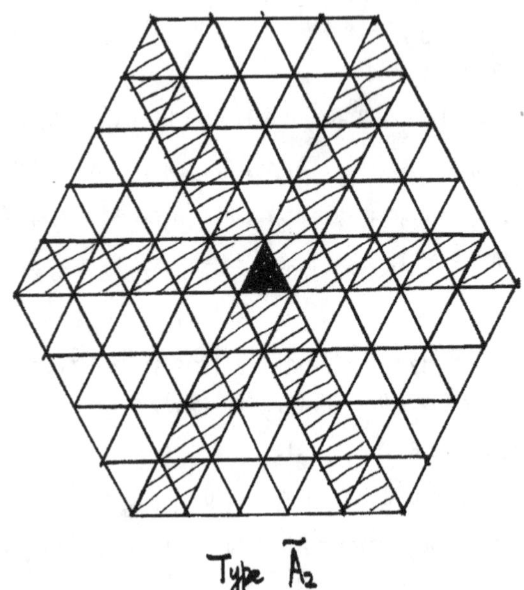

Type $\tilde{A}_2$

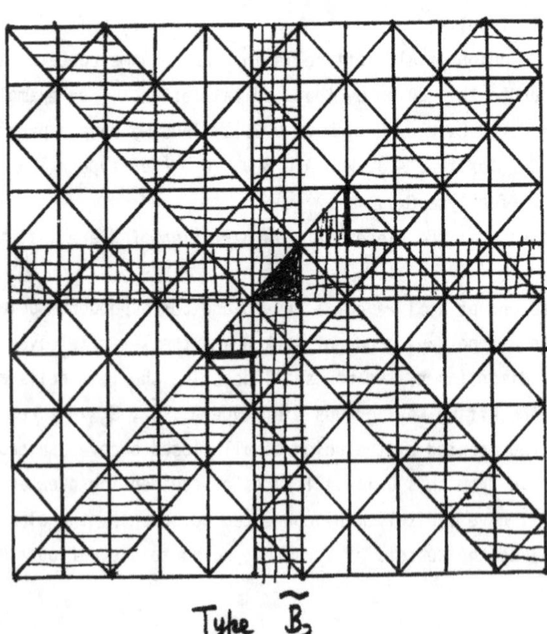

Type $\tilde{B}_2$

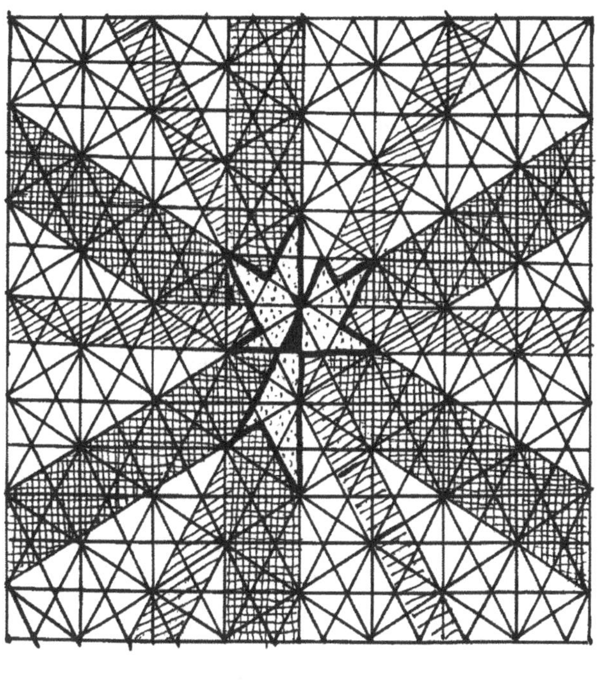

Type $\tilde{G}_2$

CHAPTER 2 : APPLICATIONS OF KAZHDAN-LUSZTIG THEORY

In Chapter 1 we defined a Coxeter group and its associated Hecke algebra. We also defined the Kazhdan-Lusztig polynomials associated with pairs of elements in a Coxeter group. These polynomials are used to define decompositions of the Coxeter group into equivalence classes called left cells, right cells and two-sided cells.

The Kazhdan-Lusztig polynomials and the cells associated to them have proved to be of importance in a number of different contexts in the Lie theory. The purpose of this chapter is to give a brief description of the way the polynomials and the cells have been used in various situations. The discussion of each such situation is necessarily quite short, but we hope that the accompanying references will give guidance to the reader who needs further details.

The different sections of the chapter are as follows. In §2.1 we describe a modification of the Kazhdan-Lusztig theory which applies to Hecke algebras with more than one parameter. This modification leads to equivalence classes called ϕ-cells. For example in the Weyl groups of type B_ℓ the left cells do not in general give irreducible representations, whereas the left ϕ-cells do.

In §2.2 we show how the 2-sided cells can be used to give a canonical isomorphism between the group algebra of a finite Coxeter group and its Hecke ring, both taken over $\mathbb{Q}[q^{\frac{1}{2}}]$.

In §2.3 we explain how the Kazhdan-Lusztig theory is related to the theory of irreducible representations of a Weyl group. We define a certain subset of the set of irreducible representations called special representations, and introduce an equivalence relation on the set of irreducible characters of W such that each equivalence class contains just one special representation. The number of equivalence classes is equal to the number of special representations, and this is also the number of 2-sided cells of the Weyl group. We also describe a process which enables us to define inductively certain representations of W called cell representations. The cell representations are not in general irreducible, but all their irreducible components are in the same equivalence class. It is conjectured that the cell representations are the representations of W which come from the left cells.

In §2.4 we state the celebrated Kazhdan-Lusztig conjecture on the multiplicities of composition factors of Verma modules, and give references for its proof.

In §2.5 we discuss the applications of the Kazhdan-Lusztig theory to the classification of the primitive ideals in the universal enveloping algebra of a semisimple Lie algebra. The primitive ideals with a given central character form a finite partially ordered set. In the case where this central character corresponds to a W-orbit of regular integral weights the number of primitive ideals with the given central character is the number of left cells of the Weyl group W.

Moreover the partial order relation on the primitive ideals is given by the partial order relation on the left cells. We also explain how the special representations of W play an important role in the theory of primitive ideals. The number of left cells in a given two-sided cell of W is the degree of the corresponding special representation of W, and the number of primitive ideals with the central character given above is the sum of the degrees of all the special representations of W.

In §2.6 we describe the application of the Kazhdan-Lusztig theory to the modular representation theory of algebraic groups and the related finite groups. The dimensions of the irreducible modules are unknown in this case but we state a conjecture of Lusztig which would determine them. This conjecture involves the Kazhdan-Lusztig polynomials for the affine Weyl group.

Finally in §2.7 we describe how the weight multiplicities in the finite dimensional irreducible modules for a semisimple Lie algebra over $\mathbb{C}$ can be given in terms of Kazhdan-Lusztig polynomials for the affine Weyl group. In the case of algebras of type A these multiplicities are closely related to the Kostka numbers, and the corresponding Kazhdan-Lusztig polynomials to the symmetric functions called Foulkes polynomials. The values of the unipotent characters of the general linear groups $GL_n(q)$ on the unipotent conjugacy classes can be described in terms of either Foulkes polynomials or Kazhdan-Lusztig polynomials.

§2.1 ϕ-CELLS OF A COXETER GROUP

Let (W,S) be a Coxeter group. We know that the left cells of W do not in general give rise to irreducible representations of W. We therefore describe a modification of the construction of the left cells, due to Lusztig [L8]. This modification gives left ϕ-cells where ϕ is a certain map from W to an infinite cyclic group Γ. If ϕ is constant we get the same theory as before. However we get a new situation by taking ϕ to have distinct values on non-conjugate elements of S. For example if W is a Weyl group of type B_ℓ ϕ can be chosen so that the left ϕ-cells all give irreducible representations of W.

We define a map $\phi: W \to \Gamma$ such that $\phi(s_{i_1} s_{i_2} \dots s_{i_r}) = \phi(s_{i_1})\,\phi(s_{i_2})\dots\phi(s_{i_r})$ for any reduced expression $s_{i_1} s_{i_2} \dots s_{i_r}$ in W. Let Γ be the infinite cyclic group generated by an element $X^{\frac{1}{2}}$ and let $\phi(s_i) = X^{m(s_i)/2}$ where $m(s_i)$ is a positive integer.

We shall set $\phi(w) = X_w^{\frac{1}{2}}$ for $w \in W$.

Let H_ϕ be the generic Hecke algebra of W with respect to ϕ. This is an algebra over $A = \mathbb{Z}[X^{\frac{1}{2}}, X^{-\frac{1}{2}}]$ with basis T_w, $w \in W$, and with multiplication given by

$$(T_s + 1)(T_s - X_s) = 0 \qquad s \in S$$

$$T_{s_1 s_2 \dots s_p} = T_{s_1} T_{s_2} \dots T_{s_p} \quad \text{if } s_1 s_2 \dots s_p \text{ is a reduced expression in W.}$$

Let $\tilde{T}_w = X_w^{-\frac{1}{2}} T_w$ for $w \in W$. The $\tilde{T}_w$ also form a basis for H_ϕ and the multiplication with respect to this basis is given by

$$(\tilde{T}_s + X_s^{-\frac{1}{2}})(\tilde{T}_s - X_s^{\frac{1}{2}}) = 0 \text{ if } s \in S$$

$$\tilde{T}_{s_{i_1} s_{i_2} \ldots s_{i_r}} = \tilde{T}_{s_{i_1}} \tilde{T}_{s_{i_2}} \ldots \tilde{T}_{s_{i_r}} \text{ if } s_{i_1} s_{i_2} \ldots s_{i_r} \text{ is a reduced expression in } W.$$

Let $a \to \bar{a}$ be the involution of the ring A defined as before. We extend it to an involution $h \to \bar{h}$ of the algebra H_ϕ by the formula

$$\overline{\sum_w a_w \tilde{T}_w} = \sum_w \overline{a_w}\, \tilde{T}_{w^{-1}}^{-1}, \quad a_w \in A.$$

Note that $\tilde{T}_s^{-1} = \tilde{T}_s + (X_s^{-\frac{1}{2}} - X_s^{\frac{1}{2}})$, $s \in S$, and so $\tilde{T}_w$ is invertible for all $w \in W$. We have

<u>Proposition 2.1.1</u> <u>Given $w \in W$, there is a unique element $C'_w \in H_\phi$ such that</u>

$$\overline{C'_w} = C'_w$$

$$C'_w = \sum_{y \leq w} \varepsilon_y \varepsilon_w \overline{P^*_{y,w}}\, \tilde{T}_y$$

<u>where $P^*_{w,w} = 1$ and, for any $y < w$, $P^*_{y,w} \in X^{-\frac{1}{2}}\mathbb{Z}[X^{-\frac{1}{2}}]$. Moreover, $X_y^{-\frac{1}{2}} X_w^{\frac{1}{2}} P^*_{y,w} \in \mathbb{Z}[X]$.</u> □

Note that in the case where $m(s_i) = 1$ for all i, we get a Hecke algebra $H = H_\phi$ exactly as defined in §1.2. Then $X_y^{-\frac{1}{2}} X_w^{\frac{1}{2}} P^*_{y,w}$ become Kazhdan-Lusztig polynomials.

It can be shown that C'_w $w \in W$, form a A-basis of H_ϕ.

Let $\underset{L,\phi}{\leq}$ be the preorder relation on W generated by the relation "$x' \underset{L,\phi}{\leq} x$ if there exists $s \in S$ such that $C'_{x'}$ appears with non-zero coefficient in $\tilde{T}_s . C'_x$ which is expressed in the C'_w-basis". We call it the left preorder. The equivalence relation associated to $\underset{L,\phi}{\leq}$ is denoted $\underset{L,\phi}{\sim}$ and the corresponding equivalence classes in W are called the left ϕ-cells of W. Given $x,y \in W$, we say that $x \underset{\Gamma,\phi}{\leq} y$ if there exists a sequence $x = x_0, x_1, \ldots, x_r = y$ of elements in W such that for $i = 0,1,\ldots,r-1$, we have either $x_i \underset{L,\phi}{\leq} x_{i+1}$ or $x_i^{-1} \underset{L,\phi}{\leq} x_{i+1}^{-1}$. The equivalence relation on W corresponding to the preorder $\underset{\Gamma,\phi}{\leq}$ is denoted $\underset{\Gamma,\phi}{\sim}$ and the corresponding equivalence classes on W are called the two-sided ϕ-cells of W.

Note that when ϕ is constant on S, these notions agree with those introduced in Chapter 1.

For any $x \in W$, we denote $I^L_{x,\phi}$ the A-submodule of H_ϕ spanned by the elements C'_y, $y \underset{L,\phi}{\leq} x$, and $\hat{I}^L_{x,\phi}$ the one spanned by the elements C'_y, $y \underset{L,\phi}{\leq} x$, $y \underset{L,\phi}{\not\sim} x$.

We define similarly $I^{\Gamma}_{x,\phi}$ and $\hat{I}^{\Gamma}_{x,\phi}$, by replacing $\underset{L,\phi}{<}, \underset{L,\phi}{\sim}$ by $\underset{\Gamma,\phi}{<}, \underset{\Gamma,\phi}{\sim}$ in the previous definition. We can show that $I^L_{x,\phi}$, $\hat{I}^L_{x,\phi}$ are left ideals of H_ϕ and that $I^{\Gamma}_{x,\phi}$, $\hat{I}^{\Gamma}_{x,\phi}$ are two-sided ideals of H_ϕ. Hence, $I^L_{x\phi}/\hat{I}^L_{x,\phi}$ is a left H_ϕ-module with a natural basis given by the images of C'_y for y in the left ϕ-cell of x. $I^{\Gamma}_{x,\phi}/\hat{I}^{\Gamma}_{x,\phi}$ is a two-sided H_ϕ-module with a natural basis given by the images of C'_y for y in the two-sided ϕ-cell of x.

<u>Example 2.1.2</u> When W is of type B_2, let $m(s_1) = 1$, $m(s_2) = c > 2$. We have

$$P^*_{s_2,s_2s_1s_2} = X^{-\frac{c+1}{2}} - X^{-\frac{c-1}{2}}, \quad P^*_{1,s_2s_1s_2} = X^{-\frac{2c+1}{2}} - X^{-\frac{2c-1}{2}}$$

$$P^*_{s_1,s_1s_2s_1} = X^{-\frac{c+1}{2}} + X^{-\frac{c-1}{2}}, \quad P^*_{1,s_1s_2s_1} = X^{-\frac{c+2}{2}} + X^{-\frac{c}{2}}$$

and $P^*_{y,w} = X^{\frac{m(y)}{2} - \frac{m(w)}{2}}$ for all other pairs $y < w$. The left ϕ-cells are $L_1 = \{1\}$, $L_2 = \{s_1\}$, $L'_4 = \{s_2, s_1s_2\}$, $L_3 = \{s_2s_1s_2\}$, $L''_4 = \{s_2s_1, s_1s_2s_1\}$, $L_5 = \{s_1s_2s_1s_2\}$. The corresponding H_ϕ-modules $I^L_{x,\phi}/\hat{I}^L_{x,\phi}$ are all irreducible with scalars extended to an algebraic closure of $\mathbb{Q}(X^{\frac{1}{2}})$. The matrices representing $\tilde{T}_{s_1}$, $\tilde{T}_{s_2}$ on the natural bases of these H_ϕ-modules are as follows.

	$\tilde{T}_{s_1}$	$\tilde{T}_{s_2}$
L_1	$(X^{\frac{1}{2}})$	$(X^{c/2})$
L_2	$(-X^{-\frac{1}{2}})$	$(X^{c/2})$
L_3	$(X^{\frac{1}{2}})$	$(-X^{-c/2})$
L_5	$(-X^{-\frac{1}{2}})$	$(-X^{-c/2})$
L'_4 or L''_4	$\begin{pmatrix} X^{\frac{1}{2}} & 0 \\ 1 & -X^{-\frac{1}{2}} \end{pmatrix}$	$\begin{pmatrix} -X^{-c/2} & X^{(c-1)/2} + X^{(1-c)/2} \\ 0 & X^{c/2} \end{pmatrix}$

Let $L_4 = L'_4 \cup L''_4$. Then the two-sided ϕ-cells are L_t, $1 < t < 5$.

This is in contrast with the situation when $m(s_1) = m(s_2) = 1$ in which case there are only four left cells: $L_{(1)} = \{1\}$, $L'_{(2)} = \{s_1, s_2s_1, s_1s_2s_1\}$, $L''_{(2)} = \{s_2, s_1s_2, s_2s_1s_2\}$, $L_{(3)} = \{s_1s_2s_1s_2\}$. There are only three two-sided cells: $L_{(t)}$, $1 < t < 3$, where we set $L_{(2)} = L'_{(2)} \cup L''_{(2)}$

Assume now that (W',S') is a Weyl group of type A_ℓ, $\ell > 3$. Let α be the unique automorphism of order 2 of (W',S'). Then the fixed point set W of α in W' is a

Coxeter group with a set of generators S corresponding to the orbits of α in S': to an orbit O, there corresponds the longest element in the subgroup generated by O. (W,S) is a Weyl group of type $B_{\ell/2}$ if ℓ is even and $B_{\ell+1/2}$ if ℓ is odd. Let ϕ be the function on W defined by $\phi(w) = X^{\ell(w)}$, where $\ell(w)$ is the length of w with respect to (W',S'). Then the restriction of ϕ to S has values $X^2,X^2,\ldots,X^2,X^3$ if ℓ is even and $X^2,X^2,\ldots,X^2,X$ if ℓ is odd. It is known that each left cell of W' contains a unique involution and it carries an irreducible representation of W.

Lusztig [L8] proved the following result on left ϕ-cells of W.

Theorem 2.1.3 Let (W',S') be the Weyl group of type A_ℓ, $\ell > 3$, and let α be the automorphism of (W',S') of order 2. Let (W,S) be the fixed point set of α in W' and ϕ the function on W defined as above. Then each left ϕ-cell of W carries an irreducible representation of W. It contains a unique involution and is the intersection of W with a left cell of W'. All the irreducible representations of W arise in this way. □

§2.2 THE CANONICAL ISOMORPHISM BETWEEN $\mathbb{Q}[X^{\frac12}]W$ AND $H_{\mathbb{Q}[X^{\frac12}]}$

As mentioned in the Introduction, although Tits [Bou] showed that the Hecke-algebra $H_{\mathbb{C}}$ is isomorphic to the group algebra $\mathbb{C}W$, no explicit isomorphism between these algebras was known until the paper of Lusztig [L4]. We shall now prove the result of Lusztig which gives an explicit construction of an isomorphism of the generic Hecke algebra over $\mathbb{Q}[X^{\frac12}]$ with the group algebra over $\mathbb{Q}[X^{\frac12}]$ of the associated Weyl group, where $X^{\frac12}$ is an indeterminate.

Let us fix a finite Weyl group (W,S) and its associated generic Hecke algebra H. Let E be the free $\mathbb{Q}[X^{\frac12}]$-module with basis $(e_w)_{w\in W}$. We know that the formulae

$$T_s e_w = \begin{cases} -e_w & \text{if } s \in \mathcal{L}(w) \\ Xe_w + X^{\frac12} \sum\limits_{\substack{y-w \\ s\in\mathcal{L}(y)}} \mu(y,w)e_y & \text{if } s \notin \mathcal{L}(w) \end{cases}$$

$$e_w T_t = \begin{cases} -e_w & \text{if } t \in R(w) \\ Xe_w + X^{\frac12} \sum\limits_{\substack{y-w \\ t\in R(y)}} \mu(y,w)e_y & \text{if } t \notin R(w) \end{cases}$$

define an (H,H)-bimodule structure on E.

We also define a left W-module structure on E by

$$s.e_w = \begin{cases} -e_w & \text{if } s \in \mathcal{L}(w) \\ e_w + \sum\limits_{\substack{y-w \\ s\in\mathcal{L}(y)}} \mu(y,w)e_y & \text{if } s \notin \mathcal{L}(w) \end{cases}$$

and a right W-module structure by

$$e_w.t = \begin{cases} -e_w & \text{if } t \in R(w) \\ e_w + \sum\limits_{\substack{y-w \\ t\in R(y)}} \mu(y,w)e_y & \text{if } t \notin R(w) \end{cases}$$

The left and right W-module structures commute with each other. However, the left H-module structure doesn't necessarily commute with the right W-module structure.

For each two-sided cell $Z \subset W$, we consider the $\mathbb{Q}[X^{\frac{1}{2}}]$-submodules E_Z and E_Z' of E, where E_Z is spanned by the e_w with $w \underset{\Gamma}{\leq} x$ for some $x \in Z$ and E_Z' by the e_w with $w \underset{\Gamma}{\leq} x$ for some $x \in Z$, $w \notin Z$.

Let $gr(E) = \underset{Z}{\oplus} (E_Z/E_Z')$, summed over all two-sided cells Z of W. It is clear that E_Z, E_Z' are both left H-submodules, left W-submodules and right W-submodules of E. So gr(E) is in a natural way a left H-module, a left W-module and a right W-module. The image $\bar{e}_w$ of the elements $e_w \in E$ form a basis of gr(E).

<u>Lemma 2.2.1</u> <u>Let W,H be as above. Then the left H-module structure and the right W-module structure on gr(E) commute.</u>

<u>Proof</u> Let $w \in W$ and let Z be the two-sided cell containing w. Let $s,t \in S$. We must prove that $(T_s\bar{e}_w)t = T_s(\bar{e}_wt)$, or equivalently that $(T_se_w)t - T_s(e_wt)$ is a linear combination of element e_y, $y \notin Z$.

Assume first that $s \in \mathcal{L}(w)$, $t \in R(w)$. Then $(T_se_w)t = T_s(e_wt) = e_w$.

Next we assume that $s \in \mathcal{L}(w)$, $t \notin R(w)$. Then $\ell(swt) = \ell(w)$ and we have $(T_se_w)t = T_s(e_wt) = -e_w - \sum\limits_{\substack{y-w \\ t\in R(y)}} \mu(y,w)e_y$, since for all y, y-w such that $t \in R(y)$, we have $s \in \mathcal{L}(y)$ by Theorem 1.5.2. The case where $s \notin \mathcal{L}(w)$, $t \in R(w)$ is entirely similar.

Finally, assume that $s \notin \mathcal{L}(w)$, $t \notin R(w)$. We have

$$(T_s e_w)t = (Xe_w + X^{\frac{1}{2}} \sum_{\substack{y-w \\ s\in\mathcal{L}(y)}} \mu(y,w)e_y)t$$

$$= Xe_w + X \sum_{\substack{y-w \\ t\in R(y)}} \mu(y,w)e_y - X^{\frac{1}{2}} \sum_{\substack{y\ w \\ s\in\mathcal{L}(y) \\ t\in R(y)}} \mu(y,w)e_y + X^{\frac{1}{2}} \sum_{\substack{y-w \\ s\in\mathcal{L}(y) \\ t\notin R(y)}} \mu(y,w)e_y$$

$$+ X^{\frac{1}{2}} \sum_{\substack{y-w \\ s\in\mathcal{L}(y) \\ t\notin R(y)}} \sum_{\substack{z-w \\ t\in R(z)}} \mu(y,w)\mu(z,y)e_z$$

Note that for the z in the last sum we have automatically $s \in \mathcal{L}(z)$.

Similarly, $T_s(e_w.t) = T_s(e_w + \sum_{\substack{y-w \\ t\in R(y)}} \mu(y,w)e_y)$

$$= Xe_w + X^{\frac{1}{2}} \sum_{\substack{y-w \\ s\in\mathcal{L}(y)}} \mu(y,w)e_y - \sum_{\substack{y-w \\ t\in R(y) \\ s\in\mathcal{L}(y)}} \mu(y,w)e_y + X \sum_{\substack{y-w \\ t\in R(y) \\ s\notin\mathcal{L}(y)}} \mu(y,w)e_y$$

$$+ X^{\frac{1}{2}} \sum_{\substack{y-w \\ t\in R(y) \\ s\notin\mathcal{L}(y)}} \sum_{\substack{z-y \\ s\in\mathcal{L}(z)}} \mu(y,w)\mu(z,y)e_z$$

Also, for the z in the last sum we have automatically $t \in R(z)$.

Subtracting, we have

$$(T_s e_w)t - T_s(e_w.t) = \sum_{\substack{y-w \\ s\in\mathcal{L}(y) \\ t\in R(y)}} (X^{\frac{1}{2}}-1)^2 \mu(y,w)e_y + X^{\frac{1}{2}} \sum_{\substack{z \\ s\in\mathcal{L}(z) \\ t\in R(z)}} \alpha_z e_z \tag{2.2.1a}$$

where $\alpha_z = \sum_{\substack{y \\ z-y-w \\ s\in\mathcal{L}(y) \\ t\notin R(y)}} \mu(y,w)\mu(z,y) - \sum_{\substack{y \\ z-y-w \\ s\notin\mathcal{L}(y) \\ t\in R(y)}} \mu(y,w)\mu(z,y)$ is an integer.

When we specialize by replacing X by 1, $(T_s e_w)t - T_s(e_w.t)$ becomes $(s.e_w).t - s.(e_w.t) = 0$ since E is a (W,W)-bimodule. The right hand side of (2.2.1a) specializes to $\sum_{\substack{z \\ s\in\mathcal{L}(z) \\ t\in R(z)}} \alpha_z e_z$. Thus, this is zero. So $\alpha_z = 0$ for all z in the sum.

Hence the right hand side of (2.2.1a) is equal to $(X^{\frac{1}{2}}-1)^2 \sum_{\substack{y-w \\ s\in\mathcal{L}(y) \\ t\in R(y)}} \mu(y,w)e_y$.

Therefore our result follows from the following lemma by which we see that all y in this sum satisfy $y \notin Z$.

<u>Lemma 2.2.2</u> <u>If W is a finite Weyl group and if we are given $y - w$ in W such that $\mathcal{L}(y) \not\subset \mathcal{L}(w)$ and $R(y) \not\subset R(w)$ then we have $y \underset{\Gamma}{\sim} w$.</u> □

The proof of Lemma 2.2.2 is based on some rather deep results on infinite dimensional representations of the complex Lie algebra whose Weyl group is W. We omit this proof and refer the readers to [BK], [Jo3], [Vo2].

<u>Lemma 2.2.3</u> <u>Let A,B be two $Q[X^{\frac{1}{2}}]$-algebras and let α be an algebra homomorphism from A to B. If for any homomorphism ψ of $\mathbb{Q}[X^{\frac{1}{2}}]$ into a field K, the specialized homomorphism $\bar{\alpha}:A_K \to B_K$ is an isomorphism then α is an isomorphism.</u>

<u>Proof</u>: Note that for any $\mathbb{Q}[X^{\frac{1}{2}}]$-algebra C and $c \in C$, if the image of c in the specialized algebra C_f of C is zero for any homomorphism f of $\mathbb{Q}[X^{\frac{1}{2}}]$ into a field, then c itself must be zero.

For each homomorphism ψ of $\mathbb{Q}[X^{\frac{1}{2}}]$ into a field K, we have the following commutative diagram

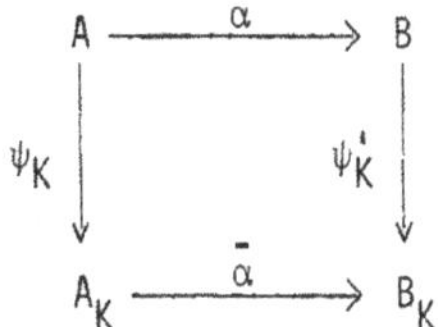

where ψ_K, ψ'_K are the specializations corresponding to ψ and hence they are both surjective. Given $b \in B$, there exists $x \in A_K$ such that $\bar{\alpha}(x) = \psi'_K(b)$. Thus there exists $a \in A$ satisfying $\psi_K(a) = x$. Hence $\psi'_K.\alpha(a) = \bar{\alpha}.\psi_K(a) = \bar{\alpha}(x) = \psi'_K(b)$ and so $\alpha(a)-b$ is in the kernel of ψ'_K for any homomorphism ψ of $\mathbb{Q}[X^{\frac{1}{2}}]$ into a field K. By the above remark, this implies that $\alpha(a) = b$ and so α is surjective. Now assume that $a' \in A$ is in the kernel of α. Then $\bar{\alpha}.\psi_K(a') = \psi'_K.\alpha(a') = 0$. By the assumption that $\bar{\alpha}$ is an isomorphism, we have $\psi_K(a') = 0$. So a' is in the kernel of ψ_K for any homomorphism ψ of $\mathbb{Q}[X^{\frac{1}{2}}]$ into a field K. Thus a' must be zero and α is injective. So α is an isomorphism. □

We can now state the result of Lusztig [L4].

<u>Theorem 2.2.4</u> <u>Assume that W is a finite Weyl group.</u>

(a) <u>There is a unique homomorphism of $\mathbb{Q}[X^{\frac{1}{2}}]$-algebras $\phi: H \to \mathbb{Q}[X^{\frac{1}{2}}]W$ such that for any $h \in H$ and any $w \in W$ we have $h\bar{e}_w = \phi(h).\bar{e}_w$.</u>

(b) Given any homomorphism ψ of $\mathbb{Q}[X^{\frac{1}{2}}]$ into a field K of characteristic o with $\psi(X) \neq 0$, the specialized homomorphism of K-algebras $\Phi_\psi : H_K \to KW$ is an isomorphism. In particular, if we take $K = \mathbb{Q}[q^{\frac{1}{2}}]$ to be the extension of $\mathbb{Q}$ of degree 1 or 2 generated by the square root $q^{\frac{1}{2}}$ of a prime power q and take ψ satisfying $\psi(X^{\frac{1}{2}}) = q^{\frac{1}{2}}$, then we get an isomorphism $\Phi_\psi : H_{\mathbb{Q}[q^{\frac{1}{2}}]} \to \mathbb{Q}[q^{\frac{1}{2}}]W$.

Proof: Let $\mathrm{End}_W\, gr(E)$ be the algebra of endomorphisms of the $\mathbb{Q}[X^{\frac{1}{2}}]$-module gr(E) which commute with the right W-module structure of gr(E). By Lemma 2.2.1, the left H-module structure on gr(E) gives rise to an algebra homomorphism $H \xrightarrow{\alpha} \mathrm{End}_W\, gr(E)$. The left W-module structure on gr(E) gives rise to an algebra homomorphism $\mathbb{Q}[X^{\frac{1}{2}}]W \xrightarrow{\beta} \mathrm{End}_W\, gr(E)$. To show (a), it is enough to show that β is an isomorphism. If so then $\Phi = \beta^{-1}.\alpha$ would be a required homomorphism. The uniqueness of Φ follows from the fact that β is an isomorphism and the condition that $h.\bar{e}_w = \Phi(h).\bar{e}_w$ for any $h \in H$ and $w \in W$. Now we shall show that β is an isomorphism. By Lemma 2.2.3, it is enough to show that for any homomorphism ψ of $\mathbb{Q}[X^{\frac{1}{2}}]$ into a field K the specialized homomorphism $\bar{\beta} : KW \to \mathrm{End}_W\, gr(E \otimes K)$ is an isomorphism. This is a composition $KW \xrightarrow{\bar{\beta}'} \mathrm{End}_W(E \otimes K) \xrightarrow{\bar{\beta}''} \mathrm{End}_W\, gr(E \otimes K)$, where $\mathrm{End}_W(E \otimes K)$ is the algebra of K-endomorphisms of $E \otimes K$ commuting with the right action of W on $E \otimes K$. We see from §1.2 and §1.5 that $E \otimes K$ is isomorphic to the group algebra KW of W as (W,W)-bimodules and hence $E \otimes K$ is the two-sided regular representation of W. So $\bar{\beta}'$ is an isomorphism. On the other hand, all KW-modules are semisimple and so $\bar{\beta}''$ is an isomorphism. This implies that β is an isomorphism and (a) follows.

Now let ψ be a homomorphism of $\mathbb{Q}[X^{\frac{1}{2}}]$ into a field K of characteristic o with $\psi(X) \neq 0$. For any $h \in H_K$ let $\hat{h}$ be the endomorphism of $E \otimes K$ defined by left multiplication by h (note that $\hat{h}$ is not necessarily in $\mathrm{End}_W(E \otimes K)$). Then $\hat{h}$ induces an endomorphism $\bar{h}$ of $gr(E \otimes K)$ which is in $\mathrm{End}_W gr(E \otimes K)$. By the above result, $\bar{h}$ can be regarded as an element of KW via the isomorphism $\bar{\beta}$. Clearly, $\bar{h} = \Phi_\psi(h)$. Thus if $h \in \ker \Phi_\psi$ then $\bar{h} = 0$ and this implies that $\hat{h}$ is a nilpotent endomorphism of $E \otimes K$. By the assumption that $\psi(X) \neq 0$, we see from §1.2 and §1.5 that $E \otimes K$ is isomorphic to H_K as the left H-modules and so $E \otimes K$ is the left regular representation of H. This implies that the map $h \to \hat{h}$ from H_K to $\mathrm{End}(E \otimes K)$ is injective. So h is also nilpotent. But H_K is semisimple. So $h = 0$ i.e. Φ_ψ is injective. By the fact that $\dim_K H_K = \dim_K KW$, Φ_ψ must be surjective and so Φ_ψ is an isomorphism. Our proof is complete. □

Examples 2.2.5

(i) Assume that $W = \langle s | s^2 = 1\rangle$ which is the Weyl group of type A_1. W has two two-sided cells $\Gamma_1 = \{1\}$, $\Gamma_2 = \{s\}$ with $\Gamma_2 \underset{\Gamma}{\leq} \Gamma_1$. So

$$\begin{cases} T_s(\bar{e}_1) = X\bar{e}_1 \\ T_s(\bar{e}_s) = -\bar{e}_s \end{cases}, \quad \begin{cases} 1\cdot\bar{e}_1 = \bar{e}_1 \\ 1.\bar{e}_s = \bar{e}_s \end{cases} \text{ and } \begin{cases} s(\bar{e}_1) = \bar{e}_1 \\ s(\bar{e}_s) = -\bar{e}_s \end{cases}$$

Assume $\Phi(T_s) = a_1.1 + a_s.s$ with $a_1, a_s \in \mathbb{Q}[X^{\frac{1}{2}}]$. Then by the condition that $T_s\cdot\bar{e}_w = \Phi(T_s)\cdot\bar{e}_w$ for all $w \in W$, we have

$$\begin{cases} a_1 + a_s = X \\ a_1 - a_s = -1 \end{cases}$$

Thus $a_1 = \frac{X-1}{2}$, $a_s = \frac{X+1}{2}$, i.e. $\Phi(T_s) = \frac{X-1}{2} + \frac{X+1}{2}s$.

(ii) The Weyl group of type A_2 has a presentation

$$W = \langle s_1, s_2 | s_1^2 = s_2^2 = (s_1s_2)^3 = 1\rangle,$$

which has three two-sided cells $\Gamma_1 = \{1\}$, $\Gamma_2 = \{s_1, s_2, s_1s_2, s_2s_1\}$ and $\Gamma_3 = \{s_1s_2s_1\}$ with $\Gamma_3 \underset{\Gamma}{\leq} \Gamma_2 \underset{\Gamma}{\leq} \Gamma_1$. So we have

$$\begin{aligned} T_{s_1}:\ & \bar{e}_1 \to X\bar{e}_1 \\ & \bar{e}_{s_1} \to -\bar{e}_{s_1} \\ & \bar{e}_{s_2} \to X\bar{e}_{s_2} + X^{\frac{1}{2}}\overline{e_{s_1s_2}} \\ & \overline{e_{s_1s_2}} \to \overline{-e_{s_1s_2}} \\ & \overline{e_{s_2s_1}} \to X\overline{e_{s_2s_1}} + X^{\frac{1}{2}}\bar{e}_{s_1} \\ & \overline{e_{s_1s_2s_1}} \to -\overline{e_{s_1s_2s_1}} \end{aligned}$$

We also have $1\cdot(\bar{e}_w) = \bar{e}_w$, $w \in W$, and

$$s_1 : \bar{e}_1 \to \bar{e}_1 \qquad s_2 : \bar{e}_1 \to \bar{e}_1$$

$$\bar{e}_{s_1} \to -\bar{e}_{s_1} \qquad \bar{e}_{s_1} \to \bar{e}_{s_1} + \bar{e}_{s_2s_1}$$

$$\bar{e}_{s_2} \to \bar{e}_{s_2} + \overline{e_{s_1s_2}} \qquad \bar{e}_{s_2} \to -\bar{e}_{s_2}$$

$$\bar{e}_{s_1s_2} \to -\bar{e}_{s_1s_2} \qquad \bar{e}_{s_1s_2} \to \bar{e}_{s_1s_2} + \bar{e}_{s_2}$$

$$\overline{e_{s_2s_1}} \to \overline{e_{s_2s_1}} + \overline{e_{s_1}} \qquad \bar{e}_{s_2s_1} \to -\bar{e}_{s_2s_1}$$

$$\overline{e_{s_1s_2s_1}} = -\overline{e_{s_1s_2s_1}} \qquad \overline{e_{s_1s_2s_1}} \to -\overline{e_{s_1s_2s_1}}$$

Let $\Phi(T_{s_1}) = \sum_{w\in W} a_w w$ with $a_w \in \mathbb{Q}[X^{\frac{1}{2}}]$. Then by the equations $T_{s_1}.\bar{e}_w = \Phi(T_{s_1}).\bar{e}_w$, $w \in W$, we get

$$\begin{pmatrix} 1 & 1 & 1 & 1 & 1 & 1 \\ 1 & -1 & 1 & 0 & -1 & 0 \\ 0 & 0 & 1 & 1 & -1 & -1 \\ 1 & 1 & -1 & -1 & 0 & 0 \\ 0 & 1 & 0 & -1 & 1 & -1 \\ 1 & -1 & -1 & 1 & 1 & -1 \end{pmatrix} \begin{pmatrix} a_1 \\ a_{s_1} \\ a_{s_2} \\ a_{s_1s_2} \\ a_{s_2s_1} \\ a_{s_1s_2s_1} \end{pmatrix} = \begin{pmatrix} X \\ -1 \\ 0 \\ X \\ X^{\frac{1}{2}} \\ -1 \end{pmatrix}$$

So we have $\Phi(T_{s_1}) = \frac{X-1}{2} + \frac{X+1}{2} s_1 + \frac{(X^{\frac{1}{2}}-1)^2}{6} (-s_2 + s_1s_2 - s_2s_1 + s_1s_2s_1)$.

By symmetry, we also have

$$\Phi(T_{s_2}) = \frac{X-1}{2} + \frac{X+1}{2} s_2 + \frac{(X^{\frac{1}{2}}-1)^2}{6}(-s_1 + s_2s_1 - s_1s_2 + s_1s_2s_1).$$

§2.3 REPRESENTATIONS OF THE WEYL GROUP

In this section we shall discuss the relationship between the decomposition of a Weyl group W into cells and the representation theory of W.

Let $\hat{W}$ be the set of irreducible representations of W. We begin by defining a subset of $\hat{W}$ introduced by Lusztig and called the set of special representations. Given any $E \in \hat{W}$ we shall define two polynomials $P_E(X)$, $D_E(X) \in \mathbb{Q}[X]$ called the fake degree and generic degree of E respectively.

We first define $P_E(X)$. Let V be the vector space affording the natural representation of W and $P = P(V)$ be the algebra of polynomial functions on V. P may be regarded in a natural way as a graded W-module. Let I be the set of W-invariants in P and I^+ the set of invariant polynomials with constant term 0. Then P/PI^+ is well known to be a W-module of dimension $|W|$ affording the regular representation of W. Since P/PI^+ is a graded W-module, this gives a grading to the regular representation of W. We define $P_E(X)$ by

$$P_E(X) = \sum_{i\geq 0} n_i X^i \in \mathbb{Z}[X]$$

where n_i is the multiplicity with which E occurs in the ith graded component of the regular representation. There is a useful formula for calculating this polynomial. It is given by

$$P_E(X) = \prod_{i=1}^{\ell} (1 - X^{d_i}) . \frac{1}{|W|} \sum_{w\in W} \frac{E(w)}{\det(1-Xw)}$$

where d_i, $1 \leq i \leq \ell$, are the degrees of the basic polynomial invariants of W [Ca2] and E(w) is the character value of E at w.

Now let G(q) be a finite Chevalley group over the field of q elements with Weyl group W. Let B(q) be a Borel subgroup of G(q), as in §1.2. Since the Hecke algebra $H_{\mathbb{C}}(q) = \mathrm{End}_{\mathbb{C}G(q)}(1_{B(q)}^{G(q)})$ is isomorphic to the group algebra $\mathbb{C}W$ there is a bijective correspondence between irreducible representations of $H_{\mathbb{C}}(q)$ and irreducible representations of W. This leads to a bijection between $\hat{W}$ and the set of irreducible characters of G(q) which occur in the induced character $1_{B(q)}^{G(q)}$. The degree of the representation of G(q) corresponding to $E \in \hat{W}$ has the form $D_E(q)$ where $D_E(X) \in \mathbb{Q}[X]$ is a polynomial with rational coefficients called the generic degree of E.

If $G(q) = SL_n(q)$ one knows that $D_E(X) = P_E(X)$ for all irreducible representations $E \in \hat{W}$. However this is not the case in general. For this reason $P_E(X)$ has been called the fake degree of E.

If we write

$$P_E(X) = \gamma_E X^{b_E} + \text{terms involving higher powers of } X$$

$$D_E(X) = \delta_E X^{a_E} + \text{terms involving higher powers of } X$$

where $\gamma_E \neq 0$, $\delta_E \neq 0$ then it was shown by Lusztig [L10] that $a_E \leq b_E$ for all E.

An irreducible representation $E \in \hat{W}$ is called a special representation if $a_E = b_E$. If W is the symmetric group S_n all irreducible representations of W are special, but this is not true for general Weyl groups.

It has been proved by Barbasch and Vogan [BV1], [BV2] that there is a 1-1 correspondence between special representations of W and 2-sided cells of W. This bijection comes from the study of primitive ideals in the universal enveloping algebra and will be discussed in §2.5.

We now describe some representations of W, not necessarily irreducible, which are conjectured to be precisely the representations coming from the left cells of W. Let W be generated as a Coxeter group by elements $s \in S$ and J be a subset of S.

Let W_J be the subgroup generated by the elements $s \in J$. For each irreducible character $E' \in \hat{W}_J$ consider the induced character

$$E'^W = \sum_{E \in \hat{W}} n_E E.$$

One knows [L10] that for each $E \in \hat{W}$ such that $n_E \neq 0$ we have $a_E \geq a_{E'}$. We define

$$\tilde{j}(E') = \sum_{\substack{E \in \hat{W} \\ a_E = a_{E'}}} n_E E.$$

and extend the definition by linearity to include the case when E' is not irreducible. Thus we truncate the sum to include only those $E \in \hat{W}$ with $a_E = a_{E'}$. We now give an inductive definition, following Lusztig [L10], of a family of representations of W called cell representations. This is the smallest set of representations satisfying the following conditions:

(i) If $W = 1$ the unit representation is a cell representation

(ii) If ϕ is a cell representation of $W_J \subset W$ then $\tilde{j}(\phi)$ and $\tilde{j}(\phi) \otimes$ sign are cell representations of W.

Cell representations have also been called constructible representations. [L10]. Lusztig [L1] gave an explicit description of the cell representations of W. Each irreducible representation of W is a component of at least one cell representation. Each cell representation contains a unique special representation as component and contains it with multiplicity 1. Two cell representations for which the special components are distinct have no common irreducible components. Thus two irreducible representations ϕ, ϕ' may be said to lie in the same family if there exist cell representations c, c' such that ϕ occurs in c, ϕ' occurs in c', and c,c' have the same special component. Thus we have a decomposition of $\hat{W}$ into families with one family for each special representation of W.

It was conjectured by Lusztig [Ca] that the cell representations of W are precisely those coming from the left cells of W by the procedure described in Chapter 1, and that two left cells lie in the same two sided cell if and only if their representations have the same special representation as component.

The situation has been clarified considerably by Barbasch and Vogan [BV1], [BV2] in their papers on primitive ideals of the enveloping algebra, which we shall discuss in §2.5.

§2.4 THE KAZHDAN-LUSZTIG CONJECTURE FOR COMPOSITION FACTORS OF VERMA MODULES

Let $\underline{g}$ be a semisimple complex Lie algebra. We wish to state a conjecture of Kazhdan and Lusztig relating Kazhdan-Lusztig polynomials with the theory of infinite dimensional representations of $\underline{g}$.

Let $\underline{h}$ be a Cartan subalgebra of $\underline{g}$, $\underline{h}^* = \mathrm{Hom}(\underline{h},\mathbb{C})$, $\Phi \subset \underline{h}^*$ the root system of $\underline{g}$, π a choice of simple roots, Φ^+ the corresponding positive root system, W the Weyl group of $\underline{g}$ in Aut($\underline{h}^*$) generated by reflections corresponding to the roots α, α runs over Φ. Then $\underline{g}$ has a decomposition

$$\underline{g} = \underline{n}_+ \oplus \underline{h} \oplus \underline{n}_-$$

where $\underline{n}_+$ is the sum of positive root spaces and $\underline{n}_-$ the sum of negative ones. Let $U = U(\underline{g})$ be the enveloping algebra of $\underline{g}$.

Elements of $\underline{h}^*$ are called weights. For any $\lambda \in \underline{h}^*$, $I_\lambda = U\underline{n}_+ + \sum_{h\in \underline{h}} U(h-\lambda(h))$ is a left ideal of U. So $V_\lambda = U/I_\lambda$ is a left $\underline{g}$-module. Let $v_\lambda = I_\lambda + \bar{1}$. Then $V_\lambda = Uv_\lambda$. We have $\underline{n}_+ v_\lambda = 0$ and $hv_\lambda = \lambda(h)v_\lambda$ for any $h \in \underline{h}$. V_λ is called the Verma module with the highest weight λ. V_λ can be decomposed as a direct sum of weight spaces under $\underline{h}$. Any weight μ of V_λ satisfies $\mu < \lambda$, where $\mu < \lambda$ means that $\lambda - \mu$ is a sum of positive roots. The multiplicity of weight λ of V_λ is 1. The sum of all submodules of V_λ with highest weights less than λ forms a unique maximal $\underline{g}$-submodule of V_λ, written K_λ. The quotient $\mathrm{V}_\lambda = V_\lambda/K_\lambda$ is an irreducible $\underline{g}$-module with the highest weight λ.

It is well known that any Verma module has a finite Jordan-Hölder series and any composition factor of the Verma module V_λ has the form V_μ for some μ with $\mu < \lambda$.

Let $Z(\underline{g})$ be the centre of U. Then $Z(\underline{g})$ is isomorphic to the polynomial ring $\mathbb{C}[X_1,\ldots,X_\ell]$ with ℓ variables X_i, $\ell = \mathrm{rank}\ \underline{g}$. Each element z of $Z(\underline{g})$ acts on V_λ by scalar multiplication. This gives a one dimensional representation of $Z(\underline{g})$ called the central character χ_λ.

The following result gives a necessary and sufficient condition for $\chi_\lambda = \chi_\mu$, $\lambda,\mu \in \underline{h}^*$.

<u>Theorem 2.4.1</u> (Harish-Chandra [Hu1]): For $\lambda,\mu \in h^*$, we have

$$\chi_\lambda = \chi_\mu \iff \lambda + \rho = w(\mu + \rho) \text{ for some } w \in W \qquad \square$$

where $\rho = \frac{1}{2} \sum_{\alpha\in\Phi^+} \alpha$.

Define a dot adtion of W on $\underline{h}^*$ by $w\cdot\lambda = w(\lambda+\rho) - \rho$ for $w \in W$ and $\lambda \in \underline{h}^*$. Then by Harish-Chandra's Theorem, we see that V_μ can occur as a composition factor of V_λ only if $\mu \in W\cdot\lambda$ and $\mu < \lambda$.

A weight $\mu \in \underline{h}^*$ is called integral if $(\lambda,\alpha^\vee) \in \mathbb{Z}$ for any $\alpha \in \Phi$, and regular if $(\lambda+\rho,\alpha^\vee) \neq 0$ for any $\alpha \in \Phi$, where ρ is half the sum of positive roots, $\alpha^\vee$ the coroot of α and (,) the Killing form on $\underline{g}$. Let P be the set of all integral weights in $\underline{h}^*$, which is an abelian group. Let $\mathbb{Z}[P]$ be the group algebra of P over $\mathbb{Z}$ with a basis $\{e_\lambda | \lambda \in P\}$. For any $\underline{g}$-module E, we define $chE = \sum_{\mu \in p} m(\mu)e_\mu$ with $m(\mu)$ the multiplicity of the weight μ in E. This is called the character of E. Let $C_1 = \{\lambda \in V | (\lambda+\rho,\alpha^\vee) < 0 \text{ for all } \alpha \in \Pi\}$ and $C_w = w \cdot C_1$, where V is the Euclidean space spanned by Φ.

We can now state the Kazhdan-Lusztig conjecture [KL1].

Kazhdan-Lusztig Conjecture 2.4.2. For any $w \in W$ and any regular integral weight $\lambda \in C_w$, $ch\, V_\lambda = \sum_{\substack{y \in W \\ y \leq w}} \varepsilon_y \varepsilon_w P_{y,w}(1)\, ch\, V_{yw^{-1}\cdot\lambda}$ where the $P_{y,w}(X)$ are Kazhdan-Lusztig polynomials and $P_{y,w}(1)$ denotes the value of $P_{y,w}(X)$ at 1.

This conjecture was proved by J.L. Brylinski and M. Kashiwara [BK] and independently by A.A. Beilinson and J.N. Bernstein [BB], using the theory of holonomic systems.

Example 2.4.3 Let $\underline{g}$ be the type A_2, $W = \langle s_1, s_2 | s_1^2 = s_2^2 = (s_1s_2)^3 = 1\rangle$ and let λ be a regular integral weight in $C_{s_2s_1}$. Then we have $chV_\lambda = chV_\lambda - ch\, V_{s_2\cdot\lambda} - chV_{s_1s_2s_1\cdot\lambda} + ch\, V_{s_1s_2\cdot\lambda}$

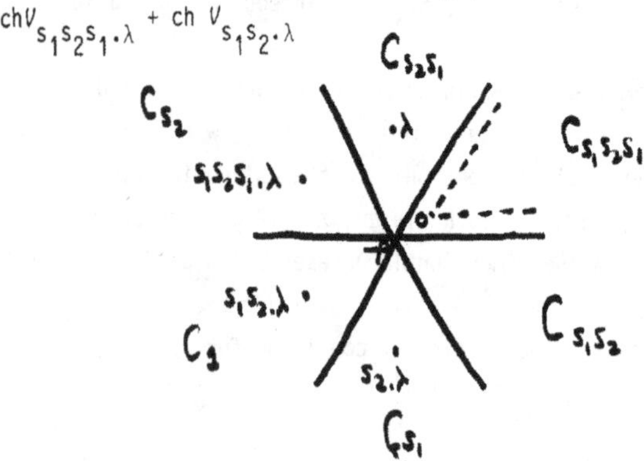

Thus the unitriangular matrix

$$(\varepsilon_y \varepsilon_w P_{y,w}(1)) \qquad y,w \in W$$

is the inverse of the matrix of decomposition numbers for the set of Verma modules whose highest weights are in a given equivalence class under the dot action of the Weyl group.

§2.5 CLASSIFICATION OF PRIMITIVE IDEALS IN UNIVERSAL ENVELOPING ALGEBRAS OF SEMISIMPLE LIE ALGEBRAS

We keep all the notation of §2.4. In particular let $\underline{g}$ be a semisimple Lie algebra over $\mathbb{C}$ and U its universal enveloping algebra. The finite dimensional irreducible representations of $\underline{g}$ (so also of U) were determined by E. Cartan [Hu1]. However the determination of the set of all irreducible representations of $\underline{g}$ does not seem to be a realistic problem, as there are so many of them. A more manageable problem is the determination of the set of primitive ideals of U. An ideal is called primitive if it is the annihilator of some irreducible representation. So every irreducible representation determines some primitive ideal, but distinct irreducible representations can determine the same primitive ideal. The determination of the primitive ideals of U has become a major goal in the theory of enveloping algebras.

Recall that in §2.4 we defined the Verma module V_λ and its irreducible quotient V_λ both with the highest weight $\lambda \in \underline{h}^*$. Let $J_\lambda = \mathrm{Ann}_U V_\lambda$. Then J_λ is a primitive ideal of U.

Duflo [Du] showed the following important result which enables us only to consider annihilators of simple highest weight modules.

Theorem 2.5.1 The J_λ, $\lambda \in \underline{h}^*$, are the only primitive ideals of U. □

For any $\lambda \neq \mu$ in $\underline{h}^*$, we always have $V_\lambda \ncong V_\mu$. But it may happen that $J_\lambda = J_\mu$. Let us consider when this can happen. We see that $z - \chi_\lambda(z)1 \in J_\lambda$ for any $z \in Z(\underline{g})$ and $\lambda \in \underline{h}^*$. So $J_\lambda = J_\mu$ implies $\chi_\lambda = \chi_\mu$. For each $\lambda \in \underline{h}^*$, we define $X_\lambda = \{J_\mu | \mu \in W\cdot\lambda\}$. Then by Theorem 2.4.1, we have $\chi_\lambda = \chi_\mu$ if and only if $\mu \in W\cdot\lambda$. In particular, X_λ is a finite set. We call X_λ fibres. Then $X = \bigcup_{\lambda \in \underline{h}^*} X_\lambda$ is the set of all primitive ideals of U.

We have the following diagram

where $\underline{Z}$ is the algebraic variety consisting of maximal ideals of $Z(\underline{g})$. It satisfies $\underline{Z} \cong \mathbb{C}^\ell.h^*/W.$ is the set of orbits of $\underline{h}^*$ under the dot action of W.

The isomorphism $\underline{h}^*/W. \to \underline{Z}$ arises as follows. $\underline{h}^*$ is an algebraic variety with coordinate ring $S[\underline{h}]$, $S[\underline{h}]$ the symmetric algebra of $\underline{h}$. W acts on $S[\underline{h}]$. The ring of invariants $S[\underline{h}]^{W.}$ is isomorphic to $Z[\underline{g}]$ by Harish-Chandra [HC]. So $S[\underline{h}]^{W.}$ is the coordinate ring **of the algebraic** variety $\underline{Z}$. Thus the isomorphism $Z[\underline{g}] \to S[\underline{h}]^{W.}$ gives rise to an isomorphism of varieties $\underline{h}^*/W. \to \underline{Z}$.

The above diagram is commutative.

Now the problem of classifying the primitive ideals of U is reduced to determining the X_λ. It is known that X_λ has a unique maximal element and also a unique minimal element with respect to inclusion. So the set X^{max} of maximal primitive ideals, one for each X_λ, is in bijection with $\underline{Z}$. Similarly for X^{min} by replacing "maximal" by "minimal".

We now compare the fibres X_λ for different weights λ.

For $\lambda \in \underline{h}^*$, let $\Phi_\lambda^+ = \{\alpha \in \Phi | (\lambda+\rho,\alpha^V) > 0 \in \mathbf{Z}\}$.

<u>Theorem 2.5.2</u> <u>(Jantzen's translation principle for Lie algebras [BJ]. Let $\lambda,\mu \in h^*$ and suppose that $\lambda-\mu$ is integral. Suppose $\Phi_\lambda^+ = \Phi_\mu^+$. Then there is an isomorphism $X_\lambda \to X_\mu$ of ordered sets with respect to inclusion.</u> □

By this theorem, if λ is integral and regular in $\underline{h}^*$ then the fibre X_λ is uniquely determined and is independent of the choice of λ.

$\lambda \in \underline{h}^*$ is called dominant if $(\lambda,\alpha^V) > 0$ for all $\alpha \in \Pi$ and antidominant if $(\lambda,\alpha^V) < 0$ for all $\alpha \in \Pi$. Now we consider the case when λ is antidominant, integral and also regular. Then $w.\lambda$ is regular and integral for any $w \in W$. Consider $X_\lambda = \{J_{w.\lambda} | w \in W\}$. We have a map $W \to X_\lambda$ by sending w to $J_{w.\lambda}$. This gives an equivalence relation on W: $w \sim w'$ in $W \Leftrightarrow J_{w.\lambda} = J_{w'.\lambda}$, which is independent of λ by Jantzen's translation principle for Lie algebras. Each equivalence class contains at least one element w with $w^2 = 1$.

Barbasch and Vogan [BV2] showed the following result for this equivalence relation on W.

<u>Theorem 2.5.3</u> <u>For any antidominant, integral, regular weight λ and $w,w' \in W$, we have</u>

$$\underline{w <_L w' \Leftrightarrow J_{w.\lambda} \supseteq J_{w'.\lambda}}$$

<u>In particular, $J_{w.\lambda} = J_{w'.\lambda} \Leftrightarrow w \sim_L w'$.</u> □

Thus the number of primitive ideals in a regular integral fibre is equal to the number of left cells in the Weyl group. Also the inclusion relations between the primitive ideals are given by the partial order relation on the left cells in W.

The proof of this result uses the proof of the Kazhdan-Lusztig conjecture for composition factors of Verma modules. Note that if λ is not integral then a similar theory can be developed by replacing W by the Weyl group of the subset of roots α satisfying $(\lambda,\alpha^V) \in \mathbf{Z}$.

Next we shall introduce the Goldie rank representations of W constructed by Joseph [JO1].

We know that for any primitive ideal J_μ of U, U/J_μ is a Noetherian prime ring whose ring of fractions is an Artinian simple ring $M_n(D)$, D a division ring. We call n the Goldie rank of J_μ. Joseph [JO1] showed that for any regular, integral weight μ in a given chamber C_w, $w \in W$, the Goldie rank of J_μ is a polynomial function in μ, say $P_w(\mu) = \text{rank } U/J_\mu$.

W acts on $\underline{h}^*$ and hence also on $P(\underline{h}^*)$ the ring of polynomial functions on $\underline{h}^*$. Joseph [J01] showed that $\sigma_w = \mathbb{C}Wp_w$ is an irreducible W-module, and that if $\mathbb{C}Wp_w \cong \mathbb{C}Wp_{w'}$ then $\mathbb{C}Wp_w = \mathbb{C}Wp_{w'}$. Let λ be any integral weight of C_1 (see §2.4). Joseph [J01] showed that $J_{w.\lambda} = J_{w'.\lambda}$ if and only if p_w is a positive multiple of $p_{w'}$. Consider the set of all w' with $\mathbb{C}Wp_{w'} = \mathbb{C}Wp_w$. For such w', we choose one $p_{w'}$ in each set of positive scalar multiples. Then these $p_{w'}$ form a basis of $\mathbb{C}Wp_w$. So $\dim \mathbb{C}Wp_w$ is the number of primitive ideals $J_{w'.\lambda}$ with $\mathbb{C}Wp_{w'} = \mathbb{C}Wp_w$. So the number of primitive ideals in X_λ is $\sum_\sigma \dim \sigma$, summed over all the Goldie rank representations of W.

Barbasch and Vogan [BV2] showed

<u>Theorem 2.5.4</u> <u>For $w, w' \in W$, we have</u>

$$\underline{\sigma_w = \sigma_{w'} \Longleftrightarrow w \underset{\Gamma}{\sim} w'} \qquad \square$$

This gives a bijection between the set of Goldie rank representations of W and the set of two-sided cells of W.

Recall that in §2.3 we defined two integers a_E, b_E for any irreducible representation E of W. We also defined a special representation E of W characterized by the property that $a_E = b_E$. Then Barbasch and Vogan [BV2] showed

<u>Theorem 2.5.5</u> <u>Each two-sided cell of W gives rise to a representation of W which has a unique irreducible component E such that a_E is minimal. The Goldie rank representations are the irreducible representations of W arising in this way. dim E is the number of left cells in the given two-sided cell. The Goldie rank representations are just the special representations of W.</u> $\square$

<u>Example 2.5.6</u> We shall now describe the situation when $\underline{g} = sl_n(\mathbb{C})$ and $W = S_n$. In this case every irreducible representation of W is a special representation, and so the situation is simpler than in the general case.

The general theory of Joseph, Barbasch and Vogan outlined above gives the following information in this case. The number of primitive ideals in a regular integral fibre is equal to the number of left cells of S_n, which is equal to the sum of the degrees of all the irreducible characters of S_n. The number of left cells in a given two sided cell of S_n is equal to the degree of one of the irreducible characters of S_n.

We can get further explicit information in this case since the irreducible characters of S_n correspond to partitions of n. (see Theorems 1.7.2 and 1.7.3). Recall that in §1.7 we defined the Robinson-Schensted map from S_n to the set of pairs (P,Q) of standard Young tableaux of the same shape. Let $w \in W$ correspond to the pair $(P(w), Q(w))$. Then w, w' lie in the same left cell if and only if $P(w) = P(w')$. Thus the left cells of S_n correspond to standard Young tableaux. So the primitive ideals in a regular integral fibre correspond to standard Young tableaux.

We also know from §1.7 that w, w' lie in the same two sided cell if and only if $P(w)$, $P(w')$ have the same shape. Thus there is one two sided cell of S_n for each partition ξ of n. The number of left cells in the two sided cell corresponding to ξ is the number of standard ξ-tableaux, and this is well known to be the degree d_ξ of the irreducible representation of S_n corresponding to ξ.

There is also an interesting connection between the two sided cells of S_n and the nilpotent orbits of $\underline{g} = sl_n(\mathbb{C})$ under the action of its adjoint group. The enveloping algebra U has a filtration

$$U^0 \subset U^1 \subset U^2 \subset \dots$$

where U^i consists of elements of degree at most i. The associated graded algebra

$$\mathrm{gr}\ U = \oplus\ U^n/U^{n-1}$$

is isomorphic to the symmetric algebra $S(\underline{g})$ of $\underline{g}$. For each J of U there is a corresponding ideal grJ in gr$U$. If J is primitive in U it can be shown that $\sqrt{\mathrm{grJ}}$ is a prime ideal in gr$U$. Now the isomorphism between $\underline{g}$ and its dual space $\underline{g}^*$ given by the Killing form leads to an isomorphism between the symmetric algebra $S(\underline{g})$ and the algebra $P(\underline{g})$ of polynomial functions on $\underline{g}$. Thus the ideal $\sqrt{\mathrm{grJ}}$ of $S(\underline{g})$ gives rise to a prime ideal of $P(\underline{g})$, and thus to an irreducible subvariety of $\underline{g}$ considered as an affine space. It turns out that this subvariety consists entirely of nilpotent elements of $\underline{g}$ and is invariant under the action of the adjoint group G of $\underline{g}$.

Let $\mathcal{O}$ be the set of nilpotent G-oribts of $\underline{g}$. There is a bijection from the set Λ_n of partitions of n to $\mathcal{O}$: for any $\xi = \{\xi_1 \geq \xi_2 \geq \dots\} \in \Lambda_n$ we let $\mathcal{O}_\xi$ be the set of nilpotent elements of $\underline{g}$ which have Jordan blocks of size $\xi_1, \xi_2, \dots$. The $\mathcal{O}_\xi$ are the elements of $\mathcal{O}$. In particular $\mathcal{O}$ is finite. Thus in the above irreducible subvariety of $\underline{g}$, there will be a unique nilpotent G-orbit which is open and dense in this subvariety. In this way we can define a map $\beta: X_\lambda \to \mathcal{O}$ by sending any primitive ideal J in X_λ to the dense orbit contained in the variety of grJ.

So we have the following diagram

$$S_n \xrightarrow{\ \alpha\ } X_\lambda \xrightarrow{\ \beta\ } \mathcal{O}$$

where $\alpha(w) = J_{w.\lambda}$ and λ is any dominant integral regular weight.

The following result was conjectured by Jantzen and proved by Joseph.

Theorem 2.5.7 (i) The map $\beta\circ\alpha : S_n \to \mathcal{O}$ is surjective. (ii) $w, w' \in S_n$ have the same image under $\beta\circ\alpha$ if and only if the standard tableaux $P(w)$, $P(w')$ have the same shape, i.e. if and only if w, w' lie in the same two sided cell.

So this gives a natural bijection between the two-sided cells of S_n and the nilpotent G-orbits on g.

§2.6 MODULAR REPRESENTATION THEORY OF ALGEBRAIC GROUPS AND RELATED FINITE GROUPS

Let G be a simply connected simple algebraic group defined over the algebraic closure K of the field $\mathbb{F}_p$ with p a prime. In this section we wish to state Lusztig's conjecture for composition factors of Weyl modules and to explain how this would give a character formula for all irreducible rational G-modules and also for finite dimensional irreducible KG^F-modules, where G^F is the group consisting of all fixed points of G under a standard Fobenius map $F:G \to G$.

First we want to define a Weyl module. Let $\underline{g}$ be the simple Lie algebra over $\mathbb{C}$ having the same root system Φ as G. Let $\underline{g}_{\mathbb{Z}} = \underline{h}_{\mathbb{Z}} \oplus \sum_{\alpha\in\Phi} \underline{g}^{\alpha}_{\mathbb{Z}}$ be a Chevalley lattice in $\underline{g}$. This means the following: (i) $\underline{h} = \underline{h}_{\mathbb{Z}} \otimes_{\mathbb{Z}} \mathbb{C}$ is a Cartan subalgebra of $\underline{g}$ on which elements of Φ are considered as linear functions and $\underline{h}_{\mathbb{Z}}$ consists of all $h \in \underline{h}$ such that $\lambda(h) \in \mathbb{Z}$ for all $\lambda \in P$, where $P \in \underline{h}^*$ is the set of integral weights. Thus if $h_\alpha \in \underline{h}$ is defined by $\lambda(h_\alpha) = (\lambda, \alpha^\vee)$ for all $\lambda \in \underline{h}^*$, then putting $h_i = h_{\alpha_i}$ one has $\underline{h}_{\mathbb{Z}} = \sum_{i=1}^{\ell} \mathbb{Z}h_i$, where $\Pi = \{\alpha_1,\ldots,\alpha_\ell\}$ is a choice of simple roots of Φ.
(ii) $\underline{g}^\alpha = \underline{g}^\alpha_{\mathbb{Z}} \otimes_{\mathbb{Z}} \mathbb{C}$ is the α-root space of $\underline{g}$ with respect to $\underline{h}$, and the generators e_α of the additive infinite-cyclic groups $\underline{g}^\alpha_{\mathbb{Z}}$ are chosen so that $[e_\alpha e_{-\alpha}] = h_\alpha$ when $\alpha \in \Phi^+$. (iii) If $\mathcal{U}_{\mathbb{Z}}$ denotes the Kostant ring of the enveloping algebra $\mathcal{U}$ of $\underline{g}$ generated by $\{\frac{e_\alpha^n}{n!}\ \alpha \in \Phi,\ n \in \mathbb{N}\}$, then $\underline{g}_{\mathbb{Z}}$ is stable under the adjoint action of $\mathcal{U}_{\mathbb{Z}}$ on $\underline{g}$. Let $\mathcal{U}_K = \mathcal{U}_{\mathbb{Z}} \otimes_{\mathbb{Z}} K$.

The Lie algebra $\underline{g}_K$ of G can be identified with $\underline{g}_{\mathbb{Z}} \otimes_{\mathbb{Z}} K$ in such a way that $\underline{h}_K = \underline{h}_{\mathbb{Z}} \otimes_{\mathbb{Z}} K$ is the Lie algebra of a maximal torus T, and identifying P with the character group X(T) of T, the subspace $\underline{g}^\alpha_K = \underline{g}^\alpha_{\mathbb{Z}} \otimes_{\mathbb{Z}} K$ of $\underline{g}_K$ becomes its α-rootspace with respect to the adjoint action of T. Thus we are identifying Φ with the root system of G with respect to T.

Now we can define a G-module called a Weyl module. Let $\lambda \in P$ be a dominant weight. Then there is a unique finite dimensional irreducible $\underline{g}$-module $V(\lambda)_{\mathbb{C}}$ with the highest weight λ. We choose a λ-weight vector v^+, obtaining a lattice $V(\lambda)_{\mathbb{Z}} = \mathcal{U}_{\mathbb{Z}} v^+$. The Weyl module $V(\lambda)$ is now constructed as $V(\lambda)_{\mathbb{Z}} \otimes_{\mathbb{Z}} K$. This vector space is first a $\mathcal{U}_K$-module, and then a G-module. The image of G under the representation on $V(\lambda)$ is generated by operators of the form $\sum_{n\geq 0} \frac{e_\alpha^n}{n!} \otimes a^n$ with $a \in K$ and $\alpha \in \Phi$.

The $\underline{g}$-module $V(\lambda)_{\mathbb{C}}$ is a direct sum of weight spaces $V(\lambda)_{\mathbb{C},\mu}$ with $\mu \prec \lambda$, and $V(\lambda)_{\mathbb{Z}}$ is the direct sum of the $V(\lambda)_{\mathbb{Z},\mu} = V(\lambda)_{\mathbb{C},\mu} \cap V(\lambda)_{\mathbb{Z}}$. We define $V(\lambda)_{\mu} = V(\lambda)_{\mathbb{Z},\mu} \otimes K$. In particular, we know $\overline{V(\lambda)}_{\mathbb{Z},\lambda} = \mathbb{Z}v^+$. Denote $x \otimes 1$ by $\bar{x}$ for any $x \in V(\lambda)_{\mathbb{Z}}$. Then $V(\lambda)_{\lambda} = K\overline{v^+}$ and $V(\lambda) = U_K\bar{v}^+$. Each submodule of $V(\lambda)$ is a direct sum of its weight spaces, so each proper submodule is contained in $\bigoplus_{\mu\neq\lambda} V(\lambda)_{\mu}$. This has to apply to the sum of all proper submodules, too, which therefore is the only maximal submodule $K(\lambda)$ of $V(\lambda)$. The quotient $L(\lambda) = V(\lambda)/K(\lambda)$ is an irreducible G-module. It is well known that the irreducible rational representations of G are the modules $L(\mu)$ where μ runs over dominant integral weights. The character $chV(\lambda)$ of $V(\lambda)$ may be written as a $\mathbb{Z}$-linear combination of characters $chL(\lambda)$ of irreducible G-modules $L(\mu)$:

$$chV(\lambda) = \sum_{\mu\in X^+(T)} [V(\lambda) : L(\mu)]chL(\mu) \qquad (*)$$

where $X^+(T)$ is the set of all dominant weights of $X(T)$, and $[V(\lambda):L(\mu)]$ is the number of times $L(\mu)$ occurs in $V(\lambda)$ as a composition factor. It is well known that $[V(\lambda):L(\mu)] \neq 0$ only if $\mu \prec \lambda$ and that $[V(\lambda):L(\lambda)] = 1$. So we may invert (*) to get

$$chL(\mu) = \sum_{\lambda\in X^+(T)} [L(\mu) : V(\lambda)]chV(\lambda) \qquad (**)$$

where $[L(\mu):V(\lambda)]$ are certain integers which need not be positive.

Lusztig's conjecture concerns the values of the integers $[L(\mu):V(\lambda)]$. In order to state the conjecture we introduce the action of a certain Coxeter group on the space $X_{\mathbb{R}} = X(T) \otimes \mathbb{R}$. Let $Y(T) = \mathrm{Hom}(K^*,T)$ be the cocharacter group of T, where K^* is the multiplicative group of K. There is a natural bilinear map $X(T) \times Y(T) \to \mathbb{Z}$ sending the pair, X,y into the integer $\langle X,y\rangle$ defined by $X(y(\lambda)) = \lambda^{\langle X,y\rangle}$ for $\lambda \in K^*$. Let $W_a(p)$ be the group of affine transformations of $X_{\mathbb{R}}$ generated by the reflections in all the affine hyperplanes of the form

$$H_{\alpha,n} = \{\lambda \in X_{\mathbb{R}} | \langle\lambda+\rho,\alpha^{\vee}\rangle = np\} \qquad \alpha \in \Phi,\ n \in \mathbb{Z}$$

where $\alpha^{\vee} \in Y(T)$ is the coroot corresponding to the root α. $W_a(p)$ is generated as a Coxeter group by the reflections in the hyperplanes $H_{\alpha_i,o}$ for simple roots α_i and $H_{\alpha_0,-1}$ where $\alpha_0^{\vee}$ is the highest coroot. ($Wa(p)$ is isomorphic to the affine Weyl group of type dual to that of G).

Let $A_1' = \{\lambda \in X_{\mathbb{R}} | -p < \langle\lambda+\rho,\alpha^{\vee}\rangle < 0 \text{ for all } \alpha \in \Phi^+\}$

and $A_w' = w.A_1'$ for all $w \in W_a(p)$. The subsets A_w' are the alcoves of $X_{\mathbb{R}}$ under the action of $W_a(p)$. An alcove A_w' is called dominant if $\langle\lambda+\rho,\alpha^{\vee}\rangle > 0$ for all $\lambda \in A_w'$ and all $\alpha \in \Phi^+$.

Let h be the Coxeter number of Φ. We can now state Lusztig's conjecture.

Lusztig's Conjecture [L2]: Assume that p is sufficiently large (we can take $p > h$). Assume that A'_w is dominant and $\lambda \in A'_w$ is an integral weight satisfying the Jantzen condition $\langle\lambda+\rho,\alpha_0^\vee\rangle < p(p-h+2)$. Then

$$\mathrm{ch}\, L(\lambda) = \sum_{\substack{y\in Wa \\ y<w \\ A'_y \text{ is dominant}}} (-1)^{\ell(w)} (-1)^{\ell(y)} P_{y,w}(1)\, \mathrm{ch}\, V(yw^{-1}.\lambda)$$

(The reader should compare this statement with that of the Kazhdan-Lusztig conjecture in §2.4).

From this one can deduce the character formula for any irreducible rational representation of G over K by making use of results of Jantzen and Steinberg as follows.

All the characters of the Weyl modules can be expressed explicitly by Weyl's character formula [Hu1]. To compute the character of any irreducible G-module $L(\lambda)$ with the highest weight $\lambda \in X(T)^+$, it is enough to compute all the coefficients $[L(\lambda):V(\mu)]$ in the expression (**).

Let $\overline{A'_1} = \{\lambda \in X_{\mathbb{R}} \mid -p < \langle\lambda+\rho,\alpha^\vee\rangle < 0 \text{ for all } \alpha \in \Phi^+\}$. Then $\overline{A'_1}$ is the closure of A'_1 in $X_{\mathbb{R}}$. Let S be the set of Coxeter generators of $Wa(p)$, i.e. the set of reflections in the walls of $\overline{A'_1}$. For any $\lambda \in A'_1$ and $w \in Wa(p)$, let $\tau(w) = \{s \in S \mid ws.\lambda < w.\lambda\}$. For any $\mu \in \overline{A'_1}$, let $S_\mu = \{s \in S \mid s.\mu = \mu\}$. We know that any alcove of $X_{\mathbb{R}}$ has the form A'_w for some $w \in Wa(p)$. For $\mu \in \overline{A'_1}$ and $w \in Wa(p)$, we say that $w.\mu$ is the upper closure of A'_w if $S_\mu \cap \tau(w) = \emptyset$. Then one can show that every $v \in X_{\mathbb{R}}$ lies in the upper closure of some alcove of $X_{\mathbb{R}}$.

It is well known that for any $\lambda,\mu \in X^+(T)$, $L(\mu)$ can be a composition factor of $V(\lambda)$ over K only if $\mu < \lambda$ and $\lambda = w.\mu$ for some $w \in Wa(p)$. Suppose $\lambda \in A'_1$, $w,w' \in Wa(p)$ with $w.\lambda,\ w'.\lambda \in X^+(T)$. Then Jantzen's translation principle for algebraic groups [Ja] asserts that for all $\mu \in \overline{A'_1}$ with $\tau(w') \cap S_\mu = \emptyset$ we have

$$L(w'.\lambda):V(w.\lambda)] = \begin{cases} [L(w'.\mu):V(w.\mu)] & \text{if } w.\mu \in X^+(T) \\ 0 & \text{otherwise} \end{cases}$$

Thus to compute the numbers $[L(\lambda):V(\mu)]$ for all pairs $\mu < \lambda$ in $X^+(T)$, it is enough to compute the numbers $[L(w'.\lambda):V(w.\lambda)]$ for only one integral weight $\lambda \in \overline{A'_1}$ and any $w,w' \in Wa(p)$ with $w.\lambda < w'.\lambda$ in $X^+(T)$. In particular, we can take λ in A'_1 when p is sufficiently large.

A weight $\mu \in X(T)$ is said to be in the restricted region X_p if $0 < \langle\mu,\alpha^\vee\rangle < p$ for each simple root α. Then $\lambda \in X^+(T)$ has a unique p-adic expression $\lambda = \lambda_0 + p\lambda_1 + \ldots \ldots + p^n\lambda_n$ so that each λ_i is in X_p.

Regarding G as an algebraic subgroup of a general linear group, we define a standard Frobenius map $F:G \to G$ by raising matrix entries to the p-th power. For any rational G-module V, let $V^{(p^r)}$ be the module V twisted by the r-th power of F. Thus $V^{(p^r)} = V$ as a vector space, but $g \in G$ acts on $V^{(p^r)}$ via $F^r(g)$. Observe that $L(\lambda)^{(p)} \cong L(p\lambda)$ by the highest weight classification. Steinberg's tensor product theorem states that $L(\lambda) \cong L(\lambda_0) \otimes L(\lambda_1)^{(p)} \otimes \ldots \otimes L(\lambda_n)^{(p^n)}$. Thus to compute the characters of $L(\lambda)$ for all dominant integral weights λ, it is enough to compute the characters of $L(\lambda)$ for all λ in the restricted region. When p is sufficiently large and $\lambda \in X_p$, the Jantzen condition $\langle\lambda+\rho,\alpha_0^\vee\rangle < p(p-h+2)$ is satisfied automatically. Therefore if Lusztig's conjecture is true then we can find the characters of all the irreducible G-modules with highest weights in $X^+(T)$.

Finally, we consider the groups of the form G^{F^n}, $n \geq 1$, where G^{F^n} is the group consisting of all F^n-fixed points of G. G^{F^n} is a finite Chevalley group. Let $X_{p^n} = \{\mu \in X(T) | 0 \leq \langle\mu,\alpha^\vee\rangle < p^n$ for all simple roots $\alpha\}$. Curtis and Steinberg [Bor] [Cu], [St] showed that the irreducible G-modules $L(\lambda)$, $\lambda \in X_{p^n}$, remain irreducible and inequivalent on restriction to G^{F^n}, and exhaust the irreducible KG^{F^n}-modules. So if Lusztig's Conjecture were proved to be true then we would get the dimensions and formal characters of all irreducible KG^{F^n}-modules.

§2.7 WEIGHT MULTIPLICITIES AND KAZHDAN-LUSZTIG POLYNOMIALS

As before, let g be a finite dimensional simple Lie algebra over $\mathbb{C}$ with Cartan subalgebra h. Let $P \subset h^*$ be the set of integral weights of g and P^+ be the subset of dominant integral weights. We recall that there is a unique finite dimensional irreducible g-module $V(\lambda)_{\mathbb{C}}$ for each $\lambda \in P^+$. The highest weight of $V(\lambda)_{\mathbb{C}}$ is λ. For any weight μ in $V(\lambda)_{\mathbb{C}}$ we define the multiplicity of μ to be the dimension of the μ weight space $V(\lambda)^{\mu}_{\mathbb{C}}$ in $V(\lambda)_{\mathbb{C}}$. We write

$$d_\mu(V(\lambda)_{\mathbb{C}}) = \dim V(\lambda)^{\mu}_{\mathbb{C}} .$$

In the present section we shall describe a result due to Lusztig which shows that the multiplicities $d_\mu(V(\lambda)_{\mathbb{C}})$ can be described as Kazhdan-Lusztig polynomials. We shall state the results without proof. The proofs can be found in [L5].

The Kazhdan-Lusztig polynomials which we need are those for an affine Weyl group. We define the action of an affine Weyl group on $h^*_{\mathbb{R}} = P \otimes \mathbb{R}$ as follows. W_a is the group of affine transformations of $h_{\mathbb{R}}^*$ generated by the reflections in all the affine hyperplanes of the form

$$L_{\alpha,n} = \{\lambda \in h_{\mathbb{R}}^*;\ \langle\lambda,\alpha^\vee\rangle = n\} \qquad \alpha \in \Phi,\ n \in \mathbb{Z}.$$

where $\alpha^\vee$ is the coroot corresponding to the root α. (Note the similarity between W_a and the group $W_a(p)$ discussed in §2.6). W_a is generated as a Coxeter group by the reflections in the hyperplanes $L_{\alpha_i,0}$ for simple roots α_i and $L_{\alpha_0,1}$ where $\alpha_0{}^{\vee}$ is the highest coroot. We can consider the Kazhdan-Lusztig polynomials of W_a with respect to this set S_a of Coxeter generators.

Now W_a can be written as a semidirect product $W_a = T_Q W$ where W is the subgroup of W_a generated by the reflections in the $L_{\alpha_i,0}$ and T_Q is the group of translations $x \to x + q$ where q lies in Q, the additive group generated by the roots. The cosets of W in W_a are therefore in natural bijective correspondence with the elements of Q. The double cosets $W\backslash W_a/W$ are therefore in natural bijective correspondence with the W-orbits on Q.

In order to be able to consider all of the irreducible representations $V(\lambda)_{\mathbb{C}}$, $\lambda \in P^+$, we must consider a slightly larger group than W_a. We know that Q is a subgroup of P of finite index. We therefore define $\tilde{W}_a$ to be the semidirect product $\tilde{W}_a = T_P W$ where T_P is the group of translations $x \to x + p$ with $p \in P$. W_a is a normal subgroup of $\tilde{W}_a$ of finite index. In fact we have a semidirect product decomposition $\tilde{W}_a = W_a\Omega$ where Ω is the set of elements of $\tilde{W}_a$ which transform the set S_a of Coxeter generators of W_a into itself. $\tilde{W}_a$ is not in general a Coxeter group. However it has the advantage that the set of double cosets $W\backslash\tilde{W}_a/W$ is in natural bijective correspondence with the set of W-orbits on P, i.e. with the set P^+ of dominant integral weights.

Now we have a length function $\ell(w)$, $w \in W_a$ and a partial order relation $w < w'$, $w,w' \in W_a$ coming from the fact that W_a is a Coxeter group. We may extend both the length function and the partial order relation to $\tilde{W}_a$ by defining

$$\ell(\gamma w) = \ell(w) \text{ if } \gamma \in \Omega\,, \quad w \in W_a$$

$$\gamma w < \gamma' w' \text{ if and only if } \gamma = \gamma' \text{ and } w < w',\ \gamma,\gamma' \in \Omega\,,\ w,w' \in W_a.$$

Using this length function and partial ordering we may define Kazhdan-Lusztig polynomials $P_{y,w}$ for $y,w \in \tilde{W}_a$ with $y < w$. These agree with the Kazhdan-Lusztig polynomials $P_{y,w}$ for W_a when $y,w \in W_a$ and satisfy $P_{\gamma y,\gamma w} = P_{y,w}$ if $y,w \in W_a$ and $\gamma \in \Omega$.

Now given $\lambda \in P$ let t_λ be the translation $x \to x + \lambda$. Then the double coset $Wt_\lambda W$ has a unique element of maximal length, which we call n_λ. Then the connection between the weight multiplicities in $V(\lambda)_{\mathbb{C}}$ and the Kazhdan-Lusztig polynomials of $\tilde{W}_a$ is given by the following theorem.

<u>Theorem 2.7.1</u> [L5]. Let μ,λ be dominant integral weights with $\mu < \lambda$. Then the multiplicity of μ in $V(\lambda)_{\mathbb{C}}$ is given by

$$d\mu(V(\lambda)_{\mathbb{C}}) = P_{n_\mu,n_\lambda}(1)$$

We shall now discuss some further results related to this in the special case when the simple Lie algebra g has type A. Let g be the Lie algebra $sl_n(\mathbb{C})$ of the special linear group $Sl_n(\mathbb{C})$. Then the dominanet integral weights of g are in bijective correspondence with partitions with at most n-1 parts. This may be seen as follows.

Let $T = \{t = \begin{pmatrix} x_1 & & 0 \\ & x_2 & \\ 0 & & x_n \end{pmatrix}$ where $x_1 x_2 \ldots x_n = 1\}$. T is a maximal torus of $Sl_n(\mathbb{C})$.

The fundamental weights of $SL_n(\mathbb{C})$ are the following functions $\omega_1,\ldots,\omega_{n-1}$ in X(T).

$$\omega_1(t) = x_1,\ \omega_2(t) = x_1 x_2, \ldots, \omega_{n-1}(t) = x_1 x_2 \ldots x_{n-1}.$$

The dominant integral weights are those of the form $\omega = m_1\omega_1 + \ldots + m_{n-1}\omega_{n-1}$ where $m_1,\ldots,m_{n-1} \in \mathbf{Z}$ satisfy $m_i \geq 0$. Thus

$$\omega(t) = x_1^{m_1+m_2+\ldots+m_{n-1}}\, x_2^{m_2+\ldots+m_{n-1}} \ldots x_{n-1}^{m_{n-1}}.$$

Let $\lambda_1 = m_1 + m_2 + \ldots + m_{n-1}$, $\lambda_2 = m_2 + \ldots + m_{n-1}$, $\ldots$ $\lambda_{n-1} = m_{n-1}$. Then $(\lambda_1,\lambda_2,\ldots,\lambda_{n-1})$ is a partition with at most n-1 parts. Conversely each partition with at most n-1 parts arises from a unique dominant integral weight. The partial order on P^+ translates to the partial order on partitions given as follows:

$(\lambda_1,\ldots,\lambda_{n-1}) \geq (\mu_1,\ldots,\mu_{n-1})$ if and only if

$$\lambda_1 \geq \mu_1$$
$$\lambda_1 + \lambda_2 \geq \mu_1 + \mu_2$$
$$\vdots$$
$$\lambda_1 + \lambda_2 + \ldots + \lambda_{n-1} \geq \mu_1 + \mu_2 + \ldots + \mu_{n-1}.$$

Now let λ,μ be partitions with at most n-1 parts and use the same notation for the corresponding dominant integral weights of $g = sl_n(\mathbb{C})$. Then the weight multiplicities $d\mu(V_\lambda(\mathbb{C}))$ can be interpreted as Kostka numbers, which play an important role in the theory of symmetric functions. The Kostka numbers can be defined as follows. We consider the ring $\mathbf{Z}[X_1 \ldots X_n]$ of polynomials in n variables with rational integer coefficients. The symmetric group S_n acts on this ring by permuting the variables, and a polynomial is said to be symmetric if it is invariant under S_n. The symmetric polynomials form a subring of $\mathbf{Z}[X_1,\ldots,X_n]$.

For each partition λ with at most n parts we introduce two symmetric functions. These are the monomial symmetric function

$$m_\lambda(X_1,\ldots,X_n) = \Sigma\, X_1^{\alpha_1} X_2^{\alpha_2} \ldots X_n^{\alpha_n}$$

summed over all distinct permutations $(\alpha_1,\ldots,\alpha_n)$ of $(\lambda_1,\ldots,\lambda_n)$ and the Schur function

$$s_\lambda(X_1,\dots,X_n) = \frac{\begin{vmatrix} X_1^{\lambda_1+n-1} & X_1^{\lambda_2+n-2} & \dots & X_n^{\lambda_n} \\ \\ X_n^{\lambda_1+n-1} & X_n^{\lambda_2+n-2} & \dots & X_n^{\lambda_n} \end{vmatrix}}{\begin{vmatrix} X_1^{n-1} & X_1^{n-2} & \dots & 1 \\ \vdots & & & \\ X_n^{n-1} & X_n^{n-2} & \dots & 1 \end{vmatrix}}$$

The Schur functions $s_\lambda(X_1,\dots,X_n)$ are the characters of the irreducible modules $V(\lambda)_{\mathbb{C}}$ for the general linear group $GL_n(\mathbb{C})$. If we put the last part $\lambda_n = 0$ we have a partition with at most n-1 parts. For such partitions the $s_\lambda(X_1,\dots,X_n)$ are the characters of the irreducible modules $V(\lambda)_{\mathbb{C}}$ for $SL_n(\mathbb{C})$ or for our simple Lie algebra $g = sl_n(\mathbb{C})$.

Now both the monomial symmetric functions $m_\lambda(X_1,\dots,X_n)$ and the Schur functions $s_\lambda(X_1,\dots,X_n)$ form $\mathbf{Z}$-bases for the ring $\mathbf{Z}[X_1,\dots,X_n]^{S_n}$ of symmetric functions. So each s_λ can be expressed as a $\mathbf{Z}$-combination of the m_μ and vice versa. We define the Kostka numbers $K_{\lambda\mu}$ by

$$s_\lambda = \sum_\mu K_{\lambda\mu} m_\mu .$$

These numbers have the following combinatorial interpretation. A semistandard tableau of shape λ is a λ-tableaux containing integers $1,2,\dots,n$, possibly with repetitions, such that the numbers in the tableau increase along each row and increase strictly down each column. If the number i appears μ_i times in the tableau we say that the tableau has content $\mu = (\mu_1,\mu_2 \dots \mu_n)$. Then $K_{\lambda\mu}$ is the number of semistandard λ-tableaux of content μ. Moreover the matrix $(K_{\lambda\mu})$ is unitriangular. (See Macdonald [M1].)

For example if n = 3 the matrix of Kostka numbers is

$\lambda\setminus\mu$	3	21	111
3	1	1	1
21	0	1	2
111	0	0	1

The relation between the weight multiplicities and the Kostka numbers is given by

$$d_\mu(V(\lambda)_{\mathbb{C}}) = K_{\lambda\mu}.$$

Now we have seen that the weight multiplicities are obtained from certain Kazhdan-Lusztig polynomials evaluated at 1. However the Kostka numbers are themselves obtained by evaluating certain polynomials at 1. These are called Foulkes polynomials, and are again defined in terms of the theory of symmetric functions. We shall now explain the connection between the Foulkes polynomials and the Kazhdan-Lusztig polynomials.

We first define for each partition λ with at most n parts a symmetric function $P_\lambda(X_1,\dots,X_n;t)$ called a Hall-Littlewood function. This is given by

$$P_\lambda(X_1,\dots,X_n;t) = \frac{1}{v_\lambda(t)} \sum_{w\in S_n} w(X_1^{\lambda_1} \dots X_n^{\lambda_n} \prod_{i<j} \frac{(X_i-tX_j)}{(X_i-X_j)})$$

where $v_\lambda(t) = v_{m_1}(t)v_{m_2}(t)\dots$ with $v_m(t) = (1+t)(1+t+t^2)\dots(1+t+\dots+t^{m-1})$ and m_i is the number of λ_j equal to i.

We see in particular that

$$P_\lambda(X_1,\dots,X_n;0) = s_\lambda(X_1,\dots,X_n)$$

$$P_\lambda(X_1,\dots,X_n;1) = m_\lambda(X_1,\dots,X_n).$$

Thus $P_\lambda(X_1,\dots,X_n;t)$ interpolates between the Schur functions and the monomial symmetric functions. The $P_\lambda(X_1,\dots,X_n;a)$ form a basis for $\mathbf{Z}[X_1,\dots,X_n]^{S_n}$ for any integer a. Thus the $s_\lambda(X_1,\dots,X_n)$ can be expressed in terms of the $P_\lambda(X_1,\dots,X_n;a)$ and vice versa. In fact there is a uniquely determined set of polynomials $K_{\lambda\mu}(t)$, called Foulkes polynomials, such that

$$s_\lambda(X_1,\dots,X_n) = \sum_\mu K_{\lambda\mu}(t)\, P_\mu(X_1,\dots,X_n;t).$$

We have in particular $K_{\lambda\mu}(0) = 1$ and $K_{\lambda\mu}(1) = K_{\lambda\mu}$. Thus the Kostka numbers are obtained from the Foulkes polynomials by specialising at 1. We also have $K_{\lambda\mu}(t) = 0$ unless $\lambda \succ \mu$.

For example if n = 3 the matrix of Foulkes polynomials $K_{\lambda\mu}(t)$ is as follows.

$\lambda \backslash \mu$	3	21	11
3	1	t	t^3
21	0	1	$t+t^2$
111	0	0	1

The relation between the Foulkes polynomials and the Kazhdan-Lusztig polynomials, as proved by Lusztig [L5] is as follows.

For each partition λ with at most n parts define $n(\lambda) = \sum_{i=1}^{n} (i-1)\lambda_i$.

<u>Theorem 2.7.3</u> Let λ,μ be two partitions with at most n-1 parts. Then

$$K_{\lambda\mu}(t) = t^{n(\lambda)-n(\mu)} P_{n_\mu,n_\lambda}(t^{-1})$$

The Foulkes polynomials $K_{\lambda\mu}(t)$ also appear in the character table of the finite general linear groups $GL_n(q)$. They appear in the part of the character table which gives the values of the unipotent characters on the unipotent conjugacy classes.

For each partition λ of n there is a unipotent representation ρ^λ of $GL_n(q)$ which may be defined as follows. Let $P_\lambda(q)$ be the parabolic subgroup of $GL_n(q)$ which is the stabilizer of a flag of subspaces of dimensions $\lambda_1, \lambda_1 + \lambda_2, \ldots, \lambda_1 + \ldots + \lambda_r$. Then ρ^λ is the irreducible representation of $GL_n(q)$ uniquely determined by the properties that it occurs in the induced representation $1_{P_\lambda(q)}^{GL_n(q)}$ but not in the induced representation $1_P^{GL_n(q)}$ for any subgroup P properly containing $P_\lambda(q)$.

For each partition μ of n there is a unipotent conjugacy class of $GL_n(q)$, consisting of all unipotent elements whose Jordan blocks have sizes given by the parts of μ.

The following theorem of Lusztig [L5] shows how the Foulkes polynomials, or Kazhdan-Lusztig polynomials, appear in this part of the character table of $GL_n(q)$.

<u>Theorem 2.7.4</u> Let λ, μ be partitions of n. Let ρ^λ be the unipotent character of $GL_n(q)$ corresponding to λ and u_μ be a unipotent element of $GL_n(q)$ of type μ. Then

$$\text{Trace } \rho^\lambda(u_\mu) = \begin{cases} q^{n(\mu)} K_{\lambda\mu}(q^{-1}) = q^{n(\lambda)} P_{n_\mu,n_\lambda}(q) & \text{if } \mu \prec \lambda \\ 0 & \text{otherwise.} \end{cases}$$

For example if n = 3 the character values Trace $\rho^\lambda(u_\mu)$ are as follows

λ \ μ	3	21	111
3	1	1	1
21	0	q	q^2+q
111	0	0	q^3

ρ^3 is the unit representation and ρ^{111} the Steinberg representation. u_{111} is the unit class and u_3 is the class of regular unipotent elements.

CHAPTER 3 : GEOMETRIC INTERPRETATIONS OF THE KAZHDAN-LUSZTIG POLYNOMIALS

§3.1 COMPLEXES OF SHEAVES ON AN ALGEBRAIC VARIETY

In this chapter we shall explain without proof how the coefficients of the Kazhdan-Lusztig polynomials for a Weyl group can be interpreted geometrically as the dimensions of certain intersection cohomology groups.

We first make some general remarks on sheaves on an algebraic variety. Let X be an irreducible algebraic variety over an algebraically closed field K. We consider X as a topological space with the Zariski topology and shall discuss sheaves on X. The sheaves we shall consider will be sheaves of vector spaces over the field $\mathbb{Q}_\ell$ of ℓ-adic numbers. Here ℓ is a prime which can be arbitrary if K has characteristic 0 but must satisfy $\ell \neq p$ if K has characteristic p. However many of the facts we shall outline are valid in the much more general context of sheaves of R-modules on X where R is any commutative ring with 1.

Given a sheaf F on X we have for any open subset U of X an associated $\mathbb{Q}_\ell$-vector space $\Gamma(U,F)$. A complex of sheaves on X is a family of sheaves F^i for $i \in \mathbb{Z}$ together with sheaf maps $d_i:F^i \to F^{i+1}$ such that $d_{i+1} \circ d_i = 0$ for all i. Thus we have

$$\cdots \to F^{-2} \xrightarrow{d_{-2}} F^{-1} \xrightarrow{d_{-1}} F^0 \xrightarrow{d_0} F^1 \xrightarrow{d_1} F^2 \xrightarrow{d_2} \cdots$$

The complex is said to be bounded if $F^i = 0$ for all i sufficiently large and for all negative i such that -i is sufficiently large. Thus only finitely many F^i are non-zero. Given such a complex of sheaves we have for any open subset U of X a sequence of linear maps

$$\cdots \quad \Gamma(U,F^{-1}) \xrightarrow{d_{-1}} \Gamma(U,F^0) \xrightarrow{d_0} \Gamma(U,F^1) \xrightarrow{d_1} \Gamma(U,F^2) \to \cdots$$

such that $\operatorname{im} d_{i-1} \subset \ker d_i$ for all i. We denote the complex of sheaves by $F^\bullet$ and define $H^i(U,F^\bullet)$ by

$$H^i(U,F^\bullet) = \ker d_i/\operatorname{im} d_{i-1}$$

Thus for each open subset U of X we have an associated $\mathbb{Q}_\ell$-vector space $H^i(U,F^\bullet)$. This collection of $\mathbb{Q}_\ell$-vector spaces forms a presheaf on X. The sheafification of this presheaf is called the ith cohomology sheaf $H^i(F^\bullet)$ of the complex $F^\bullet$.

Now suppose we have two complexes of sheaves $F^\bullet$, $G^\bullet$ on X. Suppose there exist sheaf maps $\phi_i:F^i \to G^i$ such that the following diagram is commutative.

$$\begin{array}{ccccccccccc}
\to F^{-2} & \xrightarrow{d_{-2}} & F^{-1} & \xrightarrow{d_{-1}} & F^{0} & \xrightarrow{d_{0}} & F^{1} & \xrightarrow{d_{1}} & F^{2} & \to \\
\downarrow \phi_{-2} & & \downarrow \phi_{-1} & & \downarrow \phi_{0} & & \downarrow \phi_{1} & & \downarrow \phi_{2} & \\
\to G^{-2} & \xrightarrow[d_{-2}]{} & G^{-1} & \xrightarrow[d_{-1}]{} & G^{0} & \xrightarrow[d_{0}]{} & G^{1} & \xrightarrow[d_{1}]{} & G^{2} & \to
\end{array}$$

Then we shall have an induced map ϕ_i: $H^i(F^\bullet) \to H^i(G^\bullet)$. In this situation we shall write $\phi:F^\bullet \to G^\bullet$ and call ϕ a morphism of complexes of sheaves. If ϕ has the property that each of the induced maps ϕ_i: $H^i(F^\bullet) \to H^i(G^\bullet)$ is an isomorphism then ϕ is called a quasi-isomorphism.

We now define an equivalence relation on the set of bounded complexes of sheaves on X. The equivalence classes are the smallest possible classes such that two complexes $F^\bullet$, $G^\bullet$ lie in the same equivalence class whenever there exists a quasi-isomorphism $\phi:F^\bullet \to G^\bullet$. The set of equivalence classes of bounded complexes of sheaves on X is denoted by $D^b(X)$. The elements of $D^b(X)$ are called the elements of the derived category of the category of sheaves on X.

There is an important duality map $D:D^b(X) \to D^b(X)$ called Verdier duality which is defined in [Verd].

We now discuss certain special kinds of sheaves on X. Let F be a sheaf on X and U be an open subset of X. Let $y \in U$ and F_y be the stalk of F at y. We recall that there is a natural restriction map $\Gamma(U,F) \to F_y$. F is said to be locally constant if for each $x \in X$ there is an open subset U of X containing x such that for all $y \in U$ the restriction map $\Gamma(U,F) \to F_y$ is an isomorphism. A complex of sheaves F on X is said to be cohomologically locally constant if the cohomology sheaves $H^i(F^\bullet)$ are locally constant for all i. In particular, if $F^\bullet$, $G^\bullet$ are two bounded complexes of sheaves giving rise to the same element of the derived category $D^b(X)$ their cohomology sheaves $H^i(F^\bullet)$, $H^i(G^\bullet)$ will be isomorphic and so $F^\bullet$ is cohomologically locally constant if and only if $G^\bullet$ is. Thus it is meaningful to say that an element of $D^b(X)$ is cohomologically locally constant. A complex of sheaves $F^\bullet$ is called constructible if there exists a chain

$$X_0 \subset X_1 \subset \ldots \subset X_{n-1} \subset X_n = X$$

of closed subsets X_i of X such that, for each i, the complex $F^\bullet$ is cohomologically locally constant when restricted to $X_i - X_{i-1}$. Again if $F^\bullet$, $G^\bullet$ are bounded complexes of sheaves giving the same element of the derived category $D^b(X)$, $F^\bullet$ is constructible if and only if $G^\bullet$ is. Thus we may speak of constructible elements of $D^b(X)$.

§3.2 THE INTERSECTION CHAIN COMPLEX

We now concentrate on one particular element of the derived category $D^b(X)$ which has particularly favourable properties. It has been shown by Deligne that there is a unique element $(F^\bullet) \in D^b(X)$ which satisfies the following conditions:

(i) $(F^\bullet)$ is constructible

(ii) $H^i(F^\bullet) = 0$ for $i < 0$

(iii) dim (Support $H^i(F^\bullet)) < \dim X-1-i$ for $i > 0$.

(iv) Let X_{reg} be the set of non-singular points of X. Then $F^\bullet$ restricted to X_{reg} is equivalent to the complex

$$\ldots \to 0 \to 0 \to \mathbb{Q}_\ell \to 0 \to 0 \to \ldots$$

on X_{reg}, where the constant sheaf $\mathbb{Q}_\ell$ appears in degree 0.

(v) $(F^\bullet)$ is a self dual element of $D^b(X)$ under the action of Verdier duality.

This element of $D^b(X)$ is called the intersection chain complex $IC^\bullet(X)$.

Deligne's theory is an algebraic analogue of a theory developed by Goresky and Macpherson in the context of topological spaces. Goresky and Macpherson defined, for any function of a certain kind called a perversity, a family of groups associated to any topological pseudomanifold called intersection cohomology groups. There are two extreme perversities which give rise to the homology and cohomology groups respectively. There are also two 'middle perversities'. In the case when the given topological space X is an algebraic variety over $\mathbb{C}$ these middle perversities give the same intersection cohomology groups. These are the middle intersection cohomology groups of X. These groups also arise in Deligne's theory as the hypercohomology groups of the intersection chain complex $IC^\bullet(X)$.

For this reason the cohomology sheaves $H^i(X)$ of the intersection chain complex $IC^\bullet(X)$ are called the DGM sheaves on X.

§3.3 THE CONSTRUCTION OF THE INTERSECTION CHAIN COMPLEX

We now wish to describe a construction due to Deligne [De] of the intersection chain complex on X. We first need to define an operation called truncation. Let $F^\bullet$ be a complex of sheaves on X and $k \in \mathbf{Z}$. We define the truncated complex of sheaves $\tau_{\leq k} F^\bullet$ on X by

$$(\tau_{\leq k} F^\bullet)^i = \begin{cases} F^i & \text{if } i < k \\ \ker d^i & \text{if } i = k \\ 0 & \text{if } i > k \end{cases}$$

with obvious maps $d_i : (\tau_{\leq k} F^\bullet)^i \to (\tau_{\leq k} F^\bullet)^{i+1}$. Equivalent complexes of sheaves give equivalent truncations under $\tau_{\leq k}$, so that $\tau_{\leq k}$ induces a map from the derived category $D^b(X)$ into itself. The cohomology sheaves of the truncated complex $\tau_{\leq k} F^\bullet$ are given by

$$H^i(\tau_{\leq k} F^\bullet) = \begin{cases} H^i(F^\bullet) & \text{if } i \leq k \\ 0 & \text{if } i > k \end{cases}$$

The construction of the intersection chain complex $IC^\bullet(X)$ is carried out in terms of a stratification of the algebraic vareity X. This is a chain

$$X = X^d \supset X^{d-1} \supset X^{d-2} \supset \dots \supset X^1 \supset X^0$$

where each X^i is closed in X, $X^i - X^{i-1}$ in non-singular of dimension i, and X is 'equisingular' along $X^i - X^{i-1}$ in a suitable sense. We shall not define the necessary properties precisely, but shall describe subsequently the most important special case.

Let $U_i = X - X^{d-i}$. Then we have

$$U_1 \subset U_2 \subset \dots \subset U_d \subset X$$

which is an increasing chain of open subsets of X. Let $j_i:U_i \to U_{i+1}$ be the inclusion map. For any sheaf F on U_i we shall then have the direct image sheaf j_{i*} F on U_{i+1}. There will also be a corresponding derived functor $Rj_{i*}: D^b(U_i) \to D^b(U_{i+1})$.

We begin the construction with the constant sheaf $\mathbb{Q}_\ell$ on U_1, regarded as a complex of sheaves

$$\to 0 \xrightarrow{d_{-2}} 0 \xrightarrow{d_{-1}} \mathbb{Q}_\ell \xrightarrow{d_0} 0 \xrightarrow{d_1} 0 \to$$

in which $\mathbb{Q}_\ell$ appears in degree 0. This determines an element of the derived category $D^b(U_1)$. We then form the direct image $R_{j_{1*}} \mathbb{Q}_\ell$ in $D^b(U_2)$. We truncate this with respect to $\tau_{\leq 0}$ to get $\tau_{\leq 0} R_{j_{1*}} \mathbb{Q}_\ell$ in $D^b(U_2)$. We then form the direct image in $D^b(U_3)$ and truncate again, this time by $\tau_{\leq 1}$. Continuing in this way we finally obtain an element of $D^b(X)$, and this is the intersection chain complex $IC^\bullet(X)$. Thus we have

$$IC^\bullet(X) = \tau_{\leq d-1} R_{j_{d*}} \tau_{\leq d-2} R_{j_{d-1*}} \dots \tau_{\leq 1} R_{j_{2*}} \tau_{\leq 0} R_{j_{1*}} \mathbb{Q}_\ell .$$

The cohomology sheaves $H^i(X)$ of this complex are the DGM sheaves on X.

§3.4 THE CASE OF SCHUBERT VARIETIES

In order to give the geometrical interpretation of the Kazhdan-Lusztig polynomials of a Weyl group W we must consider the special case where the variety X is a Schubert variety.

Let G be a connected reductive algebraic group over an algebraically closed field K. Let B be a Borel subgroup of G. Let $\mathcal{B} = G/B$ be the variety of Borel subgroups of G. Let W be the Weyl group of G. Then G is a disjoint union of double cosets BwB for $w \in W$. Let $\mathcal{B}_w = BwB/B$. $\mathcal{B}_w$ is called a Bruhat cell in $\mathcal{B}$. $\mathcal{B}_w$ is locally closed in $\mathcal{B}$, but not in general closed in $\mathcal{B}$. Its closure is given by

$$\overline{\mathcal{B}_w} = \bigcup_{\substack{y \in W \\ y \leq w}} \mathcal{B}_y$$

$\bar{\mathcal{B}}_w$ is called the Schubert variety corresponding to the element $w \in W$. We have $\dim \bar{\mathcal{B}}_w = \ell(w)$.

We have a stratification of $\bar{\mathcal{B}}_w$ as follows:

$$\bar{\mathcal{B}}_w = \bigcup_{\substack{y \leq w \\ \ell(y) \leq \ell(w)}} \mathcal{B}_y \supset \bigcup_{\substack{y \leq w \\ \ell(y) \leq \ell(w)-1}} \mathcal{B}_y \supset \ldots \supset \bigcup_{\substack{y \leq w \\ \ell(y) \leq 1}} \mathcal{B}_y \supset \bigcup_{\substack{y \leq w \\ \ell(y)=0}} \mathcal{B}_y$$

This is a decreasing chain of closed subsets of $\bar{\mathcal{B}}_w$. We now take the complements of these sets in $\bar{\mathcal{B}}_w$. We have:

$$\mathcal{B}_w = \bigcup_{\substack{y \leq w \\ \ell(y)=\ell(w)}} \mathcal{B}_y \subset \bigcup_{\substack{y \leq w \\ \ell(y) \geq \ell(w)-1}} \mathcal{B}_y \subset \ldots \subset \bigcup_{\substack{y \leq w \\ \ell(y) \geq 1}} \mathcal{B}_y \subset \bigcup_{\substack{y \leq w \\ \ell(y) \geq 0}} \mathcal{B}_y = \bar{\mathcal{B}}_w$$

This is an increasing chain of open subsets of $\bar{\mathcal{B}}_w$, starting with the Bruhat cell $\mathcal{B}_w$. $\mathcal{B}_w$ lies in the smooth part $(\bar{\mathcal{B}}_w)_{reg}$ of $\bar{\mathcal{B}}_w$.

The intersection chain complex can be constructed in terms of this chain of open subsets, starting with the constant sheaf $\mathbb{Q}_\ell$ on the Bruhat cell $\mathcal{B}_w$ and applying direct images and truncations alternately.

We now consider the DGM sheaves on $\bar{\mathcal{B}}_w$, i.e. the cohomology sheaves $\mathcal{H}^i(\bar{\mathcal{B}}_w)$ of the intersection chain complex $IC^\bullet(\bar{\mathcal{B}}_w)$. $\bar{\mathcal{B}}_w$ is the union of Bruhat cells $\mathcal{B}_y$ for $y \leq w$ and the stalks $\mathcal{H}^i_{p_y}(\bar{\mathcal{B}}_w)$ at all points $p_y \in \mathcal{B}_y$ are isomorphic. According to [KL2] we have

$$\mathcal{H}^{2i+1}_{p_y}(\bar{\mathcal{B}}_w) = 0 \text{ for all } y \leq w \text{ and all } i.$$

Moreover the Kazhdan-Lusztig polynomials for W are given by

$$P_{y,w}(X) = \sum_{i \geq 0} \dim \mathcal{H}^{2i}_{p_y}(\bar{\mathcal{B}}_w) X^i.$$

Thus the coefficient of X^i in $P_{y,w}(X)$ is the dimension of the local intersection cohomology group in degree 2i of $\bar{\mathcal{B}}_w$ at a point in $\mathcal{B}_y$. An alternative expression for

$P_{y,w}(X)$ is

$$P_{y,w}(X) = \sum_{i\geq 0} (-1)^i \dim H^i_{p_y}(\bar{B}_w)X^{\frac{1}{2}i}$$

in view of the fact that $H^i(\bar{B}_w) = 0$ if i is odd. Thus we see in particular that the coefficients of $P_{y,w}(X)$ are all non-negative integers.

A somewhat similar geometrical interpretation is used by Kazhdan and Lusztig to show that $P_{y,w}(X)$ also has non-negative integer coefficients when W is an affine Weyl group.

§3.5 THE CASE OF THE UNIPOTENT VARIETY

Finally, there is an interesting geometrical interpretation of the Kazhdan-Lusztig polynomials which we considered in §2.7. This time we take the algebraic variety U of all unipotent elements of $GL_n(K)$ where K is an algebraically closed field of characteristic p. We recall from §2.7 that there is a bijective correspondence between conjugacy classes of unipotent elements of $GL_n(K)$ and partitions λ of n. Let U_λ be the class of unipotent elements of type λ. Then the closure of U_λ is the unipotent variety U is given by

$$\bar{U}_\lambda = \bigcup_{\mu \leq \lambda} U_\mu$$

We consider the DGM sheaves $H^i(\bar{U}_\lambda)$. The stalks $H^i_{p_\mu}(\bar{U}_\lambda)$ for all points $p_\mu \in U_\mu$ are isomorphic.

It was shown by Lusztig [L5] that $H^{2i+1}_{p_\mu}(\bar{U}_\lambda) = 0$ for all $\mu < \lambda$ and all i. Moreover, using the notation of §2.7 we have

$$P_{n_\mu, n_\lambda}(X) = \sum_{i\geq 0} \dim H^{2i}_{p_\mu}(\bar{U}_\lambda)X^i$$

Thus the coefficient of X^i in the Kazhdan-Lusztig polynomial $P_{n_\mu, n_\lambda}(X)$ for the affine Weyl group of type $\tilde{A}_{n-1}$ is the dimension of the local intersection cohomology group in degree 2i of $\bar{U}_\lambda$ at a point of U_μ.

Using the relation between $P_{n_\mu, n_\lambda}(X)$ and the Foulkes polynomials $K_{\lambda,\mu}(X)$ described in §2.7 we have an interpretation of the Foulkes polynomials in terms of intersection cohomology. This is

$$X^{n(\mu)-n(\lambda)} K_{\lambda,\mu}(X^{-1}) = \sum_{i\geq 0} \dim H^{2i}_{p_\mu}(\bar{U}_\lambda)X^i.$$

CHAPTER 4 : THE ALGEBRAIC DESCRIPTIONS OF THE AFFINE WEYL GROUP A_n OF TYPE $\tilde{A}_{n-1}$, $n \geq 2$

In Chapter 1 we defined left, right and two-sided cells as well as RL-equivalence classes for an arbitrary Coxeter group. From now on, we shall restrict our attention only to the affine Weyl group A_n of type $\tilde{A}_{n-1}$, $n \geq 2$, and consider how to decompose this group into these equivalence classes.

In Chapters 4 to 7, we shall discuss some elementary properties of the affine Weyl group A_n. As mentioned in Chapter 1, the affine Weyl group A_n can be described in several different ways. We shall give three algebraic descriptions in Chapter 4 and one geometrical description in Chapter 6. We are indebted to Professor Lusztig for showing us the description of A_n as a set of permutations on $\mathbb{Z}$. The description of A_n as a set of affine matrices is suggested by this description of Lusztig. The matrix description of A_n will be the main tool by which we get most of our results. We list some basic definitions and terminology in §4.4 which will be used very frequently in the subsequent development. In Chapter 5 we introduce the map $\sigma: A_n \to \Lambda_n$ by Lusztig's procedure [L9] and show that the fibres of σ are invariant under star operations. Recall that in §1.7(v) all the cells of A_2 were determined. So for the sake of convenience, we shall always assume $n > 2$ after Chapter 6 inclusive, unless the contrary is specified, although most of our results are obviously valid in the case $n = 2$. We give, in Chapter 6, a geometrical description of A_n, i.e. we regard A_n as the set of alcoves of V, V a subspace of the Euclidean space $\mathbb{R}^n$ of dimension n-1. The alcove description of an affine Weyl group can be found in §1.1. The more detailed description of this can be found in the papers of D. Verma [Verm] and G. Lusztig [L3]. In contrast with their description, we give a coordinate form for each alcove. This enables us to establish the formulae to relate the geometrical description and the algebraic description of A_n. This also enables us to define, in Chapter 7, an admissible sign type of rank n. An admissible sign type can be regarded as an equivalence class of elements of A_n on one hand, and as a connected set of V on the other hand. We show in §7.3 that the cardinality of the set of all admissible sign types of rank n is finite and is equal to $(n+1)^{n-1}$. The proof of this result is quite ingenious. As a set of elements of A_n, we shall show in Chapter 18 that any admissible sign type is in some left cell. This property of an admissible sign type will be used to study the structure of cells of A_n and also to define the map $\hat{T}$ from A_n to the set of proper tabloids of rank n.

In the present chapter we shall show some basic properties of the length function $\ell(w)$ and the sets $\mathcal{L}(w)$, $\mathcal{R}(w)$ for any $w \in A_n$. In §4.3, we shall describe the left (resp. right) star operations on w in terms of permutations on $\mathbb{Z}$ and in terms of affine matrices when w lies in some $\mathcal{D}_L(s_t, s_{t+1})$ (resp. $\mathcal{D}_R(s_t, s_{t+1})$), where s_t, s_{t+1} are two Coxeter generators of A_n with $s_t\, s_{t+1}$ having order 3. The star operation is one of the most fundamental operations throughout our work.

§4.1 THREE ALGEBRAIC DESCRIPTIONS OF THE AFFINE WEYL GROUP A_n

A_n can be described in the following three different ways.

(i) By generators and relations

$$A_n = \langle s_i \mid s_i^2 = 1 \text{ and } (s_i s_j)^{m_{ij}} = 1 \text{ for } 1 \leqslant i,\ j \leqslant n \text{ and } i \neq j \rangle$$

where

$$m_{ij} = \begin{cases} 2 & \text{if } \bar{i} \neq \overline{j \pm 1} \\ 3 & \text{if } \bar{i} = \overline{j \pm 1} \end{cases}$$

with $i \to \bar{i}$ to be the natural map from $\mathbf{Z}$ to the set of the residue classes $\underline{n} = \{\bar{1}, \bar{2}, \ldots, \bar{n}\}$. We denote $\Delta = \{s_i \mid 1 \leqslant i \leqslant n\}$.

(ii) Regarded as a set of permutations on $\mathbf{Z}$.

$$A_n' = \left\{ w: \mathbf{Z} \to \mathbf{Z} \;\middle|\; \begin{array}{l} (i+n)w = (i)w + n \text{ for } i \in \mathbf{Z} \\ \sum_{t=1}^{n} (t)w = \sum_{t=1}^{n} t \end{array} \right\}$$

The relation between these two descriptions is as follows: For any i, $1 \leqslant i \leqslant n$, s_i corresponds the permutation

$$t \longrightarrow \begin{cases} t & \text{if } \bar{t} \neq \bar{i},\ \overline{i+1} \\ t-1 & \text{if } \bar{t} = \overline{i+1} \\ t+1 & \text{if } \bar{t} = \bar{i} \end{cases}$$

for $t \in \mathbf{Z}$. The above description of A_n is given by Lusztig [L9].

(iii) Regarded as the set A_n'' of all $\infty \times \infty$ affine matrices w of type $\tilde{A}_{n-1}$ which are defined as follows:

(a) The integer set $\mathbf{Z}$ is the set parametrising its rows (resp. columns). The integers parametrising its rows (resp. columns) are monotone increasing from top to bottom (resp. from left to right).

(b) The entries in each of its rows (resp. columns) are all zero except for one which is 1.

(c) Let $\{e(u,j_u) \mid u \in \mathbf{Z}\}$ be the set of its non-zero entries, where $e(u,j_u)$ lies in its (u,j_u)-position. Then $j_{u+n} = j_u + n$ for any $u \in \mathbf{Z}$ and $\sum_{u=1}^{n} j_u = \sum_{u=1}^{n} u$.

Clearly, an element $w \in A_n''$ is entirely determined by any of its non-zero entry sets $\{e(u,j_u) \mid u \in S\}$ with $S \subset \mathbf{Z}$, $|S| = n$ and $\bar{S} = \{\bar{1}, \bar{2}, \ldots, \bar{n}\}$, where $\bar{S}$ is the image of S under the natural map $i \to \bar{i}$.

For any $X, Y \in A''_n$ with $\{e_X(i,j) | i,j \in \mathbf{Z}\}$ and $\{e_Y(i',j') | i',j' \in \mathbf{Z}\}$ as their entry sets, let $Z = X \cdot Y$ be an $\infty \times \infty$ matrix satisfying condition (a) with its (i,j)-entry $\sum_{k \in \mathbf{Z}} e_X(i,k) e_Y(k,j)$ for any $i,j \in \mathbf{Z}$. It is easily shown that $Z \in A''_n$ and A''_n is a group with such a multiplication.

For $w \in A'_n$, we define an $\infty \times \infty$ matrix X_w which satisfies condition (a), (b) with $\{e(t,(t)w) | t \in \mathbf{Z}\}$ as its non-zero entries. Then $X_w \in A''_n$. So we can define a map from A'_n to A''_n by $w \to X_w$. It is obvious that such a map is a group isomorphism, where s_i, $1 \leqslant i \leqslant n$, corresponds $X_i \in A''_n$ with its non-zero entries $\{e_{X_i}(u,j_u) | u \in \mathbf{Z}\}$ satisfying

$$j_u = \begin{cases} u & \text{if } \bar{u} \neq \bar{i}, \overline{i+1} \\ u-1 & \text{if } \bar{u} = \overline{i+1} \\ u+1 & \text{if } \bar{u} = \bar{i}. \end{cases}$$

From now on, we shall identify A_n with A'_n and A''_n, and denote them all by A_n. The set $\Delta = \{s_i | 1 \leqslant i \leqslant n\}$ always denotes its (distinguished) Coxeter generators. We stipulate that $s_{i+qn} = s_i$ for any $q, i \in \mathbf{Z}$.

4.2 THE FUNCTIONS $\ell(w)$, $\mathcal{L}(w)$, $R(w)$ ON THE AFFINE WEYL GROUP A_n

Let us list some simple properties of A_n.

Lemma 4.2.1 *For $y \in A_n$, $s_i \in \Delta$,*

(i) $w = s_i \cdot y$ is obtained from y by transposing the (i+qn)-th row with the (i+1+qn)-th row for every $q \in \mathbf{Z}$.

(ii) $w = y \cdot s_i$ is obtained from y by transposing the (i+qn)-th column with the (i+1+qn)-th column for every $q \in \mathbf{Z}$.

(iii) $w = y^{-1}$ is obtained from y by transposing y.

Proof: By definition of the affine matrices and their multiplication. □

Let $\ell(\ )$ be the length function of A_n regarded as a Coxeter group. Then we have

Lemma 4.2.2 *$\ell(y) = \sum_{1 \leqslant i < j \leqslant n} \left| \left[\frac{(j)y - (i)y}{n} \right] \right|$ for $y \in A_n$, where [h] is the integer part of h for $h \in \mathbb{Q}$.*

Proof: Let us compare $\sum_{1 \leqslant i < j \leqslant n} \left| \left[\frac{(j)y-(i)y}{n} \right] \right|$ with $\sum_{1 \leqslant i < j \leqslant n} \left| \left[\frac{(j)s_t y-(i)s_t y}{n} \right] \right|$ for any

$1 \leq t \leq n$. Since

$$(h)s_t y = \begin{cases} (h)y & \text{if } \bar{h} \notin \{\bar{t}, \overline{t+1}\} \\ (h+1)y & \text{if } \bar{h} = \bar{t} \\ (h-1)y & \text{if } \bar{h} = \overline{t+1} \end{cases}$$

this implies that

$$\left|\left[\frac{(j)s_t y-(i)s_t y}{n}\right]\right| = \left|\left[\frac{(j)y-(i)y}{n}\right]\right|, \text{ if } \{\bar{i},\bar{j}\} \cap \{\bar{t},\overline{t+1}\} = \emptyset.$$

When $t \neq n$, $\displaystyle\sum_{\substack{1\leq i<j\leq n \\ |\{\bar{i},\bar{j}\}\cap\{\bar{t},\overline{t+1}\}|=1}} \left|\left[\frac{(j)y-(i)y}{n}\right]\right| = \sum_{\substack{1\leq i<j\leq n \\ |\{\bar{i},\bar{j}\}\cap\{\bar{t},\overline{t+1}\}|=1}} \left|\left[\frac{(j)s_t y-(i)s_t y}{n}\right]\right|$

When $t = n$, $1 < i < n$, we assume that $(n)y - (i)y = kn + r$ and $(i)y - (1)y = k'n + r'$ with $k, k', r, r' \in \mathbb{Z}$ and $0 \leq r, r' \leq n-1$ (actually, $1 \leq r, r' \leq n-1$ since $\overline{(n)y} \neq \overline{(i)y} \neq \overline{(1)y}$). Then

$$\left|\left[\frac{(n)s_t y-(i)s_t y}{n}\right]\right| = \left|\left[\frac{n+(1)y-(i)y}{n}\right]\right| = \left|\left[\frac{(-k')n+(n-r')}{n}\right]\right| = |-k'| = \left|\left[\frac{(i)y-(1)y}{n}\right]\right|$$

and

$$\left|\left[\frac{(i)s_t y-(1)s_t y}{n}\right]\right| = \left|\left[\frac{(i)y+n-(n)y}{n}\right]\right| = \left|\left[\frac{(-k)n+(n-r)}{n}\right]\right| = |-k| = \left|\left[\frac{(n)y-(i)y}{n}\right]\right|.$$

So we also have $\displaystyle\sum_{\substack{1\leq i<j\leq n \\ |\{\bar{i},\bar{j}\}\cap\{\bar{1},\bar{n}\}|=1}} \left|\left[\frac{(j)s_n y-(i)s_n y}{n}\right]\right|$

$$= \sum_{\substack{1\leq i<j\leq n \\ |\{\bar{i},\bar{j}\}\cap\{\bar{1},\bar{n}\}|=1}} \left|\left[\frac{(j)y-(i)y}{n}\right]\right|$$

On the other hand, when $t \neq n$, we assume that $(t+1)y - (t)y = kn + r$ with $k,r \in \mathbb{Z}$ and $1 \leq r \leq n$.

Then $\left|\left[\frac{(t+1)s_t y-(t)s_t y}{n}\right]\right| = \left|\left[\frac{(t)y-(t+1)y}{n}\right]\right| = \left|\left[\frac{(-k-1)n+(n-r)}{n}\right]\right|$

$$= |-k-1| = \begin{cases} \left|\left[\frac{(t+1)y-(t)y}{n}\right]\right| + 1 & \text{if } k \geq 0 \\ \left|\left[\frac{(t+1)y-(t)y}{n}\right]\right| - 1 & \text{if } k < 0 \end{cases} \qquad (1)$$

When $t = n$, we assume that $(n)y-(1)y = hn + u$ with $h, u \in \mathbb{Z}$ and $1 \leq u \leq n$. Then

$$|[\frac{(n)s_t y-(1)s_t y}{n}]| = |[\frac{2n+(1)y-(n)y}{n}]| = |[\frac{(1-h)n+(n-u)}{n}]|$$

$$= |1-h| = \begin{cases} |[\frac{(n)y-(1)y}{n}]| + 1 & \text{if } h \le 0 \\ |[\frac{(n)y-(1)y}{n}]| - 1 & \text{if } h > 0 \end{cases} \tag{2}$$

$$\text{So} \sum_{1\le i<j\le n} |[\frac{(j)s_t y-(i)s_t y}{n}]| = \sum_{1\le i<j\le n} |[\frac{(j)y-(i)y}{n}]| \pm 1 \tag{3}$$

It is easily seen that $\sum_{1\le i<j\le n} |[\frac{(j)y-(i)y}{n}]| = 0$ if and only if for any $1 \le i < j \le n$, the inequality $1 \le (j)y - (i)y \le n$ holds if and only if $y = 1$. (4)

Then formulae (3) and (4) imply that $\ell(y) \ge \sum_{1\le i<j\le n} |[\frac{(j)y-(i)y}{n}]|$.

To show equality, it suffices to show that for any $y \ne 1$, $\exists$ at least t, $1 \le t \le n$, such that

$$\sum_{1\le i<j\le n} |[\frac{(j)s_t y-(i)s_t y}{n}]| = \sum_{1\le i<j\le n} |[\frac{(j)y-(i)y}{n}]| - 1$$

i.e. to show that at least one of the inequalities

$(t+1)y - (t)y < 0$, for some t, $1 \le t < n$.

$(n)y - (1)y > n$

holds, or, equivalently to show that if $(t+1)y-(t)y > 0$ for all $1 \le t < n$, then $(n)y-(1)y > n$. In general, we have $(n)y-(1)y = \sum_{t=1}^{n-1} ((t+1)y-(t)y) \ge n-1$. If $(n)y-(1)y = n-1$, then $(t+1)y-(t)y = 1$ for any t, $1 \le t < n$. Hence it follows from the equation $\sum_{t=1}^{n} t = \sum_{t=1}^{n} (t)y = n\cdot(1)y + \sum_{t=1}^{n-1} t$ that $(1)y = 1$ and so $(t)y = t$ for all t, $1 \le t \le n$. But this means $y = 1$. It contradicts $y \ne 1$. Also, since $\overline{(n)y} \ne \overline{(1)y}$, we have $(n)y-(1)y \ne n$. Thus it follows that $(n)y-(1)y > n$. Our assertion is proved. □

<u>Corollary 4.2.3</u> (of the proof of Lemma 4.2.2)

<u>Let $w,y \in A_n$, $s_t \in \Delta$ with $w = s_t y$. Then</u>

(i) <u>$\ell(w) = \ell(y) + 1 \Leftrightarrow (t+1)y > (t)y$</u>

(ii) <u>$\ell(w) = \ell(y) - 1 \Leftrightarrow (t+1)y < (t)y$</u>

Proof: We write $(t+1)y - (t)y = kn + r$ with $k, r \in \mathbb{Z}$ and $1 \leq r < n$ when $t \neq n$. Then by formula (1) and Lemma 4.2.2, $\ell(w) = \ell(y) + 1 \Leftrightarrow k \geq 0 \Leftrightarrow (t+1)y-(t)y > 0$ $\Leftrightarrow (t+1)y > (t)y$.

Also, we write $(n)y-(1)y = hn + u$ with $h, u \in \mathbb{Z}$ and $1 \leq u < n$ when $t = n$. Then by formula (2) and Lemma 4.2.2, $\ell(w) = \ell(y) + 1 \Leftrightarrow h \leq 0 \Leftrightarrow (n)y-(1)y < n \Leftrightarrow$ $(n+1)y-(n)y > 0 \Leftrightarrow (n+1)y > (n)y$.

So (i) follows. Since (ii) is equivalent to (i), our proof is complete. □

Corollary 4.2.3 can be restated in terms of matrices as follows:

Corollary 4.2.3': *If w is obtained from y by transposing the $(i+qn)$-th row with the $(i+1+qn)$-th row for all $q \in \mathbb{Z}$, and if $e_y(u,j_u)$ is the non-zero entry of y lying in the (u,j_u)-position for any $u \in \mathbb{Z}$, then $\ell(w) = \ell(y) \pm 1$ and*

(i) $\ell(w) = \ell(y) + 1 \Leftrightarrow j_{i+1} > j_i$

(ii) $\ell(w) = \ell(y) - 1 \Leftrightarrow j_{i+1} < j_i$ □

The following two lemmas concern the functions $\mathcal{L}(\ )$, $\mathcal{R}(\ )$ on A_n.

Lemma 4.2.4 For any $w \in A_n$, we have

$$\mathcal{L}(w) = \{s_t \in \Delta \mid (t+1)w < (t)w\}$$

$$\mathcal{R}(w) = \{s_t \in \Delta \mid (t+1)w^{-1} < (t)w^{-1}\}$$

Proof: By definition of the functions $\mathcal{L}(\ )$, $\mathcal{R}(\ )$ (see Chapter 1) and Corollary 4.2.3. □

Lemma 4.2.5 *For any $w \in A_n$, we have $0 \leq |\mathcal{L}(w)| < n$ and $0 \leq |\mathcal{R}(w)| < n$. Moreover, the following three statements are equivalent:*

(i) $|\mathcal{L}(w)| = 0$ (ii) $|\mathcal{R}(w)| = 0$ (iii) $w = 1$

Proof: Obviously, $w = 1$ implies $|\mathcal{L}(w)| = |\mathcal{R}(w)| = 0$.

In the proof of Lemma 4.2.2, we have shown that $w \neq 1$ implies $|\mathcal{L}(w)| > 0$. And so $w \neq 1$ implies $w^{-1} \neq 1$ and then $|\mathcal{L}(w^{-1})| > 0$. It turns out $|\mathcal{R}(w)| > 0$. Therefore the latter part of the lemma has been proved. Now it is enough to show that $|\mathcal{L}(w)| \neq n$. Otherwise, we would have $(t)w > (t+1)w$ for all $1 \leq t \leq n$. So by Lemma 4.2.4.

$$(1)w > (2)w > \dots > (n)w > (n+1)w = (1)w + n, \text{ i.e. } n < 0.$$

This is impossible. Our result follows. □

§4.3 THE SUBSETS $\mathcal{D}_L(s_t)$, $\mathcal{D}_R(s_t)$ OF THE AFFINE WEYL GROUP A_n, $n > 3$.

Now we assume $n > 3$. Let us denote $\mathcal{D}_L(s_t, s_{t+1})$ and $\mathcal{D}_R(s_t, s_{t+1})$ by $\mathcal{D}_L(s_t)$ and $\mathcal{D}_R(s_t)$, respectively, for any $t \in \mathbb{Z}$. We call the map $w \to {}^*w$ in $\mathcal{D}_L(s_t)$ the left star operator on w and $y \to y^*$ in $\mathcal{D}_R(s_t)$ the right star operator on y. In terms of permutations on $\mathbb{Z}$, $w \in A_n$ lies in $\mathcal{D}_L(s_t)$ if and only if w satisfies one of the following inequalities:

(i) $(t+1)w < (t)w < (t+2)w$ (ii) $(t+1)w < (t+2)w < (t)w$

(iii) $(t)w < (t+2)w < (t+1)w$ (iv) $(t+2)w < (t)w < (t+1)w$.

Also, in terms of matrices, $w \in A_n$ lies in $\mathcal{D}_L(s_t)$ if and only if w has one of the following forms:

(i') $\begin{pmatrix} & 1 & \\ 1 & & \\ & & 1 \end{pmatrix}$ $-\ t^{th}$ rows (ii') $\begin{pmatrix} & & 1 \\ 1 & & \\ & 1 & \end{pmatrix}$ $-\ t^{th}$

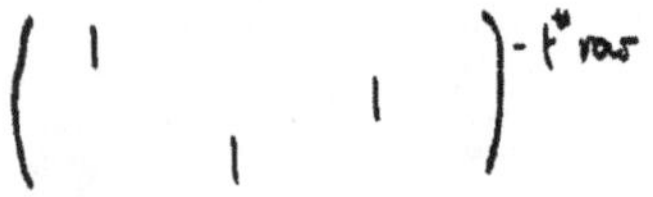

(iii') $\begin{pmatrix} 1 & & \\ & & 1 \\ & 1 & \end{pmatrix}$ $-\ t^{th}$ rows (iv') $\begin{pmatrix} & 1 & \\ & & 1 \\ 1 & & \end{pmatrix}$ $-\ t^{th}$

where form (α') is the matrix version of inequality

(α) for α = i, ii, iii, iv.

If $w \in \mathcal{D}_L(s_t)$, then w, ${}^*w$ (or ${}^*w$, w) in $\mathcal{D}_L(s_t)$ are either of forms (i'), (iv'), or of forms (iii'), (ii'), respectively.

Since $\mathcal{D}_R(s_t) = \{w \in A_n \mid w^T \in \mathcal{D}_L(s_t)\}$ by Lemma 4.2.1(iii), we have the corresponding results on $\mathcal{D}_R(s_t)$, where w^T is the transpose of the matrix w.

§4.4 SOME DEFINITIONS AND TERMINOLOGY

In this section, we shall introduce some basic definitions and terminology. Most of them will be used very frequently in the later development.

Fix an element $w \in A_n$.

(i) When we say "an entry of w", we always mean that this entry is non-zero unless the contrary is indicated. When we mention "an entry of w" (not necessarily non-zero), it means that we know both its value and its position in w. The entries are usually denoted by e in expressions such as $e(w)$, $e(i,j)$, $e_w(i,j)$, $e(i,(w))$, $e((w),j)$, etc, if it lies in the (i,j)-position of w.

(ii) Let e, e' be two entries (not necessarily non-zero) of w. Let f, f' (resp. g, g') be two rows (resp. columns) of w. We define

$$r(e,e') = \begin{cases} 0 & \text{if } e, e' \text{ lie in the same row of } w \\ \pm m & \text{otherwise} \end{cases}$$

where $m-1 \geq 0$ is the number of rows between e and e'. $r(e,e') = m$ (resp. $r(e,e') = -m$) if e' is below (resp. above) e.

We also define

$$c(e,e') = \begin{cases} 0 & \text{if } e, e' \text{ lie in the same column of } w \\ \pm \ell & \text{otherwise} \end{cases}$$

where $\ell-1 \geq 0$ is the number of columns between e and e'. $c(e,e') = \ell$ (resp. $c(e,e') = -\ell$) if e' is on the right (resp. left) of e.

Similarly, we can define $r(f,f')$, $c(g,g')$. Clearly, $r(e,e') = -r(e',e)$, $c(e,e') = -c(e',e)$, $r(e,e'') = r(e,e') + r(e',e'')$, $c(e,e'') = c(e,e') + c(e',e'')$ for any entry e" of w (not necessarily non-zero), etc.

(iii) The submatrix, consisting of any m consecutive rows of w with $m \leq n$, is called a block. Blocks will be written as A, B, C,... . We write $|A| = m$ as the size of A.

Assume that A, B are two blocks of w. Then $A \cup B$ denotes the union of A and B. In particular, when A, B are consecutive blocks of w, we see that $[A,B] = \binom{A}{B}$ is a block of w, provided that $|A| + |B| \leq n$.

(iv) For any entries e, e' of w, we say that e' is congruent to e if $r(e,e') = c(e,e') \in n\mathbb{Z}$.

For any rows f, f' (resp. columns g,g') of w, we say that f' is congruent to f (resp. g' is congruent to g) if $r(f,f') \in n\mathbb{Z}$ (resp. $c(g,g') \in n\mathbb{Z}$).

The set of all entries (resp. rows, columns) congruent to a certain entry (resp. row, column) of w is called an entry class of w (resp. a row class, a column class). Let $\bar{e}$ be a set of m entry classes of w. Then we define the size of $\bar{e}$ by $|\bar{e}| = m$.

Let $\bar{e}$ be a set of entry classes of w. Let ζ be the set of all rows and columns of w each of which contains some entry of $\bar{e}$. Then ζ is called a set of row-column classes (briefly, rc-classes) of w. We define the size of ζ, written $|\zeta|$, by the size of $\bar{e}$.

We call two blocks A, B of w congruent if the top row of B is congruent to the top row of A and $|A| = |B|$. For any block A of w with $|A| = m$, we usually denote any of its congruent blocks also by A. Moreover, let $S_A = \{e(i+u,(w)) \mid 1 \leq u \leq m\}$ be the set of entries contained in A, $S_A(q) = \{e(i+u+qn,(w)) \mid 1 \leq u \leq m\}$ for some $q \in \mathbb{Z}$. Then by abuse of terminology, we also call $S_A(q)$ a block of w and denote it by A.

(v) When we mention an entry set E of w, we always assume that any entry of E is non-zero and any two entries of E are incongruent in w. Two entry sets E, E' are regarded as the same, written E = E', if one can be obtained from the other by replacing some entries by congruent entries.

(vi) Recall that Λ_n denotes the set of partitions of n. Now we define a preorder $\geq$ on Λ_n as follows: Let $\lambda = \{\lambda_1 \geq \lambda_2 \geq \dots \geq \lambda_r\}$, $\mu = \{\mu_1 \geq \mu_2 \geq \dots \geq \mu_m\} \in \Lambda_n$. We say $\lambda \geq \mu$ if $\lambda_1 + \lambda_2 + \dots + \lambda_k \geq \mu_1 + \mu_2 + \dots + \mu_k$ for any $k \geq 1$ with the convention that $\lambda_i = \mu_j = 0$ for all $i > r$ and $j > m$, or, equivalently, if $\lambda_1 + \lambda_2 + \dots + \lambda_k \geq \mu_1 + \mu_2 + \dots + \mu_k$ for any $1 \leq k \leq r$. Clearly, this is a partial order on Λ_n. $\lambda \geq \mu$ implies $r \leq m$. For any $\lambda = \{\lambda_1 \geq \lambda_2 \geq \dots \geq \lambda_r\} \in \Lambda_n$, we call $\mu = \{\mu_1 \geq \mu_2 \geq \dots \geq \mu_m\} \in \Lambda_n$ the dual partition of λ if μ_k is the number of parts λ_j of λ with $1 \leq j \leq r$ and $\lambda_j \geq k$ for any $1 \leq k \leq m$. The map which sends any element of Λ_n to its dual is an involution of Λ_n and is order-reversing with respect to $\geq$.

(vii) Let P be a finite partially ordered set. An antichain is a subset of two by two incomparable elements of P. A chain is a subset which is totally ordered by the induced order of P. A k-chain family (resp. k-antichain family) is a subset of P which is a union of k chains (resp. k antichains). We denote by $c_k(P)$ (resp. $a_k(P)$) the maximum cardinal of k-chain families (resp. k-antichain families) of the partially ordered set P.

(viii) Let E be a set of entries of w. We define a preorder $\leq$ on E as follows. For $e(i,j)$, $e(i',j') \in E$, say $e(i',j') \leq e(i,j)$ if there exists $q \in \mathbf{Z}$ such that $i + qn \leq i'$ and $j + qn \geq j'$. It is easy to check that this preorder is well defined and is actually a partial order on E. So we can define a chain or an antichain for any entry set of w with respect to this partial order. Let $E = \{e_w(i_t,j_t) \mid 1 \leq t \leq \alpha\}$ for some $\alpha \geq 1$. Then E is a chain if and only if there exists an integer set $\{q_t \mid 1 \leq t \leq \alpha\}$ and a permutation $h_1,\dots,h_\alpha$ of $1,2,\dots,\alpha$ such that

$$i_{h_1} + q_{h_1}n < i_{h_2} + q_{h_2}n < \dots < i_{h_\alpha} + q_{h_\alpha}n \quad \text{and}$$

$$j_{h_1} + q_{h_1}n > j_{h_2} + q_{h_2}n > \dots > j_{h_\alpha} + q_{h_\alpha}n .$$

Also, E is an antichain if and only if there exists a set of integers $\{q_t \mid 1 \leq t \leq \alpha\}$ and a permutation $h_1,\dots,h_\alpha$ of $1,2,\dots,\alpha$, such that

$$i_{h_\alpha} + (q_{h_\alpha}-1)n < i_{h_1} + q_{h_1}n < i_{h_2} + q_{h_2}n < \dots < i_{h_\alpha} + q_{h_\alpha}n \quad \text{and}$$

$$j_{h_\alpha} + (q_{h_\alpha}-1)n < j_{h_1} + q_{h_1}n < j_{h_2} + q_{h_2}n < \dots < j_{h_\alpha} + q_{h_\alpha}n.$$

(ix) Suppose that $E = \{e(i_1,j_1),\dots,e(i_t,j_t)\}$ is a set of entries of w for some $t \geq 1$ such that $i_1 < \dots < i_t$ and $j_1 > \dots > j_t$ (resp. $i_1 < \dots < i_t < i_1 + n$ and

$j_1 < \dots < j_t$). Then E is called a descending (resp. increasing) chain of entries of w (briefly, a descending (resp. increasing) chain of w). Let $|E| = t$ be its size. In that case we also call $\{(i_1)w > \dots > (i_t)w$ a descending chain of w (resp. call $\{(i_1)w < \dots < (i_t)w\}$ an increasing chain of w).

Remark: (a) Note that when we say that the entry set E is a chain, we regard E as a set of entries up to congruence. But when we say that the entry set E is a descending chain, we regard E as a set of actual entries (not just up to congruence). Of course, for any chain of entries, we always can find a descending chain as its representative. Any descending chain can be obtained in this way.
(b) In contrast to what we have seen for a descending chain, there is not any relation between a chain of entries and an increasing chain. But there is some relation between an antichain and an increasing chain, although they are two different concepts. An antichain of entries is a set of entries up to congruence. But an increasing chain is a set of actual entries (not just up to congruence). For an antichain, we can always choose its entries in some block and this entry set will be automatically an increasing chain. But not any increasing chain could be obtained in this way.

(x) For a block A of w, if $e(i+1,j_1),\dots,e(i+m,j_m)$ are its entries such that $i \in \mathbb{Z}$ and $j_1 > \dots > j_m$ (resp. $j_1 < \dots < j_m$) then A is called a block of w whose entry set is a descending (resp. increasing) chain, (briefly, a DC (resp. IC) block of w). In that case, if the entries $e(i,j_0)$, $e(i+m+1,j_{m+1})$ of w satisfy $j_0 < j_1$ and $j_m < j_{m+1}$ then A is also called a maximal DC block of w (briefly, an MDC block of w). If A is contained in a block B such that $j_0 < j_1$ and $j_m < j_{m+1}$ whenever $e(i,j_0)$, $e(i+m+1,j_{m+1})$ lie in B then A is called a local MDC block of w in B. Clearly, any MDC block is also a local MDC block and any DC block is a local MDC block in itself. Similarly, we can define an MIC block, a local MIC block of w, where an MIC block is the abbreviation of a maximal IC block, etc.

(xi) For any $w \in A_n$, we say that w has the form $(A_\ell,\dots,A_1)$ at i if w has the form

$$\begin{pmatrix} \vdots \\ A_\ell(w) \\ \vdots \\ A_1(w) \end{pmatrix},$$ where $A_t(w)$, $1 \leq t \leq \ell$, are consecutive blocks of w with $|A_t(w)| = m_t$ and $\sum_{t=1}^{\ell} m_t \leq n$. i+1 is the integer labelling the first row of $A_\ell(w)$ in w. This form of w is full if $\sum_{t=1}^{\ell} m_t = n$. If we first say that w has the form $(A_\ell,\dots,A_1)$ at i and subsequently mention "the u-th entry of $A_t(w)$", it always means that this $A_t(w)$ is a single block of w lying between the (i+1)-th row and the (i+n)-th row. We denote the u-th entry of $A_t(w)$ by $e((w), j_t^u(w))$, $e((w), j_t^u(w,i))$ or $e((w), j_{A_t}^u(w))$. If all A(w)

are DC (resp. IC; MDC; MIC) blocks then we say that w has the DC (resp. IC; MDC; MIC) form $(A_\ell,\dots,A_1)$ at i. If all A(w) are local MDC (resp. local MIC) in the block $[A_\ell,\dots,A_1]$ then we say that w has local MDC (resp. local MIC) form $(A_\ell,\dots,A_1)$ at i. Sometimes, for the sake of emphasizing that the block $A_t(w)$ of w lies between the (i+1)-th row and the (i+n)-th row we denote $A_t(w)$ by $A_t(w,i)$.

(xii) In our language, for any $w \in A_n$, when we say "transposing the i-th and the j-th rows (resp. columns) of w", it always means that we transpose the (i+qn)-th and the (j+qn)-th rows (resp. columns) of w for every $q \in \mathbb{Z}$. We make the same conventions for the transformations on the blocks of w. We also make similar conventions on $\tilde{w} \in \tilde{A}_n$. (The notation $\tilde{A}_n$ will be introduced later).

(xiii) A generalized tabloid Y of rank n is, by definition, an array of n numbers 1,2,...,n into ordered columns, written $Y = (Y_1,\dots,Y_t)$, each of these columns Y_j being regarded as a subset of [n], where $[n] = \{1,2,\dots,n\}$. Let C_n (or just C) be the set of all generalized tabloids of rank n.

An element $Y = (Y_1,\dots,Y_t)$ of C is proper (resp. opposed) if $|Y_1| \geq \dots \geq |Y_t|$ (resp. $|Y_1| \leq \dots \leq |Y_t|$). Let $\hat{C}$ (resp. $\check{C}$) be the set of all proper (resp. opposed) tabloids of C.

For any $\lambda \in \Lambda_n$, an element $Y \in \check{C}$ is called a λ-tabloid if $Y = (Y_1,\dots,Y_t)$ with $\{|Y_t| \geq |Y_{t-1}| \geq \dots \geq |Y_1|\}$ being the dual partition of λ.

CHAPTER 5 : THE PARTITION OF n ASSOCIATED WITH AN ELEMENT OF THE AFFINE WEYL GROUP A_n

Let Λ_n be the set of partitions of n, $n \geq 2$. Let $\underline{\Delta}$ be the set of proper subsets of Δ. We shall define two maps: $\sigma: A_n \to \Lambda_n$ and $\pi: \underline{\Delta} \to \Lambda_n$, and show that the fibre $\sigma^{-1}(\lambda)$ is invariant under the star operations and the inverse for any $\lambda \in \Lambda_n$. These two maps, especially the first one, are at the heart of our book. In this chapter, we shall also discuss some relations between these two maps.

For any $J \in \underline{\Delta}$, W_J is by definition a standard parabolic subgroup of A_n generated by J.

Definition 5.1 A map $\pi: \underline{\Delta} \to \Lambda_n$ is defined as follows: For $J = J_1 \cup \ldots \cup J_r \in \underline{\Delta}$ with $W_J = W_{J_1} \times \ldots \times W_{J_r}$ and W_{J_j} indecomposable, $1 \leq j \leq r$, and $|J_1| \geq \ldots \geq |J_r|$, we define

$$\pi(J) = \{|J_1| + 1 \geq |J_2| + 1 \geq \ldots \geq |J_r| + 1 \geq 1 \geq \ldots \geq 1\} \in \Lambda_n.$$

It is easily seen that the map π is well defined and surjective.

Recall that in §4.4 (viii) we defined a partial order on the set of entries of an element w of A_n. To define a map $\sigma: A_n \to \Lambda_n$, we need a theorem of C. Greene, where all the notations in this theorem are as in §4.4(vii).

Theorem 5.2 (Greene's Theorem [Gr2])

Let P be a finite partially ordered set with n elements. Let q (resp. p) be the length of the largest chain (resp. antichain). We define the number $\lambda_i(P) = c_i(P) - c_{i-1}(P)$ and $\mu_i(P) = a_i(P) - a_{i-1}(P)$ (with the convention that $c_0(P) = a_0(P) = 0$). Then we have the following properties.

(i) $\lambda_1(P) \geq \ldots \geq \lambda_p(P)$. (ii) $\mu_1(P) \geq \ldots \geq \mu_q(P)$.

(iii) the partitions defined by (i) and (ii) are conjugate, where we say two partitions are conjugate if one is the dual of the other (see §4.4 (vi)). □

By using the notations of 4.4 (viii), we now can state the definition of a map $\sigma: A_n \to \Lambda_n$.

Definition 5.3 To $w \in A_n$, we associate a sequence of integers $d_1 < d_2 < \ldots < d_r = n$ as follows: d_k is the maximum cardinal of k-chain families of entries of w. Let $\lambda_1 = d_1$, $\lambda_j = d_j - d_{j-1}$ for $1 < j \leq r$. Then $\sum_{j=1}^{r} \lambda_j = n$. By Theorem 5.2, we see $\lambda_1 \geq \lambda_2 \geq \ldots \geq \lambda_r$. We define $\sigma(w) = \{\lambda_1 \geq \ldots \geq \lambda_r\}$.

In terms of a permutation on $\mathbb{Z}$, the integers d_k in the above definition can be explained as the maximum cardinal of a subset of $\mathbb{Z}$ whose elements are incongruent to each other mod n and which is a disjoint union of k subsets each of which has its natural order reversed by w. So our definition of the map σ is actually a matrix version of Lusztig's procedure [L9].

The following lemma shows that for any $\lambda \in \Lambda_n$, the fibre $\sigma^{-1}(\lambda)$ is invariant under the inverse.

Lemma 5.4 $\sigma(w) = \sigma(w^{-1})$ for any $w \in A_n$

Proof: We know from Lemma 4.2.1 (iii) that the matrix w^{-1} is the transpose of w. Since the operation of transposing a matrix keeps any descending chain and sends any two incongruent entries to incongruent ones, our result then follows immediately from the definition of the map σ. □

For $w \in A_n$, the following condition on $S = S_1 \cup \ldots \cup S_t \subset \mathbb{Z}$ is called $C_n(w,t)$: Elements of S are incongruent mod n and $(a)w > (b)w$ in S_j, $1 \leq j \leq t$, implies $a < b$. Let $E = E_1 \cup \ldots \cup E_t$ be the subset of entries of w with $E_j = \{e(a,(a)w) \mid (a)w \in S_j\}$, $1 \leq j \leq t$. Then condition $C_n(w,t)$ on $S = S_1 \cup \ldots \cup S_t$ is equivalent to the following condition on $E = E_1 \cup \ldots \cup E_t$: Elements of E are incongruent and E_j is a descending chain for any $1 \leq j \leq t$. So we can also say that $E = E_1 \cup \ldots \cup E_t$ satisfies $C_n(w,t)$ and regard condition $C_n(w,t)$ on $E = E_1 \cup \ldots \cup E_t$ as a matrix version of condition $C_n(w,t)$ on $S = S_1 \cup \ldots \cup S_t$.

Lemma 5.5 For any $w, w' \in A_n$ and $s_\alpha \in \Delta$ with $w' = s_\alpha w$ and $\ell(w') = \ell(w) + 1$, we have $\sigma(w') \geq \sigma(w)$.

Proof: If $S = S_1 \cup \ldots \cup S_t \subset \mathbf{Z}$ satisfies $C_n(w,t)$ for any $t \geq 1$, then $S' = (S_1)w^{-1}w' \cup \ldots \cup (S_t)w^{-1}w'$ satisfies $C_n(w',t)$. This implies that the integer d_t in Definition 5.2 for w' is not less than that for w, for any $t \geq 1$. So $\sigma(w') \geq \sigma(w)$. □

Corollary 5.6 If $w' = ws_\alpha$ in A_n with $s_\alpha \in \Delta$ and $\ell(w') = \ell(w) + 1$, then $\sigma(w') \geq \sigma(w)$.

Proof: This follows immediately from Lemmas 5.4, 5.5. □

The following lemma gives a relation between two maps σ and π on a given element $w \in A_n$.

Lemma 5.7 For any $w \in A_n$, $\pi(\mathcal{L}(w))$, $\pi(\mathcal{R}(w)) \leq \sigma(w)$.

Proof: By Lemma 5.4, it suffices to show that $\pi(\mathcal{L}(w)) \leq \sigma(w)$. Assume that $\mathcal{L}(w) = J = J_1 \cup \ldots \cup J_r \in \underline{\Delta}$ such that $W_J = W_{J_1} \times \ldots \times W_{J_r}$ and W_{J_j} is indecomposable,

$1 \leq j \leq r$, with $J_j = \{s_{i_j+1}, s_{i_j+2}, \ldots, s_{i_j+m_j}\}$, $m_1 \geq m_2 \geq \ldots \geq m_r$. Then $\pi(\mathcal{L}(w)) = \{m_1+1 \geq m_2+1 \geq \ldots \geq m_r+1 \geq 1 \geq \ldots \geq 1\} \in \Lambda_n$. On the other hand, for t with $1 \leq t \leq r$, let

$$S_j = \{(i_j+1)w, (i_j+2)w, \ldots, (i_j + m_j+1)w\}, \quad 1 \leq j \leq t.$$

By Lemma 4.2.4, we have

$$(i_j+1)w > (i_j+2)w > \ldots > (i_j + m_j + 1)w, \quad 1 \leq j \leq t$$

and this implies that $S = S_1 \cup \ldots \cup S_t$ satisfies $C_n(w,t)$ with $|S| = \sum_{j=1}^{t} (m_j+1)$. So $\sum_{j=1}^{t} (m_j+1) \leq \sum_{j=1}^{t} \lambda_j$ for any t with $1 \leq t \leq \min\{r,h\}$, where we assume $\sigma(w) = \{\lambda_1 \geq \ldots \geq \lambda_h\}$. But this easily implies that $\sum_{j=1}^{t} (m_j+1) \leq \sum_{j=1}^{t} \lambda_j$ for any $t \geq 1$. Therefore $\pi(\mathcal{L}(w)) \leq \sigma(w)$. □

Now we shall show that for any $\lambda \in \Lambda_n$, the fibre $\sigma^{-1}(\lambda)$ is invariant under the star operations.

<u>Lemma 5.8</u> <u>Assume that $w' = {}^*w$ in $\mathcal{D}_L(s_i)$ for some $i \in \mathbf{Z}$. Then $\sigma(w') = \sigma(w)$.</u>

<u>Proof</u>: By symmetry, it suffices to discuss the case that $(i+1)w < (i+2)w < (i)w$. In that case, $w' = s_i w$, $\ell(w') = \ell(w)-1$ and then for any $j \in \mathbf{Z}$, we have

$$(j)w' = \begin{cases} (j)w & \text{if } \bar{j} \neq \bar{i}, \overline{i+1} \\ (j+1)w & \text{if } \bar{j} = \bar{i} \\ (j-1)w & \text{if } \bar{j} = \overline{i+1} \end{cases}$$

By Lemma 5.5, it follows that $\sigma(w) \geq \sigma(w')$. We now need only show that $\sigma(w') \geq \sigma(w)$.

Let $S = S_1 \cup \ldots \cup S_t \subset \mathbf{Z}$ satisfy $C_n(w,t)$. If $\nexists\, 1 \leq j \leq t$, such that S_j contains two elements $(h_1)w$, $(h_2)w$ with $\bar{h}_1 = \bar{i}$ and $h_2 = h_1 + 1$, then $(S)w^{-1}w' = (S_1)w^{-1}w' \cup \ldots \cup (S_t)w^{-1}w'$ satisfies $C_n(w',t)$. Clearly, $|(S)w^{-1}w'| = |S|$. Now suppose that $\exists\, 1 \leq j \leq t$ such that S_j contains $(h_1)w$, $(h_2)w$ with $\bar{h}_1 = \bar{i}$ and $h_2 = h_1 + 1$. If $\nexists\, (g)w \in S$ such that $\bar{g} = \overline{i+2}$, then let

$$S'_\ell = \begin{cases} (S_\ell)w^{-1}w' & \text{if } \ell \neq j \\ ((h_1+2)w \cup S_j)w^{-1}w' - (h_1)w' & \text{if } \ell = j \end{cases}$$

where for any sets X,Y, X-Y denotes their set-theoretical difference. If $\exists (g)w \in S$ such that $\bar{g} = \overline{i+2}$, say $(g)w \in S_k$, then it is clear that $k \neq j$. Assume that

$$S_j = \{(i_{j1})w > (i_{j2})w > \ldots > (i_{j\alpha_j})w \mid i_{j1} < i_{j2} < \ldots < i_{j\alpha_j}\}$$
$$S_k = \{(i_{k1})w > (i_{k2})w > \ldots > (i_{k\alpha_k})w \mid i_{k1} < i_{k2} < \ldots < i_{k\alpha_k}\}$$

and for some $1 \leq u, u' \leq \alpha_j$, $1 \leq v \leq \alpha_k$, we have $\bar{i}_{ju} = \bar{i}$, $\bar{i}_{kv} = \overline{i+2}$ and $i_{ju'} = i_{ju}+1$. Then $u' = u+1$. By proper choice of S_k, we may assume that $i_{kv} = i_{ju} + 2$. Then let

$$S'_\ell = \begin{cases} (S_\ell)w^{-1}w' & \text{if } \ell \neq k,j \\ (i_{j1})w' > \ldots > (i_{j,u-1})w' > (i_{j,u+1})w' > (i_{kv})w' > (i_{k,v+1})w' \\ \qquad > \ldots > (i_{k\alpha_k})w'\} & \text{if } \ell = j \\ \{(i_{k1})w' > \ldots > (i_{k,v-1})w' > (i_{j,u})w' > (i_{j,u+2})w' > (i_{j,u+3})w' \\ \qquad > \ldots > (i_{j\alpha_j})w'\} & \text{if } \ell = k. \end{cases}$$

An example of this construction of S'_ℓ in terms of matrix is given in the following figure.

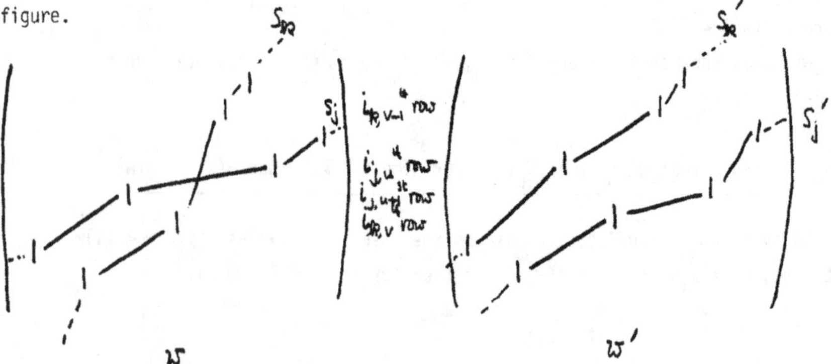

where $i_{j,u} = i_{j,u+1}-1 = i_{k,v}-2$. w' is obtained from w by transposing the $i_{j,u}$-th row with the $i_{j,u+1}$-th row.

In both cases, $S' = S'_1 \cup \ldots \cup S'_t$ satisfies $C_n(w',t)$ and $|S'| = |S|$. Now for any $S = S_1 \cup \ldots \cup S_t \subset \mathbb{Z}$, $t \geq 1$, satisfying $C_n(w,t)$, we can find $S' = S'_1 \cup \ldots \cup S'_t \subset \mathbb{Z}$ which satisfies $C_n(w',t)$ with $|S'|$ not less than $|S|$. So by the same argument as that in the proof of Lemma 5.5, we have $\sigma(w') \geq \sigma(w)$. Therefore $\sigma(w) = \sigma(w')$ as required. □

<u>Corollary 5.9</u> <u>If $w' = w^*$ in $\mathcal{D}_R(s_i)$ for some $i \in \mathbb{Z}$, then $\sigma(w') = \sigma(w)$.</u>

<u>Proof:</u> Since $w' = w^*$ in $\mathcal{D}_R(s_i)$, it implies that $w'^{-1} = {}^*(w^{-1})$ in $\mathcal{D}_L(s_i)$. By Lemma 5.8, $\sigma(w'^{-1}) = \sigma(w^{-1})$. So $\sigma(w') = \sigma(w)$ by Lemma 5.4. □

<u>Proposition 5.10</u> <u>If $w, w' \in A_n$ satisfy $w \underset{p}{\sim} w'$, then $\sigma(w) = \sigma(w')$.</u>

Proof: Since P-equivalence is generated by $w \underset{P_L}{\sim} {}^*w$ in $\mathcal{D}_L(s_i)$ and $y \underset{P_R}{\sim} y^*$ in $\mathcal{D}_R(s_j)$, where s_i, s_j run over Δ, our result follows from Lemma 5.8 and Corollary 5.9. □

Proposition 5.10 shows that for any $\lambda \in \Lambda_n$, the fibre $\sigma^{-1}(\lambda)$ is a union of some P-equivalence classes of A_n and then is also a union of some P_L-(resp. P_R-) equivalence classes of A_n.

From Lemma 5.7, we shall ask whether there exists some element $w \in \sigma^{-1}(\lambda)$ for any $\lambda \in \Lambda_n$ such that $\pi(\mathcal{L}(w)) = \sigma(w)$ (resp. $\pi(R(w)) = \sigma(w)$). The answer is affirmative. First let us show the following result.

Lemma 5.11 <u>For $w \in A_n$, we assume that $E = \{e(i_t + qn, j_t + qn) | 1 \leq t \leq \ell, q \in \mathbb{Z}\}$ is a set of entry classes of w such that $i_1 - n < i_\ell < i_{\ell-1} < \dots < i_1$ and $j_1 - n < j_\ell < j_{\ell-1} < \dots < j_1$. Let $S = \{e(i'_u, j'_u) | 1 \leq u \leq m\}$ be a descending chain of w with $i'_m < \dots < i'_1$ and $j'_m > \dots > j'_1$. Then $|E \cap S| \leq 1$.</u>

Proof: Otherwise, there exist α, β with $1 \leq \alpha < \beta \leq m$ such that $\bar{j}'_\alpha = \bar{j}_a$ and $\bar{j}'_\beta = \bar{j}_b$. Then there exist $p, q \in \mathbb{Z}$ such that $j'_\alpha = j_a + pn$, $i'_\alpha = i_a + pn$, $j'_\beta = j_b + qn$ and $i'_\beta = i_b + qn$. So the inequality $(i_a - i_b) + (p-q)n < 0$ follows from $i'_\alpha - i'_\beta < 0$ and $(j_a - j_b) + (p-q)n > 0$ from $j'_\alpha - j'_\beta > 0$, i.e. We have

$$\begin{cases} i_a - i_b < (q-p)n & (1) \\ j_a - j_b > (q-p)n & (2) \end{cases}$$

Since $0 < |i_a - i_b|, |j_a - j_b| < n$, the inequalities (1), (2) imply $q-p \geq 0$ and $q-p \leq 0$, respectively. Hence $q = p$ and we have

$$\begin{cases} i_a - i_b < 0 & (3) \\ j_a - j_b > 0 & (4) \end{cases}$$

But the inequality (3) implies $a > b$ and (4) implies $b > a$. This gives a contradiction. So our result follows. □

Fix a partition $\lambda = \{\lambda_1 \geq \lambda_2 \geq \dots \geq \lambda_r\} \in \Lambda_n$. Let $\lambda_{a_1}, \lambda_{a_2}, \dots, \lambda_{a_r}$ be a permutation of $\lambda_1, \lambda_2, \dots, \lambda_r$. Assume that $J = J_1 \cup \dots \cup J_r \underset{\neq}{\subset} \Delta$ with $J_j = \{s_{\alpha_j+1}, s_{\alpha_j+2}, \dots, s_{\alpha_j+a_j-1}\}$, $1 \leq j \leq r$, and $\alpha_j = i + \sum_{h=j+1}^{r} \lambda_{a_h}$ for some $i \in \mathbb{Z}$. Let $w_0{}^J$ be the longest element in W_J. Then $\pi(\mathcal{L}(w_0{}^J)) = \pi(R(w_0{}^J)) = \pi(J) = \lambda$.

As an affine matrix, $w_0{}^J$ has the form

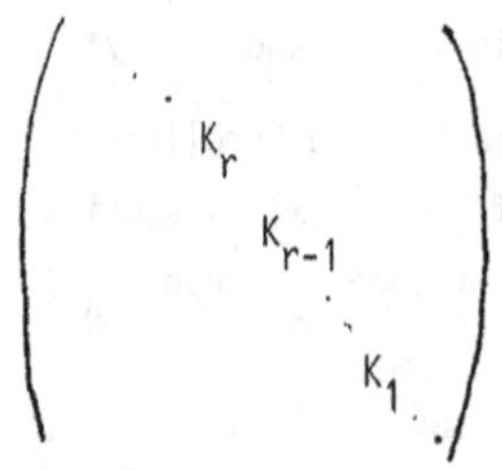

where $K_j = \begin{pmatrix} 0 & 1 \\ 1 & 0 \end{pmatrix}$ is a $\lambda_{a_j} \times \lambda_{a_j}$ diagonal block of w_0^J, $1 \leq j \leq r$. The integer labelling the first row (resp. column) of K_j in w_0^J is α_j+1. In other words, w_0^J has the MDC form $(A_r, A_{r-1}, \ldots, A_1)$ at 0 with the t-th entry $e(i_h^t, j_h^t)$ of $A_h(w_0^J)$, $1 \leq h \leq r$, $1 \leq t \leq \lambda_{a_h}$, where $i_h^t = \alpha_h + t$, $j_h^t = \alpha_{h-1} + 1 - t$.

Assume that $\mu = \{\mu_1 \geq \mu_2 \geq \ldots \geq \mu_{\lambda_1}\}$ is the dual partition of λ. Then it is clear that for any $1 \leq t \leq \lambda_1$,

$$i^t_{a_{\beta_1(t)}} - n < i^t_{a_{\beta_{\mu_t}}(t)} < i^t_{a_{\beta_{\mu_t-1}}(t)} < \ldots < i^t_{a_{\beta_1}(t)} \quad \text{and}$$

$$j^t_{a_{\beta_1}(t)} - n < j^t_{a_{\beta_{\mu_t}}(t)} < j^t_{a_{\beta_{\mu_t-1}}(t)} < \ldots < j^t_{a_{\beta_1}(t)} \quad \text{where}$$

$\beta_1(t), \beta_2(t), \ldots, \beta_{\mu_t}(t)$ is the subsequence of $1,2,\ldots,r$ such that for any h with $1 \leq h \leq \mu_t$, we have $\lambda_{a_{\beta_h}(t)} \geq t$. Let $E_t = \{e(i^t_{a_{\beta_h}(t)} + qn, j^t_{a_{\beta_h}(t)} + qn) | 1 \leq h \leq \mu_t, q \in \mathbb{Z}\}$ be the set of entry classes of w for $1 \leq t \leq \lambda_1$. Then by Lemma 5.11, the intersection of E_t with any descending chain of w_0^J has cardinal at most 1. Assume that $S = S_1 \cup \ldots \cup S_k$ is a disjoint union of k descending chains of w_0^J satisfying $C_n(w,k)$, $k \geq 1$. Then the cardinal of the intersection of E_t with S is at most $\min\{\mu_t, k\}$, for any t with $1 \leq t \leq \lambda_1$. Since $\bigcup_{t=1}^{\lambda_1} E_t$ is the full set of entry classes of w_0^J, it implies that $|S| \leq \sum_{t=1}^{\lambda_1} \min\{\mu_t, k\} = \sum_{j=1}^{k} \lambda_j$. Hence $\sigma(w_0^J) \leq \lambda$. By Lemma 5.7, this implies that $\sigma(w_0^J) = \pi(\mathcal{L}(w_0^J)) = \pi(R(w_0^J)) = \lambda$. So we have

<u>Lemma 5.12</u> <u>For any $J \in \underline{\Delta}$, let w_0^J be the longest element in W_J. Then we have</u>

$$\sigma(w_0^J) = \pi(\mathcal{L}(w_0^J)) = \pi(R(w_0^J)) = \pi(J). \quad \square$$

<u>Corollary 5.13</u> <u>The map $\sigma: A_n \to \Lambda_n$ is surjective.</u>

Proof: Since the map $\pi:\underline{\Delta} \to \Lambda_n$ is surjective, and since by Lemma 5.12, for any $J \in \underline{\Delta}$, there exists an element $w \in A_n$ such that $\sigma(w) = \pi(J)$, our conclusion follows immediately. □

Now we define another map $\delta:A_n \to \Lambda_n$ as follows.

Definition 5.14 To $w \in A_n$, we associate a sequence of integers $f_1 < f_2 < \ldots < f_m = n$ as follows, f_k is the maximum cardinal of k-antichain families of entries of w. Let $\mu_1 = f_1$, $\mu_j = f_j - f_{j-1}$ for $1 < j \leq m$. Then $\sum_{j=1}^{m} \mu_j = n$. By Theorem 5.2, we have $\mu_1 \geq \mu_2 \geq \ldots \geq \mu_m$. We define $\delta(w) = \{\mu_1 \geq \mu_2 \geq \ldots \geq \mu_m\}$.

Again by Theorem 5.2, we see that $\delta(w)$ is the dual partition of $\sigma(w)$ for any $w \in A_n$. So by the properties of the map σ, we get the corresponding properties of the map δ immediately.

Proposition 5.15

(i) For any $w \in A_n$, $\delta(w) = \delta(w^{-1})$

(ii) Assume that $w, w' \in A_n$ with $w' = s_v w$ and $\ell(w') = \ell(w)+1$ for some $s_v \in \Delta$. Then we have $\delta(w) \geq \delta(w')$. In that case, let $e_w(v,j_v)$, $e_w(v+1,j_{v+1})$ be two entries of w. Suppose $j_{v+1}-j_v > n$. (i.e. $\{e_w(v,j_v), e_w(v+1,j_{v+1})\}$ is a chain). Then we have $\delta(w) = \delta(w')$.

(iii) For any $\mu \in \Lambda_n$, the fibre $\delta^{-1}(\mu)$ is invariant under star operations.

(iv) For any $\lambda \in \Lambda_n$, we have $\sigma^{-1}(\lambda) = \delta^{-1}(\lambda^\vee)$, where $\lambda^\vee$ is the dual of λ. □

CHAPTER 6 : A GEOMETRICAL DESCRIPTION OF THE AFFINE WEYL GROUP A_n

In this chapter, we shall describe the affine Weyl group A_n as the set of alcoves of V, where V is a subspace of $\mathbb{R}^n$ of dimension n-1.

I take here the opportunity to thank Professor R.W. Carter who stimulated me to use this geometrical description of A_n in order to study more properties on the structure of cells of A_n.

§6.1 THE DESCRIPTION OF A_n AS A SET OF ALCOVES

Let $\mathbb{R}^n$, $n \geq 2$, be a Euclidean space of dimension n. Then the set $V = \{(x_1,\ldots,x_n) \in \mathbb{R}^n \mid \sum_{j=1}^{n} x_j = 0\}$ is a subspace of $\mathbb{R}^n$ of dimension n-1. For any integers k, i, j, t, m with $m > 0$, $t \in [n] = \{1,2,\ldots,n\}$ and $1 \leq i < j \leq n$, let
$H_{ij;k} = \{(x_1,\ldots,x_n) \in V \mid x_j - x_i = k\}$,
$H^m_{ij;k} = H^m_{ji;-k} = \{(x_1,\ldots,x_n) \in V \mid k < x_j - x_i < k+m\}$ and $H^m_{tt;0} = V$.
Then the connected components (alcoves) of the set $V - \bigcup_{\substack{1\leq i<j\leq n\\ k\in\mathbb{Z}}} H_{ij;k}$ are open simplices. Let $\mathbb{A}$ be the set of all such alcoves. Clearly, each alcove has the form $\bigcap_{1\leq i<j\leq n} H^1_{ij;k_{ij}}$ for some set of integers $\{k_{ij}\}_{1\leq i<j\leq n}$. In that case, sometimes we also express this alcove by $\bigcap_{1\leq i,j\leq n} H^1_{ij;k_{ij}}$ with the convention that $k_{ji} = -k_{ij}$ for any $i,j \in [n]$. We shall also make a convention that $k_{i,j\pm n} = k_{i\pm n,j} = k_{i\pm n,j\pm n} = k_{ij}$ for $i,j \in [n]$.

But given a set $\{k_{ij}\}_{1\leq i<j\leq n}$ of integers there does not always exist an alcove of the form $\bigcap_{1\leq i<j\leq n} H^1_{ij;k_{ij}}$. This fact leads us to find the conditions on the integer set $\{k_{ij}\}_{1\leq i<j\leq n}$ such that $\bigcap_{1\leq i<j\leq n} H^1_{ij;k_{ij}}$ is an alcove.

<u>Lemma 6.1.1</u> $A_1 = \bigcap_{1\leq i<j\leq n} H^1_{ij;0}$ <u>is an alcove of V.</u>

<u>Proof:</u> We see that a point $x = (x_1,\ldots,x_n)$ of $\mathbb{R}^n$ is in A_1 if and only if x satisfies the following conditions.

(i) $\sum_{i=1}^{n} x_i = 0.$ (ii) $x_1 < \ldots < x_n < x_1 + 1.$

We know that every point satisfying the above two conditions is in $\tilde{V} = V - \bigcup_{\substack{1\leq i<j\leq n\\ k\in\mathbb{Z}}} H_{ij;k}$. We also know that every point of $\tilde{V}$ is in some alcove of V and each alcove of V has the form $\bigcap_{1\leq i<j\leq n} H^1_{ij,k_{ij}}$ for some set of integers $\{k_{ij}\}_{1\leq i<j\leq n}$. So

in order to show our result, it is enough to show that A_1 is non-empty, i.e. to find a point $x \in \mathbb{R}^n$ satisfying the above two conditions. Indeed, we can take $x = (x_1,\dots,x_n)$ with $x_t = -\frac{1}{4} + \frac{t-1}{2(n-1)}$, $t \in [n]$. □

From now on, we assume that $H^m_{ij;k} = H^m_{i\pm n,j;k} = H^m_{i,j\pm n;k} = H^m_{i\pm n,j\pm n,;k}$ for any integers i,j,k,m with $i,j \in [n]$ and $m > 0$.

<u>Corollary 6.1.2</u> <u>Let A_1 be as in Lemma 6.1.1. Then $A_1 = \bigcap_{t\in[n]} H^1_{t,t+1;0}$. Moreover, A_1 has n walls which are supported by the hyperplanes $\{H_{t,t+1;0},\ 1 \le t < n;\ H_{1n;1}\}$.</u>

<u>Proof</u>: Note that $H^1_{n,n+1;0} = H^1_{1n;0}$. We see that any point $x = (x_1,\dots,x_n)$ of $\mathbb{R}^n$ is in $\bigcap_{t\in[n]} H^1_{t,t+1;0}$ if and only if x satisfies $\sum_{i=1}^{n} x_i = 0$ and $x_1 < \dots < x_n < x_1+1$. Thus the equation $A_1 = \bigcap_{t\in[n]} H^1_{t,t+1;0}$ follows by the proof of Lemma 6.1.1.

It is well known that in the Euclidean space of dimension m, any polyhedron is bounded by at least m+1 walls which are supported by some hyperplanes. This implies by the above proof that A_1 has exactly n walls. We see that for any t, $1 \le t < n$, $H_{t,t+1,;0} \cap \bar{A}_1 = \{(x_1,\dots,x_n) \in V | x_1 \le \dots \le x_t = x_{t+1} \le \dots \le x_n \le x_1 + 1\}$ has dimension n-2, where $\bar{A}_1$ is the closure of A_1 in V. This implies that $H_{t,t+1;0}$ supports a wall of A_1. By the same reason we can show that $H_{1n;1}$ supports a wall of A_1. So our result follows. □

<u>Lemma 6.1.3</u> <u>Suppose that $A_k = \bigcap_{1\le i<j\le n} H^1_{ij;k_{ij}}$ is an alcove. Then for any integers i,t,j with $1 \le i < t < j \le n$, we have $k_{it} + k_{tj} \le k_{ij} \le k_{it} + k_{tj} + 1$.</u>

<u>Proof</u>: Let $v = (x_1,\dots,x_n) \in A_k$. Then we have $k_{it} < x_t - x_i < k_{it} + 1$ and $k_{tj} < x_j - x_t < k_{ij} + 1$. So $k_{it} + k_{ij} < (x_t - x_i) + (x_j - x_t) < k_{it} + k_{ij} + 2$. i.e.

$$k_{it} + k_{tj} < x_j - x_i < k_{it} + k_{tj} + 2 \qquad (1)$$

$$\text{But we have } k_{ij} < x_j - x_i < k_{ij} + 1 \qquad (2)$$

Thus (1) and (2) together imply $k_{it} + k_{tj} < k_{ij} + 1$ and $k_{ij} < k_{it} + k_{tj} + 2$. Therefore we get $k_{it} + k_{ij} \le k_{ij} \le k_{it} + k_{tj} + 1$. □

Recall that in §4.1 we defined $A_n = \langle s_t | s_t \in \Delta\rangle$ the affine Weyl group of type $\tilde{A}_{n-1}$. We know from §1.1 that $A_n = N \rtimes P_n$, where N is the maximal normal subgroup of A_n which is free abelian of rank n-1, and P_n is a subgroup of A_n generated by the set $\{s_t | 1 \le t < n\}$ [Verm]. Let $\alpha: A_n \to P_n \cong S_n$ be the natural map with $\bar{w} = \alpha(w)$ and $\bar{s}_t = (t,t+1)$ for $t \in [n]$, where (t,t+1) is the permutation which transposes t and t+1 with the convention that $(n,n+1) = (n,1)$.

Define a right action of A_n on V in the following way: for $v = (x_1,\ldots,x_n)$, $s_t \in \Delta$ and $x,y \in A_n$,

(i)
$$(v)s_t = \begin{cases} (x_1,\ldots,x_{t-1},x_{t+1}, x_t, x_{t+2},\ldots,x_n) & \text{if } t \neq n \\ (x_n+1, x_2,\ldots,x_{n-1}, x_1-1) & \text{if } t = n \end{cases}$$

(ii) $(v)xy = ((v)x)y$.

It is easy to check that such an action of A_n on V is independent of the expression of w in A_n and so is well defined. It is plain that such an action of A_n induces a permutation on $\mathbb{A}$ in the following way: for $A_k = \bigcap_{1\leq i,j\leq n} H^1_{ij,k_{ij}} \in \mathbb{A}$, $s_t \in \Delta$ and $x,y \in A_n$,

(i) $(A_k)s_t = \bigcap_{1\leq i,j\leq n} H^1_{ij;k'_{ij}}$ with $k'_{ij} = k_{(i)\bar{s}_t,(j)\bar{s}_t} + \varepsilon^{(t)}_{ij}$ and

$$\varepsilon^{(t)}_{ij} = \begin{cases} 0, & \text{if } \{\bar{i}, \bar{j}\} \neq \{\bar{t}, \overline{t+1}\} \\ -1 & \text{if } \bar{i} = \bar{t}, \quad \bar{j} = \overline{t+1} \\ 1 & \text{if } \bar{i} = \overline{t+1}, \ \bar{j} = \bar{t}. \end{cases} \qquad (3)$$

Recall that in §4.1 $\bar{a}$ is the image of a under the natural map $\mathbb{Z} \to \underline{n}$. We keep this notation from now on.

(ii) $(A_k)xy = ((A_k)x)y$.

Note that this is exactly the right action of an affine Weyl group of type $\tilde{A}$ on a Euclidean space defined in §1.1. So this action is simply transitive on $\mathbb{A}$ [L3]. It then induces a bijection from A_n to $\mathbb{A}$ by sending w to $(A_1)w$, where $A_1 = \bigcap_{1\leq i<j\leq n} H^1_{ij;0}$, which is in $\mathbb{A}$ by Lemma 6.1.1. Usually, we write $A_w = (A_1)w = \bigcap_{1\leq i<j\leq n} H^1_{ij;k^w_{ij}}$. Clearly, $(A_x)y = A_{xy}$ for any $x,y \in A_n$.

<u>Proposition 6.1.4</u> <u>For any $w \in A_n$, we have</u>

(i) <u>$\ell(w) = \sum_{1\leq i<j\leq n} |k^w_{ij}|$.</u> (ii) <u>$R(w) = \{s_t \in \Delta | k^w_{t,t+1} < 0\}$.</u>

<u>Proof:</u> By the definition of the right action of A_n on $\mathbb{A}$, we see that for $s_t \in \Delta$, $1 \leq i < j \leq n$,

$$|k_{ij}^{ws_t}| = \begin{cases} |k_{ij}^{w}| & \text{if } \{\bar{i},\bar{j}\} \neq \{\bar{t}, \overline{t+1}\} \\ |k^{w}_{(i)\bar{s}_t,(j)\bar{s}_t}-1| & \text{if } \bar{i} = \bar{t}, \quad \bar{j} = \overline{t+1} \\ |k^{w}_{(i)\bar{s}_t,(j)\bar{s}_t}+1| & \text{if } \bar{i} = \overline{t+1}, \quad \bar{j} = \bar{t} \end{cases} \tag{4}$$

So $\sum_{1\leq i<j\leq n} |k_{ij}^{ws_t}| = \sum_{1\leq i<j\leq n} |k_{ij}^{w}| \pm 1$. Thus by the fact that $\sum_{1\leq i<j\leq n} |k_{ij}^{w}| = 0$ if and only if $w = 1$, we see $\ell(w) \geq \sum_{1\leq i<j\leq n} |k_{ij}^{w}|$. To show the equality, we claim that for any $w \neq 1$, there exists at least one element $s_r \in \Delta$ such that

$$\sum_{1\leq i<j\leq n} |k_{ij}^{ws_r}| = \sum_{1\leq i<j\leq n} |k_{ij}^{w}| - 1 \tag{5}$$

Suppose not, then for any $s_t \in \Delta$, $\sum_{1\leq i<j\leq n} |k_{ij}^{ws_t}| = \sum_{1\leq i<j\leq n} |k_{ij}^{w}| + 1$. Hence we have $|k^{w}_{t+1,t} - 1| = |k^{w}_{t+1,t}| + 1$ for all t, $1 \leq t < n$, and $|k^{w}_{n1} + 1| = |k^{w}_{n1}| + 1$. That is, $k^{w}_{t,t+1} \geq 0$ for $1 \leq t < n$ and $k^{w}_{1n} \leq 0$. But by Lemma 6.1.3, we see that for any i,j', $1 \leq i < j \leq n$, $k^{w}_{ij} \geq \sum_{u=i}^{j-1} k^{w}_{u,u+1} \geq 0$ and hence $k^{w}_{ij} \geq 0$. On the other hand, $0 \geq k^{w}_{1n} \geq k^{w}_{1i} + k^{w}_{ij} + k^{w}_{jn} \geq k^{w}_{ij}$. This implies $k^{w}_{ij} = 0$ for all i,j, $1 \leq i < j \leq n$, i.e. $w = 1$. This is a contradiction. So conclusion (5) is shown. Then there exists a sequence $x_0 = w, x_1,\ldots,x_\ell = 1$ such that for every α, $1 \leq \alpha \leq \ell$, $x_\alpha = x_{\alpha-1}s_{i_\alpha}$ with some $s_{i_\alpha} \in \Delta$ and $\sum_{1\leq i<j\leq n} |k_{ij}^{x_\alpha}| = \sum_{1\leq i<j\leq n} |k_{ij}^{x_{\alpha-1}}| -1$. Clearly, $w = s_{i_\ell} s_{i_{\ell-1}} \cdots s_{i_1}$. This implies $\ell(w) \leq \sum_{1\leq i<j\leq n} |k^{w}_{ij}|$. Hence (i) follows.

In the above proof, we see that for any $s_t \in \Delta$, $\ell(ws_t) < \ell(w)$ if and only if $\sum_{1\leq i<j\leq n} |k_{ij}^{ws_t}| = \sum_{1\leq i<j\leq n} |k_{ij}^{w}| - 1$. By (4), when $1 \leq t < n$, this is equivalent to $|k^{w}_{t+1,t} - 1| = |k^{w}_{t+1,t}| - 1$, i.e. $k^{w}_{t,t+1} < 0$. When $t = n$, this is equivalent to $|k^{w}_{n1} + 1| = |k^{w}_{n1}| - 1$, i.e. $k^{w}_{n,n+1} < 0$. Thus (ii) follows. □

Now we shall give the necessary and sufficient conditions on the integer set $\{k_{ij}\}_{1\leq i<j\leq n}$ for $\bigcap_{1\leq i<j\leq n} H^{1}_{ij,k_{ij}}$ to be an alcove of V.

Let E be the set of all $n \times n$ skew symmetric matrices over $\mathbb{Z}$. Then the map $\iota: \mathbb{A} \to E$: $\bigcap_{1\leq i,j\leq n} H^{1}_{ij,k_{ij}} \to (k_{ij})_{1\leq i,j\leq n}$ is injective. Let

$$E_0 = \{(k_{ij})_{1\leq i,j\leq n} \in E \mid k_{it} + k_{tj} \leq k_{ij} \leq k_{it} + k_{tj} + 1 \text{ for all } 1\leq i<t<j\leq n\}.$$

Then Lemma 6.1.3 asserts $\iota(\mathbb{A}) \subseteq E_0$.

Define a right action of A_n on E as follows: for $k = (k_{ij})_{1\leq i,j\leq n} \in E$, $s_t \in \Delta$ and $x,y \in A_n$,

(i) $(k)s_t = (k'_{ij})_{1\leq i,j\leq n}$ with $k'_{ij} = k_{(i)\bar{s}_t,(j)\bar{s}_t} + \varepsilon^{(t)}_{ij}$, where $\varepsilon^{(t)}_{ij}$ is defined by (3).

(ii) $(k)xy = ((k)x)y$.

As before, one can easily check that this action of A_n on E is well defined. The map ι is A_n-equivariant and so $\iota(\mathbb{A})$ is an A_n-orbit of E in E_0.

Proposition 6.1.5 $E_0 = \iota(\mathbb{A})$.

Proof: It is sufficient to show that E_0 is a single A_n-orbit. Call $k = (k_{ij})_{1\leq i,j\leq n} \in E$ a minimal element if, for any $s \in \Delta$, $\sum_{1\leq i<j\leq n} |k_{ij}| < \sum_{1\leq i<j\leq n} |k'_{ij}|$, where $k' = (k)s = (k'_{ij})_{1\leq i,j\leq n}$. It is clear that k is minimal if and only if $k_{t,t+1} > 0$ for all $t \in [n]$, where the subscripts of k are regarded in $\underline{n}$. By the same argument as that in the proof of Proposition 6.1.4, we see that E_0 contains a unique minimal element, i.e. $\iota(A_1)$. So it suffices to show that if $k \in E_0$, $s_r \in \Delta$ and $k' = (k)s_r$ then $k' \in E_0$, i.e. for any i,t,j with $1 \leq i < t < j \leq n$, $k'_{it} + k'_{tj} \leq k'_{ij} \leq k'_{it} + k'_{tj} + 1$, or equivalently,

$$k_{(i)\bar{s}_r,(t)\bar{s}_r} + k_{(t)\bar{s}_r,(j)\bar{s}_r} + \varepsilon_{it}^{(r)} + \varepsilon_{tj}^{(r)} \leq k_{(i)\bar{s}_r,(j)\bar{s}_r} + \varepsilon_{ij}^{(r)}$$
$$\leq k_{(i)\bar{s}_r,(t)\bar{s}_r} + k_{(t)\bar{s}_r,(j)\bar{s}_r} + \varepsilon_{it}^{(r)} + \varepsilon_{tj}^{(r)} + 1. \qquad (6)$$

We know that

$$\left\{\begin{array}{l} \varepsilon_{ij}^{(r)} = \begin{cases} 1 & \text{if } \{i,j\} = \{1,n\} \text{ and } r = n \\ 0 & \text{otherwise} \end{cases} \\ \\ \varepsilon_{it}^{(r)} = \begin{cases} -1 & \text{if } \{i,t\} = \{r,r+1\} \text{ and } r \neq n \\ 0 & \text{otherwise} \end{cases} \\ \\ \varepsilon_{tj}^{(r)} = \begin{cases} -1 & \text{if } \{t,j\} = \{r,r+1\} \text{ and } r \neq n \\ 0 & \text{otherwise} \end{cases} \end{array}\right. \qquad (7)$$

In general, we can easily check that for $k \in E_0$, $1 \leq \lambda < \mu < \nu \leq n$ or $1 \leq \mu < \nu < \lambda \leq n$ or $1 \leq \nu < \lambda < \mu \leq n$, the following inequalities hold

$$k_{\lambda\mu} + k_{\mu\nu} \leq k_{\lambda\nu} \leq k_{\lambda\mu} + k_{\mu\nu} + 1 \qquad (8)$$

$$k_{\nu\mu} + k_{\mu\lambda} - 1 \leq k_{\nu\lambda} \leq k_{\nu\mu} + k_{\mu\lambda} \qquad (9)$$

According to (7), when $\{\bar{r},\overline{r+1}\} \neq \{\bar{i},\bar{t}\}, \{\bar{t},\bar{j}\}, \{\bar{i},\bar{j}\}$, (6) holds by (8). When $\{\bar{r},\overline{r+1}\} = \{\bar{i},\bar{t}\}$ or $\{\bar{t},\bar{j}\}$ or $\{\bar{i},\bar{j}\}$, (6) holds by (9). Our result follows. □

Now we define a left action of A_n on $\mathbb{A}$: for $w,x,y \in A_n$ and $s_t \in \Delta$,

(i) $s_t(A_w) = A_{s_t w}$ (ii) $xy(A_w) = x(y(A_w))$.

Then it is clear that $x(A_w) = A_{xw}$ for any $x,w \in A_n$.

We know that each alcove of $\mathbb{A}$ has the form $(A_1)w$ for some $w \in A_n$. So by Corollary 6.12, any alcove of $\mathbb{A}$ has n walls. We shall label each wall of a given alcove A_w by some element of Δ as follows.

(i) When $w = 1$, A_1 has walls supported by the hyperplanes $\{H_{t,t+1;0} | 1 \leqslant t < n\} \cup \{H_{1n;1}\}$ by Corollary 6.12. We label s_t to the wall which is supported by $H_{t,t+1;0}$ for $1 \leqslant t < n$ and s_n to one supported by $H_{1n;1}$.

(ii) Inductively, suppose that we have labelled all the walls of A_w for any w with $\ell(w) < m$ for some $m > 0$. Let $w' = ws_t$ be such that $\ell(w') = \ell(w) + 1$. Then the right action of s_t on V sends A_w to $A_{w'}$ and also sends all walls of A_w to those of $A_{w'}$. Suppose that H' is the wall of $A_{w'}$ coming from the s_α-wall of A_w under s_t. Then we label H' by s_α.

We know that the right action of A_n on V induces a permutation on the set of walls of all alcoves of V. It is well known that each A_n-orbit of such walls intersects the closure of any alcove in a unique wall. Thus the label of any wall of an alcove A_w is independent of the expression of w and so is well defined.

Lemma 6.1.6. If A_w, $A_{w'}$ are two alcoves of V which share a common wall then the label of this wall for A_w is the same as for $A_{w'}$, say s_t-wall. We have $w' = s_t w$. Conversely, if $w, w' \in A_n$ have the relation $w' = s_t w$ for some $s_t \in \Delta$, then the alcoves A_w, $A_{w'}$ of V share the common s_t-wall.

Proof: Let $A_w = \bigcap_{1 \leqslant i<j \leqslant n} H^1_{ij;k^w_{ij}}$ and $A_{w'} = \bigcap_{1 \leqslant i<j \leqslant n} H^1_{ij;k^{w'}_{ij}}$. Then by our condition, we see that $k^w_{ij} = k^{w'}_{ij}$ for all pair i,j, $1 \leqslant i < j \leqslant n$, but one, say this exceptional pair is ℓ,m with $1 \leqslant \ell < m \leqslant n$, then $k^w_{\ell m} = k^{w'}_{\ell m} \pm 1$. So by Proposition 6.1.4, we have $\ell(w) = \ell(w') \pm 1$. We may assume $\ell(w) = \ell(w') - 1$ without loss of generality. Hence $|k^w_{\ell m}| = |k^{w'}_{\ell m}| - 1$. Let us apply induction on $\ell(w) \geqslant 0$. It is obvious for $\ell(w) = 0$. Now assume $\ell(w) > 0$. We claim that there exists at least one $\alpha \in [n]$ such that $s_\alpha \in R(w) \cap R(w')$. This is because $R(w) \neq \emptyset$ and $R(w) \subseteq R(w')$ by Proposition 6.1.4 (ii) and our hypothesis. Let $y = ws_\alpha$ and $y' = w's_\alpha$. Then $\ell(y) = \ell(y') - 1$ and $\ell(y) < \ell(w)$. Since A_y and $A_{y'}$ share a common wall which comes from the common wall of A_w and $A_{w'}$ under s_α, we see by inductive hypothesis that the label of this wall for Ay is the same as for Ay', say s_t-wall, and $y' = s_t y$. So the common wall of A_w and $A_{w'}$ is also labelled by s_t. Since $w = ys_t$ and $w' = y's_t$, we have $w' = s_t w$.

Conversely, assume that $w,w' \in A_n$ satisfy $w' = s_t w$ for some $s_t \in \Delta$. Let A_w be the alcove of V corresponding to w. Let A_y be the alcove differing from A_w but sharing a common s_t-wall with A_w. Then by the above proof, we have $y = s_y w = w'$. So $A_{w'}$, A_w share the common s_t-wall. □

Now we can give another description of the length function $\ell(w)$ on A_n.

<u>Proposition 6.1.7</u> <u>For any $w \in A_n$, $\ell(w)$ is the minimum number of walls of alcoves of V which separate the alcove A_w from A_1. In other words, $\ell(w)$ is the smallest number r such that there exists a sequence of alcoves $A_0 = A_w, A_1, \ldots, A_r = A_1$, where any two consecutive alcoves in this sequence share a common wall.</u>

<u>Proof</u>: Suppose that $A_0 = A_w, A_1, \ldots, A_r = A_1$ is a sequence of alcoves such that any two consecutive alcoves in this sequence share a common wall. Assume that $A_j = A_{x_j}$, $0 \le j \le r$, for $x_j \in A_n$. Then by Lemma 6.1.6, we have $x_j = s_{i_j} x_{j-1}$ for some $s_{i_j} \in \Delta$. So $w = s_{i_r} s_{i_{r-1}} \cdots s_{i_1}$. This implies that $\ell(w) \le r$ and the equality holds if we take a sequence $A_0 = A_w, A_1, \ldots, A_r = A_1$ as above such that the expression $w = s_{i_r} s_{i_{r-1}} \cdots s_{i_1}$ is a reduced form. So the proposition is proved. □

In the remaining part of this section, we shall generalize the result of Corollary 6.1.2 to any alcove of V. We shall also give a criterion for an element $s_t \in \Delta$ to be in $\mathcal{L}(w)$ for any $w \in A_n$. To do this, we need the following lemma which will be proved in the next section.

<u>Lemma 6.1.8</u> <u>Let $A_w = \bigcap_{1 \le i < j \le n} H^1_{ij;k^w_{ij}}$ for any $w \in A_n$. Then $k^{w^{-1}}_{ij} = k^w_{(j)\bar{w},(i)\bar{w}}$ for any i,j with $1 \le i < j \le n$.</u>

Let k, i, j be any integers with $1 \le i < j \le n$. Recall that at the beginning of this section we assumed that $H^1_{ji;k} = H^1_{ij;-k} = \{(x_1, \ldots, x_n) \in V \mid -k < x_j - x_i < -k+1\}$. So $H^1_{ij;k}$ (resp. $H^1_{ji;k}$) is bounded by two parallel hyperplanes $H_{ij;k}$ and $H_{ij;k+1}$ (resp. $H_{ij;k}$ and $H_{ij;-k+1}$). Let $H_{\beta\alpha;h} = H_{\alpha\beta;-h+1}$ for any integers α, β, h with $1 \le \alpha < \beta \le n$. Then $H^1_{ji;k}$ is also bounded by $H_{ji;k}$ and $H_{ji,k+1}$. So we can say that for any integers k, i, j with $i,j \in [n]$ and $i \neq j$, $H^1_{ij;k}$ is bounded by $H_{ij;k}$ and $H_{ij;k+1}$.

On the other hand, we assume that $H_{ij;k} = H_{i\pm n,j;k} = H_{i,j\pm n;k} = H_{i\pm n,j\pm n;k}$ for any integers k, i, j with $i,j \in [n]$ and $i \neq j$.

For any $w \in A_n$ and $t \in [n]$, let $H_t(w)$ be the hyperplane of V supporting the s_t-wall of the alcove A_w. Then by the definition of the s_t-wall of an alcove, we have $H_t(w) = (H_t(1))w = (H_{t,t+1;0})w$. So $H_t(w)$ has the form $H_{(t)\bar{w},(t+1)\bar{w};k_t}$ for some

$k_t \in \{k^w_{(t)\bar{w},(t+1)\bar{w}},\ k^w_{(t)\bar{w},(t+1)\bar{w}} \pm 1\}$, where we assume $A_w = \bigcap_{1\le i,j\le n} H^1_{ij;k^w_{ij}}$.

Now we wish to consider $H_t(w)$.

By Proposition 6.1.4(i) and Lemma 6.1.6, we see that $s_t \in \mathcal{L}(w)$ if and only if A_w and A_1 are on different sides of $H_t(w)$. On the other hand, by Lemma 6.1.8 and Proposition 6.1.4(ii) we see that $s_t \in \mathcal{L}(w) \iff s_t \in R(w^{-1}) \iff k^{w^{-1}}_{t,t+1} < 0 \iff k^w_{(t+1)\bar{w},(t)\bar{w}} < 0 \iff k^w_{(t)\bar{w},(t+1)\bar{w}} > 0$. Equivalently, $s_t \notin \mathcal{L}(w) \iff k^w_{(t)\bar{w},(t+1)\bar{w}} \le 0$.

First assume $s_t \in \mathcal{L}(w)$. Then $H^1_{(t)\bar{w},(t+1)\bar{w};k^w_{(t)\bar{w},(t+1)\bar{w}}}$ and A_1 are on different sides of $H_t(w)$. So $H_t(w) = H_{(t)\bar{w},(t+1)\bar{w},k^w_{(t)\bar{w},(t+1)\bar{w}}}$ by the fact that $k^w_{(t)\bar{w},(t+1)\bar{w}} > 0$.

Secondly assume $s_t \notin \mathcal{L}(w)$. Then $H^1_{(t)\bar{w},(t+1)\bar{w};k^w_{(t)\bar{w},(t+1)\bar{w}}}$ and A_1 are on the same side of $H_t(w)$. In that case, we have $k^w_{(t)\bar{w},(t+1)\bar{w}} \le 0$. If $k^w_{(t)\bar{w},(t+1)\bar{w}} < 0$ then $H_t(w) = H_{(t)\bar{w},(t+1)\bar{w};k^w_{(t)\bar{w},(t+1)\bar{w}}}$. If $k^w_{(t)\bar{w},(t+1)\bar{w}} = 0$ then $H_t(w) = H_{(t)\bar{w},(t+1)\bar{w};1}$.

So we can summarize the above results as follows.

<u>Proposition 6.1.9</u> <u>For any $w \in A_n$ and $t \in [n]$, let $H_t(w)$ be the hyperplane of V supporting the s_t-wall of the alcove A_w. Then</u>

(i)

$$H_t(w) = \begin{cases} H_{(t)\bar{w},(t+1)\bar{w};k^w_{(t)\bar{w},(t+1)\bar{w}}} & \text{if } k^w_{(t)\bar{w},(t+1)\bar{w}} \neq 0 \\ H_{(t)\bar{w},(t+1)\bar{w};1} & \text{if } k^w_{(t)\bar{w},(t+1)\bar{w}} = 0 \end{cases}$$

<u>and so $A_w = \bigcap_{t\in[n]} H^1_{(t)\bar{w},(t+1)\bar{w};k^w_{(t)\bar{w},(t+1)\bar{w}}}$.</u>

(iii) <u>$\mathcal{L}(w) = \{s_t \in \Delta \mid k^w_{(t)\bar{w},(t+1)\bar{w}} > 0\}$.</u> □

§6.2 THE RELATION BETWEEN TWO DESCRIPTIONS OF A_n

We saw in Chapter 4 that any element w of A_n can be described in terms of a permutation on $\mathbb{Z}$ which satisfies $(t+n)w = (t)w + n$ for $t \in \mathbb{Z}$ and $\sum_{h=1}^{n} (h)w = \sum_{h=1}^{n} h$. As a permutation on $\mathbb{Z}$, w can be determined by an n-tuple $((1)w^{-1}, (2)w^{-1}, \ldots, (n)w^{-1})$. Let $A_w = \bigcap_{1\le i<j\le n} H^1_{ij;k^w_{ij}}$. The following result explores the relation between these two descriptions of w.

<u>Proposition 6.2.1</u> <u>$(j)w^{-1} - (i)w^{-1} = k^w_{ij}\cdot n + r^w_{ij}$ for $1 \le i < j \le n$ and $1 \le r^w_{ij} \le n-1$.</u>

<u>Proof:</u> Apply induction on $\ell(w)$. When $\ell(w) = 0$, i.e. $w = 1$, we have $k^w_{ij} = 0$ and $(j).(1) - (i).1 = 0.n + (j-i)$. Our result is true in this case. Now assume $\ell(w) > 0$.

We may write $w = ys_t$ for some $s_t \in R(w)$. Hence

$$f = (j)w^{-1} - (i)w^{-1} = (j)s_t y^{-1} - (i)s_t y^{-1}$$

(i) Assume $t \neq n$. If $\{i,j\} \neq \{t,t+1\}$ then $(j)s_t > (i)s_t$. By inductive hypothesis, $f = k^y_{(i)\bar{s}_t,(j)\bar{s}_t}\cdot n + r^y_{(i)\bar{s}_t,(j)\bar{s}_t} = k^w_{ij}\cdot n + r^w_{ij}$, where we set $r^w_{ij} = r^y_{(i)\bar{s}_t,(j)\bar{s}_t}$. If $\{i,j\} = \{t,t+1\}$ then $(i)\bar{s}_t = t+1$ and $(j)\bar{s}_t = t$. By inductive hypothesis,

$$f = -((t+1)y^{-1} - (t)y^{-1}) = -k^y_{t,t+1}\cdot n - r^y_{t,t+1} = (-k^y_{t,t+1}-1)n + (n-r^y_{t,t+1})$$

$$= k^w_{t,t+1}\cdot n + r^w_{t,t+1}, \text{ where we set } r^w_{t,t+1} = n - r^y_{t,t+1}.$$

(ii) Assume $t = n$. If $\{i,j\} \cap \{1,n\} = \emptyset$ then $((i)s_t, (j)s_t) = (i,j)$. By inductive hypothesis,

$$f = k^y_{ij}\cdot n + r^y_{ij} = k^w_{ij}\cdot n + r^w_{ij}, \text{ where we set } r^w_{ij} = r^y_{ij}.$$

If $i = 1$ and $j \neq n$ then by inductive hypothesis,

$$f = (j)y^{-1} + n - (n)y^{-1} = n - (k^y_{jn}\cdot n + r^y_{jn}) = (-k^y_{jn})n + (n-r^y_{jn}) = k^w_{1j}\cdot n + r^w_{1j},$$

where we set $r^w_{1j} = n - r^y_{jn}$. If $i \neq 1$ and $j = n$ then by inductive hypothesis,

$$f = n + (1)y^{-1} - (i)y^{-1} = n - (k^y_{1i}\cdot n + r^y_{1i}) = (-k^y_{1i})n + (n-r^y_{1i}) = k^w_{in}\cdot n + r^w_{in},$$

where we set $r^w_{in} = n - r^y_{1i}$. If $i = 1$ and $j = n$ then by inductive hypothesis,

$$f = n + (1)y^{-1} + n-(n)y^{-1} = 2n-(k^y_{1n}\cdot n+r^y_{1n}) = (1-k^y_{1n})n+(n-r^y_{1n}) = k^w_{1n}\cdot n + r^w_{1n},$$

where we set $r^w_{1n} = n - r^y_{1n}$.

Therefore by induction, our proof is complete. □

Now we can prove Lemma 6.1.8 by using the above result.

<u>Proof of Lemma 6.1.8</u>

Assume $(t)w = q_t n + r_t$ for $t, r_t \in [n]$ and $q_t \in \mathbf{Z}$. Note $(t)\bar{w} = r_t$ and so $r_i \neq r_j$ for any $i,j \in [n]$ with $i \neq j$. By Proposition 6.2.1, we have

$$(j)w - (i)w = k^{w^{-1}}_{ij}\cdot n + r^{w^{-1}}_{ij} \text{ with } 1 \leqslant r^{w^{-1}}_{ij} \leqslant n-1.$$

(i) Assume $r_j > r_i$. Then $(j)w - (i)w = (q_j-q_i)n + (r_j-r_i)$ and so $k^{w^{-1}}_{ij} = q_j-q_i$. On the other hand, by Proposition 6.2.1,

$$(r_j)w^{-1} - (r_i)w^{-1} = ((j)\bar{w})w^{-1} - ((i)\bar{w})w^{-1} = k^w_{(i)\bar{w},(j)\bar{w}}\cdot n + r^w_{(i)\bar{w},(j)\bar{w}}\ .$$

But $(r_j)w^{-1} - (r_i)w^{-1} = ((j)w - q_jn)w^{-1} - ((i)w - q_in)w^{-1}$

$$= ((j)w)w^{-1} - q_jn - ((i)w)w^{-1} + q_in = (q_i - q_j)n + (j-i).$$

So $k^w_{(j)\bar{w},(i)\bar{w}} = -k^w_{(i)\bar{w},(j)\bar{w}} = -(q_i - q_j) = q_j - q_i = k^{w^{-1}}_{ij}$.

(ii) When $r_j < r_i$, $(j)w - (i)w = (q_j - q_i - 1)n + (n + r_j - r_i)$. So $k^{w^{-1}}_{ij} = q_j - q_i - 1$. On the other hand, by Proposition 6.2.1,

$$(r_i)w^{-1} - (r_j)w^{-1} = ((i)\bar{w})w^{-1} - ((j)\bar{w})w^{-1} = k^w_{(j)\bar{w},(i)\bar{w}} \cdot n + r^w_{(i)\bar{w},(j)\bar{w}} .$$

But

$$(r_i)w^{-1} - (r_j)w^{-1} = (i)w - q_in)w^{-1} - ((j)w - q_jn)w^{-1}$$

$$= ((i)w)w^{-1} - q_in - ((j)w)w^{-1} + q_jn = (q_j - q_i - 1)n + (n + i - j).$$

So $k^w_{(j)\bar{w},(i)\bar{w}} = q_j - q_i - 1 = k^{w^{-1}}_{ij}$. Therefore our assertion follows. □

From Lemma 6.1.8, we see that for any $w \in A_n$, the number of zeros in the set $\{k^w_{ij} | 1 \le i < j \le n\}$ is the same as that in the set $k^{w^{-1}}_{ij}$ $1 \le i < j \le n$. This fact will be useful in the next chapter.

The following result gives formulae to transform an alcove of V to the corresponding permutation on **Z**.

<u>Corollary 6.2.2</u> <u>For any $w \in A_n$, $t \in [n]$ and $1 \le i < j \le n$,</u>

(i) $(t)w^{-1} = t + \sum_{h=1}^{n} k^w_{ht}$. (ii) $(t)w = t + \sum_{h=1}^{n} k^w_{(t)\bar{w},h}$.

(iii) $r^w_{ij} = (j-i) + \sum_{h=1}^{n} (k^w_{ih} + k^w_{hj} + k^w_{ji})$.

<u>Proof</u>: (i) By Proposition 6.2.1, we have: for $1 \le i < j \le n$,

$$(j)w^{-1} - (i)w^{-1} = k^w_{ij} \cdot n + r^w_{ij} \text{ with } 1 \le r^w_{ij} < n \tag{1}$$

Then

$$(i)w^{-1} - (j)w^{-1} = k^w_{ji}\, n + r^w_{ji} - n \tag{2}$$

with the convention that $r^w_{ij} + r^w_{ji} = n$. Clearly, for any $t \in [n]$, $\sum_{t \ne h \in [n]} r^w_{th} = \frac{n(n-1)}{2}$. Now we take the sum with $i = t$ and j ranging over $[n]$ in (1) and (2):

$$\sum_{h=1}^{n} (h)w^{-1} - n \cdot (t)w^{-1} = \sum_{h=1}^{n} k^w_{th} \cdot n + \frac{n(n-1)}{2} - n(t-1) \tag{3}$$

Substitute $\sum_{h=1}^{n} (h)w^{-1} = \frac{n(n+1)}{2}$ into (3): $(-n) \cdot (t)w^{-1} = \sum_{h=1}^{n} k^w_{th} \cdot n - n - n(t-1)$.

So $(t)w^{-1} = t - \sum_{h=1}^{n} k^w_{th} = t + \sum_{h=1}^{n} k^w_{ht}$.

(ii) By (i) and Lemma 6.1.8, we have

$$(t)w = t + \sum_{h=1}^{n} k^{w^{-1}}_{ht} = t + \sum_{h=1}^{n} k^w_{(t)\bar{w},(h)\bar{w}} = t + \sum_{h=1}^{n} k^w_{(t)\bar{w},h}$$

(ii) By Proposition 6.2.1 and (i), $(j + \sum_{h=1}^{n} k^w_{hj}) - (i + \sum_{h=1}^{n} k^w_{hi}) = k^w_{ij}\cdot n + r^w_{ij}$.

Thus $r^w_{ij} = (j-i) + (\sum_{h=1}^{n} k^w_{hj} + \sum_{h=1}^{n} k^w_{ih} + k^w_{ji}\cdot n) = (j-i) + \sum_{h=1}^{n} (k^w_{ih} + k^w_{hj} + k^w_{ji})$. □

Note that for any $w \in A_n$ and $t \in [n]$ we have $((t)\bar{w}^{-1})w = q_t n+t$ for some $q_t \in \mathbb{Z}$. The following result will be very useful for the later development.

<u>Proposition 6.2.3</u> <u>For $1 \leq \alpha < \beta \leq n$ and $w \in A_n$, we have</u>

(i) $k^w_{\alpha\beta} < 0 \Leftrightarrow$ either $((\beta)\bar{w}^{-1})w - ((\alpha)\bar{w}^{-1})w > n$ or

$((\beta)\bar{w}^{-1})w - ((\alpha)\bar{w}^{-1})w > 0$ with $(\alpha)\bar{w}^{-1} > (\beta)\bar{w}^{-1}$.

(ii) $k^w_{\alpha\beta} > 0 \Leftrightarrow$ either $((\alpha)\bar{w}^{-1})w - ((\beta)\bar{w}^{-1})w > n$ or

$((\alpha)\bar{w}^{-1})w - ((\beta)\bar{w}^{-1})w > 0$ with $(\beta)\bar{w}^{-1} > (\alpha)\bar{w}^{-1}$.

(iii) $k^w_{\alpha\beta} = 0 \Leftrightarrow ((\alpha)\bar{w}^{-1} - (\beta)\bar{w}^{-1})(((\alpha)\bar{w}^{-1})w - ((\beta)\bar{w}^{-1})w) > 0$ and

$|((\alpha)\bar{w}^{-1})w - ((\beta)\bar{w}^{-1})w| < n$.

<u>Proof</u>: We have

$$\begin{cases} ((\alpha)\bar{w}^{-1})w = q_\alpha n+\alpha \\ ((\beta)\bar{w}^{-1})w = q_\beta n+\beta \end{cases} \qquad (4)$$

This implies

$$\begin{cases} (\alpha)w^{-1} = (\alpha)\bar{w}^{-1} - q_\alpha n \\ (\beta)w^{-1} = (\beta)\bar{w}^{-1} - q_\beta n \end{cases} \qquad (5)$$

(i) $k^w_{\alpha\beta} < 0 \Leftrightarrow (\beta)w^{-1} - (\alpha)w^{-1} < 0$ by Proposition 6.2.1

$\Leftrightarrow ((\beta)\bar{w}^{-1} - (\alpha)\bar{w}^{-1}) + (q_\alpha - q_\beta)n < 0$ by (5)

$\Leftrightarrow$ either $q_\alpha < q_\beta$ or $q_\alpha = q_\beta$ with $(\alpha)\bar{w}^{-1} > (\beta)\bar{w}^{-1}$

$\Leftrightarrow$ either $((\beta)\bar{w}^{-1})w - ((\alpha)\bar{w}^{-1})w > n$, or $((\beta)\bar{w}^{-1})w - ((\alpha)\bar{w}^{-1})w > 0$

with $(\alpha)\bar{w}^{-1} > (\beta)\bar{w}^{-1}$ by (4) and the condition $\alpha < \beta$.

(ii) $k^w_{\alpha\beta} > 0 \iff (\beta)w^{-1} - (\alpha)w^{-1} > n$ by Proposition 6.2.1

$\iff ((\beta)\bar{w}^{-1}-(\alpha)\bar{w}^{-1}) + (q_\alpha-q_\beta)n > n$ by (5)

$\iff$ either $q_\alpha \geq q_\beta + 2$, or $q_\alpha = q_\beta + 1$ with $(\beta)\bar{w}^{-1} > (\alpha)\bar{w}^{-1}$.

$\iff$ either $((\alpha)\bar{w}^{-1})w - ((\beta)\bar{w}^{-1})w > n$, or $((\alpha)\bar{w}^{-1})w - ((\beta)\bar{w}^{-1})w > 0$

with $(\alpha)\bar{w}^{-1} < (\beta)\bar{w}^{-1}$ by (4) and the condition $\alpha < \beta$.

(iii) $k^w_{\alpha\beta} = 0 \iff 0 < (\beta)w^{-1} - (\alpha)w^{-1} < n$ by Proposition 6.2.1.

$\iff 0 < ((\beta)\bar{w}^{-1} - (\alpha)\bar{w}^{-1}) + (q_\alpha-q_\beta)n < n$ by (5)

$\iff$ either $q_\alpha = q_\beta + 1$ with $(\alpha)\bar{w}^{-1} > (\beta)\bar{w}^{-1}$, or $q_\alpha = q_\beta$ with $(\beta)\bar{w}^{-1} > (\alpha)\bar{w}^{-1}$.

$\iff ((\alpha)\bar{w}^{-1} - (\beta)\bar{w}^{-1})((\alpha)\bar{w}^{-1})w - ((\beta)\bar{w}^{-1})w) > 0$ and

$|((\alpha)\bar{w}^{-1})w - ((\beta)\bar{w}^{-1})w| < n$ by (4) and the condition $\alpha < \beta$. □

§6.3 THE MAP $\sigma:A_n \to \Lambda_n$ DEFINED IN GEOMETRICAL TERMS

We have already defined the map $\sigma:A_n \to \Lambda_n$ by Lusztig's procedure in Chapter 5. We also know that A_n can be identified with $\mathbb{A}$. Propositions 6.2.1 and 6.2.3 exhibit the relation between the descriptions of elements of A_n in terms of alcoves and permutations on $\mathbf{Z}$. We now shall give a version of the map σ in terms of alcoves.

Any $u \in \mathbf{Z}$ can be written uniquely as the form $u = q_u n + r_u$ with $q_u \in \mathbf{Z}$ and $r_u \in [n]$.

Lemma 6.3.1 Let $w \in A_n$ and $1 \leq i < j \leq n$. Then $k^w_{ij} \neq 0$ if and only if there exists $u,v \in \mathbf{Z}$ with $r_u = j$ and $r_v = i$ such that $(u-v)((u)w^{-1}-(v)w^{-1}) < 0$. In that case, for any $\ell \in \mathbf{Z}$, let $u' = \ell n+u$ and $v' = \ell n+v$. Then $(u'-v')((u')w^{-1} - (v')w^{-1}) < 0$.

Proof: ($\Rightarrow$) By Proposition 6.2.1, we have $(j)w^{-1} - (i)w^{-1} = k^w_{ij}n + r^w_{ij}$ with $r^w_{ij} \in [n]$. If $k^w_{ij} < 0$ then $(j)w^{-1} - (i)w^{-1} < 0$ and so $(j-1)((j)w^{-1} - (i)w^{-1}) < 0$. If $k^w_{ij} > 0$ then $i' = i + k^w_{ij}\cdot n > j$ but $(j)w^{-1} - (i')w^{-1} = r^w_{ij} \geq 1 > 0$ and hence $(j-i')((j)w^{-1} - (i')w^{-1}) < 0$.

($\Leftarrow$) $(u-v)((u)w^{-1} - (v)w^{-1}) < 0$ implies $((q_u-q_v)n+(j-i))((q_u-q_v)n + ((j)w^{-1} - (i)w^{-1})) < 0$. If $u > v$ then $q_u \geq q_v$ and so $(q_u-q_v)n + ((j)w^{-1}-(i)w^{-1}) < 0$. This implies $(j)w^{-1} - (i)w^{-1} < 0$ and thus $k^w_{ij} < 0$. If $u < v$ then $q_u < q_v$ and so $(q_u-q_v)n + ((j)w^{-1} - (i)w^{-1}) > 0$, i.e. $(q_u-q_v)n + k^w_{ij}\cdot n + r^w_{ij} > 0$. Therefore $k^w_{ij} \geq q_v-q_u > 0$.

The last sentence of the lemma is obvious. □

<u>Lemma 6.3.2</u> <u>Assume $w \in A_n$ and $1 \le i_1 < \dots < i_r \le n$. Then $k^w_{i_j, i_\ell} \ne 0$ for all j, ℓ, $1 \le j < \ell \le r$ if and only if there exists $u_1, \dots, u_r \in \mathbb{Z}$ with $\{\bar{u}_1, \dots, \bar{u}_r\} = \{\bar{i}_1, \dots, \bar{i}_r\}$ such that $u_1 < \dots < u_r$ and $(u_1)w^{-1} > \dots > (u_r)w^{-1}$.</u>

<u>Proof</u>: ($\Leftarrow$) By Lemma 6.3.1.

($\Rightarrow$) When $r = 2$, the result is just as in Lemma 6.3.1.

Now assume $r \ge 3$. By Lemma 6.3.1, there exists $u_{\alpha\beta}$, $u_{\beta\alpha} \in \mathbb{Z}$ with $u_{\alpha\beta} = q_{\alpha\beta}n + i_\alpha$ and $u_{\beta\alpha} = q_{\beta\alpha}n + i_\beta$ for each pair $\alpha, \beta \in \{1, 2, \dots, r\}, \alpha \ne \beta$, such that $(u_{\alpha\beta} - u_{\beta\alpha})((u_{\alpha\beta})w^{-1} - (u_{\beta\alpha})w^{-1}) < 0$. We claim that if $u_{\alpha\beta} > u_{\beta\alpha}$ and $u_{\beta\gamma} > u_{\gamma\beta}$ then $u_{\alpha\gamma} > u_{\gamma\alpha}$. For otherwise, $u_{\alpha\gamma} < u_{\gamma\alpha}$ and this would imply

$$(q_{\alpha\beta} - q_{\beta\alpha})n + (i_\alpha - i_\beta) > 0$$

$$(q_{\beta\gamma} - q_{\gamma\beta})n + (i_\beta - i_\gamma) > 0$$

$$(q_{\gamma\alpha} - q_{\alpha\gamma})n + (i_\gamma - i_\alpha) > 0$$

So $(q_{\alpha\beta} - q_{\beta\alpha}) + (q_{\beta\gamma} - q_{\gamma\beta}) + (q_{\gamma\alpha} - q_{\alpha\gamma}) > 0$. On the other hand,

$$(u_{\alpha\beta})w^{-1} - (u_{\beta\alpha})w^{-1} < 0$$

$$(u_{\beta\gamma})w^{-1} - (u_{\gamma\beta})w^{-1} < 0$$

$$(u_{\gamma\alpha})w^{-1} - (u_{\alpha\gamma})w^{-1} < 0$$

This implies
$$(q_{\alpha\beta} - q_{\beta\alpha})n + (i_\alpha)w^{-1} - (i_\beta)w^{-1} < 0$$

$$(q_{\beta\gamma} - q_{\gamma\beta})n + (i_\beta)w^{-1} - (i_\gamma)w^{-1} < 0$$

$$(q_{\gamma\alpha} - q_{\alpha\gamma})n + (i_\gamma)w^{-1} - (i_\alpha)w^{-1} < 0$$

Hence $(q_{\alpha\beta} - q_{\beta\alpha}) + (q_{\beta\gamma} - q_{\gamma\beta}) + (q_{\gamma\alpha} - q_{\alpha\gamma}) \le 0$. This gives a contradiction.

So we can define a total order on the set $\{\tilde{i}_\alpha, 1 \le \alpha \le r\}$, for $1 \le \alpha$, $\beta \le \gamma$,

$$\tilde{i}_\alpha = \tilde{i}_\beta \quad \text{if } \alpha = \beta$$

$$\tilde{i}_\alpha > \tilde{i}_\beta \quad \text{if } u_{\alpha\beta} > u_{\beta\alpha} .$$

Now assume $\{i'_1, \dots, i'_r\} = \{i_1, \dots, i_r\}$ with $\tilde{i}'_1 < \dots < \tilde{i}'_r$.

Then $u_{i'_t, i'_{t+1}} < u_{i'_{t+1}, i'_t}$ and $(u_{i'_t, i'_{t+1}})w^{-1} > (u_{i'_{t+1}, i'_t})w^{-1}$ for $1 \le t < r$.

Let $u'_{i'_t, i'_{t+1}} = u_{i'_t, i'_{t+1}} + \sum_{j=2}^{t} (q_{i'_j, i'_{j-1}} - q_{i'_j, i'_{j+1}})n$ for $1 \le t < r$. Then it is clear that $u'_{i'_1, i'_2} < u'_{i'_2, i'_3} < \dots < u'_{i'_{r-1}, i'_r}$ and $(u'_{i'_1, i'_2})w^{-1} > (u'_{i'_2, i'_3})w^{-1} > \dots > (u'_{i'_{r-1}, i'_r})w^{-1}$.

Since $\{\overline{u'_{i'_1,i'_2}}, \ldots, \overline{u'_{i'_{r-1},i'_r}}\} = \{\overline{u_{i'_1,i'_2}}, \ldots, \overline{u_{i'_{r-1},i'_r}}\} = \{\bar{i}_1, \ldots, \bar{i}_r\}$, our proof is complete. □

Recall the definition of the map $\sigma: A_n \to \Lambda_n$ in Chapter 5. For $w \in A_n$, let $\sigma(w) = \{\lambda_1 \geq \ldots \geq \lambda_r\} \in \Lambda_n$. Then by Lemma 6.3.2, the integers $\sum_{i=1}^{\ell} \lambda_i$, $1 \leq \ell \leq r$. can be described as the maximum cardinal of subsets X of [n], where X is a disjoint union of ℓ subsets X_h, $1 \leq h \leq \ell$, of [n] such that if i,j are in X_h with $1 \leq h \leq \ell$ and $i \neq j$ then we have $k_{ij} \neq 0$.

Example 6.3.3 The alcoves corresponding to elements of A_3 are as in the following diagram, where each small triangle in this diagram represents an alcove. We label each alcove by the corresponding element of A_3 and its coordinate form. That is, suppose that

 is an alcove $\bigcap_{1 \leq i < j \leq 3} H^1_{ij,k_{ij}}$ corresponding to $w \in A_3$. Then we have

 . e.g. the alcoves corresponding to elements 1, $s_1s_2s_3$ are

, , where i denotes s_i for short in the brackets. Each wall of an alcove is labelled by one Coxeter generator of A_3 by the rule described in §6.1. (We also denote s_i by i). Thus the readers can check most of the results in the present chapter from this diagram.

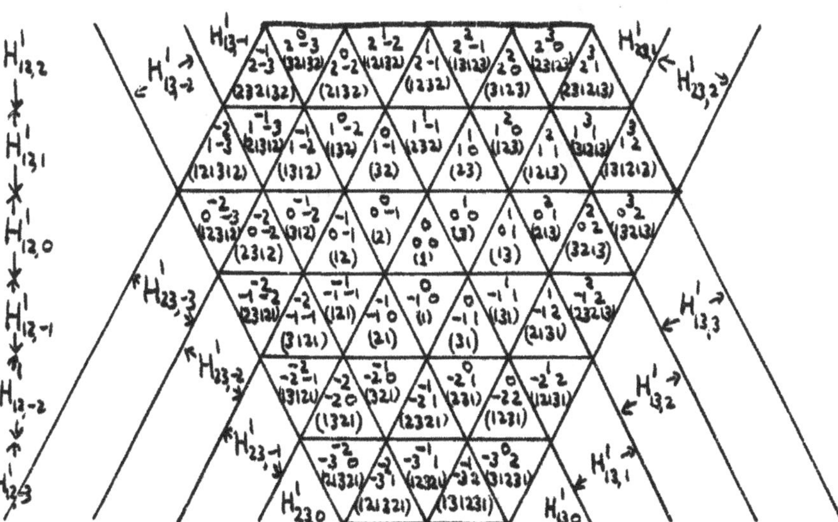

We know that $A_3 = \sigma^{-1}(\lambda_1) \cup \sigma^{-1}(\lambda_2) \cup \sigma^{-1}(\lambda_3)$ with $\lambda_1 = \{1 > 1 > 1\}$, $\lambda_2 = \{2 > 1\}$ and $\lambda_3 = \{3\}$. From the result of §1.7(vi) and the above description of the map σ, we see that the $\sigma^{-1}(\lambda_i)$, $1 < i < 3$, are two-sided cells of A_3 and that the left cells in a given two-sided cell are precisely the connected components of these fibres. Thus the left cells in A_3 are as shown in the following diagram. There is 1 left cell in the fibre $\sigma^{-1}(\lambda_1)$ which consists of the single solid alcove. There are 3 left cells in the fibre $\sigma^{-1}(\lambda_2)$ which consists of all shaded alcoves. There are 6 left cells in the fibre $\sigma^{-1}(\lambda_3)$ which consists of all unshaded alcoves.

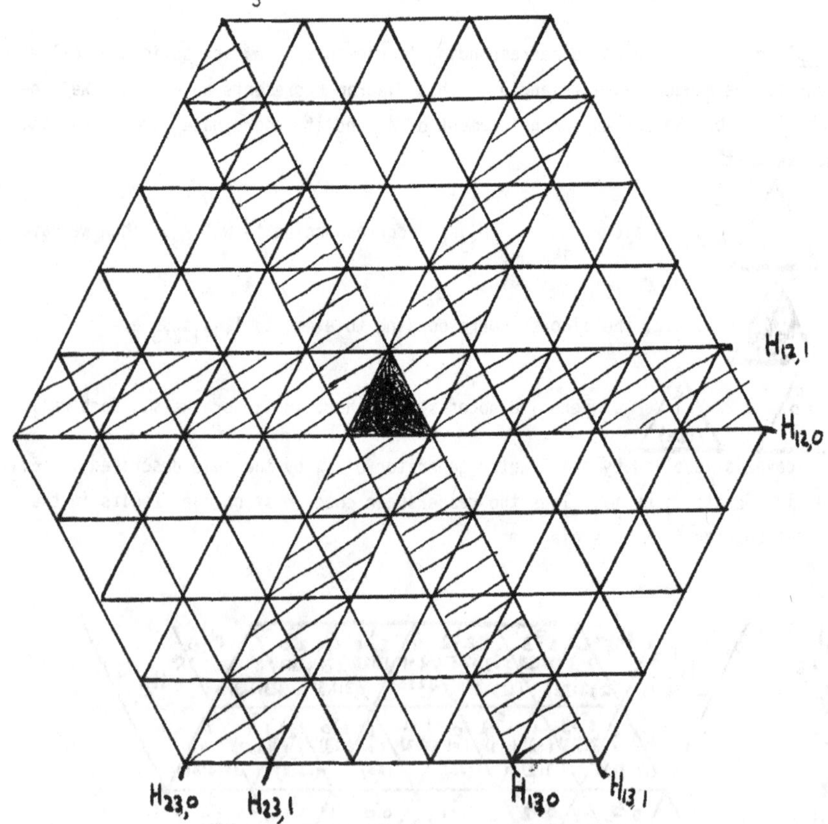

CHAPTER 7 : ADMISSIBLE SIGN TYPES OF RANK n

In the present chapter, we shall introduce the concept of an admissible sign type of rank n which is essentially an equivalence class of elements of A_n. Such a class, when regarded as a set of alcoves, forms a connected component of V in a certain sense. We shall define in §7.2 some equivalence classes, called left cells or two-sided cells on the set of admissible sign types. As we shall show in Chapter 18, these definitions are consistent with the corresponding ones on A_n. In particular, any admissible sign type is in some left cell of A_n.

The structure of an admissible sign type is quite intuitive geometrically. The cardinality of the set of all admissible sign types of rank n is finite. More precisely, we shall show in §7.3 that this number is equal to $(n+1)^{n-1}$. So it seems likely that we may have an easier way to study the properties of cells of A_n by using sign types.

I wish to thank Professor R.W. Carter who stimulated me to use these sign types by which I get all the results in this chapter and in Chapters 19 and 21.

§7.1 ADMISSIBLE SIGN TYPES AND THEIR EQUIVALENCE RELATION

Let $\bar{S}$ be the set of all sign types $X = (X_{ij})_{1\leq i<j\leq n}$ of rank n with $X_{ij} \in \{+,-,0\}$. Say a triple of signs $b^a c$ is admissible if

$$b^a c \in \{+^+ +,\ +^+ o,\ +^+ -,\ +^0 -,\ +^- -,\ o^+ +,\ o^+ o,\ o^0 o,\ o^0 -,\ o^- -,\ -^+ +,\ -^0 +,\ -^- +,\ -^0 o,\ -^- o,\ -^- -\}$$

Say $X \in \bar{S}$ is admissible if for any i,t,j with $1 \leq i < t < j \leq n$, the triple $x_{it}{}^{x_{ij}}\, x_{tj}$ is admissible Let S be the set of all admissible sign types of $\bar{S}$.

Define a map ζ: $\mathbb{A} \to \bar{S}$ such that for $A_k = \bigcap_{1\leq i<j\leq n} H^1_{ij,k_{ij}} \in \mathbb{A}$,

$\zeta(A_k) = (X_{ij})_{1\leq i<j\leq n}$ satisfies

$$X_{ij} = \begin{cases} + & \text{if } k_{ij} > 0 \\ o & \text{if } k_{ij} = 0 \\ - & \text{if } k_{ij} < 0 \end{cases}$$

Proposition 7.1.1 $\zeta(\mathbb{A}) = S$.

Proof: By Lemma 6.1.3, we see $\zeta(\mathbb{A}) \subseteq S$. Now suppose $X = (X_{ij})_{1\leq i<j\leq n} \in S$. Let $m_X = \#\{(i,j) \mid 1 \leq i < j \leq n,\ X_{ij} = -\}$. We apply induction on $m_X \geq 0$. First assume $m_X = 0$. Then by the definition of an admissible sign type, we see that $X \in S$ if and only if for any i, $1 \leq i < n$, there exists h_i with $i \leq h_i \leq n$ such that

$$X_{ij} = \begin{cases} 0 & \text{if } i < j \le h_i \\ + & \text{if } h_i < j \le n \end{cases}$$

and $h_1 \le h_2 \le \dots \le h_{n-1}$. Let $k = (k_{ij})_{1 \le i,j \le n} \in E$ (see §6.1 for the notation E) such that for $1 \le i < j \le n$, (i) $k_{ij} = 0$ if $j \le h_i$. (ii) $k_{ij} = m > 0$ if $j > h_i$ and m is the largest integer such that there exists a sequence $i_0 = i, i_1, \dots, i_m = j$ satisfying the condition that for every ℓ, $1 \le \ell \le m$, the inequality $i_\ell > h_{i_{\ell-1}}$ holds. Then we can check that $k \in E_0$ and so $A_k = \bigcap_{1 \le i < j \le n} H^1_{ij,k_{ij}} \in \mathbb{A}$ by Proposition 6.1.5. Clearly, $\zeta(A_k) = X$, i.e. $X \in \zeta(\mathbb{A})$. Now assume that $m_X > 0$. We claim that there must exist some t, $1 \le t < n$, such that $X_{t,t+1} = -$. Suppose not. This will deduce $m_X = 0$ by the admissibility of X and then contradict our hypothesis. To continue our proof, we need the following result.

Lemma 7.1.2 Assume $X = (X_{ij})_{1 \le i < j \le n} \in S$ and $X_{t,t+1} = -$ for some t, $1 \le t < n$. Let $X' = (X'_{ij})_{1 \le i < j \le n}$ be such that $X'_{ij} = X_{(i)\bar{s}_t,(j)\bar{s}_t}$ if $\{i,j\} \ne \{t,t+1\}$ and $X'_{t,t+1} = +$. Let $X'' = (X''_{ij})_{1 \le i < j \le n}$ be obtained from X' by replacing $X'_{t,t+1} = +$ by $X''_{t,t+1} = 0$. Then either X' or X'' must be in S (maybe both are in S)

Proof: First we observe that for any triple i,h,j with $1 \le i < h < j \le n$ and $(t,t+1) \ne (i,h), (h,j)$,

$$\genfrac{}{}{0pt}{}{X_{(i)\bar{s}_t,(j)\bar{s}_t}}{X_{(i)\bar{s}_t,(h)\bar{s}_t}\ X_{(h)\bar{s}_t,(j)\bar{s}_t}} \in \left\{ \genfrac{}{}{0pt}{}{X_{ij}}{X_{ih}X_{hj}}, \genfrac{}{}{0pt}{}{X_{tj}}{X_{th}X_{hj}}, \genfrac{}{}{0pt}{}{X_{t+1,j}}{X_{t+1,h}X_{hj}}, \genfrac{}{}{0pt}{}{X_{it}}{X_{ih}X_{ht}}, \genfrac{}{}{0pt}{}{X_{i,t+1}}{X_{ih}X_{h,t+1}}, \genfrac{}{}{0pt}{}{X_{ij}}{X_{it}X_{tj}}, \genfrac{}{}{0pt}{}{X_{ij}}{X_{i,t+1}X_{t+1,j}} \right\}$$

So $\genfrac{}{}{0pt}{}{X_{(i)\bar{s}_t,(j)\bar{s}_t}}{X_{(i)\bar{s}_t,(h)\bar{s}_t}\ X_{(h)\bar{s}_t,(j)\bar{s}_t}}$ are all admissible. Secondly we consider the triples

$\genfrac{}{}{0pt}{}{X_{tj}}{X_{t,t+1}\ X_{t+1,j}}$, $t+1 < j \le n$, and $\genfrac{}{}{0pt}{}{X_{i,t+1}}{X_{it}, X_{t,t+1}}$, $1 \le i < t$. Since $X_{t,t+1} = -$, we see that

$$(X_{tj}, X_{t+1,j}), (X_{i,t+1}, X_{it}) \in \{(+,+), (o,+), (-,+), (o,o), (-,o), (-,-)\}.$$

If for any i,j with $1 \le i < t$ and $t+1 < j \le n$, $(X_{tj}, X_{t+1,j}), (X_{i,t+1}, X_{it}) \in \{(+,+), (o,+), (-,o), (-,-)\}$, then we can check that the triples $\genfrac{}{}{0pt}{}{X_{t+1,j}}{X'_{t,t+1}\ X_{tj}}$, $\genfrac{}{}{0pt}{}{X_{t+1,j}}{X''_{t,t+1}\ X_{tj}}$, $\genfrac{}{}{0pt}{}{X_{it}}{X_{i,t+1}\ X'_{t,t+1}}$ and $\genfrac{}{}{0pt}{}{X_{it}}{X_{i,t+1}\ X''_{t,t+1}}$ are all admissible. So X', X'' are both in S. If for any i,j with $1 \le i < t$ and $t+1 < j \le n$,

$(X_{tj}, X_{t+1,j}), (X_{i,t+1},X_{it}) \neq (0,0)$ (resp. $\neq (-,+)$) then $\begin{smallmatrix} & X_{t+1,j} & \\ X'_{t,t+1} & & X_{tj} \end{smallmatrix}$, $\begin{smallmatrix} & X_{it} & \\ X_{i,t+1} & & X'_{t,t+1} \end{smallmatrix}$ (resp. $\begin{smallmatrix} & X_{t+1,j} & \\ X''_{t,t+1} & & X_{tj} \end{smallmatrix}$, $\begin{smallmatrix} & X_{it} & \\ X_{i,t+1} & & X''_{t,t+1} \end{smallmatrix}$) are all admissible and so X' (resp. X'') is in S. So it remains to show that if $(X_{t\ell}, X_{t+1,\ell}) = (0,0)$ for some ℓ, $t+1 < \ell \le n$, or $(X_{m,t+1}, X_{mt}) = (0,0)$ for some m, $1 \le m < t$ then $(X_{tj}, X_{t+1,j})$, $(X_{i,t+1},X_{it}) \neq (-,+)$ for any i,j with $1 \le i < t$ and $t+1 < j \le n(*)$. By symmetry, it is no loss to assume $(X_{t\ell},X_{t+1,\ell}) = (0,0)$ for some ℓ, $t+1 < \ell \le n$. If $(X_{i,t+1}, X_{it}) = (-,+)$ for some i, $1 \le i < t$ then $X_{i\ell} \in \{-,o\}$ since $\begin{smallmatrix} & X_{i\ell} & \\ X_{i,t+1} & & X_{t+1,\ell} \end{smallmatrix}$ is admissible. But on the other hand, the admissibility of the triple $\begin{smallmatrix} & X_{i\ell} & \\ X_{it} & & X_{t\ell} \end{smallmatrix}$ implies $X_{i\ell} = +$. This leads a contradiction. If $(X_{tj},X_{t+1,j}) = (-,+)$ for some j, $t+1 < j < \ell$ then the admissibility of the triples $\begin{smallmatrix} & X_{t\ell} & \\ X_{tj} & & X_{j\ell} \end{smallmatrix}$ and $\begin{smallmatrix} & X_{t+1,\ell} & \\ X_{t+1,j} & & X_{j\ell} \end{smallmatrix}$ implies $X_{j\ell} \in \{o,+\}$ and $X_{j\ell} = -$, respectively. If $(X_{tj},X_{t+1,j}) = (-,+)$ for some j, $\ell < j \le n$ then the admissibility of the triples $\begin{smallmatrix} & X_{tj} & \\ X_{t\ell} & & X_{\ell j} \end{smallmatrix}$ and $\begin{smallmatrix} & X_{t+1,j} & \\ X_{t+1,\ell} & & X_{\ell j} \end{smallmatrix}$ implies $X_{\ell j} = -$ and $X_{\ell j} \in \{o,+\}$, respectively. So both cases give a contradiction. Thus conclusion (*) follows and our proof is complete. □

Now we continue the proof of Proposition 7.1.1. By our assumption, we have $m_X > 0$ and then $X_{t,t+1} = -$ for some t, $1 \le t < n$. Let Y be one of X' and X'' defined in Lemma 7.1.2 which is in S. Then $m_Y < m_X$. By inductive hypothesis, there exists $A_w \in \mathbb{A}$ such that $\zeta(A_w) = Y$. But then we see that $\zeta(A_{ws_t}) = X$. So $\zeta(\mathbb{A}) = S$ by induction. □

Let $T = \{H_{ij,\varepsilon} \mid 1 \le i < j \le n,\ \varepsilon = 0,1\}$. Then the connected components of $V - \bigcup_{H\in T} H$ are open simplices. We see that any alcove of V lies in some connected component of $V - \bigcup_{H\in T} H$ and two alcoves correspond the same sign type if and only if they are in the same connected component of $V - \bigcup_{H\in T} H$. So the map ζ induces a bijection between the set of connected components of $V - \bigcup_{H\in T} H$ and S. Then we can identify these two sets.

<u>Example</u> Let $n = 3$. Then $V - \bigcup_{H\in T} H$ has 16 connected components. Each of them

determines a sign type $X = \begin{smallmatrix} & X_{13} & \\ X_{12} & & X_{23} \end{smallmatrix}$ as follows.

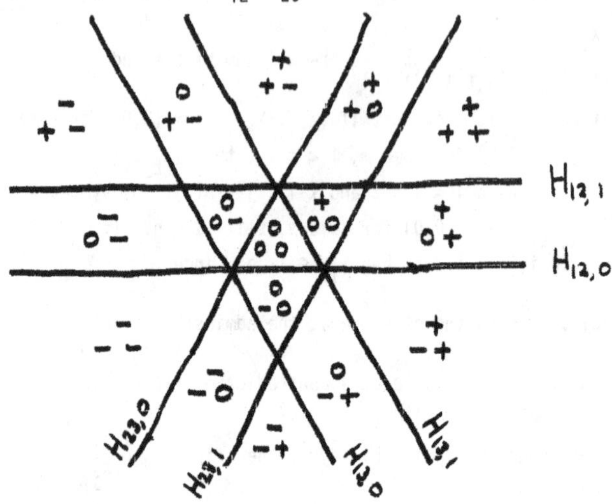

<u>Proposition 7.1.3</u> <u>For any $X \in S$ and $A_w, A_y \in \zeta^{-1}(X)$, there exists a sequence of elements $w_0 = w, w_1, \ldots, w_r = y$ in A_n such that for every h, j with $0 < h < r$ and $1 < j < r$, $A_{w_h} \in \zeta^{-1}(X)$ and $w_j = s_{i_j} w_{j-1}$ for some $s_{i_j} \in \Delta$.</u>

<u>Proof:</u> We see that each connected component of $V - \bigcup_{H \in T} H$ is convex. Our condition means that A_w, A_y are in the same connected component X of $V - \bigcup_{H \in T} H$. So there exists a sequence of alcoves $A_0 = A_w, A_1, \ldots, A_r = A_y$ in X such that for every j, $1 < j < r$, A_j and A_{j-1} share a common wall. Hence our result follows by Lemma 6.1.6. □

Recall that in §6.3 we gave a geometrical description of the map $\sigma: \mathbb{A} \to \Lambda_n$. We see that for any $X \in S$, elements of $\zeta^{-1}(X)$ have the same image under σ. So there is a unique map $\hat{\sigma}: S \to \Lambda_n$ such that the diagram

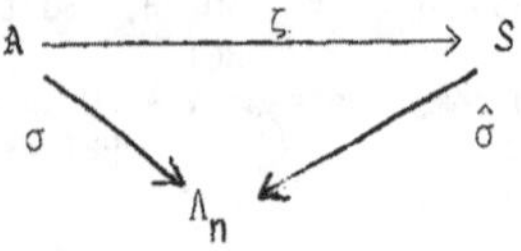

is commutative.

By the definition of the map ζ and Proposition 6.1.4(ii), for any $X \in S$, the function $R(w)$ on $\zeta^{-1}(X)$ is constant. So we can define $R(X) = R(w)$ for any $w \in \zeta^{-1}(X)$.

We see from Proposition 6.1.4 that $R(X) = \{s_t \in \Delta | X_{t,t+1} = -\}$.

§7.2 CONNECTED SETS OF A_n AND CELLS OF S

In this section, we shall introduce various connected sets of A_n. The definition of a connected (resp. left connected; right connected) set given here is a purely algebraic one rather than a geometric one, although the idea initially comes from geometry when A_n is regarded as the set of alcoves in an affine Euclidean space of dimension n-1. The reason for this is as follows. We want to discuss three kinds of sets: a left connected set, a right connected set and a connected set. But only a left connected set can be described explicitly in a geometric way. Now from the algebraic point of view, we can see that essentially a right connected set is entirely similar to a left connected set, and a connected set is only a natural generalization of the former two kinds of sets.

The connectedness is one of the important properties in the structure of cells of A_n which we shall discuss in detail in Chapter 18. At the moment, we want to use these concepts to study sign types.

Definition 7.2.1 We say a set $M \subset A_n$ is left connected if, for any $x,y \in M$, there exists a sequence $x_0 = x, x_1,\ldots,x_t = y$ in M such that for every j, $1 \leqslant j \leqslant t$, $x_j = s_{i_j} x_{j-1}$ with some $s_{i_j} \in \Delta$. Similarly, we can define a right connected set by $x_j = x_{j-1}s_{i_j}$ instead of $x_j = s_{i_j}x_{j-1}$ for each j, $1 \leqslant j \leqslant t$, as above. We say a set $N \subset A_n$ is connected if, for any $x,y \in N$, there exists a sequence $z_0 = x, z_1,\ldots,z_u = y$ in N such that for every j, $1 \leqslant j \leqslant u$, either $x_j = s_{i_j}x_{j-1}$ or $x_j = x_{j-1}s_{i_j}$ with some $s_{i_j} \in \Delta$.

Clearly, any left (resp. right) connected set of A_n is connected.

Now let us regard A_n as the set of alcoves of V. By Lemma 6.1.6, the definition of a left connected set of A_n can be restated as follows. Say $L \subset A_n$ is left connected if for any $A,B \in L$, there exists a sequence of alcoves $A_0 = A, A_1,\ldots,A_r = B$ in L for some $r \geqslant 0$ such that any two consecutive alcoves in this sequence share a common wall.

Thus we can restate Proposition 7.2.3 as follows.

Proposition 7.2.2 For any $X \in S$, $\zeta^{-1}(X)$ is a left connected set of A_n. □

Now we define a relation on S. For X, Y $\in S$, we say $X \sim Y$ if $\hat{\sigma}(X) = \hat{\sigma}(Y) = \lambda$ (see the end of §7.1 for the map $\hat{\sigma}$) for some $\lambda \in \Lambda_n$ and there exists a sequence of elements $X_0 = X, X_1,\ldots,X_r = Y$ in $\hat{\sigma}^{-1}(\lambda)$ for some $r \geqslant 0$ such that for every j, $1 \leqslant j \leqslant r$, X_j is obtained from X_{j-1} by either replacing a certain zero sign by a non-zero sign or vice versa.

This is an equivalence relation. We call each equivalence class of S a left cell. We also call $\hat{\sigma}^{-1}(\lambda)$ for any $\lambda \in \Lambda_n$ a two-sided cell. Clearly, for any $\lambda \in \Lambda_n$, $\hat{\sigma}^{-1}(\lambda)$ is a union of some left cells of S. Let $C(S)$ be the set of all left cells of S. For any $X,Y \in \varepsilon \in C(S)$, we shall write $X \underset{L}{\sim} Y$ instead of $X \sim Y$. We can define a natural map $\tau:S \to C(S)$ and a map $\bar{\sigma}:C(S) \to \Lambda_n$ such that the diagram

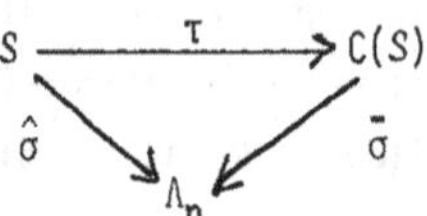

is commutative.

We shall show in Chapter 18 that the definition of a left (resp. two-sided) cell of S coincides with that of A_n under the map ζ.

We see that if X,Y are in S such that Y is obtained from X by replacing a certain zero sign by a non-zero sign or vice versa then the connected component of $V - \bigcup_{H \in T} H$ corresponding to X is adjacent to that corresponding to Y by a hyperplane of T. So there exist two alcoves $A_X \in \zeta^{-1}(X)$ and $A_Y \in \zeta^{-1}(Y)$ which share a common wall. By Lemma 6.1.6, $y = s_t x$ for some $s_t \in \Delta$. This implies that for any $\varepsilon \in C(S)$, $\zeta^{-1} \cdot \tau^{-1}(\varepsilon)$ is a left connected set of A_n in $\hat{\sigma}^{-1}(\bar{\sigma}(\varepsilon))$.

For two sets M, $N \subset A_n$ with $N \subset M$, we say that N is a maximal left connected component of M if N is left connected but $N \cup \{x\}$ is not for any $x \in M-N$.

<u>Lemma 7.2.3</u> <u>For any $\varepsilon \in C(S)$, $\zeta^{-1} \cdot \tau^{-1}(\varepsilon)$ is a maximal left connected component of $\sigma^{-1}(\bar{\sigma}(\varepsilon))$.</u>

<u>Proof</u>: By the above discussion, it is enough to show that $\zeta^{-1} \cdot \tau^{-1}(\varepsilon) \cup \{x\}$ is not left connected for any $x \in \sigma^{-1}(\bar{\sigma}(\varepsilon)) - \zeta^{-1} . \tau^{-1}(\varepsilon)$. Otherwise, there must exist some $x \in \sigma^{-1}(\bar{\sigma}(\varepsilon)) - \zeta^{-1} . \tau^{-1}(\varepsilon)$ and $y \in \zeta^{-1} \cdot \tau^{-1}(\varepsilon)$ such that x and y share a common wall. Hence either $\zeta(x) = \zeta(y)$ or $\zeta(x)$ is obtained from $\zeta(y)$ by replacing a certain non-zero sign by a zero sign or vice versa. This implies by $y \in \zeta^{-1} \cdot \tau^{-1}(\varepsilon)$ that $x \in \zeta^{-1} \cdot \tau^{-1}(\varepsilon)$ and gives a contradiction. So we get our result. □

§7.3 <u>THE CARDINALITY OF S.</u>

Recall that in §7.1 we defined the set $\bar{S}$ of sign types and the set S of admissible sign types both of which have rank n. In this section, we shall compute the cardinality $|S|$ of S. Our result will be as below.

<u>Theorem 7.3.1</u> <u>$|S| = (n+1)^{n-1}$</u>

For any $X = (X_{ij})_{1 \leq i<j \leq n} \in \bar{S}$, we make the following convention:

(i) For any i,j, $1 \leq i < j \leq n$, let X_{ji} be in the set $\{+,-,o\}$ satisfying either $\{X_{ij},X_{ji}\} = \{+,-\}$ or $\{X_{ij},X_{ji}\} = \{o,o\}$.

(ii) Let $X_{ij} = X_{i+n,j} = X_{i,j+n}$ for any $i,j \in \mathbb{Z}$.

Now suppose that $X = (X_{ij})_{1\leq i<j\leq n}$ is an element of S with $X_{t,t+1} = -$ for some t, $1 \leq t \leq n$. Let $X' = (X'_{ij})_{1\leq i<j\leq n}$ be the element of $\bar{S}$ satisfying: for any i,j, $1 \leq i < j \leq n$,

$$X'_{ij} = \begin{cases} X_{(i)\bar{s}_t,(j)\bar{s}_t} & \text{if } \{\bar{i},\bar{j}\} \neq \{\bar{t}, \overline{t+1}\} \\ 0 & \text{if } \{\bar{i},\bar{j}\} = \{\bar{t}, \overline{t+1}\} \end{cases}$$

Let $X'' = (X''_{ij})_{1\leq i<j\leq n}$ be the element of $\bar{S}$ which is obtained from X' by replacing $X'_{t,t+1} = 0$ by $X''_{t,t+1} = +$.

By Proposition 7.1.1, we have $\zeta^{-1}(X) \neq \emptyset$. Let $D_X = \{w \in A_n | ws_t \in \zeta^{-1}(X)\}$. Then $\zeta(D_X) \subseteq \{X'',X'\}$. So at least one of X' and X'' is in S (maybe both are in S).

<u>Lemma 7.3.2</u> <u>Let $X \in S$, $X',X'' \in \bar{S}$ and $D_X \subset A_n$ be as above. Then $D_X = \zeta^{-1}(\{X',X''\} \cap S)$. The map $w \to ws_t$ is a bijection from $\zeta^{-1}(X)$ (resp. $\zeta^{-1}(\{X',X''\} \cap S)$) to $\zeta^{-1}(\{X',X''\} \cap S)$ (resp. $\zeta^{-1}(X)$).</u>

<u>Proof:</u> We can check our result simply by the definition of the right action of A_n on the alcove set $\mathbb{A}$ and by the definition of the map ζ. □

We say $X = (X_{ij})_{1\leq i<j\leq n} \in S$ is dominant if $X_{ij} \in \{+,o\}$ for any i,j, $1 \leq i < j \leq n$.

<u>Lemma 7.3.3</u> <u>Suppose that $X = (X_{ij})_{1\leq i<j\leq n} \in S$ is dominant. Then there exists a unique shortest element w in $\zeta^{-1}(X)$. Let $A_y = \bigcap_{1\leq i<j\leq n} H^1_{ij,k^y_{ij}}$ for any $y \in \zeta^{-1}(X)$. Then</u>

$$\underline{k^w_{ij} = \min \{k^y_{ij} | y \in \zeta^{-1}(X)\} \text{ for any } i,j,\ 1 \leq i < j \leq n.}$$

<u>Proof:</u> For any i,j, $1 \leq i < j \leq n$ with $X_{ij} = +$, let k_{ij} be the largest integer t such that there exists a sequence of integers $\alpha_1,\alpha_2,\ldots,\alpha_{t+1}$ with $i = \alpha_1 < \alpha_2 < \ldots < \alpha_{t+1} = j$ and $X_{\alpha_u,\alpha_{u+1}} = +$ for all u, $1 \leq u \leq t$. Then by Lemma 6.1.3 and the assumption that X is dominant, we see that $k^y_{ij} \geq \sum_{u=1}^{k_{ij}} k^y_{\alpha_u,\alpha_{u+1}} \geq k_{ij}$. Let $k_{pq} = 0$ for any p,q, $1 \leq p < q \leq n$ with $X_{pq} = 0$. We shall show that $A = \bigcap_{1\leq i<j\leq n} H^1_{ij,k_{ij}}$ is in $\mathbb{A}$, i.e. show that for any integers i,h,j with $1 \leq i < h < j \leq n$,

the inequality $k_{ih} + k_{hj} \leq k_{ij} \leq k_{ih} + k_{hj} + 1$ holds. If this is proved then $w \in A_n$ with $A_w = A$ will be the unique shortest element of $\zeta^{-1}(X)$ and also w satisfies the required properties.

First we shall show $k_{ih} + k_{hj} \leq k_{ij}$. Suppose that $\alpha_1,\ldots,\alpha_{t+1}$ and $\beta_1,\ldots,\beta_{u+1}$ are two sequences satisfying the conditions that $t = k_{ih}$, $u = k_{hj}$, $i = \alpha_1<\ldots<\alpha_{t+1} = h$, $h = \beta_1 < \ldots < \beta_{u+1} = j$ and $X_{\alpha_p,\alpha_{p+1}}$, $X_{\beta_q,\beta_{q+1}} = +$ for all p,q, $1 \leq p \leq t$, $1 \leq q \leq u$. Then $\alpha_1,\ldots,\alpha_t$, $\beta_1,\ldots,\beta_{u+1}$ is a sequence with $i = \alpha_1 < \ldots < \alpha_t < \beta_1 < \ldots < \beta_{u+1} = j$ satisfying $X_{\ell m} = +$ for any two consecutive terms ℓ,m of this sequence with $\ell < m$. This implies that $k_{ij} \geq t+u = k_{ih} + k_{hj}$. Secondly we shall show $k_{ij} \leq k_{ih} + k_{hj} + 1$. We may assume $k_{ij} \geq 2$. Let $\gamma_1,\ldots,\gamma_{t+1}$ be a sequence with $t = k_{ij}$, $i = \gamma_1<\ldots<\gamma_{t+1}=j$ and $X_{\gamma_u,\gamma_{u+1}} = +$ for all u, $1 \leq u \leq t$. Let r be the largest integer satisfying $1 \leq r \leq t+1$ and $\gamma_r \leq h$. Clearly, such an integer r always exists. Let $\alpha_1,\ldots,\alpha_r$ and $\beta_1,\ldots,\beta_{t+1-r}$ be two sequences with $(\alpha_1,\ldots,\alpha_r) = (\gamma_1,\ldots,\gamma_{r-1},h)$ and $(\beta_1,\ldots,\beta_{t+1-r}) = (h,\gamma_{r+2},\ldots,\gamma_{t+1})$. Then we have $i = \alpha_1 < \ldots < \alpha_r = h$ if $r > 1$ and $h = \beta_1 < \ldots < \beta_{t+1-r} = j$ if $r < t$. Clearly, $X_{\alpha_p,\alpha_{p+1}} = X_{\beta_q,\beta_{q+1}} = +$ for all p,q, $1 \leq p \leq r-2$ and $2 \leq q \leq t-r$. We also have $X_{\alpha_{r-1},\alpha_r} = +$ if $r > 1$ and $X_{\beta_1,\beta_2} = +$ if $r < t$ by the admissibility of the triples

$$\begin{matrix} & X_{\gamma_{r-1},h} & \\ X_{\gamma_{r-1},\gamma_r} & & X_{\gamma_r,h} \end{matrix} \quad , \quad \begin{matrix} & X_{h,\gamma_{r+2}} & \\ X_{h,\gamma_{r+1}} & & X_{\gamma_{r+1},\gamma_{r+2}} \end{matrix}$$

(note $\alpha_r = \beta_1 = h$) and the assumption that $X_{\gamma_{r-1},\gamma_r} = X_{\gamma_{r+1},\gamma_{r+2}} = +$ and $X_{\gamma_r,h}$, $X_{h,\gamma_{r+1}} \in \{+,o\}$. This implies $k_{ih} \geq r-1$ and $k_{hj} \geq t-r$. Thus $k_{ih} + k_{hj} + 1 \geq t = k_{ij}$. Our proof is complete. □

For any $X = (X_{ij})_{1\leq i<j\leq n} \in \bar{S}$, we define by m_X the number of signs X_{ij}, $1\leq i<j\leq n$, with $X_{ij} = -$.

We write down the following statement for any $X \in S$.

(*) There exists a unique shortest element w in $\zeta^{-1}(X)$.
Let $A_y = \bigcap_{1\leq i<j\leq n} H^1_{ij;k^y_{ij}}$ for any $y \in \zeta^{-1}(X)$. Then $|k^w_{ij}| = \min\{|k^y_{ij}| \;\; |y \in \zeta^{-1}(X)|\}$ for any i,j, $1 \leq i < j \leq n$.

<u>Lemma 7.3.4</u> <u>Given $m \geq 0$. Suppose that statement (*) holds for any $X \in S$ with $m_X \leq m$. Let Y,Z be two elements of S with $m_y,m_z \leq m$ such that Y is obtained from Z by replacing some signs + by signs o. Let y (resp. z) be the shortest element of $\zeta^{-1}(Y)$ (resp. $\zeta^{-1}(Z)$). Then $|k^y_{ij}| = \min\{|k^x_{ij}| \;\; |x \in \zeta^{-1}(\{Y,Z\})\}$ for all i,j, $1\leq i<j\leq n$.</u>

In particular $\ell(y) < \ell(z)$.

Proof: By our assumption, we have $m_y = m_z = \ell$ for some ℓ, $0 \leq \ell \leq m$. We apply induction on $\ell \geq 0$. If $\ell = 0$ then by Lemma 7.3.3, we have $|k_{ij}^y| = \min\{|k_{ij}^x| \mid x \in \zeta^{-1}(Y)\}$ and $|k_{ij}^z| = \min\{|k_{ij}^x| \mid x \in \zeta^{-1}(Z)\}$ for any i,j, $1 \leq i < j \leq n$. By the proof of Lemma 7.3.3 and our assumption on Y and Z, we see that $0 \leq k_{ij}^y \leq k_{ij}^z$ for any i,j, $1 \leq i < j \leq n$ and that there exists at least one pair α, β, $1 \leq \alpha < \beta \leq n$, satisfying $k_{\alpha\beta}^y = 0$ and $k_{\alpha\beta}^z > 0$. So our result is true in this case.

Now assume that $\ell > 0$. By the admissibility of Y, there must exist some t, $1 \leq t < n$, such that $Y_{t,t+1} = -$. Thus we also have $Z_{t,t+1} = -$, where we assume $Y = (Y_{ij})_{1 \leq i < j \leq n}$ and $Z = (Z_{ij})_{1 \leq i < j \leq n}$. Let $Y' = (Y'_{ij})_{1 \leq i < j \leq n}$ be the element of $\bar{S}$ such that for any i,j, $1 \leq i < j \leq n$,

$$Y'_{ij} = \begin{cases} Y_{(i)\bar{s}_t,(j)\bar{s}_t} & \text{if } (i,j) \neq (t,t+1) \\ 0 & \text{if } (i,j) = (t,t+1) \end{cases}$$

Let $Y'' = (Y''_{ij})_{1 \leq i < j \leq n}$ be the element of $\bar{S}$ which is obtained from Y' by replacing $Y'_{t,t+1} = 0$ by $Y''_{t,t+1} = +$. We define Z', $Z'' \in \bar{S}$ from Z in the same way as Y', Y'' from Y. Then by our assumption on Y and Z, we see that Y' is obtained from Y'' (resp. Z', Z'') by replacing some signs + by signs o. We also see that Y'' is obtained from Z'' by replacing some signs + by signs o.

If $Y' \in S$ then by Lemma 7.3.2 there must exist some $x \in \zeta^{-1}(Y)$ with $k_{t,t+1}^x = -1$. By our assumption that statement (*) holds for Y we see that $y' = ys_t \in \zeta^{-1}(Y')$ with $\ell(y') = \ell(y) - 1$. Moreover, y' must be the shortest element of $\zeta^{-1}(Y')$ by Lemma 7.3.2. Again by Lemma 7.3.2 $z' = zs_t \in \zeta^{-1}(\{Z',Z''\} \cap S)$ with $\ell(z') = \ell(z) - 1$. By inductive hypothesis, we have $|k_{ij}^{y'}| = \min\{|k_{ij}^x| \mid x \in \zeta^{-1}(\{Y',Y'',Z',Z''\} \cap S)\}$ for all i,j, $1 \leq i < j \leq n$ and $\ell(y') < \ell(x)$ for any $x \in \zeta^{-1}(\{Y',Y'',Z',Z''\} \cap S)$. In particular, $\ell(y) < \ell(z')$. This implies by Lemma 7.3.2 that $|k_{ij}^y| = \min\{|k_{ij}^x| \mid x \in \zeta^{-1}(\{Y,Z\})\}$ for all i,j, $1 \leq i < j \leq n$ and $\ell(y) \leq \ell(z)$. Our result is true in this case, too.

In the case that Y', $Z' \notin S$, we can show our result by a similar argument as above. The only difference is that now we have $y' = ys_t \in \zeta^{-1}(Y'')$ and $z' = zs_t \in \zeta^{-1}(Z'')$.

So it remains to show that the case that $Y' \notin S$ and $Z' \in S$ will never happen, or equivalently, to show that $Z' \in S$ implies $Y' \in S$.

Now assume $Z' \in S$. We shall show $Y' \in S$. By the fact that $Z'_{t,t+1} = 0$ and the admissibility of Z' we have that for any i,j with $1 \leq i < t$ and $t+1 < j \leq n$,

$$(Z'_{i,t}, Z'_{i,t+1}),\ (Z'_{t+1,j}, Z'_{t,j}) \in \{(+,+),\ (o,o),\ (-,-),\ (o,+),\ (-,o)\}$$

We know that $\{Y',Y''\} \cap S \neq \emptyset$. To show that $Y' \in S$, we need only check that for any i,j with $1 \leq i < t$ and $t+1 < j \leq n$,

$$(Y'_{it}, y'_{i,t+1}), (Y'_{t+1,j}, Y'_{t,j}) \in \{(+,+), (o,o), (-,-), (o,+), (-,o)\}.$$

Remember that Y' is obtained from Z' by replacing some signs + by signs o. Thus the only possibility to make Y' not in S is that there exists a pair $(Y'_{i,t}, Y'_{i,t+1}) = (+,o)$ for some i, $1 \leq i < t$, or a pair $(Y'_{t+1,j}, Y'_{t,j}) = (+,o)$ for some j, $t+1 < j \leq n$. But this would imply $(Y''_{i,t}, Y''_{i,t+1}) = (+,o)$ or $(Y''_{t+1,j}, Y''_{t,j}) = (+,o)$, i.e. $Y'' \notin S$, which contradicts $\{Y',Y''\} \cap S \neq \emptyset$. Therefore our results follows. □

Proposition 7.3.5 Statement (*) is true for any $X \in S$.

Proof: We apply induction on $m_X \geq 0$. If $m_X = 0$ then Lemma 7.3.3 verifies our assertion. Now assume $m_X > 0$. By the admissibility of $X = (X_{ij})_{1\leq i<j\leq n}$ there must exist some t, $1 \leq t < n$, satisfying $X_{t,t+1} = -$. Let $X' = (X'_{ij})_{1\leq i<j\leq n}$ and $X'' = (X''_{ij})_{1\leq i<j\leq n}$ be the two elements of $\bar{S}$ with respect to X defined as before. Then $m_{X'}, m_{X''} < m_X$ and X' is obtained from X'' by replacing $X''_{t,t+1} = +$ by $X'_{t,t+1} = 0$. We have $|\{X',X''\} \cap S| \geq 1$ in general. If $|\{X',X''\} \cap S| = 1$ then by inductive hypothesis statement (*) is true for $\{X',X''\} \cap S$. Hence by Lemma 7.3.2 statement (*) is also true for X. If $|\{X',X''\} \cap S| = 2$, i.e. $X',X'' \in S$ then by inductive hypothesis statement (*) is true for both X' and X''. Let x', x'' be the shortest elements of $\zeta^{-1}(X')$, $\zeta^{-1}(X'')$, respectively. Then by Lemma 7.3.4 we have $|k^{x'}_{ij}| = \min\{|k^{y}_{ij}| \;|y \in \zeta^{-1}(\{X',X''\})$ for all i,j, $1 \leq i < j \leq n$ and $\ell(x') < \ell(x'')$. So by Lemma 7.3.2 $x = x's_t$ is the shortest element of $\zeta^{-1}(X)$ satisfying the condition that $|k^{x}_{ij}| = \min\{|k^{y}_{ij}| \;\; |y \in \zeta^{-1}(X)\}$. That is, statement (*) is also true for X.

Therefore our assertion follows by induction. □

Proposition 7.3.5 also gives us a criterion for an element $w \in \zeta^{-1}(X)$ to be the shortest element of $\zeta^{-1}(X)$.

Corollary 7.3.6 For any $X \in S$ and $y \in \zeta^{-1}(X)$, let $A_y = \bigcap_{1\leq i<j\leq n} H^1_{ij;k^y_{ij}}$. Then an element $w \in \zeta^{-1}(X)$ is the shortest element of $\zeta^{-1}(X)$ if and only if $|k^{w}_{ij}| = \min\{|k^{y}_{ij}| \;\; |y \in \zeta^{-1}(X)\}$ for any i,j, $1 \leq i < j \leq n$. □

Now we shall give another criterion for an element to be the shortest element of $\zeta^{-1}(X)$ for any $x \in S$.

Let $w', w \in A_n$ and $s_t \in \Delta$ satisfy $w' = s_t w$ and $\ell(w') = \ell(w) - 1$. Then by the definition of the left action of A_n on the alcove set $\mathbf{A}$ we have $k^{w'}_{ij} = k^{w}_{ij}$ for all pairs i,j, $1 \leq i < j \leq n$, but one. Let this exceptional pair be α,β, $1 \leq \alpha < \beta \leq n$. Then we have $|k^{w'}_{\alpha\beta}| = |k^{w}_{\alpha\beta}| - 1$ and $k^{w'}_{\alpha\beta} = k^{w}_{\alpha\beta} \pm 1$. Now assume that w is the shortest element of $\zeta^{-1}(\zeta(w))$. Then by Proposition 7.3.5, we must have $k^{w}_{\alpha\beta} = \pm 1$ and $k^{w'}_{\alpha\beta} = 0$.

In particular, $w' \notin \zeta^{-1}(\zeta(w))$.

Proposition 7.3.7 Let $X \in S$ and $w \in \zeta^{-1}(X)$. Then w is the shortest element of $\zeta^{-1}(X)$ if and only if for any $s_t \in \mathcal{L}(w)$, we have $s_t w \notin \zeta^{-1}(X)$.

Proof: Let w be the shortest element of $\zeta^{-1}(X)$. By the above discussion, it is sufficient to show that if $y \in \zeta^{-1}(X)$ with $y \neq w$ then there must exist some $s_t \in \mathcal{L}(y)$ such that $s_t y \in \zeta^{-1}(X)$. Now assume $y \in \zeta^{-1}(X)$ with $y \neq w$. Then by Proposition 7.3.5 we have: for any i,j, $1 \leqslant i < j \leqslant n$,

$$k^y_{ij} \geqslant k^w_{ij} > 0 \quad \text{if } X_{ij} = +$$

$$k^y_{ij} \leqslant k^w_{ij} < 0 \quad \text{if } X_{ij} = -$$

$$k^y_{ij} = k^w_{ij} = 0 \quad \text{if } X_{ij} = 0$$

and the set $D_{y,w} = \{(\alpha,\beta) \mid 1 \leqslant \alpha < \beta \leqslant n,\ k^y_{\alpha\beta} \neq k^w_{\alpha\beta}\}$ is non-empty. Let $D^+_{y,w} = \{(\alpha,\beta) \in D_{y,w} \mid X_{\alpha\beta} = +\}$ and $D^-_{y,w} = \{(\alpha,\beta) \in D_{y,w} \mid X_{\alpha\beta} = -\}$. Let
$D_1 = \bigcap_{\substack{1\leqslant i<j\leqslant n\\ (i,j)\notin D_{y,w}}} H^1_{ij,k^w_{ij}}$, $D_2 = \bigcap_{(\alpha,\beta)\in D^+_{y,w}} H^{k^y_{\alpha\beta}+1-k^w_{\alpha\beta}}_{\alpha\beta;k^w_{\alpha\beta}}$ and $D_3 = \bigcap_{(\alpha,\beta)\in D^-_{y,w}} H^{k^w_{\alpha\beta}+1-k^y_{\alpha\beta}}_{\alpha\beta,k^y_{\alpha\beta}}$
with the convention that $D_i = V$ if the set of indices for the corresponding intersection is empty. Let $D = D_1 \cap D_2 \cap D_3$. We have $A_w, A_y \subset D \subset X$ regarded as sets of vectors of V. On the other hand, we see that for an alcove $A \in \mathbb{A}$, either $A \subset D$ or $A \cap D = \emptyset$. So D can be regarded as a set of all elements x of A_n with $A_x \subset D$. Thus w (resp. y) is the shortest (resp. longest) element in D. Since D is a convex set of V, which contains more than one alcove of V, there must exist some alcove A_x in D with $x \neq y$ such that A_x and A_y share a common wall. That is, $x = sy$ for some $s \in \Delta$ by Lemma 6.1.6. Clearly, we have $s \in \mathcal{L}(y)$ and $x \in \zeta^{-1}(X)$. So our result follows. □

Let $E(S) = E_n(S) = \{w \in A_n \mid w \text{ is the shortest element of } \zeta^{-1}(\zeta(w))\}$. Let $E(S)^{-1} = E_n(S)^{-1} = \{w \mid w^{-1} \in E_n(S)\}$. For any $w \in A_n$, we see by Propositions 7.3.7 and 6.1.4 that $w \in E(S)^{-1}$ if and only if $k^w_{t,t+1} = -1$ for any $s_t \in \mathcal{R}(w)$ if and only if $k^w_{t,t+1} \geqslant -1$ for any t, $1 \leqslant t \leqslant n$, where we assume $A_w = \bigcap_{1\leqslant i<j\leqslant n} H^1_{ij,k^w_{ij}}$.

We define $H^+_{ij,k} = \{(x_1,\ldots,x_n) \in V \mid x_j - x_i > k\}$ and $H^-_{ij;k} = \{(x_1,\ldots,x_n) \in V \mid x_j - x_i < k\}$ for any $k,i,j \in \mathbb{Z}$ with $1 \leqslant i < j \leqslant n$. Then regarded as a set of alcoves of V, $E(S)^{-1}$ is the set of all alcoves of V contained in $H_n = (\bigcap_{1\leqslant t<n} H^+_{t,t+1;-1}) \cap H^-_{1n;2}$. In particular, this implies by the convexity of H_n that $E(S)^{-1}$ is a left connected set of A_n (see §7.2).

Let us consider an example: when $n = 3$, $E_3(S)^{-1}$ is the set of all alcoves in the fully shadowed area of the following figure

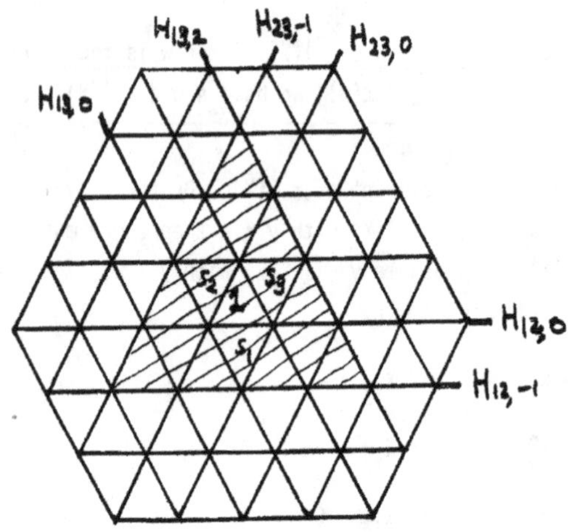

From this figure, we see that $H_3 = H^+_{12;-1} \cap H^+_{23;-1} \cap H^-_{13;2}$ is a triangle (i.e. the fully shadowed area) similar to the alcove A_1 (i.e. the alcove labelled by 1). The scale of A_1 to H_3 is one-fourth and so the area of H_3 is $4^{\dim V}$ times to that of A_1. Then H_3 contains $4^{\dim V}$ alcoves altogether.

Recall the notation $H^m_{ij;k} = \{(x_1,\ldots,x_n) \in V \mid k < x_j - x_i < k+m\}$ in §6.1 for any integers m,k,i,j with $m > 0$ and $1 \le i < j \le n$.

<u>Lemma 7.3.8</u> Let $H^{n+1}_n = (\bigcap_{1 \le t < n} H^{n+1}_{t,t+1;-1}) \cap H^{n+1}_{1n,-(n-1)}$ for $n \ge 2$. Then $H_n = H^{n+1}_n$.

<u>Proof</u>: Since $H^{n+1}_{t,t+1;-1} \subset H^+_{t,t+1;-1}$, $1 \le t < n$, and $H^{n+1}_{1n;-(n-1)} \subset H^-_{1n;2}$, the inclusion $H_n \supseteq H^{n+1}_n$ is obvious.

Now let $x = (x_1,\ldots,x_n) \in H_n$. To show $x \in H^{n+1}_n$, it is enough to show that $x_{t+1} - x_t < n$ for any t, $1 \le t < n$ and $x_n - x_1 > -(n-1)$.

For $1 \le t < n$, we have

$$x_{t+1} - x_t = -\sum_{i=1}^{t-1} (x_{i+1} - x_i) - \sum_{j=t+1}^{n-1} (x_{j+1} - x_j) + (x_n - x_1) < (t-1) + (n-t-1) + 2 = n.$$

We also have $x_n - x_1 = \sum_{i=1}^{n-1} (x_{i+1} - x_i) > -(n-1)$. So the inclusion $H_n \subseteq H^{n+1}_n$ is shown and hence $H_n = H^{n+1}_n$. □

<u>Lemma 7.3.9</u> <u>Let $m > 0$ be an integer and let an integer set $k = \{k_t \mid 1 \le t \le n\}$ satisfy the condition that $k_n = \sum_{i=1}^{n-1} k_i$. Let $H^m_{n,k} = (\bigcap_{1 \le t < n} H^m_{t,t+1;k_t}) \cap H^m_{1n;k_n}$.</u>

<u>Then $H^m_{n,k}$ contains exactly $m^{\dim V}$ alcoves of V.</u>

Proof: We see $A_1 = H^1_{n,K_0}$ by Corollary 6.1.2, where $K_0 = (k_1,\ldots,k_n) = (0,\ldots,0)$ satisfies the condition $\sum_{i=1}^{n-1} k_i = k_n$. Then for any $m > 0$, H^m_{n,K_0} is similar to A_1 in geometrical shape and the scale of A_1 is one m-th of that of H^m_{n,K_0}. So the volume of H^m_{n,K_0} is $m^{\dim V}$ times that of A_1. This implies that H^m_{n,K_0} contains exactly $m^{\dim V}$ alcoves of V. Now we take any other integer set $K = \{k_t | 1 \leq t \leq n\}$ with $k_n = \sum_{i=1}^{n-1} k_i$. Let $x = (x_1,\ldots,x_n)$ with $x_t = -\sum_{i=t}^{n-1} k_i + \sum_{i=1}^{n-1} \frac{ik_i}{n}$ for $1 \leq t \leq n$. Then $\sum_{\alpha=1}^{n} x_\alpha = 0$ and so $x \in V$. Moreover, we have $x_{t+1}-x_t = k_t$ for all t, $1 \leq t < n$ and $x_n - x_1 = \sum_{\alpha=1}^{n} k_\alpha = k_n$. Let T_x be the translation on V which sends the point $(0,\ldots,0)$ to x. Then T_x also sends H^m_{n,K_0} to $H^m_{n,K}$. Hence $H^m_{n,K}$ also contains $m^{\dim V}$ alcoves of V by the condition that k_t, $1 \leq t \leq n$, are all integers. □

<u>Corollary 7.3.10 Let H_n^{n+1} be defined as in Lemma 7.3.8. Then H_n^{n+1} contains exactly $(n+1)^{n-1}$ alcoves of V.</u>

Proof: Let $K = (-1,\ldots,-1, -(n-1))$. Then $H_n^{n+1} = H_{n,K}^{n+1}$. Since $\dim V = n-1$, our result follows by Lemma 7.3.9. □

Proof of Theorem 7.3.1

By Proposition 7.3.5, it is enough to show $|E(S)| = (n+1)^{n-1}$, or equivalently, to show $|E(S)^{-1}| = (n+1)^{n-1}$. But this follows by Lemma 7.3.8 and Corollary 7.3.10. □

Remark 7.3.11 Let Φ be the root system of type A_ℓ, $\ell = n-1$, and let h be the Coxeter number of Φ. Then Theorem 7.3.1 can be reformulated by $|S| = (h+1)^\ell$. This formula is also valid for other types of affine Weyl groups, where the definition of an admissible sign type for these groups is similar to that for A_n. The last assertion was conjectured by R.W. Carter and now has been proved by the author.

Example Let n = 4. Then there are 125 sign types

$$X = \begin{matrix} & & X_{14} & & \\ & X_{13} & & X_{24} & \\ X_{12} & & X_{23} & & X_{34} \end{matrix}$$

in S altogether. They are divided into 5 two-sided cells, i.e. $\hat{\sigma}^{-1}(\{1>1>1>1\})$, $\hat{\sigma}^{-1}(\{2>1>1\})$, $\hat{\sigma}^{-1}(\{2>2\})$, $\hat{\sigma}^{-1}(\{3>1\})$ and $\hat{\sigma}^{-1}(\{4\})$.

$\hat{\sigma}^{-1}(\{1 > 1 > 1 > 1\}) = \begin{smallmatrix} & 0 & \\ 0 & & 0 \\ 0 & 0 & 0 \end{smallmatrix}$. $\hat{\sigma}^{-1}(\{2 > 1 > 1\})$ consists of 20 sign types and they are divided into 4 left cells:

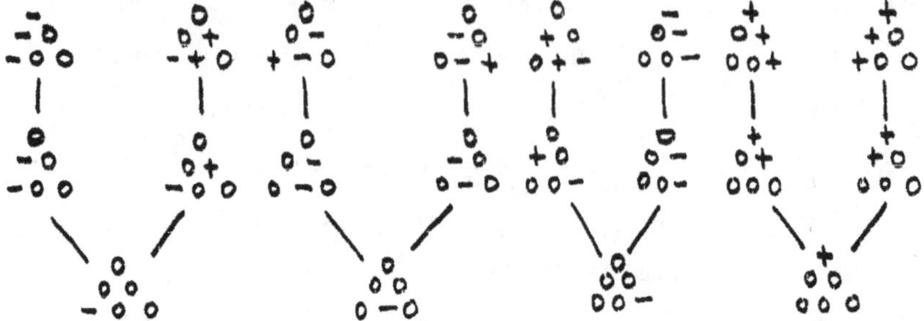

Each of them forms a connected diagram as above, where two sign types are joined by a bond if and only if one is obtained from the other by either replacing a certain zero sign by a non-zero sign or vice versa, (the same for the following cases)

$\hat{\sigma}^{-1}(\{2 > 2\})$ consists of 16 sign types and they are divided into 6 left cells:

$\hat{\sigma}^{-1}(\{3 > 1\})$ consists of 64 sign types. They are divided into 12 left cells:

$\hat{\sigma}^{-1}(\{4\})$ consists of 24 sign types. Each of them forms a left cell:

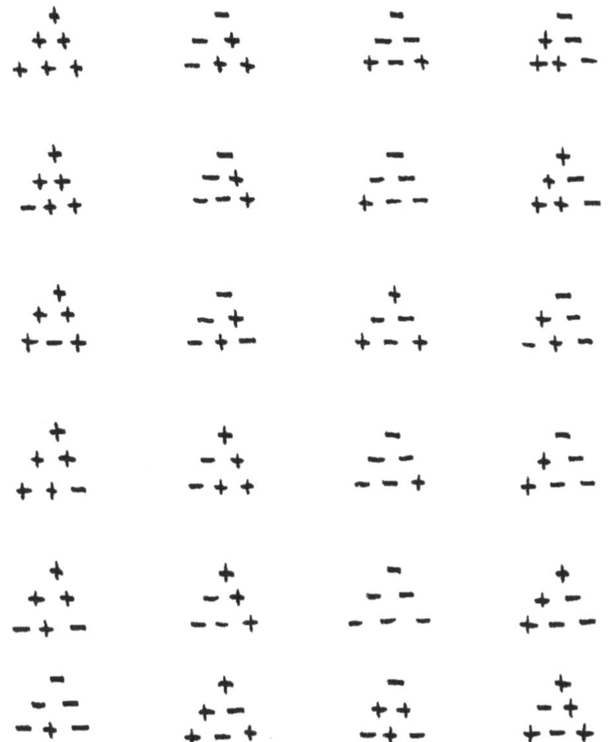

CHAPTER 8 : ITERATED STAR OPERATIONS AND INTERCHANGING OPERATIONS ON BLOCKS

We shall prove most of our main results in Chapters 8 to 18. The approach to these results is purely algebraic except for Proposition 18.2.2 which involves admissible sign types.

We define, in Chapter 8, iterated star operations on an element w of A_n and interchanging operations on blocks of w. They are all successions of left star operations. These operations are very important in the proof of our main results.

The proof of the results on the decomposition of A_n into cells starts from Chapter 9 and is divided into four main steps.

First we show in Chapters 9 and 10 that for any $\lambda \in \Lambda_n$, $\sigma^{-1}(\lambda)$ is a union of RL-equivalence classes of A_n and for any $w \in \sigma^{-1}(\lambda)$, there exists an element y of N_λ with $y \underset{P_L}{\sim} w$, where N_λ is the set of left normalized elements (often simply called normalized elements) of $\sigma^{-1}(\lambda)$.

The second step is the most difficult part in the whole of our proof and includes Chapters 11 to 14. In this step, we define a map $T: N_\lambda \to \mathcal{C}_\lambda$ for some fixed $\lambda \in \Lambda_n$ and then show that for $y, w \in N_\lambda$, $y \underset{L}{\sim} w$ if and only if $T(y) = T(w)$. The proof of this result is achieved by considering a new kind of operation on an element w of N_λ called a raising operation on a layer of w, which is not in general a succession of star operations but gives an element in the same left cell as w. Since $|\mathcal{C}_\lambda| = \frac{n!}{\prod_{j=1}^{m} \mu_j!}$ with $\mu = \{\mu_1 \geq \ldots \geq \mu_m\}$ the dual partition of λ, this confirms Lusztig's first conjecture [L9] which says that $\sigma^{-1}(\lambda)$ consists of $\frac{n!}{\prod_{j=1}^{m} \mu_j!}$ left (resp. right) cells of A_n for any $\lambda \in \Lambda_n$.

Thirdly, we show in Chapter 15 that $\sigma^{-1}(\lambda)$ is just a single RL-equivalence class of A_n for any $\lambda \in \Lambda_n$.

Finally, by applying Theorem 1.5.3 [L12] to the affine Weyl group A_n, we verify Lusztig's second conjecture [L9] in Chapter 17 which says that $\sigma^{-1}(\lambda)$ is a two-sided cell of A_n for any $\lambda \in \Lambda_n$.

On the other hand, in Chapter 16, we give another characterization of any left cell of A_n in terms of the generalized right τ-invariant and conclude that any two elements y, w of A_n lie in the same left cell if and only if they have the same generalized right τ-invariant. Let P be any standard parabolic subgroup of A_n isomorphic to the symmetric group S_n. We show in Chapters 16 and 17 that the intersection of P with $\sigma^{-1}(\lambda)$ for any $\lambda \in \Lambda_n$ is non-empty and is just a two-sided cell of P and this verifies Lusztig's third conjecture in the case $W_a = A_n$. This conjecture [L2] states that each two-sided cell in W_a meets some proper standard parabolic subgroup of W_a, where W_a is any affine Weyl group. We also show that the intersection of P with any left (resp. right) cell of A_n is either empty or a left (resp. right) cell of P.

In Chapter 18, we show that left star operations commute with right star operations on an element of A_n. We also show that each two-sided cell or P-equivalence class in A_n is a connected set and that each left (resp. right) cell of P_L-(resp. P_R-) equivalence class in A_n is a left (resp. right) connected set. Moreover, each left (resp. right) cell in A_n is a maximal left (resp. right) connected component in the two-sided cell containing it. The map $\zeta: A_n \to S$ (see §7.1) induces a bijection between the set of two-sided (resp. left) cells of A_n and those of S. We show that the intersection of a left cell with a right cell in the same non-identity two-sided cell of A_n is an infinite set. Finally we show that the intersection of a left cell with a right cell in any Coxeter group is either empty or a discrete set, in a natural sense.

§8.1 ITERATED STAR OPERATIONS

We wish to determine the left cell of A_n containing a given element w. We know from Theorem 1.6.3 that each P_L-equivalence class of A_n lies in some left cell of A_n. But any two elements of a P_L-equivalence class of A_n can be transformed from one to another by a succession of left star operations. It will turn out that we can perform various interchanging operations on blocks of w which are successions of left star operations and so give us elements in the same left cell as w. Although we cannot in general obtain all elements in the same left cell as w in this way, the interchanging operations on blocks will be crucial in our subsequent determination of the left cells.

To define interchanging operations on blocks of an element of A_n, we shall first introduce iterated star operations.

Assume that $w \in A_n$ has a DC form (A) at $i \in \mathbb{Z}$ with $|A(w)| = r$, $1 \leqslant r < n$. Assume that the entry $e(i,j(w))$ of w satisfies $j(w) < j_A^1(w)$, where $e((w),j_A^h(w))$ is the h-th entry of $A(w)$ for $1 \leqslant h \leqslant r$. Let $k = \max\{h \mid 1 \leqslant h \leqslant r,\ j_A^h(w) > j(w)\}$. Then there exists a sequence of elements $x_0 = w, x_1, \ldots, x_{r-1}$, in A_n such that for every t, $1 \leqslant t \leqslant r-1$, we have $x_t = {}^*x_{t-1}$ in $\mathcal{D}_L(s_{i+t-1})$. In particular, $w' = x_{r-1}$ has a DC form (A) at i-1 with $|A(w')| = r$ such that

$$j_A^u(w') = \begin{cases} j_A^u(w) & \text{for } 1 \leqslant u \leqslant r,\ u \neq k \\ j(w) & \text{for } u = k \end{cases}$$

where $e((w'), j_A^h(w'))$ is the h-th entry of $A(w')$ for $1 \leqslant h \leqslant r$ and the entry $e(i+r, j(w'))$ of w' satisfies $j(w') = j_A^k(w)$.

<u>Definition 8.1.1</u> For $w, w' \in A_n$, $1 \leqslant r < n$ and $i \in \mathbb{Z}$, we write $w' \xleftarrow{*(i+1,r)} w$, if

(i) w has a DC form (A) at i with $|A(w)| = r$. (ii) $\exists$ a sequence of elements $x_0 = w, x_1, \ldots, x_{r-1} = w'$ in A_n such that for every $1 \leq h \leq r-1$, we have $x_h = {}^*x_{h-1}$ in $\mathcal{D}_L(s_{i+h-1})$.

For $w, w' \in A_n$, $i \in \mathbb{Z}$ and $1 \leq r < r+m \leq n$, we write $w' \xleftarrow{*(i+1,r,m)} w$, if $\exists$ a sequence of elements $x_0 = w, x_1, \ldots, x_m = w'$ in A_n such that for every $1 \leq h \leq m$, we have $x_h \xleftarrow{*(i+2-h,r)} x_{h-1}$. We write $w \xrightarrow{*(i+1,r,m)} w'$ if $w \xleftarrow{*(i+1-m,r,m)} w'$.

Here is an example for $w \in A_n$ which has DC form (A) $i \in \mathbb{Z}$ with $|A(w)| = 3$.

i^{th} row, $j(w)^{th}$ column, $A(w)$, $j_A^2(w)^{th}$ column

Then there exists w' with $w' \xleftarrow{*(i+1,3)} w$ as follows:

i^{th} row, $A(w')$, $j(w')^{th}$ column

where $j(w') = j_A^2(w)$.

<u>Remark 8.1.2</u> (i) If w has a DC form (A) at $i \in \mathbb{Z}$ with $|A(w)| = r < n$, then we can easily check that w' with $w' \xleftarrow{*(i+1,r)} w$ exists if and only if $j_0 < j_A^1(w)$. Also, w" with $w \xrightarrow{*(i+1,r)} w''$ exists if and only if $j_A^r(w) < j_{r+1}$, where $e(i,j_0)$, $e(i+r+1,j_{r+1})$ are the entries of w.

(ii) Given $w \in A_n$, $i \in \mathbb{Z}$, $r,m > 0$ with $r + m \leq n$, an element w' $w' \xleftarrow{*(i+1,r,m)} w$ or $w \xrightarrow{*(i+1,r,m)} w'$ is unique if it exists.

(iii) In general, the expression $w \xrightarrow{*(i+1,r,m)} w'$ is equivalent to $w \xleftarrow{*(i+1-m,r,m)} w'$ but not to $w' \xleftarrow{*(i+1,r,m)} w$.

§8.2 SOME RESULTS ON ITERATED STAR OPERATIONS

In this section, the first two lemmas give a necessary and sufficient condition on an element w of A_n for which the iterated star operations on w can be carried out. The last lemma states a property of the iterated star operations.

Lemma 8.2.1 Assume that $w \in A_n$ has a DC form (A) at $i \in \mathbb{Z}$. Let m satisfy $1 \leq m \leq n-|A|$. Then there exists w' with $w \xrightarrow{*(i+1,|A|,m)} w'$ if and only if $A(w)$ is a longest descending chain in $A'(w)$, where $A'(x)$ is the block of x consisting of rows from the $(i+1)$-th to the $(i + |A| + m)$-th for $x \in A_n$.

Proof: ($\Leftarrow$) We apply induction on $m \geq 1$. The result is obvious for $m = 1$. When $m > 1$, since $A(w)$ is a longest descending chain in the block $[A(w), B(w)]$, where $B(w)$ is the $(i + |A| + 1)$-th row of w, there exists w_1 such that $w \xrightarrow{*(i+1,|A|)} w_1$. By the proof of Lemma 5.8, $A(w_1)$ is a longest descending chain in $A'(w_1)$ and so also in $A''(w_1)$, where $A(w_1)$ (resp. $A''(w_1)$) is the block of w_1 consisting of rows from the $(i+2)$-th to the $(i+1 + |A|)$-th (resp. from the $(i+2)$-th to the $(i + |A| + m)$-th). By inductive hypothesis, there exists w' such that $w_1 \xrightarrow{*(i+2,|A|,m-1)} w'$. So we have $w \xrightarrow{*(i+1,|A|,m)} w'$.

($\Rightarrow$) We also apply induction on $m \geq 1$. It is true for $m = 1$. Now assume $m > 1$. Since w' exists, it implies that there exists w_1 such that $w \xrightarrow{*(i+1,|A|)} w_1$ and w_1 such that $w \xrightarrow{*(i+1,|A|)} w_1$ and $w_1 \xrightarrow{*(i+2,|A|,m-1)} w'$. By inductive hypothesis, $A(w_1)$ is a longest descending chain in the block $[A(w_1), C(w_1)]$, where $C(w_1)$ is the block of w_1 consisting of rows from the $(i + |A| + 2)$-th to the $(i + |A| + m)$-th. By Lemma 5.8, to reach our goal, it suffices to show that $A(w_1)$ is also a longest descending chain in $A'(w_1)$, or equivalently to show that for any descending chain $D \subseteq A'(w_1)$, there exists a descending chain $D' \subseteq [A(w_1), C(w_1)]$ such that $|D'| = |D|$. Now suppose we are given a fixed descending chain $D \subseteq A'(w_1)$. If $D \subseteq [A(w_1), C(w_1)]$, then let $D' = D$. If $D \nsubseteq [A(w_1), C(w_1)]$, let $D = \{e_{w_1}(i_t,j_t) \mid 1 \leq t \leq \ell,\ i_1 < \dots < i,\ j_1 > \dots > j_\ell\}$. Then $i_1 = i + 1$. Let $e_{w_1}(i_0,j_0)$ be the first entry of $A(w_1)$. Since $i_0 = i_1+1 > i_1$ and $j_0 > j_1$, this implies that $e_{w_1}(i_0,j_0) \notin D$. But by $i_1 < i_2$, $i_0 = i_1+1$ and $i_2 \neq i_1+1$, we have $i_0 < i_2$. Clearly $j_0 > j_1 > j_2$. Let $D' = (\{e_{w_1}(i_0,j_0)\} \cup D) - \{e_{w_1}(i_1,j_1)\}$. Then $D' \subseteq [A(w_1), C(w_1)]$ is as required.□

Lemma 8.2.2 Assume that $w \in A_n$ has a DC form (A) at $i \in \mathbb{Z}$. Suppose m satisfies $1 \leq m \leq n-|A|$. Then there exists w' with $w' \xleftarrow{*(i+1,|A|,m)} w$ if and only if $A(w)$ is a longest descending chain in $A'(w)$ where $A'(x)$ is the block of x consisting of rows from the $(i+1-m)$-th to the $(i+|A|)$-th for $x \in A_n$.

Proof: ($\Rightarrow$) We know that the existence of $w' \in A_n$ with $w' \xleftarrow{*(i+1,|A|,m)} w$ is equivalent to the existence of $w' \in A_n$ with $w' \xrightarrow{*(i+1-m,|A|,m)} w$ and that when w' does exist, w' has a DC form (A) at i-m with $|A(w')| = |A(w)|$. Now we start with w'. Then since w with $w' \xrightarrow{*(i+1-m,|A|,m)} w$ exists, we see by Lemma 8.2.1 that $A(w')$ is a longest descending chain in $A'(w')$. So $A(w)$ is a longest descending chain in $A'(w)$ by the proof of Lemma 5.8.

($\Leftarrow$) Imitate the proof of the corresponding part of Lemma 6.2.1. □

Lemma 8.2.3 Assume that $w \in A_n$ has a DC form (A) at $i \in \mathbb{Z}$ with $|A(w)| = r$, $1 \leq r \leq n-2$. Let $e(i-1,j_1(w))$, $e(i,j_2(w))$ be entries of w. If $w' \in A_n$ with $w' \xleftarrow{*(i+1,r,2)} w$ exists, let $e(i+r-1, j_1(w'))$, $e(i+r, j_2(w'))$ be entries of w'. Then

(i) $j_1(w) > j_2(w) \Leftrightarrow j_1(w') > j_2(w')$

(ii) $j_1(w) < j_2(w) \Leftrightarrow j_1(w') < j_2(w')$.

Proof: It is enough to show the implication in the direction "$\Rightarrow$" for both cases.

(i) Let $a_2 = \max \{h | 1 \leq h \leq r,\ j_A^h(w) > j_2(w)\}$

$a_1 = \max \{h | 1 \leq h < a_2,\ j_A^h(w) > j_1(w)\}$.

Then since w' exists, this implies that a_1, a_2 both exist with $j_1(w') = j_A^{a_1}(w)$ and $j_2(w') = j_A^{a_2}(w)$. Since $a_1 < a_2$, we have $j_1(w') > j_2(w')$.

(ii) We may assume that $r \geq 2$, since otherwise, the result is obvious. Let $a = \max \{h | 1 \leq h \leq r,\ j_A^h(w) > j_2(w)\}$. Then $j_2(w') = j_A^a(w)$. Since $j_1(w') = \min \{j_2(w),\ j_A^h(w) | 1 \leq h \leq r,\ j_A^h(w) > j_1(w)\}$, it follows that $j_1(w') \leq j_2(w) < j_2(w')$. □

§8.3 THE INTERCHANGING OPERATIONS $\rho_{A_2}^{A_1}$ AND $\theta_{A_1}^{A_2}$.

In this section, we shall first define the interchanging operations $\rho_{A_2}^{A_1}$ and $\theta_{A_1}^{A_2}$ on an element $w \in A_n$ and then give a necessary and sufficient condition for which the interchanging operations on w can be carried out. We also give the formulae to calculate these operations. One special case when w has a local MDC form which is quasi-normal (resp. normal) for the first k layers, $k \geq 1$, is most interesting for us. In that case, ρ operations on w have a very good behaviour.

We assume in this section that $w \in A_n$ always has a local MDC form (A_2,A_1) at $i \in \mathbb{Z}$ with $|A_t| = m_t$, $t = 1,2$, unless the contrary is specified.

Definition 8.3.1 Set $w' = \rho_{A_2}^{A_1}(w)$ if $\exists\, w'$ satisfying $w \xrightarrow{*(i+1,m_2,m_1)} w'$ and $w'' = \theta_{A_1}^{A_2}(w)$ if $\exists\, w''$ satisfying $w'' \xleftarrow{*(i+m_2+1,m_1,m_2)} w$.

Note that such elements w', w'' do not always exist in general. But when w' (resp. w'') does exist, then by Lemma 8.2.3, w' (resp. w'') has a local MDC form (A_1',A_2') at i with $|A_t'| = m_t$, $t = 1,2$. We shall always assume that w' (resp. w'') has a local MDC form (A_1,A_2) at i unless the contrary is specified. Clearly, in that case, $w = \theta_{A_2}^{A_1}\, \rho_{A_2}^{A_1}(w)$ (resp. $w = \rho_{A_1}^{A_2}\, \theta_{A_1}^{A_2}(w)$). For convenience we admit blocks occurring in $\rho_{A_2}^{A_1}$ or $\theta_{A_1}^{A_2}$ which have the size 0. We make a convention that $w = \rho_{A_2}^{A_1}(w)$ and $w = \theta_{A_1}^{A_2}(w)$ when either $|A_1| = 0$ or $|A_2| = 0$.

When w' does exist, we define a sequence $\xi(w,\rho_{A_2}^{A_1})$:

$x_{0,m_2-1} = w,\ x_{11}, x_{12}, \ldots, x_{1,m_2-1}, x_{21}, x_{22}, \ldots, x_{2,m_2-1}, \ldots, x_{m_1,1}, x_{m_1,2}, \ldots, x_{m_1,m_2-1}$ such that $x_{hj} = {}^*x_{h,j-1}$ in $\mathcal{D}_L(s_{i+h-1+m_2-j})$ for h,j with $1 \leq h \leq m_1$ and $1 < j \leq m_2-1$, and $x_{h+1,1} = {}^*x_{h,m_2-1}$ in $\mathcal{D}_L(s_{i+h-1+m_2})$ for $0 \leq h < m_1$. Then $\xi(w,\rho_{A_2}^{A_1})$ exists with $w' = x_{m_1,m_2-1}$. We call $\xi(w,\rho_{A_2}^{A_1})$ the sequence corresponding to $\rho_{A_2}^{A_1}$ on w. Similarly, we can define the sequence $\xi(w,\theta_{A_1}^{A_2})$ corresponding to $\theta_{A_1}^{A_2}$ on w when w'' exists.

For example, let w have a local MDC form (A_2,A_1) at $i \in \mathbb{Z}$ with $|A_2| = 3$ and $|A_1| = 2$ as follows:

$$\begin{array}{l} (i+1)^{st}\text{ row} \\ \\ \\ \\ \\ \end{array} \begin{pmatrix} & & & 1 & \\ & & 1 & & \\ 1 & & & & \\ & & & & 1 \\ & 1 & & & \end{pmatrix} \begin{array}{l} \left.\begin{array}{l} \\ \\ \\ \end{array}\right\} A_2(w) \\ \left.\begin{array}{l} \\ \\ \end{array}\right\} A_1(w) \end{array}$$

Then $w' = \rho_{A_2}^{A_1}(w)$ has the following form

$$\begin{array}{l} (i+1)^{st}\text{ row} \\ \\ \\ \\ \\ \end{array} \begin{pmatrix} & & & 1 & \\ 1 & & & & \\ & & & & 1 \\ & & 1 & & \\ & 1 & & & \end{pmatrix} \begin{array}{l} \left.\begin{array}{l} \\ \\ \end{array}\right\} A_1(w') \\ \left.\begin{array}{l} \\ \\ \\ \end{array}\right\} A_2(w') \end{array}$$

and the sequence corresponding to $\rho_{A_2}^{A_1}$ on w is $\xi(w,\rho_{A_2}^{A_1})$: $x_{0,2} = w$, x_{11}, x_{12}, x_{21}, $x_{22} = w'$, where $x_{11} = {}^*x_{02}$ in $\mathcal{D}_L(s_{i+2})$, $x_{12} = {}^*x_{11}$ in $\mathcal{D}_L(s_{i+1})$, $x_{21} = {}^*x_{12}$ in $\mathcal{D}_L(s_{i+3})$ and $x_{22} = {}^*x_{21}$ in $\mathcal{D}_L(s_{i+2})$.

In the above example, we also have $w = \theta_{A_2}^{A_1}(w')$ and the sequence corresponding to $\theta_{A_2}^{A_1}$ on w' is obtained from the sequence $\xi(w, \rho_{A_2}^{A_1})$ just by reversing its order.

Now we shall give the formulae for calculation of $\rho_{A_2}^{A_1}(w)$ and $\theta_{A_1}^{A_2}(w)$, and also give a necessary and sufficient condition on the existence of $\rho_{A_2}^{A_1}(w)$ or $\theta_{A_1}^{A_2}(w)$.

Let

$$(8.3.2)\quad \begin{cases} \alpha_0 = 0 \\ \alpha_u = \min\,\{h \mid \alpha_{u-1} < h \leq m_2,\ j_2^h(w) < j_1^u(w)\},\ 1 \leq u \leq m_1 \end{cases}$$

and

$$(8.3.3)\quad \begin{cases} \beta_{m_2+1} = m_1+1 \\ \beta_u = \max\,\{h \mid 1 \leq h < \beta_{u+1},\ j_1^h(w) > j_2^u(w)\},\ 1 \leq u \leq m_2 \end{cases}$$

<u>Lemma 8.3.4</u> <u>When $|A_2(w)| \neq 0$, we have</u>

$$w' = \rho_{A_2}^{A_1}(w) \text{ exists} \iff \text{all } \alpha_u,\ 1 \leq u \leq m_1, \text{ exist.}$$

<u>When they do exist, we get</u>

$$\begin{cases} j_1^u(w') = j_2^{\alpha_u}(w) & \text{if } 1 \leq u \leq m_1 \\ j_2^v(w') = \begin{cases} j_2^v(w) & \text{if } 1 \leq v \leq m_2,\ v \notin \{\alpha_1,\dots,\alpha_{m_1}\} \\ j_1^h(w) & \text{if } v = \alpha_h \text{ for some } 1 \leq h \leq m_1 \end{cases} \end{cases}$$

<u>Proof</u>: By definition of $\rho_{A_2}^{A_1}$. □

<u>Lemma 8.3.5</u> <u>When $|A_2(w)| \neq 0$, we have</u>

$$w' = \rho_{A_2}^{A_1}(w) \text{ exists} \iff A_2(w) \text{ is a longest descending chain in the block } [A_2(w), A_1(w)].$$

<u>Proof</u>: By Lemma 8.2.1 and the definition of $\rho_{A_2}^{A_1}$. □

Lemma 8.3.6 When $|A_1(w)| \neq 0$, we have $w'' = \theta_{A_1}^{A_2}(w)$ exists $\Leftrightarrow$ all β_u, $1 \leq u \leq m_2$, exist.

When they do exist, we get

$$\begin{cases} j_2^u(w'') = j_1^{\beta_u}(w) & \text{if } 1 \leq u \leq m_2 \\ j_1^v(w'') = \begin{cases} j_1^v(w) & \text{if } 1 \leq v \leq m_1,\ v \notin \{\beta_1,\ldots,\beta_{m_2}\} \\ j_2^h(w) & \text{if } v = \beta_h \text{ for some } 1 \leq h \leq m_2 \end{cases} \end{cases}$$

Proof: By definition of $\theta_{A_1}^{A_2}$. □

Lemma 8.3.7 When $|A_1(w)| \neq 0$, we have

$w'' = \theta_{A_1}^{A_2}(w)$ exists $\Leftrightarrow$ $A_1(w)$ is a longest descending chain in the block $[A_2(w), A_1(w)]$.

Proof: By Lemma 8.2.2 and the definition of $\theta_{A_1}^{A_2}$. □

In the remainder of this section, we shall introduce the concept of being quasi-normal (resp. normal) for the first k layers of a local MDC form of w. Before doing this, we give some definitions.

Definition 8.3.8 For any ℓ, $i \in \mathbb{Z}$ with $\ell > 0$ and $1 \leq i \leq \ell$, suppose that either $a_i \in \mathbb{Z}$ or a_i does not exist. We define $(a_1,\ldots,a_\ell)^{(om)} \in \bigcup_{t=0}^{\ell} \mathbb{Z}^t$ as follows:

(i) $(a_1,\ldots,\hat{a}_j,\ldots,a_\ell)^{(om)} = (a_1,\ldots,a_\ell)^{(om)}$, if a_j does not exist for some $1 \leq j \leq \ell$, where the notation $\hat{a}_j$ means that we omit the term a_j.

(ii) $(a_1,\ldots,a_\ell)^{(om)} = (a_1,\ldots,a_\ell)$, if all a's exist with the convention that $Z^0 = \emptyset$.

Definition 8.3.9 For any ℓ, $i \in \mathbb{Z}$ with $\ell > 0$ and $1 \leq i \leq \ell$, suppose that either $a_i \in \mathbb{Z}$ or a_i does not exist. We define $(a_1 < \ldots < a_\ell)^{(om)}$ as follows:

(i) Let $(a_1 < \ldots < a_\ell)^{(om)}$ be $(a_1 < \ldots < \hat{a}_j < \ldots < a_\ell)^{(om)}$, if a_j does not exist for some $1 \leq j \leq \ell$.

(ii) Let $(a_1 < \dots < a_\ell)^{(om)}$ be $a_1 < \dots < a_\ell$, if all a's exist with the convention that $(a_1 < \dots < a_\ell)^{(om)}$ is an identity when $\ell \le 1$.

Suppose that w has a local MDC form $(A_\ell,\dots,A_1)$ at $i \in \mathbb{Z}$. For any $u \ge 1$, the u-th layer with respect to this local MDC form is, by definition, the set of entries of w consisting of all the u-th entries of $A_t(w)$, $1 \le t \le \ell$, and their congruent entries.

Definition 8.3.10 Suppose that w has a local MDC form $(A_\ell,\dots,A_1)$ at $i \in \mathbb{Z}$. Let k satisfy $1 \le k \le \max\{|A_t| \,|\, 1 \le t \le \ell\}$. If for any h, $1 \le h \le k$, $(j_\ell^h(w) < j_{\ell-1}^h(w) < \dots < j_1^h(w))^{(om)}$ holds, then we say that w has a local MDC form $(A_\ell,\dots,A_1)$ at i which is quasi-normal for the first k layers. If for any h, $1 \le h \le k$, $(j_1^h(w) - n < j_\ell^h(w) < j_{\ell-1}^h(w) < \dots < j_1^h(w))^{(om)}$ holds, then we say that w has a local MDC form $(A_\ell,\dots,A_1)$ at i which is normal for the first k layers. In particular, in the above definition, if $k = \max\{|A_t| \,|\, 1 \le t \le \ell\}$, we shall say that w has a local MDC form $(A_\ell,\dots,A_1)$ at i which is quasi-normal (resp. normal).

Now assume that w has a local MDC form (A_2,A_1) at i which is quasi-normal for the first k layers with $k = \min\{m_1,m_2\}$. Then in Formula (8.3.2), we have $\alpha_v = v$ for all $1 \le v \le k$. Let $\alpha'_u = \alpha_{u+k} - k$ for any $0 \le u \le m_1 - k$. Then we have

$$(8.3.11) \qquad \begin{cases} \alpha'_0 = 0 \\ \alpha'_u = \min\{h \,|\, \alpha'_{u-1} < h \le m_2 - k,\ j_2^{k+h}(w) < j_1^{k+u}(w)\},\ 1 \le u \le m_1 - k \end{cases}$$

Corollary 8.3.12 Assume that $w \in A_n$ has a local MDC form (A_2,A_1) at $i \in \mathbb{Z}$ with $|A_2| \ne 0$ which is quasi-normal for the first k layers for some $k \ge 0$. Then

$$w' = \rho_{A_2}^{A_1}(w) \text{ exists} \iff \text{all } \alpha_u,\ k+1 \le u \le m_1, \text{ exist.}$$

When they do exist, we have

$$\begin{cases} (j_1^h(w'),\ j_2^h(w')) = (j_2^h(w),\ j_1^h(w)), & \text{if } 1 \le h \le k \\ j_1^u(w') = j_2^{\alpha_u}(w), & \text{if } k+1 \le u \le m_1 \\ j_2^v(w') = \begin{cases} j_2^v(w), & \text{if } k+1 \le v \le m_1 \text{ and } v \notin \{\alpha_{k+1},\dots,\alpha_{m_1}\} \\ j_1^h(w), & \text{if } v = \alpha_h \text{ for some } k+1 \le h \le m_1 \end{cases} \end{cases}$$

Proof: This follows from Lemma 8.3.4. □

We can regard Lemma 8.3.4 as a special case of Corollary 8.3.12 when $k = 0$.

§8.4 MORE GENERAL INTERCHANGING OPERATIONS

In this section, we assume that $w \in A_n$ has a local MDC form $(A_\ell,\dots,A_1)$ at $i \in \mathbb{Z}$. We shall extend the results of §8.3 to more general cases.

First let us observe an example: Assume that $w \in A_n$ has a local MDC form (A_4,A_3,A_2,A_1) at $i \in \mathbb{Z}$.

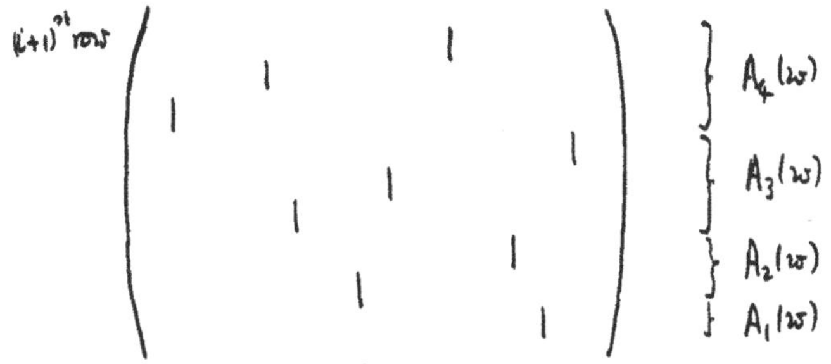

Then there exist elements x_1, x_2, x_2', x_3 and x_4 such that $x_1 = \rho_{A_3}^{A_2}(w)$, $x_2 = \rho_{A_3}^{A_1}(x_1)$, $x_2' = \rho_{A_4}^{A_2}(x_1)$, $x_3 = \rho_{A_4}^{A_2}(x_2) = \rho_{A_3}^{A_1}(x_2')$ and $x_4 = \rho_{A_4}^{A_1}(x_3)$ are as follows:

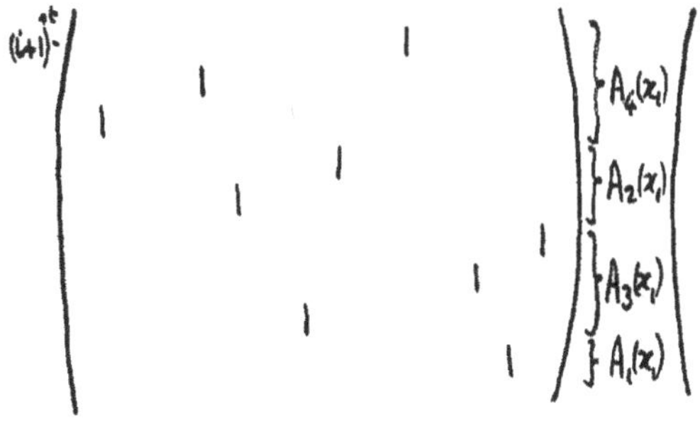

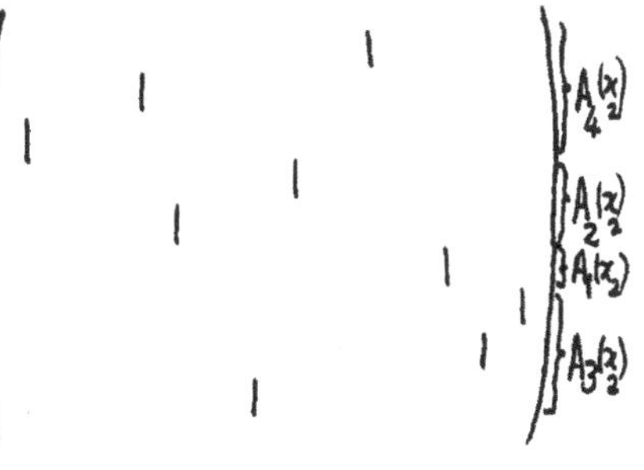

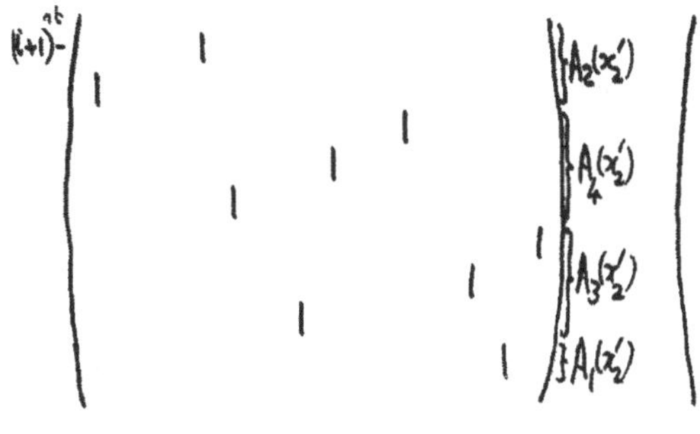

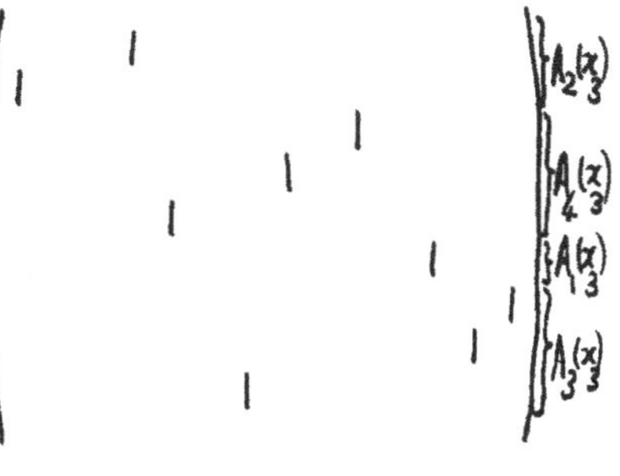

$(\ell+1)^{st}$ row

$A_2(x_4)$
$A_1(x_4)$
$A_4(x_4)$
$A_3(x_4)$

x_4

So $x_4 = \rho^{A_1}_{A_4}\,\rho^{A_2}_{A_4}\,\rho^{A_1}_{A_3}\,\rho^{A_2}_{A_3}(w) = \rho^{A_1}_{A_4}\,\rho^{A_1}_{A_3}\,\rho^{A_2}_{A_4}\,\rho^{A_2}_{A_3}(w)$.

<u>Definition 8.4.1</u> Set

$$\begin{cases} \rho^{A_1}_{A_2,A_3,\dots,A_\ell}(w) = \rho^{A_1}_{A_\ell}\,\rho^{A_1}_{A_{\ell-1}}\cdots\rho^{A_1}_{A_2}(w) \\ \rho^{A_{\ell-1},A_{\ell-2},\dots,A_1}_{A_\ell}(w) = \rho^{A_1}_{A_\ell}\,\rho^{A_2}_{A_\ell}\cdots\rho^{A_{\ell-1}}_{A_\ell}(w) \\ \rho^{A_j,\dots,A_1}_{A_{j+1},\dots,A_\ell}(w) = \rho^{A_1}_{A_{j+1},\dots,A_\ell}\,\rho^{A_2}_{A_{j+1},\dots,A_\ell}\cdots\rho^{A_j}_{A_{j+1},\dots,A_\ell}(w), \quad 1 < j < \ell. \end{cases}$$

$$\begin{cases} \theta^{A_2,A_3,\dots,A_\ell}_{A_1}(w) = \theta^{A_\ell}_{A_1}\,\theta^{A_{\ell-1}}_{A_1}\cdots\theta^{A_2}_{A_1}(w) \\ \theta^{A_\ell}_{A_{\ell-1},A_{\ell-2},\dots,A_1}(w) = \theta^{A_\ell}_{A_1}\,\theta^{A_\ell}_{A_2}\cdots\theta^{A_\ell}_{A_{\ell-1}}(w) \\ \theta^{A_{j+1},\dots,A_\ell}_{A_j,\dots,A_1}(w) = \theta^{A_\ell}_{A_j,\dots,A_1}\,\theta^{A_{\ell-1}}_{A_j,\dots,A_1}\cdots\theta^{A_{j+1}}_{A_j,\dots,A_1}(w), \quad 1 < j < \ell. \end{cases}$$

when the right hand sides exist.

<u>Lemma 8.4.2</u> When they exist, we have: for any $1 < j < \ell$,

$$\underline{\rho^{A_1}_{A_{j+1},\dots,A_\ell}\,\rho^{A_2}_{A_{j+1},\dots,A_\ell}\cdots\rho^{A_j}_{A_{j+1},\dots,A_\ell}(w)}$$

$$= \underline{\rho^{A_j,\dots,A_1}_{A_\ell}\,\rho^{A_j,\dots,A_1}_{A_{\ell-1}}\cdots\rho^{A_j,\dots,A_1}_{A_{j+1}}(w)}$$

$$\frac{\theta^{A_\ell}_{A_j,\ldots,A_1}\ \theta^{A_{\ell-1}}_{A_j,\ldots,A_1}\ \cdots\ \theta^{A_{j+1}}_{A_j,\ldots,A_1}(w)}{}$$

$$= \underline{\theta^{A_{j+1},\ldots,A_\ell}_{A_1}\ \theta^{A_{j+1},\ldots,A_\ell}_{A_2}\ \cdots\ \theta^{A_{j+1},\ldots,A_\ell}_{A_j}(w)}$$

<u>Proof</u>: By definition of ρ, θ it suffices to show that when they exist, $\rho^{A_{j_1}}_{A_{i_1}}\ \rho^{A_{j_2}}_{A_{i_2}}(w) = \rho^{A_{j_2}}_{A_{i_2}}\ \rho^{A_{j_1}}_{A_{i_1}}(w)$ and $\theta^{A_{i_1}}_{A_{j_1}}\ \theta^{A_{i_2}}_{A_{j_2}}(w) = \theta^{A_{i_2}}_{A_{j_2}}\ \theta^{A_{i_1}}_{A_{j_1}}(w)$ hold, where $j_1, j_2, i_1, i_2 \in \{1,2,\ldots,\ell\}$ are distinct. But this is obvious. □

As in the above example, we can rewrite x_4 by $\rho^{A_2,A_1}_{A_4}\ \rho^{A_2,A_1}_{A_3}(w)$ or by $\rho^{A_1}_{A_3,A_4}\ \rho^{A_2}_{A_3,A_4}(w)$.

The above lemma tells us that the expression of $\rho^{A_j,\ldots,A_1}_{A_{j+1},\ldots,A_\ell}$ (or $\theta^{A_{j+1},\ldots,A_\ell}_{A_j,\ldots,A_1}$) in terms of ρ^A_B (or θ^A_B) is not unique. So for the sake of definiteness, we shall define the sequence corresponding to $\rho^{A_j,\ldots,A_1}_{A_{j+1},\ldots,A_\ell}$ (resp. $\theta^{A_{j+1},\ldots,A_\ell}_{A_j,\ldots,A_1}$) on w in the following way. These sequences will be used in Chapter 12 to define a family of much longer and also more important sequences.

If $w' = \rho^{A_1}_{A_2,\ldots,A_\ell}(w)$ exists, then there exist sequences corresponding to $\rho^{A_1}_{A_j}$ on $\rho^{A_1}_{A_2,\ldots,A_{j-1}}(w)$ for $2 < j < \ell$ (For the definition of these sequences, see §8.3) with the convention that $\rho^{A_1}_{A_2,\ldots,A_{j-1}}$ is the identity map on w for j = 2. We call $\xi(w,\rho^{A_1}_{A_2,\ldots,A_\ell})$ the sequence corresponding to $\rho^{A_1}_{A_2,\ldots,A_\ell}$ on w if it is obtained by linking all the above sequences together. Furthermore, we call $\xi(w, \rho^{A_j,\ldots,A_1}_{A_{j+1},\ldots,A_\ell})$ the sequence corresponding to $\rho^{A_j,\ldots,A_1}_{A_{j+1},\ldots,A_\ell}$ on w if it is obtained by linking all the sequences
$\xi(\rho^{A_j,\ldots,A_t}_{A_{j+1},\ldots,A_\ell}(w), \rho^{A_{t-1}}_{A_{j+1},\ldots,A_\ell})$, $1 < t < j+1$, together, with the convention that $\rho^{A_j,\ldots,A_t}_{A_{j+1},\ldots,A_\ell}(w) = w$ for t = j+1. We can define $\xi(w, \theta^{A_{j+1},\ldots,A_\ell}_{A_j,\ldots,A_1})$ as the sequence corresponding to $\theta^{A_{j+1},\ldots,A_\ell}_{A_j,\ldots,A_1}$ on w to be obtained from the sequence $\xi(\theta^{A_{j+1},\ldots,A_\ell}_{A_j,\ldots,A_1}(w), \rho^{A_1,\ldots,A_j}_{A_\ell,\ldots,A_{j+1}})$ by reversing its order.

The following two lemmas generalize Lemmas 8.3.5 and 8.3.7.

Lemma 8.4.3 Assume that $w \in A_n$ has a local MDC form $(A_\ell,\dots,A_1)$ at $i \in \mathbb{Z}$. Then $w' = \rho_{A_\ell}^{A_{\ell-1},A_{\ell-2},\dots,A_1}(w)$ exists if and only if $A_\ell(w)$ is a longest descending chain in the block $[A_\ell(w),\dots,A_1(w)]$.

Proof: Since $w \xrightarrow{*(i+1,\ |A_\ell|,\ \sum_{t=1}^{\ell-1}|A_t|)} w'$, this follows from Lemma 8.2.1. □

Lemma 8.4.4 Assume that $w \in A_n$ has a local MDC form $(A_\ell,\dots,A_1)$ at $i \in \mathbb{Z}$. Then $w' = \rho_{A_1}^{A_2,A_3,\dots,A_\ell}(w)$ exists if and only if $A_1(w)$ is a longest descending chain in the block $[A_\ell(w),\dots,A_1(w)]$.

Proof: ($\Rightarrow$) The existence of w' implies that $w = \rho_{A_1}^{A_\ell,A_{\ell-1},\dots,A_2}\ \theta\ \rho_{A_1}^{A_2,A_3,\dots,A_\ell}(w)$ $= \rho_{A_1}^{A_\ell,A_{\ell-1},\dots,A_2}(w')$. By Lemma 8.4.3, it follows that $A_1(w')$ is a longest descending chain in the block $[A_\ell(w'),\dots,A_1(w')]$ and then $A_1(w)$ is a longest descending chain in the block $[A_\ell(w),\dots,A_1(w)]$ by Lemma 5.8.

($\Leftarrow$) This is proved as in the proof of Lemma 8.2.1. □

The next lemma and its corollary show that ρ operations preserve the property of being quasi-normal (resp. normal) for the first k layers of a local MDC form.

Lemma 8.4.5 Assume that $w \in A_n$ has a local MDC form $(A_{1\ell_1},\dots,A_{11},A_{2\ell_2},\dots,A_{21},$ $A_{3\ell_3},\dots,A_{31},A_{4\ell_4},\dots,A_{41})$ at $i \in \mathbb{Z}$ which is quasi-normal (resp. normal) for the first k layers. If there exists $w' = \rho_{A_{21},\dots,A_{2\ell_2}}^{A_{3\ell_3},\dots,A_{31}}(w)$, then for any h with $1 \leq h \leq k$, we have $(j^h_{1\ell_1}(w'),\dots,j^h_{11}(w'),\ j^h_{3\ell_3}(w'),\dots,j^h_{31}(w'),\ j^h_{2\ell_2}(w'),\dots,$ $j^h_{21}(w'),\ j^h_{4\ell_4}(w'),\dots,j^h_{41}(w'))^{(om)} = (j^h_{1\ell_1}(w),\dots,j^h_{11}(w),\ j^h_{2\ell_2}(w),\dots,\ j^h_{21}(w),$ $j^h_{3\ell_3}(w),\dots,j^h_{31}(w),\ j^h_{4\ell_4}(w),\dots,j^h_{41}(w))^{(om)}$. In particular, w' has a local MDC form $(A_{1\ell_1},\dots,A_{11},\ A_{3\ell_3},\dots,A_{31},\ A_{2\ell_2},\dots,A_{21},\ A_{4\ell_4},\dots,A_{41})$ at i which is quasi-normal (resp. normal) for the first k layers.

Proof: By repeatedly applying Lemma 8.2.3 and Corollary 8.3.12. □

Remark 8.4.6 (i) Assume that $w \in A_n$ has a full DC form $(A_\ell,\dots,A_1)$ at $i \in \mathbb{Z}$ and assume that there exists $w' = \rho_{A_{i_1},\dots,A_{i_\alpha}}^{A_{i'_1},\dots,A_{i'_\beta}}(w)$. Consider the block $A_t(w')$. This is defined up to congruence mod n. But when we mention the u-th entry $e((w'), j_t^u(w'))$ (or $e((w'), j_t^u(w',i')))$ of $A_t(w')$, we mean that the block $A_t(w')$ involved here has been specified precisely by the following rule;

(1) $A_t(w')$ lies between the (i+1)-th row and the (i+n)-th row of w' if $1 \notin \{i_1,\dots,i_\alpha\} \cup \{i'_1,\dots,i'_\beta\}$.

(2) $A_t(w')$ lies between the $(i + \sum_{t=1}^{\beta} |A_{i'_t}| + 1)$-th row and the $(i + \sum_{t=1}^{\beta} |A_{i'_t}| + n)$-th row if $1 \in \{i_1,\dots,i_\alpha\}$.

(3) $A_t(w')$ lies between the $(i - \sum_{t=1}^{\alpha} |A_{i_t}| + 1)$-th row and the $(i - \sum_{t=1}^{\alpha} |A_{i_t}| + n)$-th row if $1 \in \{i'_1,\dots,i'_\beta\}$.

(ii) Assume that $w \in A_n$ has a full DC form $(A_\ell,\dots,A_1)$ at $i \in \mathbb{Z}$ and assume that there exists $w'' = \theta_{A_{i_1},\dots,A_{i_\alpha}}^{A_{i'_1},\dots,A_{i'_\beta}}(w)$. This time when we mention the u-th entry $e((w''), j_t^u(w''))$ (or $e((w''), j_t^u(w'',i'')))$ of $A_t(w'')$, we specify $A_t(w'')$ involved here by the following rule:

(1) $A_t(w'')$ lies between the (i+1)-th row and the (i+n)-th row of w'' if $1 \notin \{i_1,\dots,i_\alpha\} \cup \{i'_1,\dots,i'_\beta\}$.

(2) $A_t(w'')$ lies between the $(i - \sum_{t=1}^{\beta} |A_{i'_t}| + 1)$-th row and the $(i - \sum_{t=1}^{\beta} |A_{i'_t}| + n)$-th row if $1 \in \{i_1,\dots,i_\alpha\}$.

(3) $A_t(w'')$ lies between the $(i + \sum_{t=1}^{\alpha} |A_{i_t}| + 1)$-th row and the $(i + \sum_{t=1}^{\alpha} |A_{i_t}| + n)$-th row if $1 \in \{i'_1,\dots,i'_\beta\}$.

Corollary 8.4.7 <u>Assume that $w \in A_n$ has a full MDC form $(A_\ell,\dots,A_1)$ at $i \in \mathbb{Z}$ which is normal for the first k layers. If there exists $w' = \rho_{A_1,\dots,A_v}^{A_\ell,\dots,A_{v+1}}(w)$ for some v, $1 \leq v < \ell$, then w' has the full MDC form $(A_\ell,\dots,A_1)$ at $i + \sum_{t=v+1}^{\ell} |A_t|$ which is</u>

<u>normal for the first k layers such that for any h, $1 \leq h \leq k$,</u>

$$\underline{(j_\ell^h(w'),\ j_{\ell-1}^h(w'),\dots,j_1^h(w'))^{(om)} = (j_v^h(w),\ j_{v-1}^h(w),\dots,j_1^h(w),\ j_\ell^h(w) + n,}$$

$$\underline{j_{\ell-1}^h(w) + n,\dots,j_{v+1}^h(w) + n)^{(om)}} \qquad (8.4.8).$$

<u>Proof</u>: Formula (8.4.8) follows from Lemma 8.4.5 since w has an MDC form $(A_v,\dots,A_1,A_\ell,\dots,A_{v+1})$ at $i + \sum_{t=v+1}^{\ell} |A_t|$ which is normal for the first k layers. This implies that w' has a full MDC form $(A_\ell,\dots,A_1)$ at $i + \sum_{t=v+1}^{\ell} |A_t|$ which is normal for the first k layers. □

CHAPTER 9 : THE SUBSET $\sigma^{-1}(\lambda)$ OF THE AFFINE WEYL GROUP A_n

In Chapter 5, we have defined a map σ from the affine Weyl group A_n to the set of partitions of n. In this chapter, we consider the fibre $\sigma^{-1}(\lambda)$ corresponding to a given partition λ of n. Our main aim is to show that each element w in this fibre can be transformed by a succession of left star operations into an element y of a rather special form. y has the property that the blocks in a full MDC form have size $\lambda_1,\ldots,\lambda_r$, where $\lambda_1,\ldots,\lambda_r$ are the parts of λ with $\lambda_1 \geqslant \ldots \geqslant \lambda_r$.

This result will have two useful consequences. In the first place, it enables us to show that the fibre $\sigma^{-1}(\lambda)$ is a union of RL-equivalence classes. This is done in §9.4. In the second place, we shall use this algorithm as the initial stage of a longer process for passing from w to a simpler type of element which will be described in the next chapter.

It is easily seen that $\sigma^{-1}(\{1 \geqslant \ldots \geqslant 1\}) = \{\mathbb{1}\}$ is a two-sided cell of A_n. So we may assume that $\lambda \neq \{1 \geqslant \ldots \geqslant 1\}$.

§9.1 TWO SIMPLE LEMMAS ON ITERATED STAR OPERATIONS

Lemma 9.1.1 Assume that $w \in A_n$ has a DC form (A_2,A_1) at $i \in \mathbb{Z}$, where $A_1(w)$ is a longest descending chain in the block $[A_2(w), A_1(w)]$. Then there exists an integer k with $k = \max\{h \mid 1 \leqslant h \leqslant |A_1|,\ j_2^1(w) < j_1^h(w)$ and $w' = \theta_{A_1}^{A_2}(w)$ such that w' has the DC form (A_1,A_2) at i with $j_2^1(w') = j_1^k(w)$.

Proof: The existence of the element $w' \in A_n$ and the integer k follows from Lemma 8.3.7 and the condition that $A_1(w)$ is a longest descending chain in the block $[A_2(w), A_1(w)]$. Then formula (8.3.3) and Lemma 8.3.6 imply $j_2^1(w') = j_1^k(w)$. □

Lemma 9.1.2 Assume that $w \in A_n$ has a DC form (A) at $i \in \mathbb{Z}$ with $|A(w)| = \ell$, and suppose that the entry set $\{e(i+\ell+u, j_u(w)) \mid 0 \leqslant u \leqslant m\}$ of w with $m + \ell \leqslant n$ satisfies $j_A^{\ell}(w) = j_0(w) < j_1(w) < \ldots < j_m(w)$. Then there exists $w' \in A_n$ with $w \xrightarrow{*(i+1,\ell,m)} w'$ such that w' has the DC form (A) at i+m with $j_A^{\ell}(w') = j_k(w)$, where $k = \max\{h \mid 0 \leqslant h \leqslant m,$ $j_h(w) < j_A^{\ell-1}(w)\}$. Let $\{e(i+1+u, j_u(w')) \mid 0 \leqslant u \leqslant m\}$ be the entry set of w'. Then $j_0(w') < j_{\ell}(w') < \ldots < j_m(w') = j_A^{1}(w')$ and $j_h(w') = j_h(w)$ for $0 \leqslant h < k$.

Proof: The existence of w' follows from 8.2.1. The other conclusions follow from Lemma 8.2.3 (ii) and by repeatedly applying formula (8.3.2) with $m_1 = 1$, where m_1 is as in (8.3.2). □

§9.2 THE SUBSET F OF THE AFFINE WEYL GROUP A_n

We shall describe the process of passing from an arbitrary element $w \in \sigma^{-1}(\lambda)$ to an element y of the required form by a succession of left star operations in two stages. In the first stage, we shall show that we can pass from w to an element in the subset F defined below. In the second stage, we shall show that starting from an element of $\sigma^{-1}(\lambda) \cap F$, we can pass to an element y of the required form.

We define $F = \bigcup_{i=0}^{n-1} F_i$ with $F_i = \{w \in A_n | (i+1)w < \ldots < (i+n)w\}$ for $0 \leq i \leq n-1$.

Lemma 9.2.1 *For any $w \in A_n$, there exists $y \in F$ such that $y \underset{P_L}{\sim} w$.*

Proof: It suffices to show that there exists y with $y \underset{P_L}{\sim} w$ such that for some $t \in \mathbb{Z}$, we have $(t+1)y < \ldots < (t+n)y$. We may assume $w \neq 1$ since otherwise the result is trivial.

By Lemmas 4.2.4 and 4.2.5, there exists $i \in \mathbb{Z}$ such that $(i+1)w < (i+2)w < \ldots < (i+j)w$ for some $2 \leq j \leq n$. We apply induction on $\ell = n-j \geq 0$. The result is trivial for $\ell = 0$. Now assume $\ell > 0$. If $(i+j+1)w > (i+j)w$, then the induction can be applied. Otherwise, assume that for some $0 \leq u < j$.

$$(i+1)w < (i+2)w < \ldots < (i+u)w < (i+j+1)w < (i+u+1)w < \ldots < (i+j)w.$$

Let $z = w$, then $z_0 \in \mathcal{D}_L(s_{i+j+1})$. Let $z_1 = {}^*z_0$ in $\mathcal{D}_L(s_{i+j-1})$, then $z_1 \in \mathcal{D}_L(s_{i+j-2})$ if $j \geq 3$. In general, we have $z_u \in \mathcal{D}_L(s_{i+j-u}) \cap \mathcal{D}_L(s_{i+j-u-1})$ for $1 \leq u \leq j-2$, where $z_u = {}^*z_{u-1}$ is in $\mathcal{D}_L(s_{i+j-u})$ for $1 \leq u \leq j-1$. Let $y_1 = z_{j-1}$. Then

$$\begin{cases} (i+v)w = (i+v+1)y_1 & \text{if } 1 \leq v \leq j,\ v \neq u+1 \\ (i+j+1)w = (i+u+2)y_1 & \\ (i+u+1)w = (i+1)y_1 & \\ (h)w = (h)y_1 & \text{for } \bar{h} \notin \{\overline{i+1}, \overline{i+2}, \ldots, \overline{i+j+1}\} \end{cases}$$

In particular, we have

$$\begin{matrix} (i+2)y_1 & < (i+3)y_1 & < \ldots < & (i+u+1)y_1 & < (i+u+2)y_1 & < (i+u+3)y_1 & < \ldots < & (i+j+1)y_1 \\ \| & \| & & \| & \| & \| & & \| \\ (i+1)w & < (i+2)w & < \ldots < & (i+u)w & < (i+j+1)w & < (i+u+2)w & < \ldots < & (i+j)w. \end{matrix}$$

Let us take an example to illustrate the above result. Assume that $n > 5$ and w has the form

$$\begin{matrix} (i+1)^{th}\text{ row} \\ (i+u)^{th}\text{ row} \\ \\ (i+j)^{th}\text{ row} \\ \\ \end{matrix} \begin{pmatrix} 1 & & & & \\ & 1 & & & \\ & & & 1 & \\ & & & & 1 \\ & & 1 & & \end{pmatrix}$$

$$w$$

where $u = 2$ and $j = 4$. Then y_1 has the form

$$\begin{matrix} (i+1)^{th}\text{ row} \\ \\ \\ \\ \\ \end{matrix} \begin{pmatrix} & & & 1 & \\ 1 & & & & \\ & 1 & & & \\ & & 1 & & \\ & & & & 1 \end{pmatrix}$$

$$y_1$$

with $(i+1)y_1 = (i+3)w$ and

$$\begin{matrix} (i+2)y_1 & < & (i+3)w & < & (i+4)y_1 & < & (i+5)y_1 \\ \| & & \| & & \| & & \| \\ (i+1)w & < & (i+2)w & < & (i+5)w & < & (i+4)w. \end{matrix}$$

Now let us continue to prove Lemma 9.2.1. Clearly, $\sum_{v=2}^{j+1} (i+v)y_1 < \sum_{v=1}^{j} (i+v)w$ and $(i+j+1)y_1 < (i+j)w$. If $(i+j+2)y_1 > (i+j+1)y_1$, then our assertion follows by inductive hypothesis. Otherwise, the same procedure can be repeated. We get $y_1, y_2, \ldots,$ where for each $c > 1$,

$$(i+c+1)y_c < (i+c+2)y_c < \ldots < (i+j+c)y_c, \quad \sum_{v=c+1}^{j+c} (i+v)y_c < \sum_{v=c}^{j+c-1} (i+v)y_{c-1},$$

and $(i+y+c)y_c < (i+j)w$. We claim that there exists $c > 1$ such that $(i+j+c+1)y_c > (i+j+c)y_c$ and then the induction can be applied. Otherwise, for any $c > 1$, $(i+j+c+1)y_c < (i+j+c)y_c$ and then we have

$$\sum_{v=1}^{j} (i+v)w > \sum_{v=2}^{j+1} (i+v)y_1 > \sum_{v=3}^{j+2} (i+v)y_2 > \ldots > \sum_{v=c+1}^{j+c} (i+v)y_c > \ldots .$$

By the conditions that $\sum_{v=1}^{n} (v)x = \sum_{v=1}^{n} v$ and $(v+n)x = (v)x + n$ for any $x \in A_n$, this implies that $\lim_{c\to\infty} \sum_{v=c+1}^{c+n} (i+v)y_c = \infty$. So there exist integers c, h with c sufficiently large and $i+j+c < h \leq i+n+c$ such that $(h)y_c > (i+j)w$. Hence

$$(i+j+(h-1-i-j)+1)y_{h-1-i-j} = (h)y_{h-1-i-j} = (h)y_c > (i+j)w$$

$$\geq (i+j+(h-1-i-j))y_{h-1-i-j} = (h-1)y_{h-1-i-j}.$$

This gives a contradiction. □

For $w \in F_0$, we define an m chain set for w to be a set of m disjoint sequences $\{i_{uv}, 1 \leq v \leq \alpha_u\}$ for $1 \leq u \leq m$ such that $i_{uv} \in \{1,\ldots,n\}$ and i_{uv} are all distinct, and for every u,v with $1 \leq u \leq m$ and $1 < v \leq \alpha_u$, we have $(i_{uv})w - (i_{u,v-1})w > n$. We call an m chain set for w saturated if it satisfies the further conditions:

(i) For every u,v with $1 \leq u \leq m$ and $1 \leq v < \alpha_u$,

$$i_{uv} = \max\,\{h \mid 1 \leq h \leq n,\ (i_{u,v+1})w-(h)w > n,\ h \notin \{i_{pq},\ 1 \leq p < u,\ 1 \leq q \leq \alpha_p\}\}$$

(ii) $$\begin{cases} i_{1\alpha_1} = n \\ i_{u\alpha_u} = \max\,\{h \mid 1 \leq h \leq n,\ h \notin \{i_{pq},\ 1 \leq p < u,\ 1 \leq q \leq \alpha_p\}\},\ 1 < u \leq m. \end{cases}$$

(iii) $\{h \mid 1 \leq h \leq n,\ (i_{u,1})w-(h)w > n,\ h \notin \{i_{pq},\ 1 \leq p < u,\ 1 \leq q \leq \alpha_p\}\} = \emptyset,\ 1 \leq u \leq m$

(iv) $\displaystyle\sum_{j=1}^{m} \alpha_u = n.$

Clearly, in this case, both m and α_u, $1 \leq u \leq m$, are entirely determined by w. So we call such an m chain set the saturated chain set for w and denote it by $N^0(w)$. Obviously, $\alpha_1 \geq \ldots \geq \alpha_m$, so $N^0(w)$ determines a partition $\{\alpha_1 \geq \ldots \geq \alpha_m\}$ of n. In such a way, we get a map $F_0 \xrightarrow{\phi_0} \Lambda_n$ by sending w to $\{\alpha_1 \geq \ldots \geq \alpha_m\}$.

For example, assume $n = 5$ and that $w \in F_0$ has the following form:

1st column

1st row
$$\begin{pmatrix} 1 & & & & \\ & 1 & & & \\ & & 1 & & \\ & & & 1 & \\ & & & & 1 \end{pmatrix}$$

Then $\{1 < 3 < 5\} \cup \{2 < 4\}$ is the saturated chain set and so $\phi_0(w) = \{3 \geq 2\} \in \Lambda_5$.

Let Φ be the automorphism of A_n such that $\Phi(s_i) = s_{i+1}$ for all $i \in \mathbb{Z}$. Then Φ induces a bijective map from F to itself by restriction to F which sends F_i onto F_{i+1}, $0 \leq i \leq n-1$, with the convention that $F_n = F_0$. So we also get a map $F_i \xrightarrow{\phi_i} \Lambda_n$ such that the diagram

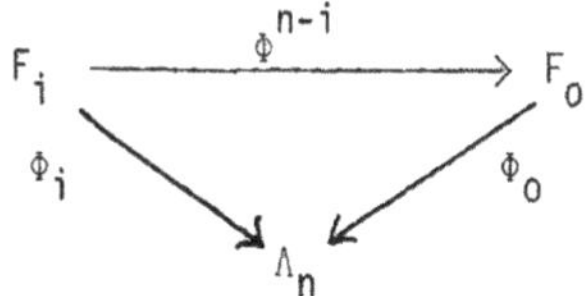

commutes.

Lemma 9.2.2 For any $w \in F_o \cap \sigma^{-1}(\lambda)$, we have $\phi_o(w) = \lambda$.

Proof: Let $N(w) = \bigcup_{u=1}^{m} \{i_{u1} < i_{u2} < \ldots < i_{u\alpha_u}\}$ be the saturated chain set for w. Let $i'_{uv} = i_{uv} - (v-1)n$ for $1 \leq u \leq m$, $1 \leq v \leq \alpha_u$. Then $i'_{u1} > \ldots > i'_{u,\alpha_u}$ and $(i'_{u1})w < \ldots < (i'_{u,\alpha_u})w$ for any $1 \leq u \leq m$. So $m \geq r$ and $\sum_{t=1}^{\ell} \alpha_t \leq \sum_{t=1}^{\ell} \lambda_t$ for any $1 \leq \ell \leq r$. Now it suffices to show that for any $1 \leq \ell \leq r$, $\sum_{t=1}^{\ell} \lambda_t \leq \sum_{t=1}^{\ell} \alpha_t$.

First we claim that for any $i = pn + a$, $j = qn + b$ with $1 \leq a, b \leq n$ and $p,q \in \mathbb{Z}$, if $i < j$ and $(i)w > (j)w$, we must have $a > b$, $p < q$ and $(a)w - (b)w > n$. For

$(i)w > (j)w \Rightarrow (a)w > (q-p)n + (b)w$

$\Rightarrow (a)w - (b)w > (q-p)n \geq 0$ since $i < j$

$\Rightarrow (a)w > (b)w$

$\Rightarrow a > b$ by our hypothesis on w

$\Rightarrow p < q$ since $i < j$

$\Rightarrow (a)w - (b)w > n$.

Now suppose $S = S_1 \cup \ldots \cup S_\ell \subset \mathbb{Z}$ satisfies $C_n(w,\ell)$ with $S_t = \{(h_{t1})w > \ldots > (h_{t\beta_t})w | h_{t1} < \ldots < h_{t\beta_t}\}$ for $1 \leq t \leq \ell$ and let h'_{tj} be defined by $\bar{h}'_{tj} = \bar{h}_{tj}$ with $1 \leq h'_{th} \leq n$. Then, by the above claim, we have $(h'_{tj})w - (h'_{t,j+1})w > n$ and $h'_{tj} > h'_{t,j+1}$ for $1 \leq t \leq \ell$ and $1 \leq j < \beta_t$. Let $k_{tu} = h'_{t,\beta_t+1-u}$ for $1 \leq t \leq \ell$ and $1 \leq u \leq \beta_t$. Then $K = \bigcup_{t=1}^{\ell} \{k_{t1} < k_{t2} < \ldots < k_{t\beta_t}\}$ is an ℓ chain set for w and we must show $\sum_{t=1}^{\ell} \beta_t \leq \sum_{t=1}^{\ell} \alpha_t$. If n is not a term of any chain of K, then by replacing k_{1,β_1} by n, we also get an ℓ chain set for w. If n is a term of some chain of K, we may without loss of generality assume that $k_{1\beta_1} = n$. We now define a number h such that $i_{1,\alpha_1+1-j} = k_{1,\beta_1+1-j}$, $1 \leq j < h$ for some $1 < h \leq \alpha_1 + 1$ with h as large as possible. We shall now modify the set K as follows. If $h \leq \alpha_1$, then when

i_{1,α_1+1-h} is a term of some chain of K, we replace k_{1,β_1+1-h} by i_{1,α_1+1-h} in the case $\beta_1 > h$, or we replace $\{k_{11} < k_{12} < \ldots < k_{1h}\}$ by $\{i_{1,\alpha_1+1-h} < i_{1,\alpha_1+2-h} < \ldots < i_{1,\alpha_1}\}$ in the case $\beta_1 = h-1$. If $i_{1,\alpha_1+1-h} = k_{uv}$ for some $1 < u \leq \ell$, $1 \leq v \leq \beta_u$, we replace $\{k_{u1} < \ldots < k_{uv} < k_{1,\beta_1+2-h} < \ldots < k_{1,\beta_1}\}$ by $\{k_{11} < \ldots < k_{1,h-1+v}\}$, $\{k_{11} < \ldots < k_{1,\beta_1+1-h} < k_{u,v+1} < \ldots < k_{u,\beta_u}\}$ by $\{k_{u1} < \ldots < k_{u,\beta_1+\beta_u+1-h-v}\}$, $h-1+v$ by β_1 and $\beta_1 + \beta_u+1-h-v$ by β_u, we also get an ℓ chain set for w. If $h = \alpha_1+1$, then by definition of the saturated chain set for w, we have $\beta_1 = \alpha_1$. In such a way we can make $\beta_1 = \alpha_1$ and replace $\{k_{11} < \ldots < k_{1,\beta_1}\}$ by $\{i_{11} < \ldots < i_{1\alpha_1}\}$. The cardinal of this new K is equal to or greater than that of the original one. Similarly, we can make $\beta_v = \alpha_v$ and $\{k_{v1} < \ldots < k_{v\beta_v} = \{i_{v1} < \ldots < i_{v\alpha_v}\}$ for all $1 < u \leq \ell$. But now we have $|K| = \sum_{t=1}^{\ell} \alpha_t$ and so the cardinal of the initial K is not greater than $\sum_{t=1}^{\ell} \alpha_t$. Our assertion then follows. □

Corollary 9.2.3 *For any $0 \leq i \leq n-1$, $w \in F_i \cap \sigma^{-1}(\lambda)$, we have $\phi_i(w) = \lambda$.*

Proof: $\phi_i(w) = \phi_0(\phi^{n-i}(w))$. Since $w \in \sigma^{-1}(\lambda)$ implies $\phi^{n-i}(w) \in \sigma^{-1}(\lambda)$, our conclusion follows by Lemma 9.2.2. □

§9.3 THE SUBSET H_λ OF $\sigma^{-1}(\lambda)$

Let H_λ be the set of all elements w of $\sigma^{-1}(\lambda)$ which have a full MDC form $(A_r,\ldots,A_1)$ at i for some $i \in \mathbf{Z}$ with $|A_h| = \lambda_h$, $1 \leq h \leq r$. Such an MDC form of $w \in H_\lambda$ is called a standard MDC form of w. By Lemma 4.2.4, $w \in H_\lambda$ implies that $\pi(\mathcal{L}(w)) = \lambda$.

Suppose $n = 5$, $\lambda = \{3 > 2\}$. Then the following element $w \in A_5$ lies in the set H_λ.

$$(i+1)^{\text{th}}\text{ row}\quad \begin{pmatrix} & & & 1 & \\ & 1 & & & \\ & & & & 1 \\ & & 1 & & \\ 1 & & & & \end{pmatrix} \begin{matrix} \}\, A_2(w) \\ \\ \}\, A_1(w) \\ \\ \end{matrix}$$

$$w$$

Lemma 9.3.1 *Let $w \in F \cap \sigma^{-1}(\lambda)$. Then there exists $y \in A_n$ with $y \underset{P_L}{\sim} w$ such that for some $j \in \mathbf{Z}$,*

$(j+1)y < (j+2)y < \ldots < (j+n-\lambda_1+1)y > (j+n-\lambda_1+2)y > \ldots > (j+n)y$.

Proof: This is trivial for $w = 1$. Now assume $w \neq 1$. Then we have $(i+n)w - (i+1)w > n$ with $(i+1)w < \dots < (i+n)w$ for some $0 \leq i \leq n-1$. Let $i+1 \leq i_1 < i_2 < \dots < i_\alpha = i+n$ such that for each $1 < t \leq \alpha$, $(i_t)w - (i_{t-1})w > n$ but $(i_t)w - (i_{t-1}+1)w < n$. By Corollary 9.2.3, we have $\alpha = \lambda_1$. We know that $(i)w > (i+1)w$. By Lemma 9.1.2, there exists $x_1 \in A_n$ with $w \xrightarrow{*(i,2,n-2)} x_1$ satisfying $(i_{\alpha-1})w = (i+n-1)x_1 < (i+n-2)x_1$ and $(i)x_1 < (i+1)x_1 < \dots < (i+n-2)x_1$ with $(k-1)x_1 = (k)w$ for all $i+1 \leq k < i_{\alpha-1}$. If $\alpha \geq 3$, then $(i)x_1 = (i+1)w < (i_{\alpha-1}-n)w = (i-1)x_1 < (i-2)x_1$. Also by Lemma 9.1.2, there exists $x_2 \in A_n$ with $x_1 \xrightarrow{*(i-2,3,n-3)} x_2$ satisfying

$$(i_{\alpha-2})w = (i_{\alpha-2}-1)x_1 = (i + n - \sum_{t=1}^{2} t)x_2 < (i + n-1 - \sum_{t=1}^{2} t)x_2$$

$$< (i + n-2 - \sum_{t=1}^{2} t)x_2 > (i+n-3 - \sum_{t=1}^{2} t)x_2 > \dots > (i+1 - \sum_{t=1}^{2} t)x_2$$

with $(k-2)x_2 = (k)x_1$ for all $i \leq k < i_{\alpha-2}-1$. That is, $(k - \sum_{t=1}^{2} t)x_2 = (k)w$ for all $i+1 \leq k < i_{\alpha-2}$. In general, suppose we have $x_1,\dots,x_\ell$ with $\ell \leq \alpha -2$ such that for each $1 < u \leq \ell$, the element x_u with

$$x_{u-1} \xrightarrow{*(i+1 - \sum_{t=1}^{u} t,\ u+1,\ n-u-1)} x_u$$

satisfies $(i_{\alpha-u})w = (i+n - \sum_{t=1}^{u} t)x_u < (i+n-1 - \sum_{t=1}^{u} t)x_u < \dots < (i+n-u - \sum_{t=1}^{u} t)x_u$

$> (i+n-u-1) - \sum_{t=1}^{u} t)x_u > \dots > (i+1 - \sum_{t=1}^{u} t)x_u$ with $(k - \sum_{t=1}^{u} t)x_u = k(w)$ for all $i + 1 \leq k < i_{\alpha-u}$. Then

$$(i + 1 - \sum_{t=1}^{\ell} t)x_\ell = (i+1)w < (i_{\alpha-\ell}-n)w = (i - \sum_{t=1}^{\ell} t)x_\ell < (i-1 - \sum_{t=1}^{\ell} t)x_\ell < \dots <$$

$(i+1 - \sum_{t=1}^{\ell+1} t)x_\ell$ and $(i_{\alpha-\ell-1} - \sum_{t=1}^{\ell} t)x_\ell = (i_{\alpha-\ell-1})w$ is the greatest number among the set $\{(v - \sum_{t=1}^{\ell} t)x_\ell \mid i+1 \leq v < i_{\alpha-\ell}\}$ which is smaller than $(i_{\alpha-\ell}-n - \sum_{t=1}^{\ell} t)x_\ell = (i_{\alpha-\ell}-n)w$. So by Lemma 9.1.2, there exists $x_{\ell+1} \in A_n$ with

$$x_\ell \xrightarrow{*(i+1 - \sum_{t=1}^{\ell+1} t,\ \ell+2,\ n-\ell-2)} x_{\ell+1}$$ satisfying

$$(i_{\alpha-\ell-1})w = (i_{\alpha-\ell-1} - \sum_{t=1}^{\ell} t)x_{\ell} = (i+n - \sum_{t=1}^{\ell+1} t)x_{\ell+1} < (i+n-1 - \sum_{t=1}^{\ell+1} t)x_{\ell+1} < \dots$$

$$< (i+n-\ell-1 - \sum_{t=1}^{\ell+1} t)x_{\ell+1} > (i+n-\ell-2 - \sum_{t=1}^{\ell+1} t)x_{\ell+1} > \dots > (i+1 - \sum_{t=1}^{\ell+1} t)x_{\ell+1}$$

with $(k - \sum_{t=1}^{\ell+1} t)x_{\ell+1} = (k)w$ for all $i+1 \leq k < i_{\alpha-\ell-1}$. Clearly, if $\ell < \alpha-3$, then

$$(i+1 - \sum_{t=1}^{\ell+1} t)x_{\ell+1} = (i+1)w < (i_{\alpha-\ell-1}-n)w = (i - \sum_{t=1}^{\ell+1} t)x_{\ell+1} < (i-1 - \sum_{t=1}^{\ell+1} t)x_{\ell+1} < \dots$$

$< (i+1 - \sum_{t=1}^{\ell+2} t)x_{\ell+1}$ and the same procedure can be repeated. Finally, we get $y = x_{\alpha-2}$ and take $j = i - \sum_{t=1}^{\alpha-1} t$. Then y has the required property. □

Suppose $n = 5$ and that $w \in A_5$ is the following matrix.

(i+1)th row

w

Then $w \in F \cap \sigma^{-1}(\{3 > 2\})$. There exists x_1 with $w \xrightarrow{*(i,2,3)} x_1$ which is as follows.

(i-2)th row

x_1

with $(i)x_1 = (i+1)w$. So x_1 here plays the role of y in Lemma 9.3.1.

Assume $w \in \sigma^{-1}(\lambda)$ such that for some $i, m \in \mathbb{Z}$ with $1 \leq m < r$, we have

$$\begin{cases} (i+1)w < \dots < (i+n+1 - \sum_{t=1}^{m} \lambda_t)w \\ (i+n+1 - \sum_{t=1}^{\ell} \lambda_t)w > \dots > (i+n - \sum_{t=1}^{\ell-1} \lambda_t)w, \ 1 \le \ell \le m \end{cases}$$

Let $i+1 \le i_1 < i_2 < \dots < i_\alpha = i+n - \sum_{t=1}^{m} \lambda_t$ such that

$$\begin{cases} \beta_{m+1,j} = i_j, \ 1 \le j \le \alpha, \text{ all exist} \\ (\beta_{\ell j})w = \min \{(i+n - \sum_{t=1}^{\ell} \lambda_t + v)w \mid 1 \le v \le \lambda_\ell, (i+n - \sum_{t=1}^{\ell} \lambda_t + v)w > (\beta_{\ell+1,j})w\}, \\ \qquad 1 \le j \le \alpha \\ \qquad 1 \le \ell \le m \\ (i_j)w = \max \{(k)w \mid i+1 \le k < i_{j+1}, (\beta_{1,j+1})w - (k)w > n\}, \ 1 \le j \le \alpha \\ \{(k)w \mid i+1 \le k < i_1, (\beta_{1,1})w - (k)w > n\} = \emptyset. \end{cases}$$

Since $w \in \sigma^{-1}(\lambda)$, it follows that once $\beta_{m+1,j}$ exists, all $\beta_{\ell j}$, $1 \le \ell \le m$, automatically exist.

Using the above example and replacing x_1 by w, we have

$$(i+1)^{st}\text{ row} \begin{pmatrix} 1 & & & & \\ & & & 1 & \\ & & & & 1 \\ & & 1 & & \\ & 1 & & & \end{pmatrix}$$

$$w$$

with $m = 1$ and $\alpha = 2$. Thus $\beta_{2,1} = i+1$, $\beta_{2,2} = i+2$, $\beta_{1,1} = i+5$ and $\beta_{1,2} = i+3$.

We shall prove that

Lemma 9.3.2 Let $w \in \sigma^{-1}(\lambda)$ be as above. Then $\alpha = \lambda_{m+1}$.

Before doing this, let us show some simple results.

Lemma 9.3.3 Let $w \in \sigma^{-1}(\lambda)$ be as in Lemma 9.3.2. Then $0 < (\beta_{1j})w - (i_j)w < n$ for any $1 \le j \le \alpha$.

Proof: It suffices to show that $(\beta_{1j})w - (i_j)w < n$ for any $1 \leq j \leq \alpha$. If $(\beta_{1j})w - (i_j)w > n$ for some $1 \leq j \leq \alpha$, let

$$S_u = \{(i+n+1 - \sum_{t=1}^{u} \lambda_t)w > (i+n+2 - \sum_{t=1}^{u} \lambda_t)w > \ldots >$$

$$(\beta_{uj})w > (\beta_{u-1,j}+1)w > \ldots > (i+n - \sum_{t=1}^{u-2} \lambda_t)w\} \text{ for } 1 < u \leq m$$

$$S_1 = \{(i+n+1 - \lambda_1)w > (i+n+2-\lambda_1)w > \ldots >$$

$$(\beta_{1j})w > (i_j+n)w > (\beta_{mj}+1+n)w > (\beta_{mj}+2+n)w > \ldots > (i+2n - \sum_{t=1}^{m-1} \lambda_t)w\}.$$

Then $S = S_1 \cup \ldots \cup S_m$ satisfies $C_n(w,m)$ with $|S| = \sum_{t=1}^{m} \lambda_t + 1$. This contradicts $w \in \sigma^{-1}(\lambda)$. Also, $(\beta_{1j})w - (i_j)w \neq n$ since $(\overline{\beta_{1j}})w \neq (\overline{i_j})w$. Our result follows. □

<u>Corollary 9.3.4</u> <u>Let $w \in \sigma^{-1}(\lambda)$ be as in Lemma 9.3.2. Then for $1 \leq \ell, \ell' \leq m$ and $1 \leq j, j' \leq \alpha$, we have</u>

$$\beta_{\ell j} = \beta_{\ell' j'} \Leftrightarrow \ell = \ell' \text{ and } j = j'.$$

Proof: $(\Leftarrow)$ This is obvious.

$(\Rightarrow)$ It is clear that $\beta_{\ell j} = \beta_{\ell' j'}$ implies $\ell = \ell'$. So it suffices to show that for any $1 \leq \ell \leq m$ and $1 \leq j < \alpha$, $(\beta_{\ell j})w < (\beta_{\ell,j+1})w$ holds. In general, we have $(\beta_{\ell j})w \leq (\beta_{\ell,j+1})w$. If $(\beta_{\ell j})w = (\beta_{\ell,j+1})w$ for some $1 \leq \ell \leq m$, $1 \leq j < \alpha$, then we have $(\beta_{1j})w = (\beta_{1,j+1})w$. But $(\beta_{1j})w - (i_j)w < n$ and $(\beta_{1,j+1})w - (i_j)w > n$. This gives a contradiction. □

<u>Lemma 9.3.5</u> <u>Let $w \in \sigma^{-1}(\lambda)$ be as in Lemma 9.3.2. Then $\alpha \leq \lambda_{m+1}$.</u>

Proof: We shall define sets $S_1, S_2, \ldots, S_m, S_{m+1}$ as follows. From $1 \leq u \leq m$, let

$$S_u = \{(i+n+1 - \sum_{t=1}^{u} \lambda_t)w > (i+n+2 - \sum_{t=1}^{u} \lambda_t)w > \ldots > (\beta_{u\alpha})w$$

$$> (\beta_{u-1,\alpha}+1)w > (\beta_{u-1,\alpha}+2)w > \ldots > (\beta_{k,\alpha-u+k})w$$

$$> (\beta_{k-1,\alpha-u+k}+1)w > (\beta_{k-1,\alpha-u+k})w > \ldots > (\beta_{k',\alpha-u+k'})w$$

$$> (\beta_{k'-1,\alpha-u+k'-1})w > (\beta_{k'-2,\alpha-u+k'-1}+1)w > (\beta_{k'-2,\alpha-u+k'-1}+2)w > \ldots \}$$

$$S_{m+1} = \{(\beta_{m+1,\alpha})w > (\beta_{m,\alpha}+1)w > (\beta_{m,\alpha}+2)w > \ldots > (\beta_{k,\alpha-m-1+k})w$$

$$> (\beta_{k-1,\alpha-m-1+k}+1)w > (\beta_{k-1,\alpha-m-1+k}+2)w > \ldots > (\beta_{k',\alpha-m-1+k'})w$$

$$> (\beta_{k'-1,\alpha-m-1+k'-1})w > (\beta_{k'-2,\alpha-m-1+k'-1}+1)w > (\beta_{k'-2,\alpha-m-1+k'-1}+2)w > \ldots\}$$

where $k \not\equiv 1$, $k' \equiv 1 \pmod{m+1}$ and $\beta_{f,g} = \beta_{f+m+1,g+m+1} + n$. It remains to specify the last term of each S_v. Let $\alpha-v = k(m+1) + q$, $1 \leq q \leq m+1$, $h \in \mathbf{Z}$. Then the last term of S_v, $1 \leq v \leq m+1$, is

$$\begin{cases} (i+(h+2)n - \sum_{t-1}^{m-q} \lambda_t)w \text{ if } \beta_{m+1-q,1} \neq i+n - \sum_{t=1}^{m-q} \lambda_t \text{ and } 1 \leq q \leq m. \\ (\beta_{m+2-q,1} + (h+1)n)w \text{ otherwise.} \end{cases}$$

Then $S = S_1 \cup \ldots \cup S_{m+1}$ satisfies $C_n(w,m+1)$ with

$$\bar{S} = \{\overline{(i+n+1 - \sum_{t=1}^{m} \lambda_t)w},\ \overline{(i+n+2 - \sum_{t=1}^{m} \lambda_t)w}, \ldots, \overline{(i+n)w}, \overline{i_j}, 1 \leq j \leq \alpha\}\ .$$

It follows that $|S| = \sum_{t=1}^{m} \lambda_t + \alpha$. So $\alpha \leq \lambda_{m+1}$ by $w \in \sigma^{-1}(\lambda)$. □

An example of $S = S_1 \cup \ldots \cup S_{m+1}$ in the proof of Lemma 9.3.5 is as follows. Suppose $n = 10$, $\lambda = \{4 > 3 > 3\}$ and that $w \in \sigma^{-1}(\lambda)$ is as below.

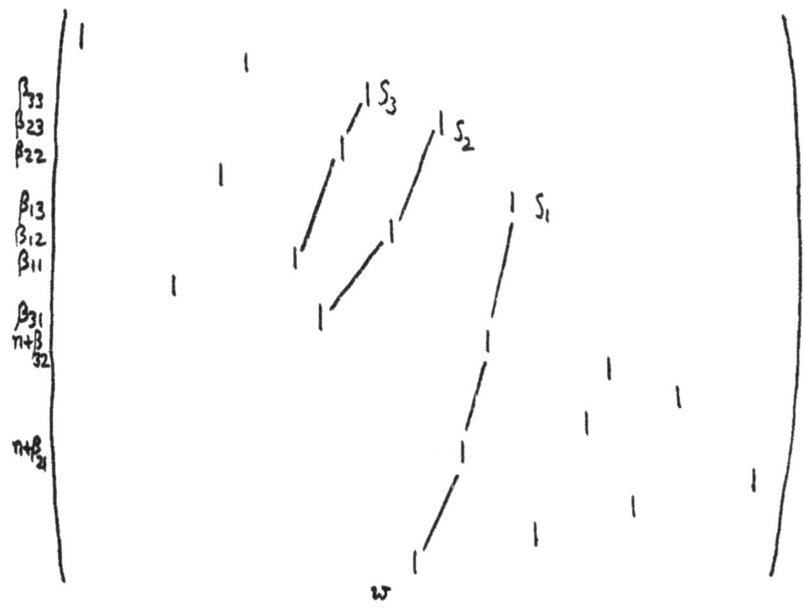

Then w is as in Lemma 9.3.5 with $m = 2$ and $\alpha = 3$. The sets S_1, S_2, S_3 are as in this diagram which consist of all entries of w occurring as the vertices of the corresponding broken lines. We see that the lowest vertex of each broken line is either the last term of some maximal descending chain of w or lies in some congruent row of the $\beta_{\ell,j}$-th row of w for some j,ℓ with $1 \leq j \leq \alpha$ and $1 \leq \ell \leq m+1$.

Proof of Lemma 9.3.2

Let $S = S_1 \cup \ldots \cup S_{m+1}$ be as in the proof of Lemma 9.3.5. It suffices to show that if $S' = S_1' \cup \ldots \cup S_{m+1}'$ satisfies $C_n(w,m+1)$, then $|S'| \leq \sum_{t=1}^{m} \lambda_t + \alpha$. We know that

$$\{(i+1)w, (i+2)w, \ldots, (i+n - \sum_{t=1}^{m} \lambda_t)w\} = \bigcup_{1 \leq t \leq \alpha} \{(j)w \mid i_{t-1} < j \leq i_t\}$$

with the convention that $i_0 = i$. For each $1 \leq t \leq \alpha$, let $U_t = \{(j)w \mid i_{t-1} < j \leq i_t\} \cup \{(\beta_{ht})w \mid 1 \leq h \leq m\}$. By Corollary 9.3.4, $U_t \cap U_{t'} = \emptyset$ for any t, t' with $1 \leq t, t' \leq \alpha$ and $t \neq t'$. It is clear that $\{(\overline{i+j})w \mid 1 \leq j \leq n\} - \bigcup_{t=1}^{\alpha} \bar{U}_t \subseteq \bar{S}$ and $|\bar{U}_t \cap \bar{S}| = m+1$ for any $1 \leq t \leq \alpha$, where $\bar{X} = \{\bar{x} \mid x \in X\}$. So it suffices to show that $|\bar{U}_t \cap \bar{S}'| \leq m+1$ for any $1 \leq t \leq \alpha$. But

$$U_t = \{(i_{t-1}+1)w < (i_{t-1}+2)w < \ldots < (i_t)w < (\beta_{m,t})w < (\beta_{m-1,t})w < \ldots < (\beta_{1,t})w\}$$

where $i+1 \leq i_{t-1}+1 < i_{t-1}+2 < \ldots < i_t < \beta_{m,t} < \beta_{m-1,t} < \ldots < \beta_{1,t} \leq i+n$ and $(\beta_{1t})w - (i_{t-1}+1)w < n$. By Lemma 5.11, our conclusion follows. □

<u>Lemma 9.3.6</u> <u>Let $w \in \sigma^{-1}(\lambda)$ be as in Lemma 9.3.2. Then there exists $y \in A_n$ with $y \underset{P_L}{\sim} w$ such that for some $j \in \mathbf{Z}$,</u>

$$\begin{cases} \underline{(j+1)w < (j+2)w < \ldots < (j+n+1 - \sum_{t=1}^{m+1} \lambda_t)w} \\ \underline{(j+n+1 - \sum_{t=1}^{\ell} \lambda_t)w > \ldots > (j+n - \sum_{t=1}^{\ell-1} \lambda_t)w, \ 1 \leq \ell \leq m+1} \end{cases}$$

<u>Proof</u>: If $\alpha = 1$, then $y = w$ is as required. Now assume $\alpha > 1$. By Lemma 8.2.2, there exists a sequence of elements $x_{10} = w, x_{11}, \ldots, x_{1m}$ such that for each $1 \leq \ell \leq m$, we have

$$x_{1\ell} \xleftarrow{*(i+n+1-\sum_{t=1}^{m+1-\ell}\lambda_t,\lambda_{m+1-\ell})} x_{1,\ell-1}.$$ By Lemma 9.1.1, $(i)x_{1m} = (\beta_{1\alpha})w-n >$ $(i_{\alpha-1})w \geq (i+1)w = (i+1)x_{1m}$. So by Lemma 9.1.2, there exists $x_{1,m+1}$ with $x_{1m} \xrightarrow{*(i,2,n-\sum_{t=1}^{m}\lambda_t-2)} x_{1,m+1}$ having the following properties:

(i) $(i_{\alpha-1})w = (i+n-1-\sum_{t=1}^{m}\lambda_t)x_{1,m+1} > (i+n-2-\sum_{t=1}^{m}\lambda_t)x_{1,m+1}$.

(ii) $(i+n-2-\sum_{t=1}^{m}\lambda_t)x_{1,m+1} > (i+n-3-\sum_{t=1}^{m}\lambda_t)x_{1,m+1} > \ldots > (i)x_{1,m+1}$

with $(k-1)x_{1,m+1} = (k)w$ for all k with $i+1 \leq k < i_{\alpha-1}$ or $\beta_{\ell\alpha} < k \leq i+n-\sum_{t=1}^{\ell-1}\lambda_t, 1\leq\ell\leq m$.

(iii) $(i+n-\sum_{t=1}^{\ell}\lambda_t)x_{1,m+1} > \ldots > (i+n-1-\sum_{t=1}^{\ell-1}\lambda_t)x_{1,m+1}$ for $1 \leq \ell \leq m$.

If $\alpha > 2$, then there exists elements $x_{20} = x_{1,m+1}, x_{21},\ldots,x_{2m}$ such that for each $1 \leq \ell \leq m$, we have $x_{2\ell} \xleftarrow{*(i+n-\sum_{t=1}^{m+1-\ell}\lambda_t,\ m+1-\ell,\ 2)} x_{2,\ell-1}$. By Lemma 9.1.1,

$(i-2)x_{2m} > (i-1)x_{2m} = (\beta_{1,\alpha-1})w-n > (i_{\alpha-2})w \geq (i+1)w = (i)x_{1,m+1} = (i)x_{2,m}$.

So by Lemma 9.1.2, there exists $x_{2,m+1}$ with $x_{2,m} \xrightarrow{*(i-2,3,n-\sum_{t=1}^{m}\lambda_t-3)} x_{2,m+1}$ satisfying the properties as follows:

(i) $(i_{\alpha-2})w = (i+n-3-\sum_{t=1}^{m}\lambda_t)x_{2,m+1} < (i+n-4-\sum_{t=1}^{m}\lambda_t)x_{2,m+1}$

$$< (i+n-5-\sum_{t=1}^{m}\lambda_t)x_{2,m+1}.$$

(ii) $(i+n-5-\sum_{t=1}^{m}\lambda_t)x_{2,m+1} > (i+n-6-\sum_{t=1}^{m}\lambda_t)x_{2,m+1} > \ldots > (i-2)x_{2,m+1}$

with $(k-\sum_{t=1}^{2}t)x_{2,m+1} = (k-1)x_{1,m+1} = (k)w$ for all k with $i+1 \leq k < i_{\alpha-2}$ or $\beta_{\ell,\alpha-1} < k \leq i+n-\sum_{t=1}^{\ell-1}\lambda_t,\ 1 \leq \ell \leq m$.

(iii) $(i+n-2-\sum_{t=1}^{\ell}\lambda_t)x_{2,m+1} > \ldots > (i+n-3-\sum_{t=1}^{\ell-1}\lambda_t)x_{2,m+1}$ for $1 \leq \ell \leq m$.

In general, if we have got $x_{h,m+1}$, $h < \alpha-1$, satisfying

(i) $(i_{\alpha-h})w = (i+n-\sum_{t=1}^{h} t - \sum_{t=1}^{m}\lambda_t)x_{h,m+1} < (i+n-1-\sum_{t=1}^{h} t - \sum_{t=1}^{m}\lambda_t)x_{h,m+1}$

$< \dots < (i+n-h-\sum_{t=1}^{h} t - \sum_{t=1}^{m}\lambda_t)x_{h,m+1}.$

(ii) $(i+n-h-\sum_{t=1}^{h} t - \sum_{t=1}^{m}\lambda_t)x_{h,m+1} > (i+n-h-1-\sum_{t=1}^{h} t - \sum_{t=1}^{m}\lambda_t)x_{h,m+1}$

$> \dots > (i+1-\sum_{t=1}^{h} t)x_{h,m+1}$ with $(k-\sum_{t=1}^{h} t)x_{h,m+1} = k(w)$ for all k with

$i+1 \leq k < i_{\alpha-h}$ or $\beta_{\ell,\alpha-h+1} < k \leq i+n-\sum_{t=1}^{\ell-1}\lambda_t$, $1 \leq \ell \leq m$.

(iii) $(i+n+1-\sum_{t=1}^{h} t - \sum_{t=1}^{\ell}\lambda_t)x_{h,m+1} > \dots > (i+n-\sum_{t=1}^{h} t - \sum_{t=1}^{\ell-1}\lambda_t)x_{h,m+1}$

for $1 \leq \ell \leq m$. Then there exists elements $x_{h+1,o} = x_{h,m+1}, x_{h+1,1},\dots,x_{h+1,m}$ such that for each $1 \leq \ell \leq m$, we have

$$x_{h+1,\ell} \xleftarrow{*(i+n+1-\sum_{t=1}^{h} t - \sum_{t=1}^{m+1-\ell}\lambda_t,\ \lambda_{m+1-\ell},\ h+1)} x_{h+1,\ell-1}.$$

By Lemma 9.1.1, we see that

$$(i+1-\sum_{t=1}^{h+1} t)x_{h+1,m} > (i+2-\sum_{t=1}^{h+1} t)x_{h+1,m} > \dots > (i-\sum_{t=1}^{h} t)x_{h+1,m}$$

$$= (\beta_{1,\alpha-h})w-n > (i_{\alpha-h-1})w \geq (i+1)w = (i+1-\sum_{t=1}^{h} t)x_{h+1,m}.$$

So by Lemma 9.1.2, there exists $x_{h+1,m+1}$ with

$$x_{h+1,m} \xrightarrow{*(i+1-\sum_{t=1}^{h+1} t,\ h+2,\ n-\sum_{t=1}^{m}\lambda_t-h-2)} x_{h+1,m+1}$$

satisfying the properties as below:

(i) $(i_{\alpha-h-1})w = (i+n-\sum_{t=1}^{h+1} t - \sum_{t=1}^{m}\lambda_t)x_{h+1,m+1} < (i+n-1-\sum_{t=1}^{h+1} t -$

$\sum_{t=1}^{m}\lambda_t)x_{h+1,m+1} < \dots < (i+n-h-1-\sum_{t=1}^{h+1} t - \sum_{t=1}^{m}\lambda_t)x_{h+1,m+1}.$

(ii) $(i+n-h-1 - \sum_{t=1}^{h+1} t - \sum_{t=1}^{m} \lambda_t)x_{h+1,m+1} > (i+n-h-2 - \sum_{t=1}^{h+1} t - \sum_{t=1}^{m} \lambda_t)x_{h+1,m+1}$

$> \ldots > (i+1 - \sum_{t=1}^{h+1} t)x_{h+1,m+1}$ with $(k - \sum_{t=1}^{h+1} t)x_{h+1,m+1} = (k)w$ for all k

with $i+1 \leq k < i_{\alpha-h-1}$ or $\beta_{\ell,\alpha-h} < k \leq i+n - \sum_{t=1}^{\ell-1} \lambda_t$, $1 \leq \ell \leq m$.

(iii) $(i+n+1 - \sum_{t=1}^{h+1} t - \sum_{t=1}^{\ell} \lambda_t)x_{h+1,m+1} > \ldots > (i+n - \sum_{t=1}^{h+1} t - \sum_{t=1}^{\ell-1} \lambda_t)x_{h+1,m+1}$

for $1 \leq \ell \leq m$. So if $h + 1 < \alpha-1$, such a recurring procedure can be carried on. Finally, we get $y = x_{\alpha-1,m+1}$. By Lemma 9.3.2, y is as required. □

As in the example for the proof of Lemma 9.3.5, let $x_{10} = w$, x_{11}, x_{12}, x_{13} be a sequence such that

$$x_{11} \xleftarrow{*(i+4,\lambda_2)} x_{10}, \quad x_{12} \xleftarrow{*(i+7,\lambda_1)} x_{11}, \quad x_{12} \xrightarrow{*(i,2)} x_{13}.$$

We have x_{13} as follows:

(i+1)

Then let $x_{20} = x_{13}$, x_{21}, x_{22}, x_{23} be a sequence such that

$$x_{21} \xleftarrow{*(i+3,\lambda_2,2)} x_{20}, \quad x_{22} \xleftarrow{*(i+6,\lambda_1,2)} x_{21}, \quad x_{22} \xrightarrow{*(i-2,3,0)} x_{23}.$$

We get the following x_{23}

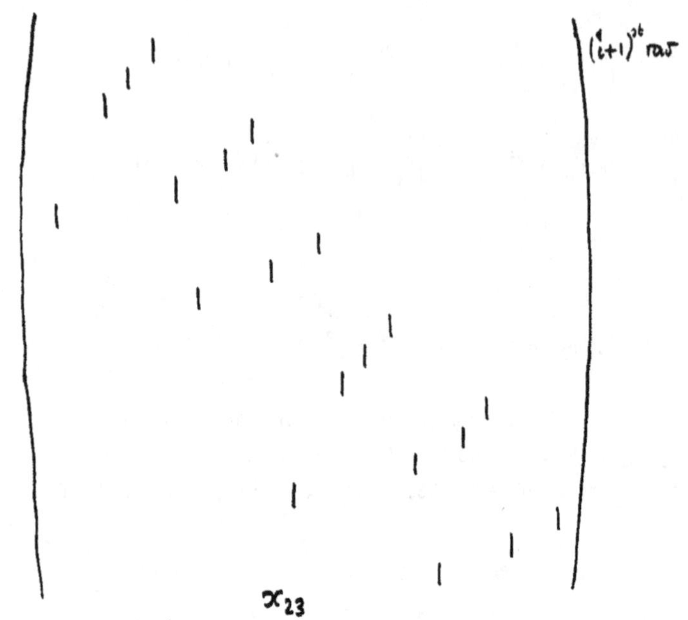

which is required.

Now we can prove our main result in this chapter.

<u>Proposition 9.3.7</u> <u>For any $w \in \sigma^{-1}(\lambda)$, there exists $y \in H_\lambda$ with $y \underset{P_L}{\sim} w$.</u>

<u>Proof</u>: By Lemma 9.2.1, there exists $x \in F$ such that $x \underset{P_L}{\sim} w$. By Lemma 5.8 $x \in F \cap \sigma^{-1}(\lambda)$. It follows from Lemma 9.3.1 and 9.3.6 that there exists y with $y \underset{P_L}{\sim} x$ such that for some $j \in \mathbf{Z}$ and all $1 \leqslant \ell \leqslant r$, we have

$$(j+1+\sum_{t=\ell+1}^{r} \lambda_t)y > (j+2+\sum_{t=\ell+1}^{r} \lambda_t)y > \ldots > (j+\sum_{t=\ell}^{r} \lambda_t)y. \quad \text{i.e. } y \in H_\lambda. \quad \square$$

§9.4 <u>$\sigma^{-1}(\lambda)$ IS A UNION OF RL-EQUIVALENCE CLASSES</u>

Recall the notations $\tilde{\Gamma}$, $\underset{L}{\sim}$, $\underset{R}{\sim}$ and $\underset{RL}{\sim}$ introduced in Chapter 1. Now we shall show an important result on the RL-equivalence relation.

<u>Proposition 9.4.1</u> <u>For $y, w \in A_n$, $y \underset{RL}{\sim} w$ implies $\sigma(y) = \sigma(w)$.</u>

<u>Proof</u>: It suffices to prove that $y \underset{R}{\sim} w$ implies $\sigma(y) = \sigma(w)$. By Proposition 9.3.7, there exists a sequence of elements $y_0 = y, y_1, \ldots, y_r$ in A_n such that for each $1 \leqslant j \leqslant r$, $y_j = {}^*y_{j-1}$ in $\mathcal{D}_L(s_{i_j})$ for some $s_{i_j} \in \Delta$ and $\pi(\mathcal{L}(y_r)) = \sigma(y_r) = \sigma(y)$. Since $w \underset{R}{\sim} y$, by Theorems 1.5.2(ii) and 1.6.2(i), there also exists a sequence of elements $w_0 = w, w_1, \ldots, w_r$ in A_n such that for each $1 \leqslant j \leqslant r$, $w_j = {}^*w_{j-1}$ in $\mathcal{D}_L(s_{i_j})$

and for each $0 < k < r$, $\mathcal{L}(w_k) = \mathcal{L}(y_k)$. In particular, $\pi(\mathcal{L}(w_r)) = \pi(\mathcal{L}(y_r)) = \sigma(y)$. By Lemma 5.7, $\sigma(w) = \sigma(w_r) \geq \pi(\mathcal{L}(w_r)) = \sigma(y)$. By symmetry, $\sigma(y) \geq \sigma(w)$. Therefore, we get $\sigma(y) = \sigma(w)$. □

By Proposition 9.4.1, we see that for any $\lambda \in \Lambda_n$, $\sigma^{-1}(\lambda)$ is a union of some RL-equivalence classes of A_n.

CHAPTER 10 : THE SET N_λ OF NORMALIZED ELEMENTS OF $\sigma^{-1}(\lambda)$

In Chapter 9, we have shown that any element $w \in \sigma^{-1}(\lambda)$ is P_L-equivalent to some element of H_λ. Now we shall prove that any element $w \in H_\lambda$ is P_L-equivalent to some normalized element defined below. A normalized element has a quite simple form and good properties which will be useful to us later.

In 9.3, we have introduced the set H_λ. Here we give a lemma for H_λ. Let $\lambda = \{\lambda_1 \geq \dots \geq \lambda_r\} \in \Lambda_n$ and let $\mu = \{\mu_1 \geq \dots \geq \mu_m\} \in \Lambda_n$ be the dual partition of λ.

<u>Lemma 10.1</u> <u>Assume that $w \in H_\lambda$ has a standard MDC form $(A_r,\dots,A_1)$ at $i \in \mathbb{Z}$. Then for any $1 \leq t \leq \lambda_1$,</u>

$$\underline{j^t_{\mu_t}(w) < j^t_{\mu_t-1}(w) < \dots < j^t_1(w)}$$

<u>Proof:</u> Otherwise, there exists some t, h with $1 \leq t \leq \lambda_1$ and $1 \leq h < \mu_t$ such that $j^t_h(w) < j^t_{h+1}(w)$. Let

$$S_v = \{e((w),\ j^u_v(w)) \mid 1 \leq u \leq \lambda_v\},\ 1 \leq v < h$$

$$S_h = \{e((w),\ j^u_{h+1}(w)),\ e((w),\ j^{u'}_h(w)) \mid 1 \leq u \leq t,\ t \leq u' \leq \lambda_h\}$$

Then $S = S_1 \cup \dots \cup S_h$ satisfies $C_n(w,h)$ with $|S| = \sum_{u=1}^{h} \lambda_u + 1$. This contradicts $w \in \sigma^{-1}(\lambda)$. So our proof is complete. □

<u>Definition 10.2</u> Let N_λ be the set of all elements $w \in H_\lambda$, called normalized elements of $\sigma^{-1}(\lambda)$, such that w has a standard MDC form $(A_r,\dots,A_1)$ at $i \in \mathbf{Z}$ which is normal. In that case, if for any $1 < u \leq \lambda_1$, either $j^u_t(w) > j^{u-1}_{t+1}(w)$ for some $1 \leq t < \mu_u$ or $j^u_{\mu_u}(w) > j^{u-1}_1(w)-n$, then w is called a principal normalized element of $\sigma^{-1}(\lambda)$. Let $\bar{N}_\lambda$ be the set of all principal normalized elements of $\sigma^{-1}(\lambda)$.

Strictly speaking, we shall call an element defined above a left normalized element (or a left principal normalized element), since we can also define a right normalized element (or a right principal normalized element) to be an element which is obtained by transposing a left normalized element (or a left principal normalized element). But in this book, we shall only consider left normalized elements and left principal normalized elements. So the word "left" here is usually omitted.

Now we shall give some examples of elements in N_λ or in $\bar{N}_\lambda$.

(i) Assume that $J^i = J_1^i \cup \dots \cup J_r^i \in \underline{\Delta}$ with $J_j^i = \{s_{\alpha_j+1}, s_{\alpha_j+2}, \dots, s_{\alpha_j+\lambda_j-1}\}$, $1 \leqslant j \leqslant r$ and $\alpha_j = i + \sum_{h=j+1}^{r} \lambda_h$ for some $i \in \mathbf{Z}$. Let $w_0^{J^i}$ be the longest element in W_{J^i}. Then by Lemma 5.12, we have $w_0^{J^i} \in \sigma^{-1}(\lambda)$. So by observing its matrix, we can see that $w_0^{J^i}$ has a full MDC form $(A_r, A_{r-1}, \dots, A_1)$ at i with $|A_j(w^{J^i}_0)| = \lambda_j$ for any $1 \leqslant j \leqslant r$. We also see that the t-th entry of $A_h(w_0^{J^i})$ is $e((w_0^{J^i}), j_h^t(w_0^{J^i})) = e(i + \sum_{u=h+1}^{r} \lambda_u + t, i + \sum_{u=h}^{r} \lambda_u + 1 - t)$ for any h, t with $1 \leqslant h \leqslant r$ and $1 \leqslant t \leqslant \lambda_h$. We can check that for any $1 \leqslant t \leqslant \lambda_1$, $j_1^t(w_0^{J^i}) - n < j^t_{\mu_t}(w_0^{J^i}) < j^t_{\mu_t-1}(w_0^{J^i}) < \dots < j_1^t(w_0^{J^i})$. Moreover, for $1 \leqslant t < \lambda_1$, if $\mu_t \geqslant 2$, then we have $j_1^{t+1}(w_0^{J^i}) = i + \sum_{u=1}^{r} \lambda_u + 1 - (t+1) > i + \sum_{u=2}^{r} \lambda_u + 1-t = j_2^t(w_0^{J^i})$; if $\mu_t = 1$, then we have $j_1^{t+1}(w_0^{J^i}) = j_1^t(w_0^{J^i}) - 1 > j_1^t(w_0^{J^i}) - n$. So $w_0^{J^i} \in \bar{N}_\lambda$. The elements $w_0^{J^i}$, $0 \leqslant i < n$, in $\bar{N}_\lambda$ will play an important role in Chapter 14.

(ii) If $\lambda = \{n\}$, then we have $N_\lambda = H_\lambda$. An element $w \in N_\lambda$ lies in $\bar{N}_\lambda$ if and only if w has a standard MDC form (A) at $i \in \mathbf{Z}$ with $j_A^t(w) - j_A^{t+1}(w) < n$ for all $1 \leqslant t < n$.

We know that for any $1 \leqslant t, t' \leqslant n$ with $t \neq t'$, $\overline{j_A^t(w)} \neq \overline{j_A^{t'}(w)}$. On the other hand, if we are given a permutation of $\bar{1}, \bar{2}, \dots, \bar{n}$, say $\bar{i}_1, \bar{i}_2, \dots, \bar{i}_n$, then there exists a unique element $w \in A_n$ which has an MDC form (A) at i for some $i \in \mathbf{Z}$ with $|A(w)| = n$ such that $\overline{j_A^u(w)} = \bar{i}_u$ and $j_A^t(w) - j_A^{t+1}(w) < n$ for all u, t with $1 \leqslant u \leqslant n$ and $1 \leqslant t < n$. This implies that $|\bar{N}_{\{n\}}| = n!$. Later on, we shall see that $|\bar{N}_{\{n\}}|$ is just the number of left cells in $\sigma^{-1}(\{n\})$. But this result cannot be extended to the general case when $\lambda \in \Lambda_n$ is arbitrary.

(iii) If $\lambda = \{2 \geqslant 1 \geqslant \dots \quad 1\}$, then $H_\lambda = \sigma^{-1}(\lambda)$. Suppose $w \in \sigma^{-1}(\lambda)$ with $s_t \in \mathcal{L}(w)$. Then $w \in N_\lambda$ if and only if $(t+2)w > (t)w$. Also, $w \in \bar{N}_\lambda$ if and only if $w \in N_\lambda$ with $\ell(w) < n$.

(iv) Let us consider the case n = 3. There are three partitions $\lambda_1 = \{1 \geqslant 1 \geqslant 1\}$, $\lambda_2 = \{2 \geqslant 1\}$ and $\lambda_3 = \{3\}$ of 3. We have $\sigma^{-1}(\lambda_1) = H_{\lambda_1} = N_{\lambda_1} = \bar{N}_{\lambda_1} = 1$. Secondly, H_{λ_2} is the set of all shaded alcoves in the following diagram:

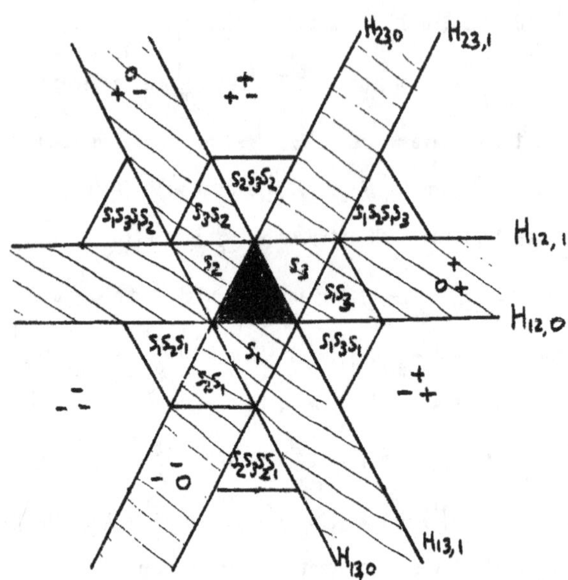

$N_{\lambda_2} = \zeta^{-1}\left({}^{\ o}_{x\ -}\right) \cup \zeta^{-1}\left({}^{\ -}_{-\ o}\right) \cup \zeta^{-1}\left({}^{\ +}_{o\ +}\right) \cup \{s_1,s_2,s_3\}$ and $\bar{N}_{\lambda_2} = \{s_1,s_2,s_3,s_2s_1, s_3s_2,$ $s_1s_3\}$, where $X = {}^{\ \ X_{13}}_{X_{12}\ X_{23}}$ are sign types. Finally, let $H^{-1}_{\lambda_3} = \{w \in A_3 | w^{-1} \in H_{\lambda_3}\}$ and $N^{-1}_{\lambda_3} = \{w \in A_3 | w^{-1} \in N_{\lambda_3}\}$. Then $H^{-1}_{\lambda_3} = N^{-1}_{\lambda_3} = \zeta^{-1}\left({}^{\ +}_{+\ -}\right) \cup \zeta^{-1}\left({}^{\ +}_{-\ +}\right) \cup \zeta^{-1}\left({}^{\ -}_{-\ -}\right)$

and $\bar{N}_{\lambda_3} = \{s_1s_2s_1,\ s_2s_3s_2,\ s_1s_3s_1,\ s_1s_3s_1s_2,\ s_1s_2s_1s_3,\ s_2s_3s_2s_1\}$

Remark In the last example, we see that for any $\lambda \in \Lambda_n$ with $n = 3$ and for any element $w \in \bar{N}_\lambda$, w is the shortest element of $\zeta^{-1}(\zeta(w))$. By using Proposition 7.3.7 we can show that the above conclusion is also true for any $n > 2$.

Lemma 10.3 For $\ell > 2$, assume that $w \in \sigma^{-1}(\lambda)$ has an MDC form $(A_{\ell-1},\dots,A_1,A_\ell)$ at $i \in \mathbb{Z}$ with the following properties:

(i) $|A_t(w)| = \lambda_t$ for $1 < t < \ell$, $|A_\ell(w)| > k+1$ for some $0 < k < \lambda_\ell$.

(ii) For all $1 < h < k$, $j^h_\ell(w) > j^h_1(w) > j^h_2(w) > \dots > j^h_{\ell-1}(w)$.

Then there exists $w' = \rho^{A_\ell}_{A_1,\dots,A_{\ell-1}}(w)$ such that w' has the MDC form $(A_\ell,\dots,A_1)$ at i with

$$\begin{cases} j^{k+1}_\ell(w) - n < j^{k+1}_\ell(w') < j^{k+1}_\ell(w) \\ j^{k+1}_t(w') > j^{k+1}_t(w), \text{ for all } 1 < t < \ell-1. \end{cases}$$

Proof: The existence of w' follows from Lemma 8.3.5.

Let
$$\begin{cases} i_1 = \min\{h \mid k < h \leq \lambda_1,\ j_1^h(w) < j_\ell^{k+1}(w)\} \\ i_u = \min\{h \mid k < h \leq \lambda_u,\ j_u^h(w) < j_{u-1}^{i_{u-1}}(w)\},\ 1 < u \leq \ell-1 \end{cases}$$

Then
$$\begin{cases} j_\ell^{k+1}(w') = j_{\ell-1}^{i_{\ell-1}}(w) \\ j_t^{k+1}(w') = \max\{j_t^{k+1}(w),\ j_{t-1}^{i_{t-1}}(w)\},\ 1 \leq t \leq \ell-1. \end{cases}$$

So it follows that $j_\ell^{k+1}(w') < j_\ell^{k+1}(w)$ and $j_t^{k+1}(w') \geq j_t^{k+1}(w)$ for $1 \leq t \leq \ell-1$. Suppose $j_\ell^{k+1}(w) - n > j_\ell^{k+1}(w')$. i.e. $j_\ell^{k+1}(w) - n > j_{\ell-1}^{i_{\ell-1}}(w)$. Let

$$S_t = \{e((w), j_{t+1}^u(w)),\ e((w), j_t^v(w)) \mid 1 \leq u < i_{t+1},\ i_t \leq v \leq \lambda_t\},\ 1 \leq t < \ell-1$$

$$S_{\ell-1} = \{e((w), j_1^u(w)),\ e((w), j_\ell^{k+1}(w)),\ e((w), j_{\ell-1}^v(w)+n) \mid 1 \leq u < i_1,\ i_{\ell-1} \leq v \leq \lambda_{\ell-1}\}.$$

Then $S = S_1 \cup \ldots \cup S_{\ell-1}$ satisfies $C_n(w, \ell-1)$. But $|S| = \sum_{t=1}^{\ell-1} \lambda_t + 1$. It contradicts $w \in \sigma^{-1}(\lambda)$. So $j_\ell^{k+1}(w) - n < j_\ell^{k+1}(w')$ because $\overline{j_\ell^{k+1}(w)} \neq \overline{j_\ell^{k+1}(w')}$. □

Proposition 10.4 For any $w \in H_\lambda$, there exists $y \in N_\lambda$ such that $y \underset{\rho_L}{\sim} w$.

Proof: Suppose that $w \in H_\lambda$ has a standard MDC form $(A_r, \ldots, A_1)$ at i and suppose that k is the largest number with $0 \leq k \leq \lambda_1$ such that for all h with $1 \leq h \leq k$, we have $(j_1^h(w) - n < j_r^h(w) < j_{r-1}^h(w) < \ldots < j_1^h(w))^{(om)}$. If $k = \lambda_1$, then $w \in N_\lambda$ and the result is trivial. Now assume $k < \lambda_1$. Then there exists v such that v is the smallest number satisfying $1 < v \leq r$ and $j_1^{k+1}(w) - j_v^{k+1}(w) > n$. By Lemmas 8.4.3 and 5.8, there exists $w' = \rho_{A_1, \ldots, A_{v-1}}^{A_r, \ldots, A_v}(w)$ in H_λ. By Corollary 8.4.7 we have

$$(j_1^h(w') - n < j_r^h(w') < j_{r-1}^h(w') < \ldots < j_1^h(w'))^{(om)} \text{ for all } 1 \leq h \leq k \qquad (1)$$

Assume that for some $v \leq m \leq r$, $\lambda_m \geq k+1$ but $\lambda_{m+1} < k+1$ (such an m obviously exists). Let

$$w_1 = \rho_{A_1, \ldots, A_{v-1}}^{A_r, \ldots, A_{m+1}}(w),\quad w_2 = \rho_{A_1, \ldots, A_{v-1}}^{A_m}(w_1), \ldots,$$

$$w_{m+2-v} = \rho_{A_1, \ldots, A_{v-1}}^{A_v}(w_{m+1-v}) = w'.$$

Then

$$(j_1^h(w_1),\ j_2^h(w_1),\dots,j_{v-1}^h(w_1))^{(om)} = (j_1^h(w),\ j_2^h(w),\dots,j_{v-1}^h(w))^{(om)}$$

$$(j_v^h(w_1),\ j_{v+1}^h(w_1),\dots,j_m^h(w_1))^{(om)} = (j_v^h(w),\ j_{v+1}^h(w),\dots,j_m^h(w))^{(om)},\quad k+1 \le h \le \lambda_1$$

$$(j_v^g(w_1) + n,\ j_{v+1}^g(w_1)+n,\dots,j_m^g(w_1)+n,\ j_1^g(w_1),\ j_2^g(w_1),\dots,j_{v-1}^g(w_1),$$

$$j_{m+1}^g(w_1),\ j_{m+2}^g(w_1),\dots,j_r^g(w_1))^{(om)}$$

$$= (j_v^g(w) + n,\ j_{v+1}^g(w)+n,\dots,j_r^g(w)+n,\ j_1^g(w), j_2^g(w),\dots,j_{v-1}^g(w))^{(om)},\quad 1 \le g \le k.$$

So for $1 \le g \le k$,

$$(j_v^g(w_1)+n)-n < j_r^g(w_1) < j_{r-1}^g(w_1) < \dots < j_{m+1}^g(w_1) < j_{v-1}^g(w_1) < j_{v-2}^g(w_1)$$

$$< \dots < j_1^g(w_1) < j_m^g(w_1) + n < j_{m-1}^g(w_1) + n < \dots < j_v^g(w_1) + n)^{(om)}$$

By Lemma 10.3, we have

$$\begin{cases} (j_t^{k+1}(w) + n)-n < j_t^{k+1}(w') < j_t^{k+1}(w) + n & \text{for } v \le t \le m \\ j_{t'}^{k+1}(w') \ge j_{t'}^{k+1}(w) & \text{for } 1 \le t' \le v-1 \end{cases}$$

But it is clear that $j_1^{k+1}(w') = j_1^{k+1}(w)$ by our hypothesis. So this together with Lemma 10.3 implies that

$$\begin{cases} 0 < j_1^{k+1}(w') - j_{t'}^{k+1}(w') \le j_1^{k+1}(w) - j_{t'}^{k+1}(w) & \text{for } 1 < t' \le v-1 \\ 0 < j_1^{k+1}(w') - j_t^{k+1}(w') < j_1^{k+1}(w) - j_t^{k+1}(w) & \text{for } v \le t \le m \end{cases}$$

In particular, $0 < j_1^{k+1}(w') - j_v^{k+1}(w') < j_1^{k+1}(w) - j_v^{k+1}(w)$. (2)

Combining (1), (2), we see that whenever $j_1^{k+1}(w) - j_v^{k+1}(w) > n$, we can always find $w' \in H_\lambda$ by a succession of ρ operations such that w' has a standard MDC form $(A_r,\dots,A_1)$ at i' for some $i' \in \mathbf{Z}$ which is normal for the first k layers with $j_1^{k+1}(w') - j_t^{k+1}(w') < n$ for all $1 < t < v$ and $0 < j_1^{k+1}(w') - j_v^{k+1}(w') \lneqq j_1^{k+1}(w) - j_v^{k+1}(w)$. Applying induction on the difference $j_1^{k+1}(w) - j_v^{k+1}(w) > n$, we can finally find $w' \in H_\lambda$ which has all the above properties and also satisfies $0 < j_1^{k+1}(w') - j_v^{k+1}(w') < n$ (note that $j_1^{k+1}(w') \ne j_v^{k+1}(w')$ always holds). If we denote such an element w' still by w, then the number v for w defined above gets strictly larger. By applying induction on the number $r-v > 0$, we can find $x \in H_\lambda$ with $x \underset{P_L}{\sim} w$ such that x has a standard MDC form which is normal for the first $k+1$ layers. Finally, by applying induction on the number $\lambda_1 - k > 0$, we can get the required $y \in N_\lambda$. □

We shall finish this chapter by giving an element of H_λ to see how to normalize it into N_λ:

Assume that $n = 5$, $\lambda = \{3 > 2\}$ and $w \in H_\lambda$ has a standard MDC form (A_2, A_1) at $i \in \mathbb{Z}$ as follows:

j_1th column

$(i+1)$st row

$A_2(w)$

$A_1(w)$

w

where j is some integer. Now $j_1^1(w) - j_2^1(w) = 8 > 5$, let $x = \rho_{A_1}^{A_2}{}^{2}(w)$. Then $x \in H_\lambda$ is as follows:

j_1th column

$(i+3)$rd row

$A_2(x)$

$A_1(x)$

x

We see that x has a standard MDC form (A_2, A_1) at $i + 2$ which is normal for the first layer. Now $j_1^2(x) - j_2^2(x) = 8 > 5$, let $y = \rho_{A_1}^{A_2}{}^{2}(x)$. Then from the following matrix, we see that $y \in H_\lambda$ is a normalized element.

j_1th column

$(i+5)$th row

$A_2(y)$

$A_1(y)$

y

CHAPTER 11 : THE ORBIT SPACE $\tilde{A}_n$ OF THE AFFINE WEYL GROUP A_n

In this chapter, we shall introduce the concept of an unlabelled affine matrix which is essentially a ϕ-orbit of A_n, where ϕ is an automorphism on A_n defined in §9.2. Let $\tilde{A}_n$ be the set of all ϕ-orbits of A_n. Then there exists a natural map $\eta: A_n \to \tilde{A}_n$, most of the results on A_n can be carried over to $\tilde{A}_n$ under this map with slight modification. We shall define an operation of deletion on some special kind of elements $\tilde{w} \in \tilde{A}_n$ which sends $\tilde{w}$ into $\tilde{A}_m$ for some $m < n$. The commutativity between interchanging operations and deletion will enable us to simplify the proofs in the subsequent chapters.

§11.1 DEFINITION OF $\tilde{A}_n$

Let $\tilde{A}_n$ be the set of all matrices $\tilde{w}$ which satisfy the following conditions:

(i) $\tilde{w}$ is an $\infty \times \infty$ matrix with no bounded edge in any direction.

(ii) Each row (resp. column) of $\tilde{w}$ contains a unique non-zero entry and this entry is 1.

(iii) If e, e' are two entries of $\tilde{w}$ with $r(e,e') = c(e,e') \in n\mathbf{Z}$, then e is non-zero if and only if e' is non-zero, where the functions r(,), c(,) are defined in the same way as in A_n.

So $\tilde{w} \in \tilde{A}_n$ can be determined by any of its n consecutive rows (or columns).

For $\tilde{y}, \tilde{y}' \in \tilde{A}_n$, we say $\tilde{y} = \tilde{y}'$ if there exists entries e_0, e_0' (not necessarily non-zero) of $\tilde{y}$, $\tilde{y}'$, respectively, such that for any entries e, e' of $\tilde{y}$, $\tilde{y}'$, respectively, with $r(e_0,e) = r(e_0',e')$ and $c(e_0,e) = c(e_0',e')$, we have that e is non-zero if and only if e' is non-zero.

For any $\tilde{w} \in \tilde{A}_n$, we say an entry e (not necessarily non-zero) of $\tilde{w}$ is diagonal if there exists non-zero entries $e_1, e_2, \ldots, e_n$ of $\tilde{w}$ such that $r(e,e_j) = j$ for any $1 \leq j \leq n$ and $\sum_{j=1}^{n} c(e,e_j) = \sum_{j=1}^{n} j$.

Lemma 11.1.1 <u>There exists a unique diagonal entry in each row (resp. column) of $\tilde{w}$ for any $\tilde{w} \in \tilde{A}_n$.</u>

<u>Proof</u>: By symmetry, it suffices to show that there exists a unique diagonal entry in each row of $\tilde{w}$. Now let us fix a row of $\tilde{w}$ and take any entry e (not necessarily non-zero) in it. Assume that $e_1, \ldots, e_n$ are non-zero entries of $\tilde{w}$ such that $r(e,e_t) = t$ for $1 \leq t \leq n$ and $\sum_{t=1}^{n} c(e,e_t) = m$. Since $c(e,e_t)$'s are incongruent mod n, there exists some $q \in \mathbf{Z}$ such that $m = \sum_{t=1}^{n} t + qn$. Let e' be the entry of $\tilde{w}$ satisfying $r(e,e') = 0$ and $c(e,e') = q$. Then $r(e',e_t) = t$ for $1 \leq t \leq n$ and

$\sum_{t=1}^{n} c(e',e_t) = \sum_{t=1}^{n} t$. So e' is the diagonal entry in the given row. The uniqueness of e' is obvious. □

By the property that if e, e' are two non-zero entries of $\tilde{w}$ with $r(e,e') \in n\mathbf{Z}$ then $c(e,e') = r(e,e')$, we can easily see that if e, e' are two entries of $\tilde{w}$ (not necessarily non-zero) with $c(e,e') = r(e,e')$ then e and e' are either both diagonal or both not. So the set of all diagonal entries of $\tilde{w}$ forms a diagonal line of $\tilde{w}$ which is uniquely determined.

§11.2 THE MAP $\eta: A_n \to \tilde{A}_n$.

For any $w \in A_n$, let $\eta(w)$ be the matrix which comes from w by forgetting the integers labelling the rows and columns of w. We can check that $\eta(w) \in \tilde{A}_n$. This defines a map $\eta: A_n \to \tilde{A}_n$.

Lemma 11.2.1 η is surjective.

Proof: For any $\tilde{w} \in \tilde{A}_n$, there exists a diagonal entry, say e, of $\tilde{w}$ by Lemma 11.1.1. We give integer labels to all rows and columns of $\tilde{w}$ such that e lies in the (0,0)-position. Then we get an element of A_n. Clearly, the image of this element under η is $\tilde{w}$ and so η is surjective. □

§11.3 THE PARTITION ASSOCIATED WITH AN ELEMENT OF $\tilde{A}_n$

Let Φ be the automorphism of A_n such that $\Phi(s_i) = s_{i+1}$ for any $i \in \mathbf{Z}$. Then the following lemma is clear.

Lemma 11.3.1 For any $\tilde{w} \in \tilde{A}_n$, $\eta^{-1}(\tilde{w})$ is just a Φ-orbit in A_n. □

Remark 11.3.2 (i) By Lemma 11.3.1, $\tilde{A}_n$ can be regarded as the set of Φ-orbits of A_n.

(ii) One can show that $|\eta^{-1}(\tilde{w})| \,\big|\, n$, for any $\tilde{w} \in \tilde{A}_n$. Fix $d \in \mathbf{Z}^+$ with $d|n$, let $A_n(d) = \{w \in A_n \,\big|\, |\eta^{-1}(\eta(w))| \,\big|\, d\}$. Then there exists a map $P_d: A_n(d) \to A_d$ by regarding any $w \in A_n(d)$ as an element of A_d. One can check that P_d is bijective.

For $w \in A_n$, assume that $A = \{e_w(i_t,j_t) \mid 1 \leq t \leq m,\ i_1 < \ldots < i_m,\ j_1 < \ldots < j_m\}$ is any descending chain of w, then $A' = \{e_{\Phi(w)}(i_t+1,\ j_t+1) \mid 1 \leq t \leq m\}$ is the corresponding descending chain of $\Phi(w)$. By this fact, one can easily see that if $w \in \sigma^{-1}(\lambda)$ then $\Phi(w) \in \sigma^{-1}(\lambda)$. So we can define a map $\tilde{\sigma}: \tilde{A}_n \to \Lambda_n$ by $\tilde{\sigma}(\tilde{w}) = \sigma(y)$ for $\tilde{w} \in \tilde{A}_n$ and any $y \in \eta^{-1}(\tilde{w})$. Then we get a commutative diagram as follows:

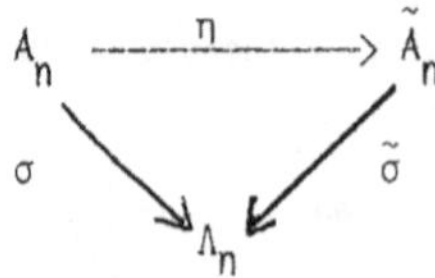

§11.4 THE FUNCTIONS $\ell(\tilde{w})$, $\mathcal{L}(\tilde{w})$, $R(\tilde{w})$ AND STAR OPERATIONS IN $\tilde{A}_n$.

Definition 11.4.1 We say that $J = J_1 \cup \ldots \cup J_t \in \underline{\Delta}$ is a standard decomposition of J into indecomposable subsets if there exist integers i, m_v, k_v for $1 \leq v \leq t$ with $0 < m_v < k_v$ and $\sum_{v=1}^{t} k_v = n$ such that for every $1 \leq h \leq t$,

$J_h = \{s_{\alpha_u} \mid 1 \leq u \leq m_h,\ \alpha_u = i + \sum_{v=1}^{h-1} k_v + u\}$. Call $(m_1,k_1), (m_2,k_2),\ldots,(m_t,k_t)$ the sequence associated with such a standard decomposition of J and call t the number of components of such a standard decomposition of J.

For example, suppose n = 5 and $J = \{s_1,s_2,s_4\}$. Then $J = \{s_1,s_2\} \cup \{s_4\}$is a standard decomposition of J with the associated sequence (2,3), (1,2).

For any $J \in \underline{\Delta}$, we can see that the sequence associated with a standard decomposition of J is uniquely determined by J up to a cyclic permutation, where a sequence $x_1',x_2',\ldots,x_t'$ is said to be a cyclic permutation of a sequence $x_1,x_2,\ldots,x_t$ if there exists an integer h, $1 \leq h \leq t$, such that $(x_1',x_2',\ldots,x_t') = (x_h,x_{h+1},\ldots,x_t,x_1,x_2,\ldots,x_{h-1})$.

We say J, $J' \in \underline{\Delta}$ are similar, written $J \sim J'$, if there exist standard decompositions of J, J' which have the same associated sequence.

For example, when n = 5, $J = \{s_1,s_2\} \cup \{s_4\}$ and $J' = \{s_2,s_3\} \cup \{s_5\}$ have the same associated sequence (2,3), (1,2) and so $J \sim J'$.

This is an equivalence relation. Let $C\ell(\underline{\Delta})$ be the set of all such equivalence classes of $\underline{\Delta}$. Then there exists a natural map from $\underline{\Delta}$ to $C\ell(\underline{\Delta})$ by sending J to $\tilde{J}$, where $J \in C\ell(\underline{\Delta})$ is the class containing J.

For any $\tilde{J}$, $\tilde{J}' \in C\ell(\underline{\Delta})$, we say $\tilde{J} \subseteq \tilde{J}'$ if there exists $J_0 \in \tilde{J}$ and $J_0' \in \tilde{J}'$ such that $J_0 \subseteq J_0'$. Clearly, this defines a partial order on $C\ell(\underline{\Delta})$.

By Lemma 11.3.1, for any $\tilde{w} \in \tilde{A}_n$, elements y in $\eta^{-1}(\tilde{w})$ have the same length and the sets $\mathcal{L}(y)$ (resp. $R(y)$) are all similar. So one can define $\ell(\tilde{w}) = \ell(y)$, $\mathcal{L}(\tilde{w}) = \widetilde{\mathcal{L}(y)}$ and $R(\tilde{w}) = \widetilde{R(y)}$ for any $y \in \eta^{-1}(\tilde{w})$. If $y \in \mathcal{D}_L(s_t)$ for some $1 \leq t \leq n$, let $y' = {}^*y$ in $\mathcal{D}_L(s_t)$ and let $\tilde{y}' = \eta(y')$. We see that the element $\tilde{y}'$ is independent of the choice of $y \in \eta^{-1}(\tilde{w})$ in the following sense: Suppose $x \in \eta^{-1}(\tilde{w})$ with $x = \Phi^h(y)$. Then $x \in \mathcal{D}_L(s_{t+h})$. Let $x' = {}^*x$ in $\mathcal{D}_L(s_{t+h})$ and let $\tilde{x}' = \eta(x')$. Then $\tilde{x}' = \tilde{y}'$. So we can say that $\tilde{y}'$ is obtained from $\tilde{w}$ by a left star operation. Similarly one can define a right star operation on $\tilde{w}$. So one can define the equivalence relations $\underset{P_L}{\sim}$, $\underset{P_R}{\sim}$ and $\underset{P}{\sim}$ on $\tilde{A}_n$ in a natural way. One can also define a descending chain, a block, a DC block, an MDC block, a local MDC block, an entry (resp. row, column, row-column) class, etc, of an element $\tilde{w} \in \tilde{A}_n$, and then define ρ, θ operations on $\tilde{w}$ and the sequence $\xi(\tilde{w},\rho)$ (resp. $\xi(\tilde{w},\theta)$) corresponding to ρ (resp. θ) on $\tilde{w}$ in a natural way. For example, we say a non-zero entry set $\{e_1,\ldots,e_t\}$ of $\tilde{w}$ is a descending chain of $\tilde{w}$ if for any $1 \le i < j \le t$, we have $r(e_i,e_j) \cdot c(e_i,e_j) < 0$. We say a submatrix consisting of any m, $m \le n$, consecutive rows of $\tilde{w}$ is a block of $\tilde{w}$. We say $\tilde{w}$ has DC form $(A_\ell,\ldots,A_1)$ if $A_\ell,\ldots,A_1$ are consecutive DC blocks of $\tilde{w}$ downwards with $\sum_{t=1}^{\ell} |A_t| \le n$. In that case, e_{tu} (or $e_{tu}(\tilde{w})$) usually denotes the u-th entry of $A_t(\tilde{w})$.

§11.5 INTERCHANGING OPERATIONS ON BLOCKS IN $\tilde{A}_n$.

Since $\tilde{A}_n$ can be regarded as the set of ϕ-orbits of A_n, we can get the analogues for $\tilde{A}_n$ of most of the results on A_n immediately.

Assume that $\tilde{w} \in \tilde{A}_n$ has local MDC form (A_2,A_1). Set

$$\begin{cases} i_0 = 0 \\ i_u = \min\{h \mid c(e_{2h},e_{1u}) > 0,\ i_{u-1} < h \le |A_2|\},\ 1 \le u \le |A_1| \end{cases} \tag{11.5.1}$$

(Compare with formula (8.3.2))

Lemma 11.5.2 (Compare with Lemma 8.3.4)

Assume that $\tilde{w} \in \tilde{A}_n$ has local MDC form (A_2,A_1).

Then $\rho_{A_2}^{A_1}(\tilde{w})$ exists $\Leftrightarrow$ all i_u, $1 \le u \le |A_1|$, exist. When they exist, let $f_{tu}(\tilde{w})$ be the row of $\tilde{w}$ containing $e_{tu}(\tilde{w})$. Then $\tilde{w}' = \rho_{A_2}^{A_1}(\tilde{w})$ is obtained from $\tilde{w}$ by permuting the rows in the block $[A_2(\tilde{w}), A_1(w)]$ as follows.

$$f_{1u}(\tilde{w}') \text{ comes from } f_{2,i_u}(\tilde{w}) \quad \text{for } 1 \le u \le |A_1|$$

$$f_{2v}(\tilde{w}') \text{ comes from } \begin{cases} f_{2v}(\tilde{w}) \text{ if } v \notin \{i_u \mid 1 \le u \le |A_1|\} \\ f_{1u}(\tilde{w}) \text{ if } v = i_u \text{ for some } 1 \le u \le |A_1| \end{cases} \tag{11.5.3}$$

where $f_{tu}(\tilde{w}')$ is the row of $\tilde{w}'$ containing $e_{tu}(\tilde{w}')$.

So $\tilde{w}'$ has local MDC form (A_1,A_2). □

Lemma 11.5.4 (Compare with Lemma 8.3.5)

Assume that $\tilde{w} \in \tilde{A}_n$ has local MDC form (A_2,A_1).

Then $\tilde{w}' = \rho_{A_2}^{A_1}(\tilde{w})$ exists $\Leftrightarrow$ $A_2(\tilde{w})$ is a longest descending chain in the block $[A_2(\tilde{w}), A_1(\tilde{w})]$. □

Lemma 11.5.5 (Compare with Lemma 8.3.7)

Assume that $\tilde{w} \in \tilde{A}_n$ has local MDC form (A_2,A_1).

Then $\tilde{w}' = \theta_{A_1}^{A_2}(\tilde{w})$ exists $\Leftrightarrow$ $A_1(\tilde{w})$ is a longest descending chain in the block $[A_2(\tilde{w}), A_1(\tilde{w})]$. □

In formula (11.5.1), if $\tilde{w}$ has local MDC form (A_2,A_1) which is quasi-normal for the first k layers, then $i_h = h$ for $1 \leq h \leq k$. When they exist, let $i_{u+k} = k + i'_{ku}$ for $1 \leq u \leq |A_1| - k$, where

$$i'_{ku} = \min\{h \mid c(e_{2,k+h}, e_{1,k+u}) > 0,\ i'_{k,u-1} < h \leq |A_2| - k\} \tag{11.5.6}$$

with the convention that $i'_{ko} = 0$. (Compare with formula (8.3.11))

Lemma 11.5.7 (Compare with Corollary 8.3.12)

Assume that $\tilde{w} \in \tilde{A}_n$ has local MDC form (A_2,A_1) which is quasi-normal for the first k layers. Then

$$\tilde{w}' = \rho_{A_2}^{A_1}(\tilde{w}) \text{ exists} \Leftrightarrow \text{all } i'_{ku},\ 1 \leq u \leq |A_1| - k, \text{ exist.} \quad \square$$

When all i'_{ku}, $1 \leq u \leq |A_1| - k$, do exist, we have $|A_2| \geq |A_1|$. If $k \leq |A_1|$, then

$$\begin{cases} \left.\begin{array}{l} f_{1h}(\tilde{w}') \text{ comes from } f_{2h}(\tilde{w}) \\ f_{2h}(\tilde{w}') \text{ comes from } f_{1h}(\tilde{w}) \end{array}\right\} \text{ for } 1 \leq h \leq k \\ f_{1,k+u}(\tilde{w}') \text{ comes from } f_{2,i_{k+u}}(\tilde{w}) \text{ for } 1 \leq u \leq |A_1| - k \\ f_{2,k+v}(\tilde{w}') \text{ comes from } \begin{cases} f_{2,k+v}(\tilde{w}) \text{ for } 1 \leq v \leq |A_1| - k,\ v \notin \{i'_{ku} \mid 1 \leq u \leq |A_1| - k\} \\ f_{1,k+u}(\tilde{w}) \text{ for } v = i'_{ku}, \text{ some } 1 \leq u \leq |A_1| - k \end{cases} \end{cases}$$

If $k > |A_1|$, then

$$\left|\begin{array}{ll} f_{1h}(\tilde{w}') \text{ comes from } f_{2h}(\tilde{w}) \\ f_{2h}(\tilde{w}') \text{ comes from } f_{1h}(\tilde{w}) \end{array}\right\} \text{ for } 1 \leq h \leq |A_1|$$
$$f_{2t}(\tilde{w}') \text{ comes from } f_{2t}(\tilde{w}) \quad \text{for } |A_1| < t \leq |A_2|$$

We see that the choice of i'_{ku}'s is only dependent on the set of inequalities

$$\{c(e_{2,k+h}, e_{1,k+h'}) \gtrless 0 \mid 1 \leq h \leq |A_2| - k,\ 1 \leq h' \leq |A_1| - k\} \qquad (11.5.9)$$

in the case that $\tilde{w}$ has local MDC form (A_2, A_1) which is quasi-normal for the first k layers.

Lemma 11.5.10 (Compare with Corollary 8.4.3)

Assume that $\tilde{w} \in \tilde{A}_n$ has local MDC form $(A_\ell, \dots, A_1)$. Let $\tilde{w}' = \rho_{A_\ell}^{A_{\ell-1},\dots,A_1}(\tilde{w})$.

Then $\tilde{w}'$ exists $\Longleftrightarrow$ $A_\ell(\tilde{w})$ is a longest descending chain in the block $[A_\ell(\tilde{w}), \dots, A_1(\tilde{w})]$.

□

Lemma 11.5.11 (Compare with Corollary 8.4.4)

Assume that $\tilde{w} \in \tilde{A}_n$ has local MDC form $(A_\ell, \dots, A_1)$. Let $\tilde{w}' = \theta_{A_1}^{A_2,\dots,A_\ell}(\tilde{w})$.

Then $\tilde{w}'$ exists $\Longleftrightarrow$ $A_1(\tilde{w})$ is a longest descending chain in the block $[A_\ell(\tilde{w}), \dots, A_1(\tilde{w})]$.

□

Lemma 11.5.12 (Compare with Lemma 8.4.5)

Assume that $\tilde{w} \in \tilde{A}_n$ has local MDC form $(A_{1\ell_1}, \dots, A_{11}, A_{2\ell_2}, \dots, A_{21}, A_{3\ell_3}, \dots, A_{31}, A_{4\ell_4}, \dots, A_{41})$ which is quasi-normal (resp. normal) for the first k layers. If there exists $\tilde{w}' = \rho_{A_{21},\dots,A_{2\ell_2}}^{A_{3\ell_3},\dots,A_{31}}(\tilde{w})$, then $\tilde{w}'$ has local MDC form $(A_{1\ell_1}, \dots, A_{11}, A_{3\ell_3}, \dots, A_{31}, A_{2\ell_2}, \dots, A_{21}, A_{4\ell_4}, \dots, A_{41})$ which is quasi-normal (resp. normal) for the first k layers. Let $e_{tu}(\tilde{w})$ (resp. $e_{tu}(\tilde{w}')$) be the uth entry of the t-th block in such a form of $\tilde{w}$ (resp. $\tilde{w}'$). For $k \leq u \leq k$, let $i_1, \dots, i_t$ with $1 \leq i_1 < \dots < i_t \leq \sum_{\alpha=1}^{4} \ell_\alpha$ (resp. $i'_1, \dots, i'_{t'}$ with $1 \leq i'_1 < \dots < i'_{t'} \leq \sum_{\alpha=1}^{4} \ell_\alpha$) be be the integer set such that $e_{vu}(\tilde{w})$ exists if and only if $v \in \{i_1,\ldots,i_t\}$ (resp. $e_{vu}(\tilde{w}')$ exists if and only if $v \in \{i'_1,\ldots,i'_{t'}\}$). Then $t = t'$ and for any $1 \le \alpha$, $\beta \le t$, we have $c(e_{i_{\alpha},u}(\tilde{w}),\, e_{i_{\beta},u}(\tilde{w})) = c(e_{i'_{\alpha},u}(\tilde{w}'),\, e_{i'_{\beta},u}(\tilde{w}'))$. □

Lemma 11.5.13 (Compare with Corollary 8.4.7)

Assume that $\tilde{w} \in \tilde{A}_n$ has full MDC form $(A_v,\ldots,A_1,\, A_\ell,\ldots,A_{v+1})$ which is normal for the first k layers. If there exists $\tilde{w}' = \rho^{A_\ell,\ldots,A_{v+1}}_{A_1,\ldots,A_v}(\tilde{w})$ for some $1 \le v < \ell$, then $\tilde{w}'$ has full MDC form $(A_\ell,\ldots,A_1)$ which is normal for the first k layers. □

Lemma 11.5.14 (Compare with Lemma 5.11)

For $\tilde{w} \in \tilde{A}_n$, let $E = \{e_u \mid 1 \le u \le \ell\} \cup \{e \mid r(e,e_u) = c(e,e_u) \in n\mathbf{Z}$, for some $1 \le u \le \ell\}$ be a set of entry classes of $\tilde{w}$ such that $0 < r(e_i,e_j) < n$ and $0 < c(e_i,e_j) < n$ for any $1 \le i < j \le \ell$. Let S be a descending chain on $\tilde{w}$. Then $|E \cap S| \le 1$. □

§11.6 TOTALLY ORDERED SETS WITH A DISTANCE FUNCTION

Our purpose of introducing a totally ordered set with a distance function is to establish the commutativity of interchanging operations with deletion.

We say that R is a totally ordered set with a distance function if R is a non-empty set with a map $R \times R \to \mathbf{Z}$: $(\alpha,\beta) \xrightarrow{\langle\ ,\ \rangle} \langle\alpha,\beta\rangle$ satisfying: for any α, β, $\nu \in R$, we have

$$\begin{cases} \text{(i)} & \langle\alpha,\beta\rangle = 0 \Longleftrightarrow \alpha = \beta \\ \text{(ii)} & \langle\alpha,\nu\rangle = \langle\alpha,\beta\rangle + \langle\beta,\nu\rangle \end{cases} \tag{11.6.1}$$

We can also write $R = \{R, \langle\ ,\ \rangle\}$. Clearly, for any $\alpha,\beta \in R$, the equation $\langle\alpha,\beta\rangle = -\langle\beta,\alpha\rangle$ always holds.

We say $\alpha \le \beta$ if $\langle\alpha,\beta\rangle \ge 0$. This is a total order in R. We say a sequence $\xi{:}\alpha_1,\ldots,\alpha_\ell$ in R is descending if $\langle\alpha_i,\alpha_j\rangle < 0$ for any $1 \le i < j \le \ell$. Set $|\xi| = \ell$.

Let $\tilde{R}$ be the set of all descending sequences in R. Let $\tilde{R}^2_0$ be the set of all ordered pairs (ξ,ζ) with $\xi{:}\alpha_1,\ldots,\alpha_\ell$ and $\zeta{:}\beta_1,\ldots,\beta_m$ belonging to $\tilde{R}$ and $\alpha_t \ne \beta_u$ for any $1 \le t \le \ell$, $1 \le u \le m$. Let

$$\begin{cases} i_0 = 0 \\ i_u = \min\,\{h \mid \langle\alpha_h,\beta_u\rangle > 0,\ i_{u-1} < h \le \ell\}, \quad 1 \le u \le m \end{cases} \tag{11.6.2}$$

Clearly, there exists a maximal integer k, $0 < k < m$, such that $i_0,\ldots,i_k$ exist. We call the sequence $i_0,\ldots,i_k$ the ρ sequence associated with (ξ,ζ). When $k = m$, such a sequence is called a full ρ sequence. Let $\tilde{R}_1^2 \subseteq \tilde{R}_0^2$ be the set of all pairs (ξ,ζ) which have a full ρ sequence.

For any $(\xi,\zeta) \in \tilde{R}_1^2$, let (ξ',ζ') be the ordered pair with $\xi':\alpha_1',\ldots,\alpha_\ell'$ and $\zeta':\beta_1',\ldots,\beta_m'$ such that

$$\begin{cases} \underline{\beta_u' = \alpha_{i_u},\ 1 < u < m} \\ \alpha_v' = \begin{cases} \underline{\alpha_v,\ v \notin \{i_u,\ 1 < u < m\}} \\ \underline{\beta_u,\ v = i_u \text{ for some } 1 < u < m} \end{cases} \end{cases} \tag{11.6.3}$$

Clearly, $(\alpha_1',\ldots,\alpha_\ell',\ \beta_1',\ldots,\beta_m')$ is a permutation of $(\alpha_1,\ldots,\alpha_\ell,\ \beta_1,\ldots,\beta_m)$ and $(\xi',\zeta') \in \tilde{R}_0^2$. So we have defined a map $\rho:\tilde{R}_1^2 \to \tilde{R}_0^2$ by $\rho(\xi,\zeta) = (\xi',\zeta')$. Let ϕ be a map from $\{\alpha_t,\beta_u \mid 1 < t < \ell,\ 1 < u < m\}$ to $\{1,2,\ldots,\ell+m\}$ such that $\phi(\alpha_t) = t$ and $\phi(\beta_u) = \ell+u$. Set

$$\tilde{\rho}_R(\xi,\zeta) = \begin{pmatrix} 1,\ldots, & \ell, & \ell+1,\ldots, & \ell+m \\ \phi(\alpha_1'),\ldots, & \phi(\alpha_\ell'), & \phi(\beta_1'),\ldots, & \phi(\beta_m') \end{pmatrix}$$

Then $\tilde{\rho}_R$ is a map from $\tilde{R}_1^2$ to $\bigcup_{n\geq 2}^{\infty} S_n$, where S_n is the permutation group on $\{1,2,\ldots,n\}$.

<u>Definition 11.6.4</u> Assume that R, R' are two totally ordered sets with distance functions. Let $(\xi,\zeta) \in \tilde{R}_1^2$, $(\xi',\zeta') \in \tilde{R}_1'^2$. We say $(\xi,\zeta) \sim (\xi'\ \xi')$ if $|\xi| = |\xi|'$, $|\zeta| = |\zeta'|$ and $\tilde{\rho}_R(\xi,\zeta) = \tilde{\rho}_{R'}(\xi',\zeta')$.

Clearly, $(\xi,\zeta) \sim (\xi',\zeta')$ if and only if their ρ sequences are the same.

<u>Lemma 11.6.5</u> <u>Assume that $\{R, \langle\ ,\ \rangle_R\}$, $\{R', \langle\ ,\ \rangle_{R'}\}$ are two totally ordered sets with distance functions. Let $(\xi,\zeta) \in \tilde{R}_0^2, (\xi',\zeta') \in \tilde{R}_0'^2$ with $\xi:\alpha_1,\ldots,\alpha_\ell$; $\zeta:\beta_1,\ldots,\beta_m$; $\xi':\alpha_1',\ldots,\alpha_\ell'$ and $\zeta':\beta_1',\ldots,\beta_m'$. If for any t, u with $1 < t < \ell$ and $1 < u < m$, we have $\langle\alpha_t,\beta_u\rangle_R \cdot \langle\alpha_t',\beta_u'\rangle_{R'} > 0$, then $(\xi,\zeta) \in \tilde{R}_1^2$ if and only if $(\xi',\zeta') \in \tilde{R}_1'^2$. When $(\xi,\zeta) \in \tilde{R}_1^2$ and $(\xi',\zeta') \in \tilde{R}_1'^2$, we have $(\xi,\zeta) \sim (\xi',\zeta')$.</u>

<u>Proof</u>: Let $I:i_0,\ldots,i_k$ and $I':i_0',\ldots,i_{k'}'$ be the ρ sequences of (ξ,ζ) and (ξ',ζ'), respectively. Then to reach our goal, it suffices to show that $k = k'$ and

$(i_0, \ldots, i_k) = (i'_0, \ldots, i'_{k'})$. Clearly, $i_0 = i'_0 = 0$. In general, assume $i'_h = i_h$ for some u, $u < m$ and all h, $0 \leq h \leq u$. We have $\langle\alpha_h, \beta_{u+1}\rangle_R \cdot \langle\alpha'_h, \beta'_{u+1}\rangle_{R'} > 0$ for all h, $i_u < h \leq \ell$ and $\langle\alpha_t, \beta_{u+1}\rangle_R < \langle\alpha_{t+1}, \beta_{u+1}\rangle_{R'}$ $\langle\alpha'_t, \beta'_{u+1}\rangle_{R'} < \langle\alpha'_{t+1}, \beta'_{u+1}\rangle_{R'}$ for any t, $i_u < t < \ell$. By formula 11.6.2, this implies that either i_{u+1}, u'_{u+1} both do not exist or $i_{u+1} = i'_{u+1}$. By induction on u, we have $k' = k$ and $(i_0, \ldots, i_k) = (i'_0, \ldots, i'_k)$. □

Example 11.6.6 For any block, say A, of $w \in A_n$ with $|A| \leq n$, let ξ_A be the set of integers labelling the columns of w which contain some entry of A. For any j, $j' \in \xi_A$, we define a map $d:(j,j') \to j'-j$. Then $\{\xi_A, d(\ ,\ )\}$ is a totally ordered set with a distance function. We have $\tilde{\xi}_A = \{\xi: j_1, j_2, \ldots, j_\alpha | j_1 > j_2 > \ldots > j_\alpha, j_i \in \xi_A\}$. $(\tilde{\xi}_A)^2_0$ is the set of all pairs (ξ,ζ) of $\tilde{\xi}_A$ such that all terms of ξ,ζ are distinct. $(\tilde{\xi}_A)^2_1$ is the set of all pairs $(\xi_2,\xi_1) \in (\tilde{\xi}_A)^2_0$ with $\xi_t: j_{t1}, j_{t2}, \ldots, j_{tm_t}$, $t = 1,2$ such that for any $1 \leq u \leq m_2$, $|\{h | j_{2u} - j_{1h} > 0,\ 1 \leq h \leq m_1\}| \leq m_2 - u$. In particular, if A_2, A_1 are two consecutive DC subblocks of A with A_2 lying above A_1, let $\xi_t: j^1_t(w), \ldots, j^{m_t}_t(w)$, $t = 1,2$, where $|A_t| = m_t$ and $e((w), j^u_y(w))$ is the u-th entry of A_t. Then

$$\rho^{A_1}_{A_2}(w) \text{ exists} \Leftrightarrow (\xi_2,\xi_1) \in (\tilde{\xi}_A)^2_1$$

by definition of ρ operation on w. If $w' = \rho^{A_1}_{A_2}(w)$ does exist, let $\xi'_t: j'_t(w'), j^2_t(w'), \ldots, j^{m_t}_t(w')$, $t = 1,2$. Then it is clear that $(\xi'_2,\xi'_1) \in (\tilde{\xi}_A)^2_0$ satisfies formula 11.6.3 by replacing (ξ,ζ) and (ξ',ζ') by (ξ_2,ξ_1) and (ξ'_2,ξ'_1), respectively, i.e. $\rho(\xi_2,\xi_1) = (\xi'_2,\xi'_1)$.

Example 11.6.7 For any block, say A, of $\tilde{w} \in \tilde{A}_n$ with $|A| \leq n$, let ζ_A be the set of entries of A with the map $(e,e') \xrightarrow{c(\ ,\ )} c(e,e')$. Then $\{\xi_A, c(\ ,\ )\}$ is a totally ordered set with a distance function. We have $\tilde{\zeta}_A = \{\zeta: e_{i_1}, \ldots, e_{i_\alpha} | c(e_{i_t}, e_{i_u}) < 0$ for any $t < u\}$. $(\tilde{\xi}_A)^2_0$ is the set of all pairs (ξ,ζ) of $\tilde{\xi}_A$ such that all terms of ξ,ζ are distinct. $(\tilde{\xi}_A)^2_1$ is the set of all pairs $(\zeta_2,\zeta_1) \in (\tilde{\zeta}_A)^2_0$ with $\zeta_t: e_{t1}, \ldots, e_{tm_t}$, $t = 1,2$, such that for any $1 \leq u \leq m_2$, $|\{h | c(e_{1h}, e_{2u}) > 0,$ $1 \leq h \leq m_1\}| \leq m_2 - u$. In particular, if A_2, A_1 are two consecutive DC subblocks of A with A_2 lying above A_1, let (ζ_2,ζ_1) be as above with $|A_t| = m_t$ and e_{tu} the u-th entry of A_t. Then

$$\rho^{A_1}_{A_2}(\tilde{w}) \text{ exists} \Leftrightarrow (\zeta_2,\zeta_1) \in (\tilde{\zeta}_A)^2_1$$

by definition of ρ operation on $\tilde{w}$. If $\tilde{w}' = \rho_{A_2}^{A_1}(\tilde{w})$ does exist, assume that $f_{tu}(\tilde{w}')$ is the row of $\tilde{w}'$ containing the u-th entry of $A_t(\tilde{w}')$ and assume that $f_{tu}(\tilde{w}')$ comes from the row $f_{tu}(\tilde{w})$ of $\tilde{w}$ under the operation $\rho_{A_2}^{A_1}$. Let e'_{tu} be the entry of $f_{tu}(\tilde{w})$. Then $e'_{tu} \in \zeta_A$. Let $\zeta_t: e'_{t1}, \ldots, e'_{tm_t}$, $t = 1,2$. Then we see that $\rho(\zeta_2, \zeta_1) = (\zeta'_2, \zeta'_1) \in (\tilde{\zeta}_A)^2_o$.

§11.7 DELETION OPERATIONS IN $\tilde{A}_n$

For a given set E of entry classes of $\tilde{w} \in \tilde{A}_n$, we define the corresponding set ζ_E of rc-classes of $\tilde{w}$ to be the set of all rows and columns of $\tilde{w}$ which contain some entry of E. The size of ζ_E is defined to be the size of E, and is written $|\zeta_E|$. Every rc-class of $\tilde{w}$ is defined in this way.

Let ζ be a set of rc-classes of $\tilde{w} \in \tilde{A}_n$ with $0 \leq |\zeta| = m < n$ and let $\tilde{v}$ be the matrix obtained from $\tilde{w}$ by deleting all rows and columns of ζ. Then $\tilde{v} \in \tilde{A}_{n-m}$. In that case, let e, e' be any two entries of $\tilde{w}$ not lying in ζ. Then the corresponding entries e_o, e'_o of e, e' in $\tilde{v}$ satisfy

$$\begin{cases} \underline{r_{\tilde{v}}(e_o,e'_o) = r_{\tilde{w}}(e,e') - a_{\tilde{w},\zeta}(e,e')} \\ \underline{c_{\tilde{v}}(e_o,e'_o) = c_{\tilde{w}}(e,e') - b_{\tilde{w},\zeta}(e,e')} \end{cases} \qquad (11.7.1)$$

where $a_{\tilde{w},\zeta}(e,e')$ (resp. $b_{\tilde{w},\zeta}(e,e')$) is the number of rows (resp. columns) of ζ lying between e and e'.

Let $\ell = n-m$. Then we have the following results.

Lemma 11.7.2 (i) $\underline{r_{\tilde{v}}(e_o,e'_o) > 0 \iff r_{\tilde{w}}(e,e') > 0}$

$\underline{r_{\tilde{v}}(e_o,e'_o) < 0 \iff r_{\tilde{w}}(e,e') < 0}$

(ii) $\underline{\text{For any } q \in \mathbb{Z},\ r_{\tilde{v}}(e_o,e'_o) = q\ell \iff r_{\tilde{w}}(e,e') = qn}$

(iii) $\underline{0 \leq r_{\tilde{v}}(e_o,e'_o) < \ell \iff 0 \leq r_{\tilde{w}}(e,e') < n}$

(iv) $\underline{\text{Let } r_{\tilde{v}}(e_o,e'_o) = q'\ell + r',\ r_{\tilde{w}}(e,e') = qn + r \text{ with } q, q', r, r' \in \mathbb{Z},}$

$\underline{0 \leq r' < \ell \text{ and } 0 \leq r < n. \text{ Then } q = q'.}$

<u>In (i) - (iv), by replacing $r_{\tilde{v}}(\ ,\ )$, $r_{\tilde{w}}(\ ,\ )$ by $c_{\tilde{v}}(\ ,\ )$, $c_{\tilde{w}}(\ ,\ )$, we have the corresponding results.</u>

Proof: (i) - (iii) are obvious. For (iv), let e" be the entry of $\tilde{w}$ such that $r_{\tilde{w}}(e,e") = qn$. Then e" does not lie in ζ. So there exists the corresponding entry $e_0"$ of e" in $\tilde{v}$. By (ii), we have $r_{\tilde{v}}(e_0,e_0") = q\ell$. Since $r_{\tilde{w}}(e",e') = r_{\tilde{w}}(e",e) + r_{\tilde{w}}(e,e') = r_{\tilde{w}}(e,e') - r_{\tilde{w}}(e,e") = (qn+r) - qn = r$, we have $0 \leq r_{\tilde{w}}(e",e') < n$. Hence by (iii), we have $0 \leq r_{\tilde{v}}(e_0",e_0') < \ell$. So $r_{\tilde{v}}(e_0,e_0') = r_{\tilde{v}}(e_0,e_0") + r_{\tilde{v}}(e_0",e_0') = q\ell + r_{\tilde{v}}(e_0",e_0')$. By uniqueness of such an expression, it follows that $q = q'$. □

When $\tilde{v}$ is obtained from $\tilde{w}$ deleting ζ, we can also say that a block of $\tilde{v}$ is obtained from a block, say A, of $\tilde{w}$ by deleting all rows and columns in ζ. We usually denote such a block of $\tilde{v}$ also by A. For the sake of distinction, we write $A(\tilde{w})$, $A(\tilde{v})$, respectively.

§11.8 COMMUTATIVITY OF INTERCHANGING OPERATIONS WITH DELETION

Assume that $\tilde{w} \in \tilde{A}_n$ has a local MDC form (A_2,A_1) which is quasi-normal for the first k layers. Let $\tilde{v} \in \tilde{A}_{n-2k}$ be obtained from $\tilde{w}$ by deleting ζ, where ζ is the set of rc-classes of $\tilde{w}$ containing all entries $e_{tu}(\tilde{w})$, $1 \leq t \leq 2$, $1 \leq u \leq k$ with $|\zeta| = 2k$.

Lemma 11.8.1 $\tilde{w}' = \rho_{A_2}^{A_1}(\tilde{w})$ exists $\Longleftrightarrow$ $\tilde{v}' = \rho_{A_2}^{A_1}(\tilde{v})$ exists.

When they do exist, let ζ' be the set of rc-classes of $\tilde{w}'$ containing all $e_{tu}(\tilde{w}')$, $t = 1,2$, $1 \leq u \leq k$ with $|\zeta'| = 2k$. Let $\tilde{v}" \in \tilde{A}_{n-2k}$ be obtained from $\tilde{w}'$ by deleting ζ'. Then $\tilde{v}" = \tilde{v}'$.

Proof: Clearly $R = \{\{e_{t,k+u}(\tilde{w})|_{1\leq u\leq m_t-k}^{t=1,2}\}, c_{\tilde{w}}(\ ,\ )\}$, $R' = \{\{e_{tu}(\tilde{v})|_{1\leq u\leq m_t-k}^{t=1,2}\}, c_v(\ ,\ )\}$ are two totally ordered sets with distance functions, where $m_t = |A_t(\tilde{w})|$, $t = 1,2$. Let

$$\begin{cases} \zeta_t(w) : e_{t,k+1}(\tilde{w}),\dots,e_{t,m_t}(\tilde{w}) \\ \zeta_t(\tilde{v}) : e_{t1}(v),\dots,e_{t,m_t-k}(\tilde{v}) \end{cases} \quad \text{for } t = 1,2.$$

Then by Lemma 11.5.7, we have

$$\tilde{w}' = \rho_{A_2}^{A_1}(\tilde{w}) \text{ exists} \Longleftrightarrow (\zeta_2(\tilde{w}), \zeta_1(\tilde{w})) \in \tilde{R}_1^2$$

$$\tilde{v}' = \rho_{A_2}^{A_1}(\tilde{v}) \text{ exists} \Longleftrightarrow (\zeta_2(\tilde{v}), \zeta_1(\tilde{v})) \in \tilde{R}'^2_1.$$

Since for any u, u' with $1 \le u \le m_2-k$, $1 \le u' \le m_1-k$, we have $c_{\tilde{w}}(e_{2,k+u}(\tilde{w}), e_{1,k+u'}(\tilde{w})) \cdot c_{\tilde{v}}(e_{2,u}(\tilde{v}), e_{1,u'}(\tilde{v})) > 0$, we see by Lemma 11.6.5 that $(\zeta_2(\tilde{w}), \zeta_1(\tilde{w})) \in \tilde{R}_1^2$ if and only if $(\zeta_2(\tilde{v}),\zeta_1(v)) \in \tilde{R}'^2_1$. It follows that $\tilde{w}'$ exists if and only if $\tilde{v}'$ exists. When they both exist, also by Lemma 11.6.5, we have $(\zeta_2(\tilde{w}), \zeta_1(\tilde{w})) \sim (\zeta_2(\tilde{v}), \zeta_1(\tilde{v}))$. So by Formula 11.5.3, we get $\tilde{v}'' = \tilde{v}'$. □

Assume that w has an MDC form $(A_\ell,\ldots,A_1)$ which is normal for the first k layers with $k \le \min\{|A_t| \mid 1 \le t \le \ell\}$. Let $\zeta(\tilde{w})$ be the set of rc-classes containing all e_{tu}, $1 \le t \le \ell$, $1 \le u \le k$ with $|\zeta(\tilde{w})| = \ell k$. Let $\tilde{v} \in \tilde{A}_{n-\ell k}$ be obtained from $\tilde{w}$ by deleting $\zeta(\tilde{w})$. Then $\tilde{v}$ has the corresponding DC form $(A_\ell,\ldots,A_1)$ with $|A_t(\tilde{v})| = |A_t(\tilde{w})|-k$ (where we admit some $|A_t(\tilde{v})| = 0$) for $1 \le t \le \ell$ and the u-th entry of $A_t(\tilde{v})$ comes from the (k+u)-th entry of $A_t(\tilde{w})$.

Lemma 11.8.1 implies the following corollary.

<u>Corollary 11.8.2</u> <u>Let $\tilde{w}$, $\tilde{v}$ be as above. Then for any $1 \le j < \ell$, $\tilde{W}' = \rho_{A_{j+1}}^{A_j}(\tilde{w})$ exists $\Leftrightarrow$ $\tilde{v}' = \rho_{A_{j+1}}^{A_j}(\tilde{v})$ exists. When they both exist, let $\zeta(\tilde{w}')$ be the set of rc-classes of $\tilde{w}'$ containing all $e_{tu}(\tilde{w}')$, $1 \le t \le \ell$, $1 \le u \le k$ with $|\zeta(\tilde{w}')| = \ell k$. Let $\tilde{v}'' \in \tilde{A}_{n-\ell k}$ be obtained from $\tilde{w}'$ by deleting $\zeta(\tilde{w}')$. Then $\tilde{v}'' = \tilde{v}'$.</u>

<u>Proof</u>: Let $\zeta_{j,j+1}(\tilde{w})$ be the set of rc-classes of $\tilde{w}$ containing all $e_{tu}(\tilde{w})$, $t = j$, $j+1$, $1 \le u \le k$ with $|\zeta_{j,j+1}(\tilde{w})| = 2k$. Let $\tilde{u} \in \tilde{A}_{n-2k}$ be obtained from $\tilde{w}$ by deleting $\zeta_{j,j+1}(\tilde{w})$. Then $\tilde{v} \in \tilde{A}_{n-\ell k}$ is obtained from $\tilde{u}$ by deleting $\zeta_{\widehat{j,j+1}}(\tilde{u})$, where $\zeta_{\widehat{j,j+1}}(\tilde{u})$ is the set of rc-classes of $\tilde{u}$ containing all $e_{tu}(\tilde{u})$, $1 \le t \le \ell$, $t \ne j$, $j+1$, $1 \le u \le k$, with $|\zeta_{\widehat{j,j+1}}(\tilde{u})| = (\ell-2)k$. By Lemma 11.8.1, $\tilde{y} = \rho_{A_{j+1}}^{A_j}(\tilde{w})$ exists $\Leftrightarrow$ $\tilde{u}_1 = \rho_{A_{j+1}}^{A_j}(\tilde{u})$ exists. When they both exist, let $\zeta_{j,j+1}(\tilde{y})$ be the set of rc-classes of $\tilde{y}$ containing all $e_{tu}(\tilde{y})$, $t = j$, $j+1$, $1 \le u \le k$ with $|\zeta_{j,j+1}(\tilde{y})| = 2k$. Then also by Lemma 11.8.1, $\tilde{u}_1$ is obtained from $\tilde{y}$ by deleting $\zeta_{j,j+1}(\tilde{y})$, i.e. the diagram

$$\begin{array}{ccc} \tilde{w} & \xrightarrow{\text{deleting } \zeta_{j,j+1}(\tilde{w})} & \tilde{u} \\ {\scriptstyle \rho_{A_{j+1}}^{A_j}}\downarrow & & \downarrow{\scriptstyle \rho_{A_{j+1}}^{A_j}} \\ \tilde{y} & \xrightarrow{\text{deleting } \zeta_{j,j+1}(\tilde{y})} & \tilde{u}_1 \end{array}$$

commutes. But by the same argument as that in the proof of Lemma 11.8.1, we can show that

$$\tilde{u}_1 = \rho^{A_j}_{A_{j+1}}(\tilde{u}) \text{ exists} \Leftrightarrow \tilde{x} = \rho^{A_j}_{A_{j+1}}(\tilde{v}) \text{ exists.}$$

When the both exist, the diagram

$$\begin{array}{ccc} \tilde{u} & \xrightarrow{\text{deleting } \zeta_{\hat{j},\hat{j+1}}(\tilde{u})} & \tilde{v} \\ \rho^{A_j}_{A_{j+1}} \downarrow & & \downarrow \rho^{A_j}_{A_{j+1}} \\ \tilde{u}_1 & \xrightarrow{\text{deleting } \zeta_{\hat{j},\hat{j+1}}(\tilde{u}_1)} & \tilde{x} \end{array}$$

commutes. Combining the above two results, we reach our goal. □

§11.9 COMMUTATIVITY OF INTERCHANGING OPERATIONS WITH THE MAP η.

Assume that $w \in A_n$ has a DC form (A_2,A_1) at $i \in \mathbb{Z}$. We assume that the corresponding DC form of $\tilde{w} = \eta(w)$ is also (A_2,A_1). Then by definition of ρ, θ operations on both w and $\tilde{w}$, the following lemma is immediate.

Lemma 11.9.1 Assume that $w \in A_n$ has a DC form (A_2,A_1) at $i \in \mathbb{Z}$. Then

$$w' = \rho^{A_1}_{A_2}(w) \text{ exists} \Leftrightarrow \eta(w)' = \rho^{A_1}_{A_2}(\eta(w)) \text{ exists.}$$

Also, $w'' = \theta^{A_2}_{A_1}(w)$ exists $\Leftrightarrow \eta(w)'' = \theta^{A_2}_{A_1}(\eta(w))$ exists.

When they all exist, the diagrams

$$\begin{array}{ccc} w & \xrightarrow{\eta} & \eta(w) \\ \rho^{A_1}_{A_2} \downarrow & & \downarrow \rho^{A_1}_{A_2} \\ w' & \xrightarrow{\eta} & \eta(w') \end{array}$$

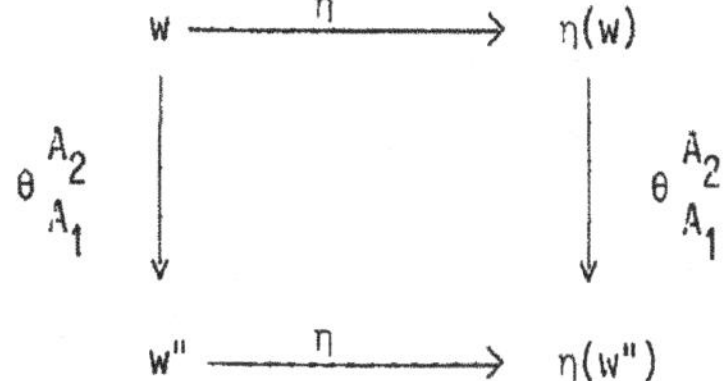

commute. □

CHAPTER 12 : THE SEQUENCE $\xi(w,k)$ BEGINNING WITH AN ELEMENT OF N_λ.

In the present chapter, the partition $\lambda = \{\lambda_1 \geq \ldots \geq \lambda_r\} \in \Lambda_n$ is fixed. $\mu = \{\mu_1 \geq \ldots \geq \mu_m\}$ is the dual partition of λ. To avoid the trivial case, we assume $\lambda \neq \{1 \geq \ldots \geq 1\}$.

In Chapter 13, we shall introduce a raising operation on an element w of N_λ and show that if w' is obtained from w by a raising operation and $w' \in N_\lambda$ then w' lies in the same left cell as w. In order to do this, we shall make use of a certain sequence $\xi(w,k)$ which is discussed in this chapter.

§12.1 A DESCRIPTION OF N_λ.

<u>Lemma 12.1.1</u> <u>Assume that $w \in A_n$ has a full MDC form $(A_r,\ldots,A_1)$ at $i \in \mathbf{Z}$ which is normal with $|A_t(w)| = \lambda_t$, $1 \leq t \leq r$. Then $w \in \sigma^{-1}(\lambda)$ and so $w \in N_\lambda$.</u>

<u>Proof:</u> Assume that $w \in \sigma^{-1}(\lambda')$ for some $\lambda' = \{\lambda_1' \geq \ldots \geq \lambda_{r'}'\} \in \Lambda_n$. It suffices to show $\lambda' = \lambda$. We claim $\lambda' \geq \lambda$. For, let $S_t = \{e((w), j_t^u(w)) | 1 \leq u \leq \lambda_t\}$, $1 \leq t \leq r$. Then $S = S_1 \cup \ldots \cup S_r$ satisfies $C_n(w,r)$ with $|S_1 \cup \ldots \cup S_t| = \sum_{h=1}^{t} \lambda_h$ for any $1 \leq t \leq r$.

On the other hand, if for some $1 \leq t' \leq r'$, $S' = S_1' \cup \ldots \cup S_{t'}'$ satisfies $C_n(w,t')$, let $\bar{e}_u$ be the u-th layer of w, $1 \leq u \leq \lambda_1$. Then by Lemma 5.11, $|\bar{e}_u \cap S_v'| \leq 1$. So $|(\bigcup_{v=1}^{t'} S_v') \cap \bar{e}_u| \leq \min\{\mu_u, t'\}$ for $1 \leq u \leq \lambda_1$. This implies $|S'| \leq \sum_{h=1}^{t'} \lambda_h$. So $\lambda' = \lambda$ and $w \in \sigma^{-1}(\lambda)$. The latter part of this lemma is immediate. □

<u>Remark 12.1.2</u> The above lemma gives a sufficient condition for an element of A_n to belong to N_λ. Obviously, it is also a necessary condition.

§12.2 A SEQUENCE $\xi(w,r)$ BEGINNING WITH AN ELEMENT OF H_λ.

We begin with the case $k = r$ since this is considerably easier than the general case and the general case will make use of this special case.

Now assume that $\lambda_r > 1$.

<u>Lemma 12.2.1</u> <u>Assume that $w \in \sigma^{-1}(\lambda)$ has a full DC form $(A_{r+1},\ldots,A_1)$ at $i \in \mathbf{Z}$ with $|A_t(w)| = \lambda_t$, $1 \leq t < r$, $|A_r(w)| = \lambda_r - 1$ and $|A_{r+1}(w)| = 1$. Suppose that for some $1 \leq m \leq \lambda_r$, $j_r^m(w) < j_{r+1}^1(w) < j_r^{m-1}(w)$ with the convention that $j_r^0(w) = \infty$,</u>

$j_r^{\lambda_r}(w) = -\infty$. Then there exists $w' = \rho^{A_{r+1}}_{A_1,\ldots,A_{r-1}}(w)$ having the full DC form $(A_r, A_{r+1}, A_{r-1},\ldots,A_1)$ at $i+1$ with $j_r^m(w') < j_{r+1}^1(w) < j_{r+1}^1(w') < j_r^{m-1}(w')$.

Proof: The existence of w' follows by Lemma 8.3.5 and Lemma 5.8. Let

$$i_1 = \min\{h \mid 1 \le h \le \lambda_1,\ j_1^h(w) < j_{r+1}^1(w) + n\}$$

$$i_u = \min\{h \mid 1 \le h \le \lambda_u,\ j_u^h(w) < j_{u-1}^{i_{u-1}}(w)\},\ 1 < u \le r-1$$

Then by repeatedly applying Lemma 8.3.4, all $i_1,\ldots,i_{r-1}$ exist and $j_{r+1}^1(w') = j_{r-1}^{i_{r-1}}(w)$. If $j_{r-1}^{i_{r-1}}(w) > j_r^{m-1}(w)$, then $j_r^{m-1}(w) + n > j_{r+1}^1(w) + n > j_1^{i_1}(w) > j_2^{i_2}(w) > \ldots > j_{r-1}^{i_{r-1}}(w) > j_r^{m-1}(w)$. By Lemma 5.11, we need at least $(r+1)$ descending chains of w to contain an entry set $\{e((w),j_v) \mid 1 \le v \le r+1\}$ with $\{\bar{j}_v \mid 1 \le v \le r+1\} = \{j_v^{i_v}(w),\ 1 \le v \le r-1;\ \overline{j_{r+1}^1(w)},\ \overline{j_r^{m-1}(w)}\}$. This contradicts $w \in \sigma^{-1}(\lambda)$. So $j_{r-1}^{i_{r-1}}(w) < j_r^{m-1}(w)$ (since $\overline{j_{r-1}^{i_{r-1}}(w)} \neq \overline{j_r^{m-1}(w)}$.) i.e. $j_{r+1}^1(w') < j_r^{m-1}(w')$.

On the other hand, if $j_{r-1}^{i_{r-1}}(w) < j_{r+1}^1(w)$, let

$$\{S_1 = \{e((w),j_1^u(w)) \mid 1 \le u < i_1\} \cup \{e((w),j_{r+1}^1(w)+n)\} \cup \{e((w),j_{r+1}^v(w)+n) \mid i_{r-1} \le v \le \lambda_{r-1}\}$$

$$S_h = \{e((w),j_h^u(w)) \mid 1 \le u < i_h\} \cup \{e((w),j_{h-1}^v(w)) \mid i_{h-1} \le v \le \lambda_{h-1}\},\ 1 < h \le r-1.$$

Then $S = S_1 \cup \ldots \cup S_{r-1}$ satisfies $C_n(w,r-1)$ with $|S| = \sum_{h=1}^{r-1} \lambda_h + 1$. This also contradicts $w \in \sigma^{-1}(\lambda)$. So $j_{r-1}^{i_{r-1}}(w) > j_{r+1}^1(w)$ (also since $\overline{j_{r-1}^{i_{r-1}}(w)} \neq \overline{j_{r+1}^1(w)}$). i.e. $j_{r+1}^1(w') > j_{r+1}^1(w)$. Since $j_{r+1}^1(w) > j_r^m(w) = j_r^m(w')$, it follows that $j_r^m(w') < j_{r+1}^1(w) < j_{r+1}^1(w') < j_r^{m-1}(w')$. □

For $1 \le m \le \lambda_r$, let $H^m_{\lambda,r}$ be the set of all elements $w \in \sigma^{-1}(\lambda)$ which have a full DC form $(A_{r+1},\ldots,A_1)$ at i for some $i \in \mathbb{Z}$ with $|A_t(w)| = \lambda_t$, $1 \le t < r$, $|A_r(w)| = \lambda_r - 1$, $|A_{r+1}(w)| = 1$ and $j_r^m(w) < j_{r+1}^1(w) < j_r^{m-1}(w)$. Then it is clear that $H_\lambda = H^1_{\lambda,r}$.

Lemma 12.2.2 For any $w \in H^m_{\lambda,r}$, $1 \le m < \lambda_r$, there exists $x \in H^{m+1}_{\lambda,r}$ such that $x = \rho^{A_{r+1}}_{A_1,\ldots,A_r}(w)$.

Proof: As in Lemma 12.2.1, there exists $w' = {}^{A_{r+1}}_{A_1,\ldots,A_{r-1}}(w)$ such that $j_r^m(w') < j_{r+1}^1(w') < j_r^{m-1}(w')$. So there exists $x = \rho_{A_r}^{A_{r+1}}(w')$. Obviously, $x \in H_{\lambda,r}^{m+1}$, as required. □

By Lemma 12.2.2, for any $w \in H_\lambda$ which has a standard MDC form $(A_r,\ldots,A_1)$ at $i \in \mathbf{Z}$, let A_r^o (resp. A_r^1) be the block consisting of the first row (resp. the last λ_r-1 rows) of A_r. Then there exists a sequence of elements $w_o = w, w_1,\ldots,w_{\lambda_r}$ such that for every $1 \leqslant j < \lambda_r$, $w_j = \rho_{A_1,\ldots,A_{r-1},A_r^1}^{A_r^o}(w_{j-1}) \in H_{\lambda,r}^{j+1}$,

$w_{\lambda_r} = \rho_{A_1,\ldots,A_{r-1}}^{A_r^o}(w_{\lambda_r-1}) \in H_\lambda$.

Let $\xi(w,r)$ be the sequence obtained by linking all the sequences $\xi(w_{j-1}, \rho_{A_1,\ldots,A_{r-1},A_r^1}^{A_r^o})$, $1 \leqslant j < \lambda_r$, $\xi(w_{\lambda_r-1}\ \rho_{A_1,\ldots,A_{r-1}}^{A_r^o})$ together. (These sequences have been defined in §8.4). We can see that when $\lambda_1 \neq \lambda_r$, such a sequence exists and is unique, but when $\lambda_1 = \lambda_r$ and $r > 1$, there may exist r different such sequences beginning with w because we can name any MDC block of w by A_r. But in either case, the length of such a sequence is independent of the choice of w in H_λ and of the choice of the standard MDC forms.

The above results still hold even when $\lambda_r = 1$ (the trivial case). In that case, the sequence $\xi(w,r)$ is just $\xi(w, \rho_{A_1,\ldots,A_{r-1}}^{A_r^o})$. So we need make no restriction on λ_r.

§12.3 THE DELETION MAP $d(\lambda,m)$

In the remainder of this chapter, we shall make use of the deletion operation introduced in Chapter 11 to relate the construction of the sequence $\xi(w,k)$ to the special case $k = r$ which has just been considered. In this section, we shall define a deletion operator $d(\lambda,m)$ on a special subset $\tilde{F}(\lambda,m)$ of $\tilde{\sigma}_n^{-1}(\lambda)$ and give some result for this operator, where $\tilde{\sigma}_n$ is just $\tilde{\sigma}$ on $\tilde{A}_n$.

We define $\tilde{F}(\lambda,m)$ to be the set of all elements $\tilde{w} \in \tilde{\sigma}_n^{-1}(\lambda)$ which have the full MDC form $(A_r,\ldots,A_1)$ which is normal for the first m layers with $|A_h(\tilde{w})| \geqslant m$ for all $1 \leqslant h \leqslant r$. Let $\tau = \{\lambda_1-m \geqslant \ldots \geqslant \lambda_r-m\} \in \Lambda_{n-rm}$. We define a map $d(\lambda,m):\tilde{F}(\lambda,m) \to \tilde{A}_{n-rm}$ by deleting $\zeta(\tilde{w},m)$ from $\tilde{w} \in \tilde{F}(\lambda,m)$, where $\zeta(\tilde{w},m)$ is the set of rc-classes of $\tilde{w}$ containing the first m entries of all MDC blocks $\tilde{w}$ with $|\zeta(\tilde{w},m)| = rm$.

Lemma 12.3.1 Im $d(\lambda,m) \subseteq \tilde{\sigma}_{n-rm}^{-1}(\tau)$.

To show this lemma, we need the following result.

Lemma 12.3.2 Assume that $w \in \sigma^{-1}(\lambda)$ has a full DC form $(A_\ell,\ldots,A_1)$ at i. Then for any t, $1 \leq t \leq r$, there exists $S = S_1 \cup \ldots \cup S_t$ satisfying $C_n(w,t)$ and also satisfying the following conditions.

(i) Let $|S_h| = m_h$, $1 \leq h \leq t$. Then $\sum_{h=1}^{t} m_h = \sum_{h=1}^{t} \lambda_h$.

(ii) Let $S_h = \{e(i_{hk},j_{hk}) | 1 \leq k \leq m_h,\ i_{h_1} < \ldots < i_{hm_h},\ j_{h_1} > \ldots > j_{hm_h}\}$.

Then $\overline{j_{h1}} = \overline{j^1_{v_h}}(w)$ for some $1 \leq v_h \leq \ell$.

Proof: Let $E = \{\overline{j^1_v(w)} | 1 \leq v \leq \ell\} \subseteq \underline{n}$. Since $w \in \sigma^{-1}(\lambda)$, there exists $S = S_1 \cup \ldots \cup S_t$ satisfying $C_n(w,t)$ such that $|S| = \sum_{h=1}^{t} \lambda_h$ and $S_h = \{e(i_{hk},j_{hk}) | 1 \leq k \leq m_h,\ i_{h1} < \ldots < i_{hm_h},\ j_{h1} > \ldots > j_{hm_h}\}$, $1 \leq h \leq t$. If $\{\overline{j_{h1}} \,|1 \leq h \leq t\} \subseteq E$, then S is as required. Otherwise, there exists $\overline{j_{h1}} \notin E$, say $\overline{j_{h1}} = \overline{j^u_v(w)}$ for some $1 \leq v \leq \ell$, $u > 1$. Without loss of generaltiy, we may assume that $j_{h1} = j^u_v(w)$. Let

$$\begin{cases} S^1_h = S_h \cup \{e((w),\ j^\alpha_v(w)) | 1 \leq \alpha < u\} \\ S^1_k = S_k - \{e(i,j) | \bar{j} \in \{\overline{j^\alpha_v(w)} | 1 \leq \alpha < u\}\},\ 1 \leq k \leq t,\ k \neq h. \end{cases}$$

Then $S^1 = S^1_1 \cup \ldots \cup S^1_t$ satisfies $C_n(w,t)$ with $|S^1| \geq |S|$. But $|S^1| \leq \sum_{h=1}^{t} \lambda_h$ in general since $w \in \sigma^{-1}(\lambda)$. So we must have $|S^1| = |S|$.

For S, let $a(S,h) \geq 0$ be the largest number such that $e(i_{h\alpha},j_{h\alpha}) = e((w),j^\alpha_v(w))$ for all $1 \leq \alpha \leq a(S,h)$ and some $1 \leq v \leq \ell$. Let $a(S) = \sum_{h=1}^{t} a(S,h)$. Similarly, we can define $a(S^1)$ for S^1. Then it is clear that $n \geq a(S^1) > a(S)$. If S^1 is still not as required, then by the same procedure, we can get S^2, S^3,... in turn with $a(S^1) < a(S^2) < \ldots$. Since $a(S^\alpha) \leq n$ for all $\alpha \geq 1$. This implies that after a finite number of steps, we can get the required disjoint union of descending chains of w. □

Lemma 12.3.3 Assume that $\tilde{w} \in \tilde{\sigma}^{-1}(\lambda)$ has a full DC form $(A_\ell,\ldots,A_1)$. Then for any $1 \leq t \leq r$, there exists $S = S_1 \cup \ldots \cup S_t$ satisfying $C_n(\tilde{w},t)$ and also satisfying the following conditions.

(i) Let $|S_h| = m_h$, $1 \leq h \leq t$. Then $\sum_{h=1}^{t} m_h = \sum_{h=1}^{t} \lambda_h$.

(ii) Let $S_h = \{e_{(hk)} \mid 1 \leq k \leq m_h,\ r(e_{(hk_1)}, e_{(hk_2)}) > 0,$

$c(e_{(hk_1)}, e_{(hk_2)}) < 0$, for any $1 \leq k_1 < k_2 \leq m_h\}$

Then $r(e_{(h1)}, e_{v_h 1}) = C(e_{(h1)}, e_{v_h 1}) \in n\mathbb{Z}$ for some $1 \leq v_h \leq \ell$, where e_{uv} is the v-th entry of A_u.

Proof: This is only a version of Lemma 12.3.2 in $\tilde{A}_n$. □

Proof of Lemma 12.3.1 Let $\tilde{y} \in \operatorname{Im} d(\lambda,m)$ with $\tilde{y} \in \tilde{\sigma}_{n-rm}^{-1}(\tau')$. Let $\tilde{w} \in \tilde{F}(\lambda,m)$ be such that $\tilde{y} = d(\lambda,m)(\tilde{w})$ and $\tilde{w}$ has a full MDC form $(A_r,\ldots,A_1)$ which is normal for the first m layers. Let $e_{uv}(\tilde{w})$ be the v-th entry of $A_u(\tilde{w})$, for u,v with $1 \leq u \leq r$, $1 \leq v \leq |A_u(\tilde{w})|$. First we claim that $\tau' \geq \tau$. For, suppose that $S = S_1 \cup \ldots \cup S_t$ satisfies $C_n(\tilde{w},t)$ with $|S| = \sum_{h=1}^{t} \lambda_h$, $1 \leq t \leq r$. Let S' be obtained from S by deleting $\zeta(\tilde{w},m)$. Then by Lemma 11.5.14, $|S'| \geq |S| - tm = \sum_{h=1}^{t} \tau_h$. Since S' satisfies $C_{n-rm}(\tilde{y},t')$ with $t' \leq t$, we must have $\tau' \geq \tau$, where $t' = t$ if $\tau_t \geq 1$ or $t' = \max\{h \mid 1 \leq h \leq r,\ \tau_h \geq 1\}$ if $\tau_t = 0$.

Secondly, we claim that $\tau' \leq \tau$. For, by Lemma 12.3.3, for any $1 \leq t \leq r$, there exists $S' = S_1' \cup \ldots \cup S_t'$ satisfying $C_{n-mr}(\tilde{y},t)$ and all conditions obtained from those in Lemma 12.3.3 by replacing λ, $\tilde{w}$ by τ', $\tilde{y}$. This implies that for any $1 \leq t \leq r$, there exists $S = S_1 \cup \ldots \cup S_t$ satisfying $C_n(\tilde{w},t)$ such that

(i) no entry of S lies in $\zeta(\tilde{w},m)$.

(ii) $|S| = \sum_{h=1}^{t} \tau_h'$.

(iii) Let $S_h = \{e_{(hk)} \mid 1 \leq k \leq |S_h|,\ r(e_{(hk_1)}, e_{(hk_2)}) > 0,$

$c(e_{(hk_1)}, e_{(hk_2)}) < 0$ for $1 \leq k_1 < k_2 \leq |S_h|\}$.

Then $r(e_{(h1)}, e_{u_h,m+1}(\tilde{w})) = c(e_{(h1)}, e_{u_h,m+1}(\tilde{w})) \in n\mathbb{Z}$ for some $u_h \in \{1,\ldots,r\}$.

Without loss of generality, we may assume $e_{(h1)} = e_{u_h,m+1}(\tilde{w})$ for $1 \leq h \leq t$. Let

$$S_h'' = S_h \cup \{e_{u_h,\alpha}(\tilde{w}) \mid 1 \leq \alpha \leq m\},\ 1 \leq h \leq t.$$

Then $S'' = S_1'' \cup \ldots \cup S_t''$ satisfies $C_n(\tilde{w},t)$ with $|S''| = \sum_{h=1}^{t} \tau_h' + tm$.

It follows by $\tilde{w} \in \tilde{\sigma}_n^{-1}(\lambda)$ that $|S''| \leq \sum_{h=1}^{t} \lambda_h$. Hence $\sum_{h=1}^{t} \tau'_h \leq \sum_{h=1}^{t} \tau_h$. So $\tau' \leq \tau$ and then $\tau = \tau'$, i.e. $\tilde{y} \in \tilde{\sigma}_{n-rm}^{-1}(\tau)$. Im $d(\lambda,m) \subseteq \tilde{\sigma}_{n-rm}^{-1}(\tau)$. □

§12.4 THE SUBSET $\tilde{H}_{\lambda,k}$ OF $\tilde{\sigma}^{-1}(\lambda)$

In this section, we introduce a certain subset $\tilde{H}_{\lambda,k}$ of $\tilde{A}_n$ and show that certain combinations of ρ operations can be applied to any given element $\tilde{w} \in \tilde{H}_{\lambda,k}$. This will subsequently enable us to construct the sequence $\xi(\tilde{w},k)$.

In this section and §12.5, §12.6, we shall assume that $1 \leq k < r$ and $\lambda_k > \lambda_{k+1}$.

Let $k = i_0 < i_1 < \ldots < i_t = r$ and $i_{t+1} = r+1$ be such that for every $0 \leq h \leq t$, $\lambda_{i_h} > \lambda_{i_h+1} = \ldots = \lambda_{i_{h+1}}$. Let

$$\alpha_{\lambda,h} = \lambda_{i_h} - \lambda_{i_{h+1}},\ \tau_a^h = \lambda_a - \lambda_{i_{h+1}},\ \beta_{\lambda,h} = n - \sum_{j=1}^{\lambda_{i_{h+1}}} \mu_j$$

for h, a with $0 \leq h \leq t$ and $1 \leq a \leq i_h$. We have $\tau^h = \{\tau_1^h \geq \ldots \geq \tau_{i_h}^h\} \in \Lambda_{\beta_{\lambda,h}}$.

Let $\tilde{N}_\tau = \eta(N_\tau)$ for any $\tau \in \Lambda_m$ with $m \geq 1$. Given $\tilde{y} \in \tilde{N}_{\tau^h}$ with $1 \leq h < t$, we construct an element $\tilde{w}$ as follows.

Assume that $\tilde{y}$ has the standard MDC form $(B_{i_h},\ldots,B_1)$. Let $f_{pq}(\tilde{y})$ be the row of $\tilde{y}$ containing the entry $e_{pq}(\tilde{y})$. For each u, $0 \leq u < h$, let $B_{i_u}^0(\tilde{y})$ (resp. $B_{i_u}^1(\tilde{y})$) be the block consisting of the first $\tau_{i_{u+1}}^h$ (resp. the last $\tau_{i_u}^h - \tau_{i_{u+1}}^h - 1$) rows of $B_{i_u}(\tilde{y})$. We permute the blocks from $[B_{i_{u+1}-1}, B_{i_{u+1}-2},\ldots,B_{i_u+1},B_{i_u}^0, f_{i_u,\tau_{i_{u+1}}^h+1}]$ to $[f_{i_u,\tau_{i_{u+1}}^h+1}, B_{i_{u+1}-1},\ldots,B_{i_u+1},B_{i_u}^0]$ in the matrix $\tilde{y}$ for each u, $0 \leq u < h$, and get $\tilde{w}$.

Let $\tilde{H}_{\tau^h,k}$ be the set of all elements $\tilde{w}$ of $\tilde{A}_{\beta_{\lambda,h}}$ obtained by the above procedure. We see that for any $\tilde{w} \in \tilde{H}_{\tau^h,k}$ the corresponding $\tilde{y} \in \tilde{N}_{\tau^h}$ must be uniquely determined and also the standard MDC form $(B_{i_h},\ldots,B_1)$ of $\tilde{y}$ is uniquely determined.

Let $\tilde{w} \in \tilde{H}_{\tau^h,k}$ have full MDC form $(A_{i_h},\ldots,A_1)$. We say that this MDC form of $\tilde{w}$ is semistandard if it comes from a standard MDC form $(B_{i_h},\ldots,B_1)$ of $\tilde{y} \in \tilde{N}_{\tau^h}$ by the above procedure, where

$$A_v(\tilde w) \text{ comes from } \begin{cases} B_v(\tilde y) \text{ if } 1 \le v \le i_h,\ v \notin \{i_0,\dots,i_h\}. \\ [B^0_{i_u}(\tilde y),\ B^1_{i_u}(\tilde y),\ f_{i_{u-1},\tau^h_{i_u}+1}(\tilde y)] \text{ if } v = i_u,\ 0 < u < h. \\ [B^0_{i_0}(\tilde y),\ B^1_{i_0}(\tilde y)] \text{ if } v = i_0. \\ [B_{i_h}(\tilde y),\ f_{i_{h-1},\ \tau^h_{i_h}+1}(\tilde y)] \text{ if } v = i_h. \end{cases}$$

Clearly, the semistandard MDC form of $\tilde w$ is uniquely determined.

Let $\tilde H_{\tau^0,k} = \tilde N_{\tau^0}$. Clearly, $\tilde H_{\tau^h,k} \subseteq \tilde F(\tau^h, \tau^h_{i_h})$ for all $0 \le h \le t$ and $\tilde H_{\lambda,k} = \tilde H_{\tau^t,k}$.

Let us give an example for an element $\tilde w \in \tilde A_n$ which lies in $\tilde H_{\lambda,k}$ and is obtained from $\tilde y \in \tilde N_\lambda$ in the above way, where $n = 9$, $k = 1$, $\lambda = \{5 > 2 > 2\}$.

$\tilde y$ $\qquad\qquad$ $\tilde w$

<u>Lemma 12.4.2</u> $\tilde H_{\lambda,k} \subseteq \tilde\sigma^{-1}(\lambda)$.

<u>Proof</u>: Take any $\tilde w \in \tilde H_{\lambda,k}$. In the matrix $\tilde w$, let

$$\tilde e_{uv} = \begin{cases} e_{uv} & \text{for (i) } 1 \le u \le r,\ u \notin \{i_1,\dots,i_{t-1}\} \text{ (ii) } u=i_h \\ & \text{and } 1 \le v \le \lambda_{i_{h+1}} \text{ with } 0 \le h \le t-1. \\ e_{u,v-1} & \text{for (i) } u=i_h,\ 1 \le h < t,\ \lambda_{i_{h+1}}+1 < v < \lambda_{i_h}+1 \text{ (ii) } u=i_0, \\ & \lambda_{i_1}+1 < v \le \lambda_{i_0}. \end{cases}$$

Then $0 < c(\tilde e_{uv}, \tilde e_{u'v}) < n$ for $1 \le u' < u \le r$ and $1 \le v \le \lambda_1$ when $\tilde e_{uv}$, $\tilde e_{u'v}$ both exist.

Suppose that $\tilde{w} \in \tilde{\sigma}^{-1}(\lambda')$, $\lambda' = \{\lambda_1' \geq \dots \geq \lambda_r'\}$. Let $S = S_1 \cup \dots \cup S_h$ satisfy $C_n(\tilde{w},h)$. Let $\tilde{e}_v$ be the set of entry classes containing all $\tilde{e}_{uv}$, $1 \leq u \leq r$, which exist, and let $\mu_v = |\tilde{e}_v|$. Then by Lemma 11.5.14, $|S \cap \tilde{e}_v| \leq \min\{\mu_v, h\}$. This implies

$$|S| = \sum_{v=1}^{\lambda_1} |S \cap \tilde{e}_v| \leq \sum_{i=1}^{h} \lambda_i$$

So $\lambda' \leq \lambda$.

On the other hand, let $S_v' = \{\text{all entries in } A_v(\tilde{w})\}$,

$$1 \leq v < k,\ S_k' = \Big(\bigcup_{h=1}^{t-1} \{\tilde{e}_{i_h,v}(\tilde{w}) | \lambda_{i_{h+1}} + 1 < v \leq \lambda_{i_h} + 1\}\Big) \cup$$

$$\{\tilde{e}_{i_t,v} | 1 \leq v \leq \lambda_{i_t} + 1\} \cup \{\tilde{e}_{i_0,v} | \lambda_{i_1} + 1 < v \leq \lambda_{i_0}\},$$

$S_h' = \{\text{all entries in } A_{h-1}(\tilde{w})\} - S_k'$, $k < h \leq r$. Then $S' = S_1' \cup \dots \cup S_r'$ satisfies $C_n(\tilde{w},r)$ with $|S_h'| = \lambda_h$, $1 \leq h \leq r$. So $\lambda' \geq \lambda$. Therefore $\lambda = \lambda'$ and $\tilde{w} \in \tilde{\sigma}^{-1}(\lambda)$, i.e. $\tilde{H}_{\lambda,k} \subseteq \tilde{\sigma}^{-1}(\lambda)$. □

For any $\tilde{w} \in \tilde{H}_{\lambda,k}$ with semi-standard MDC form $(A_r,\dots,A_1)$, let $\tilde{v}_1 = d(\tau^t, \tau^t_{i_t})(\tilde{w})$. Then $\tilde{v}_1$ has the full MDC form $(A_{i_{t-1}},\dots,A_1)$ satisfying all conditions for $\tilde{H}_{\tau^{t-1},k}$, where $A_{i_{t-1}}(\tilde{v}_1)$ is the MDC block of $\tilde{v}_1$ coming from $[A_{i_t}(\tilde{w}),\dots,A_{i_{t-1}}(\tilde{w})]$ by deletion and $A_u(\tilde{v}_1)$ comes from $A_u(\tilde{w})$ for $1 \leq u < i_{t-1}$. So $\tilde{v}_1 \in \tilde{H}_{\tau^{t-1},k}$, $d(\tau^t,\tau^t_{i_t})(\tilde{H}_{\lambda,k}) \subseteq \tilde{H}_{\tau^{t-1},k}$. In general, for any $1 < h \leq t$, $\tilde{v}_{t+1-h} = d(\tau^h,\tau^h_{i_h})\cdot d(\tau^{h+1},\tau^{h+1}_{i_{h+1}})\dots d(\tau^t,\tau^t_{i_t})(\tilde{w})$ exists and has the full MDC form $(A_{i_{h-1}},\dots,A_1)$ satisfying all conditions for $\tilde{H}_{\tau^{h-1},k}$. Then $\tilde{v}_{t+1-h} \in \tilde{H}_{\tau^{h-1},k}$, where $A_{i_{h-1}}(\tilde{v}_{t+1-h})$ comes from $[A_{i_h}(\tilde{v}_{t-h}),\dots,A_{i_{h-1}}(\tilde{v}_{t-h})]$ and $A_u(\tilde{v}_{t+1-h})$ comes from $A_u(\tilde{v}_{t-h})$, $1 \leq u < i_{h-1}$. Finally, there exists $\tilde{v}_t = d(\tau',\tau'_{i_1})(\tilde{v}_{t-1})$ having the full MDC form $([A_{k+1},A_k], A_{k-1},\dots,A_1)$ satisfying all conditions for $\tilde{N}_{\tau^0}$, where $A_{k+1}(\tilde{v}_t)$ comes from $[A_{i_1}(\tilde{v}_{t-1}),\dots,A_{k+1}(\tilde{v}_{t-1})]$ and $A_u(\tilde{v}_t)$ comes from $\tilde{A}_u(\tilde{v}_{t-1})$, $1 \leq u \leq k$. Thus we have

$$\tilde{H}_{\tau^t,k} \xrightarrow{d(\tau^t,\tau^t_{i_t})} \tilde{H}_{\tau^{t-1},k} \xrightarrow{d(\tau^{t-1},\tau^{t-1}_{i_{t-1}})} \dots \xrightarrow{d(\tau^2,\tau^2_{i_2})} \tilde{H}_{\tau^1,k} \xrightarrow{d(\tau^1,\tau^1_{i_1})} \tilde{H}_{\tau^0,k}$$

Under the map $\tilde{w} \to d(\tau^h,\tau^h_{i_h})\cdot d(\tau^{h+1},\tau^{h+1}_{i_{h+1}})\dots d(\tau^t,\tau^t_{i_t})(\tilde{w})$ the blocks of $\tilde{w}$ are deleted into the following ones.

$$[A_r(\tilde{w}),\ldots,A_{k+1}(\tilde{w})] \rightarrow \begin{cases} [A_{i_{h-1}}(\tilde{v}_{t+1-h}),\ldots,A_{k+1}(\tilde{v}_{t+1-h})] & h > 1 \\ A_{k+1}(\tilde{v}_t) & h = 1 \end{cases}$$

$$[A_r(\tilde{w}),\ldots,A_{i_1}(\tilde{w})] \rightarrow \begin{cases} [A_{i_{h-1}}(\tilde{v}_{t+1-h}),\ldots,A_{i_1}(\tilde{v}_{t+1-h})] & h > 1 \\ A_{k+1}(\tilde{v}_t) & h = 1 \end{cases}$$

$$[A_r(\tilde{w}),\ldots,A_{i_1+1}(\tilde{w})] \rightarrow \begin{cases} [A_{i_{h-1}}(\tilde{v}_{t+1-h}),\ldots,A_{i_1+1}(\tilde{v}_{t+1-h})] & h > 2 \\ A_{i_1,1}(\tilde{v}_{t-1}) & h = 2 \\ \emptyset & h = 1 \end{cases}$$

$$A^o_{i_1}(\tilde{w}) \longrightarrow \begin{cases} A^o_{i_1}(\tilde{v}_{t+1-h}) & h \geq 2 \\ A^*_{i_1}(\tilde{v}_{t-1}) & h = 2 \\ \emptyset & h = 1 \end{cases}$$

$$A^!_{i_1}(\tilde{w}) \longrightarrow \begin{cases} A^!_{i_1}(\tilde{v}_{t+1-h}) & h > 1 \\ A_{k+1}(\tilde{v}_t) & h = 1 \end{cases}$$

$$A_{i_1}(\tilde{w}) \longrightarrow \begin{cases} A_{i_1}(\tilde{v}_{t+1-h}) & h > 2 \\ A^!_{i_1,|A_{i_1}|-1}(\tilde{v}_{t-1}) & h = 2 \\ A_{k+1}(\tilde{v}_t) & h = 1 \end{cases}$$

$$A_u(\tilde{w}) \longrightarrow A_u(\tilde{v}_{t+1-h})$$

$(1 < u < k)$

where $A_{i_1,u}$ (resp. $A^!_{i_1,u}$) is the block consisting of the first u (resp. the last u) rows of A_{i_1} for $0 < u < |A_{i_1}|$. $A^o_{i_1} = A_{i_1,|A_{i_1}|-1}$. $A^*_{i_1}$ is the block obtained from A_{i_1} by omitting the first row and the last row. $\emptyset$ is the block of size 0. $A^!_{i_1} = A^!_{i_1,1}$.

Now let us observe an example to illustrate what the above statement means

Suppose $\tilde{w} \in \tilde{H}_{\lambda,1}$ with $\lambda = \{6 > 4 > 2\} \in \Lambda_{12}$ which has the following semi-standard MDC form (A_3, A_2, A_1).

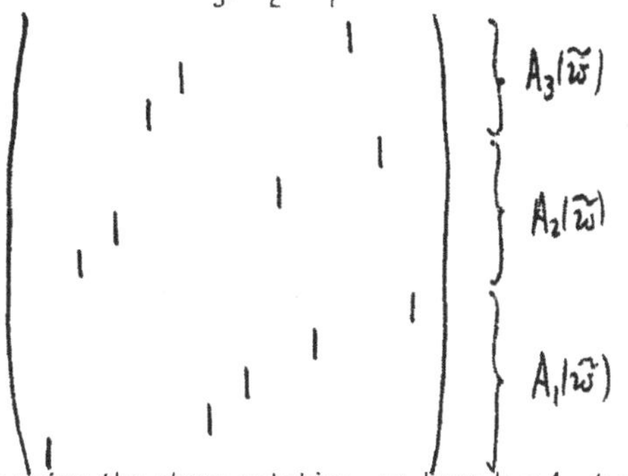

$A_3(\tilde{w})$

$A_2(\tilde{w})$

$A_1(\tilde{w})$

So by using the above notation, we have $k = 1$, $t = 2$, $i_1 = 2$. Let $\tau^1 = \{4>2\} \in \Lambda_6$ and $\tau^0 = \{2\} \in \Lambda_2$. Then $\tilde{v}_1 = d(\tau^1,2)(\tilde{w}) \in \tilde{H}_{\tau^1,1}$ and $\tilde{v}_2 = d(\tau^0,2)(\tilde{v}_1) \in \tilde{H}_{\tau^0,1} = \tilde{N}_{\tau^0}$ are as follows.

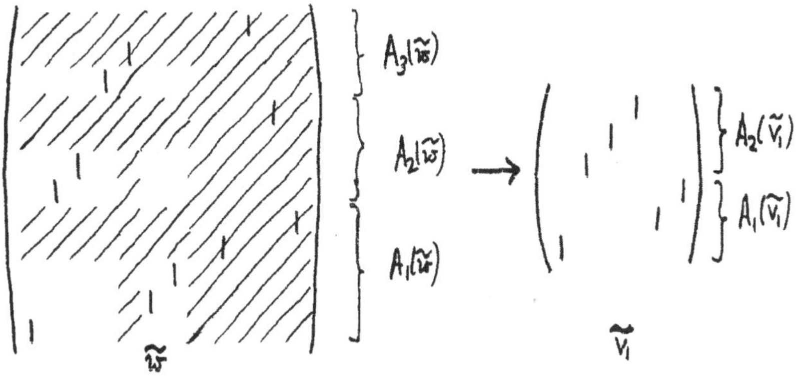

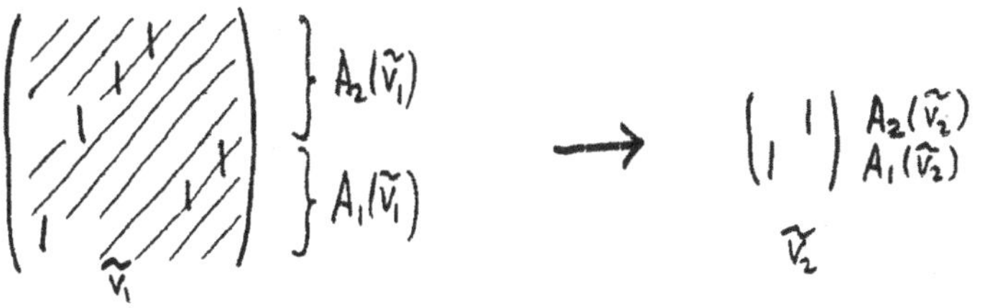

$A_3(\tilde{w})$ $A_2(\tilde{w})$ $A_1(\tilde{w})$ $\tilde{w}$ → $A_2(\tilde{v}_1)$ $A_1(\tilde{v}_1)$ $\tilde{v}_1$

$A_2(\tilde{v}_1)$ $A_1(\tilde{v}_1)$ $\tilde{v}_1$ → $A_2(\tilde{v}_2)$ $A_1(\tilde{v}_2)$ $\tilde{v}_2$

where each matrix on the right is obtained from the corresponding one on the left by leaving out all rows and columns fully shaded.

We see that the blocks of $\tilde{w}$ here are deleted into the following ones.

$$[A_3(\tilde{w}),\ A_2(\tilde{w})] \xrightarrow{d(\tau^1,2)} A_2(\tilde{v}_1) \xrightarrow{d(\tau^0,2)} A_2(\tilde{v}_2)$$

$$A_1(\tilde{w}) \xrightarrow{d(\tau^1,2)} A_1(\tilde{v}_1) \xrightarrow{d(\tau^0,2)} A_1(\tilde{v}_2)$$

$$A_3(\tilde{w}) \xrightarrow{d(\tau^1,2)} A_{2,1}(\tilde{v}_1) \xrightarrow{d(\tau^0,2)} \emptyset$$

$$A_2(\tilde{w}) \xrightarrow{d(\tau^1,2)} A'_{2,2}(\tilde{v}_1) \xrightarrow{d(\tau^0,2)} A_2(\tilde{v}_2)$$

$$A_2^0(w) \xrightarrow{d(\tau^1,2)} A_2^*(\tilde{v}_1) \xrightarrow{d(\tau^0,2)} \emptyset$$

$$A_2^1(\tilde{w}) \xrightarrow{d(\tau^1,2)} A_2^1(\tilde{v}_1) \xrightarrow{d(\tau^0,2)} A_2(\tilde{v}_2)$$

Now we return to discuss the general cases. By Corollary 11.8.2, we have

$$\rho^{A_{i_1}^0}_{A_k}\ \rho^{A_{i_1}}_{A_1,\dots,A_{k-1}}\ \rho^{A_r,\dots,A_{i_1+1}}_{A_1,\dots,A_k}\ (\rho^{A_r,\dots,A_{k+1}}_{A_1,\dots,A_k})^{\alpha_{\lambda,0}^{-1}}\ (\tilde{w}) \text{ exists}$$

$$\Leftrightarrow\ \rho^{A_{i_1}^0}_{A_k}\ \rho^{A_{i_1}}_{A_1,\dots,A_{k-1}}\ \rho^{A_{i_{h-1}},\dots,A_{i_1+1}}_{A_1,\dots,A_k}(\rho^{A_{i_{h-1}},\dots,A_{k+1}}_{A_1,\dots,A_k})^{\alpha_{\lambda,0}^{-1}}\ (\tilde{v}_{t+1-h})$$
$$\text{exists, } 2 < h < t.$$

$$\Leftrightarrow\ \rho^{A_{i_1}^*}_{A_k}\ \rho^{A'_{i_1},|A_{i_1}|-1}_{A_1,\dots,A_{k-1}}\ \rho^{A_{i_1},1}_{A_1,\dots,A_k}\ (\rho^{A_{i_1},\dots,A_{k+1}}_{A_1,\dots,A_k})^{\alpha_{\lambda,0}^{-1}}\ (\tilde{v}_{t-1})$$

$$= \rho^{A_{i_1}^0}_{A_k}\ \rho^{A_{i_1}}_{A_1,\dots,A_{k-1}}\ (\rho^{A_{i_1},\dots,A_{k+1}}_{A_1,\dots,A_k})^{\alpha_{\lambda,0}^{-1}}\ (\tilde{v}_{t-1}) \text{ exists} \tag{1}$$

Also,

$$\rho^{A_r,\dots,A_{i_1}}_{A_1,\dots,A_k}\ (\rho^{A_r,\dots,A_{k+1}}_{A_1,\dots,A_k})^{\alpha_{\lambda,0}^{-1}}(\tilde{w}) \text{ exists}$$

$$\Leftrightarrow\ \rho^{A_{i_{h-1}},\dots,A_{i_1}}_{A_1,\dots,A_k}\ (\rho^{A_{i_{h-1}},\dots,A_{k+1}}_{A_1,\dots,A_k})^{\alpha_{\lambda,0}^{-1}}\ (\tilde{v}_{t+1-h}) \text{ exists } 1 < h < t$$

$$\Leftrightarrow\ (\rho^{A_{k+1}}_{A_1,\dots,A_k})^{\alpha_{\lambda,0}}(\tilde{v}_t) \text{ exists.} \tag{2}$$

Also,

$$\rho^{A_{i_1}}_{A_1,\ldots,A_{k-1}} (\rho^{A_{i_1},\ldots,A_{k+1}}_{A_1,\ldots,A_k})^{\alpha_\lambda,0^{-1}} (\tilde{v}_{t-1}) \text{ exists}$$

$$\Longleftrightarrow \rho^{A_{k+1}}_{A_1,\ldots,A_{k-1}} (\rho^{A_{k+1}}_{A_1,\ldots,A_k})^{\alpha_\lambda,0^{-1}} (\tilde{v}_t) \text{ exists} \tag{3}$$

Since $\tilde{v}_{t,1} = \rho^{A_{k+1}}_{A_1,\ldots,A_{k-1}} (\rho^{A_{k+1}}_{A_1,\ldots,A_k})^{\alpha_\lambda,0^{-1}} (\tilde{v}_t)$ is the last term of the sequence $\xi(\tilde{v}_t,k)$ beginning with $\tilde{v}_t \in H_{\tau^0}$, we see that $\tilde{v}_{t,1}$ exists but $\rho^{A_{k+1}}_{A_k}(\tilde{v}_{t,1})$ doesn't. It follows from (2), (3) that

$$\tilde{v}_{t-1,1} = \rho^{A_{i_1}}_{A_1,\ldots,A_{k-1}} (\rho^{A_{i_1},\ldots,A_{k+1}}_{A_1,\ldots,A_k})^{\alpha_\lambda,0^{-1}} (\tilde{v}_{t-1}) \text{ exists but } \rho^{A_{i_1}}_{A_k}(\tilde{v}_{t-1,1})$$

doesn't, where we stipulate $\tilde{v}_0 = \tilde{w}$.

We claim that $\rho^{A^0_{i_1}}_{A_k}(\tilde{v}_{t-1,1})$ exists. Otherwise, there exists u with $0 < u < |A_{i_1}|-1$ such that $\tilde{v}_{t-1,2} = \rho^{A_{i_1},u}_{A_k}(\tilde{v}_{t-1,1})$ exists but $\rho^{A_{i_1},u+1}_{A_k}(\tilde{v}_{t-1,1})$ doesn't. Hence $\tilde{v}_{t-1,2}$ has the DC form $(A_{i_1}-1,\ldots,A_{k+1}, A_{i_1,u}, [A_k,A'_{i_1,|A_{i_1}|-u}], A_{k-1},\ldots,A_1)$ with $|[A_k(\tilde{v}_{t-1,2}), A'_{i_1,|A_{i_1}|-u}(\tilde{v}_{t-1,2})]| > \tau^1_k + 1$ and then

$$\sum_{u=1}^{k-1} |A_u(\tilde{v}_{t-1,2})| + |[A_k(\tilde{v}_{t-1,2}), A'_{i_1,|A_{i_1}|-u}(\tilde{v}_{t-1,2})]| > \sum_{u=1}^{k} \tau^1_u + 1.$$

But by Lemmas 12.3.1 and 5.8 $\tilde{v}_{t-1,2} \in \tilde{\sigma}^{-2}_{\beta_{\lambda,1}}(\tau^1)$. This gives a contradiction.

Therefore, by (1), (2), we have

$$\tilde{w}' = \rho^{A^0_{i_1}}_{A_k} \rho^{A_{i_1}}_{A_1,\ldots,A_{k-1}} \rho^{A_r,\ldots,A_{i_1+1}}_{A_1,\ldots,A_k} (\rho^{A_r,\ldots,A_{k+1}}_{A_1,\ldots,A_k})^{\alpha_\lambda,0^{-1}} (\tilde{w}) \text{ exists}$$

but $\rho^{A^1_{i_1}}_{A_k}(\tilde{w}')$ doesn't. This means that $\tilde{w}'$ has the full MDC form

$$(A_{i_1}-1,\ldots,A_{k+1}, A_r,\ldots,A_{i_1+1}, A^0_{i_1}, [A_k,A^1_{i_1}], A_{k-1},\ldots,A_1)$$

where when $t > 1$,

$$\begin{cases} |A_u| = \lambda_u, & 1 < u < r, \quad u \neq k, i_1 \\ |A^o_{i_1}| = \lambda_{i_1} - 1 & \\ |[A_k, A^1_{i_1}]| = \lambda_k & \\ |A_r| = \lambda_r + 1 & \end{cases}$$

When $t = 1$, $\begin{cases} |A_u| = \lambda_u, & 1 < u < i_1, \; u \neq k \\ |A^o_{i_1}| = \lambda_{i_1} & \\ |[A_k, A^1_{i_1}]| = \lambda_k. & \end{cases}$

§12.5 THE SEQUENCE $\xi(\tilde{w},k)$ BEGINNING WITH $\tilde{w} \in \tilde{N}_\lambda$.

For any $\tilde{w} \in \tilde{N}_\lambda$, assume that $\tilde{w}$ has the standard MDC form $(A_r,\ldots,A_1)$. We split $A_k(\tilde{w})$ into two blocks $A^o_k(\tilde{w})$ and $A^1_k(\tilde{w})$, where $A^o_k(\tilde{w})$ (resp. $A^1_k(\tilde{w})$) consists of the first row (resp. the last λ_k-1 rows) of $A_k(\tilde{w})$. By Lemmas 11.5.10 and 11.5.12, there exists $\tilde{w}' = \rho^{A_r,\ldots,A_{k+1}}_{A_1,\ldots,A_{k-1}}(\tilde{w})$ having full MDC form $(A_k, A_r, A_{r-1},\ldots,A_{k+1},A_{k-1},\ldots,A_1)$ which is normal. We claim that there also exists $\tilde{w}'' = \rho^{A_r,\ldots,A_{k+1}}_{A^1_k}(\tilde{w}')$ since $\lambda_k > \lambda_h$, $k+1 < h < r$ and $\tilde{w}'$ has local MDC form $(A^1_k, A_r,\ldots,A_{k+1})$ which is quasi-normal for the first λ_k-1 layers. Then $\tilde{w}''$ has full MDC form $([A^o_k,A_r], A_{r-1},\ldots,A_{k+1},A^1_k,A_{k-1},\ldots,A_1)$ satisfying all conditions for $\tilde{H}_{\lambda,k}$. i.e. $\tilde{w}'' \in \tilde{H}_{\lambda,k}$.

Combining the above result with §12.4 and noting that $[A^o_k,A_r]$, A^1_k play the same roles as A_r, A_k in 12.4, we see that there exists

$$\tilde{y} = \rho^{A^o_{i_1}}_{A^1_k} \rho^{A_{i_1}}_{A_1,\ldots,A_{k-1}} \rho^{A^o_k,A_r,\ldots,A_{i_1+1}}_{A_1,\ldots,A_{k-1},A^1_k} \left(\rho^{A^o_k,A_r,\ldots,A_{k+1}}_{A_1,\ldots,A_{k-1},A^1_k}\right)^{\alpha_{\lambda,0}-1}$$

$$\rho^{A_r,\ldots,A_{k+1}}_{A_1,\ldots,A_{k-1},A^1_k}(\tilde{w}) = \rho^{A^o_{i_1}}_{A^1_k} \rho^{A_{i_1}}_{A_1,\ldots,A_{k-1}} \rho^{A_r,\ldots,A_{i_1+1}}_{A_1,\ldots,A_{k-1},A^1_k}$$

$$\left(\rho^{A_r,\ldots,A_{k+1},A^o_k}_{A_1,\ldots,A_{k-1},A^1_k}\right)^{\alpha_{\lambda,0}}(\tilde{w}) \tag{4}$$

but there doesn't exist $\rho^{A^1_{i_1}}_{A^1_k}(\tilde{y})$. So there exists a sequence $\tilde{w}_0 = \tilde{w}, \tilde{w}_1,\ldots,\tilde{w}_{\alpha_{\lambda,0}+3} = \tilde{y}$ such that

$$\begin{cases} \tilde{w}_u = \rho^{A_r,\dots,A_{k+1},A_k^0}_{A_1,\dots,A_{k-1},A_k^1}(\tilde{w}_{u-1}) & \text{for } 1 \leq u \leq \alpha_{\lambda,0} \\ \tilde{w}_{\alpha_{\lambda,0}+1} = \rho^{A_r,\dots,A_{i_1+1}}_{A_1,\dots,A_{k-1},A_k^1}(\tilde{w}_{\alpha_{\lambda,0}}) & \\ \tilde{w}_{\alpha_{\lambda,0}+2} = \rho^{A_{i_1}}_{A_1,\dots,A_{k-1}}(\tilde{w}_{\alpha_{\lambda,0}+1}) & \\ \tilde{w}_{\alpha_{\lambda,0}+3} = \rho^{A_{i_1}^0}_{A_k^1}(\tilde{w}_{\alpha_{\lambda,0}+2}) & \end{cases}$$

By linking all the sequences $\xi(\tilde{w}_u, \rho^{A_r,\dots,A_{k+1},A_k^0}_{A_1,\dots,A_{k-1},A_k^1})$, $0 \leq u < \alpha_{\lambda,0}$,

$\xi(\tilde{w}_{\alpha_{\lambda,0}}, \rho^{A_r,\dots,A_{i_1+1}}_{A_1,\dots,A_{k-1},A_k^1})$, $\xi(\tilde{w}_{\alpha_{\lambda,0}+1}, \rho^{A_{i_1}}_{A_1,\dots,A_{k-1}})$ and

$\xi(\tilde{w}_{\alpha_{\lambda,0}+2}, \rho^{A_{i_1}^0}_{A_k^1})$ together (for the definition of these sequences, see §8.4, §11.4), we get the sequence $\xi(\tilde{w},k)$ beginning with $\tilde{w} \in N_\lambda$.

<u>Lemma 12.5.2</u> <u>The length of $\xi(\tilde{w},k)$ is independent of the choice of $\tilde{w}$ in $\tilde{N}_\lambda$.</u>

<u>Proof</u> This follows from formula (4). □

Let $\tilde{x}_0 = \tilde{w}$, $\tilde{x}_1,\dots,\tilde{x}_{\alpha_{\lambda,0}+1}$, $\tilde{z}$ be such that for $1 \leq j \leq \alpha_{\lambda,0} + 1$,

$$\begin{cases} \tilde{x}_j = \rho^{A_{i_1}^0}_{A_k^1}\, \rho^{A_{i_1}}_{A_1,\dots,A_{k-1}}\, \rho^{A_r,\dots,A_{i_1+1}}_{A_1,\dots,A_{k-1},A_k^1} \left(\rho^{A_r,\dots,A_{k+1},A_k^0}_{A_1,\dots,A_{k-1},A_k^1}\right)^{j-1}(\tilde{w}) \\ \tilde{z} = \rho^{A_r,\dots,A_{k+1},A_k^0}_{A_1,\dots,A_{k-1},A_k^1}(\tilde{w}). \end{cases}$$

Then all $\tilde{x}_j$, $0 \leq j \leq \alpha_{\lambda,0} + 1$, $\tilde{z}$ lie in $\xi(\tilde{w},k)$. Let us consider full MDC forms for these elements.

(i) Suppose $t > 1$.

Since $\rho^{A_r}_{A_1,\ldots,A_{k-1},A_k^1}(\tilde{w})$ has full MDC form $(A_{r-1},\ldots,A_{k+1}, [A_k^0,A_r],$ $A_k^1,A_{k-1},\ldots,A_1)$ and

$$\tilde{x}_j = \rho^{A^0_{i_1}}_{A^1_k}\ \rho^{A_{i_1}}_{A_1,\ldots,A_{k-1}}\ \rho^{A_{r-1},\ldots,A_{i_1+1}}_{A_1,\ldots,A_{k-1},A_k^1}\ (\rho^{A_{r-1},\ldots,A_{k+1},[A_k^0,A_r]}_{A_1,\ldots,A_{k-1},A_k^1})^{j-1}$$

$\rho^{A_r}_{A_1,\ldots,A_k^1}(\tilde{w})$ for $1 < j < \alpha_{\lambda,0} + 1$, we see that $[A_k^0(\tilde{x}_j), A_r(\tilde{x}_j)]$ is an MDC block of $\tilde{x}_j$ for $1 < j < \alpha_{\lambda,0} + 1$. Since $\rho^{A^1_{i_1}}_{A^1_k}(\tilde{x}_j)$ exists for $1 < j < \alpha_{\lambda,0}$ but $\rho^{A^1_{i_1}}_{A^1_k}(\tilde{x}_{\alpha_{\lambda,0}+1})$ doesn't, we see that $[A_k^1(\tilde{x}_j), A^1_{i_1}(\tilde{x}_j)]$ is not a DC block of $\tilde{x}_j$ for $1 < j < \alpha_{\lambda,0}$ but $[A_k^1(\tilde{x}_{\alpha_{\lambda,0}+1}), A^1_{i_1}(\tilde{x}_{\alpha_{\lambda,0}+1})]$ is a DC block of $\tilde{x}_{\alpha_{\lambda,0}+1}$. So by Lemma 8.2.3, $\tilde{x}_j$ has full MDC form

$(A_{i_1-1},\ldots,A_{k+1}, [A_k^0,A_r],A_{r-1},\ldots,A_{i_1+1},A^0_{i_1},A^1_k,A^1_{i_1}, A_{k-1},\ldots,A_1)$ for $1 < j < \alpha_{\lambda,0}$ and $\tilde{x}_{\alpha_{\lambda,0}+1}$ has full MDC form

$(A_{i_1-1},\ldots,A_{k+1},[A_k^0,A_r], A_{r-1},\ldots,A_{i_1+1}, A^0_{i_1},[A^1_k,A^1_{i_1}],A_{k-1},\ldots,A_1)$

(ii) Suppose $t = 1$.

The results on $[A_k^1,A^1_{i_1}]$ are the same as that in the case $t > 1$.

Since $\rho^{A^0_{i_1}}_{A_1,\ldots,A_{k-1},A_k^1}(\tilde{w})$ has full MDC form

$(A^1_{i_1},A_{i_1-1},\ldots,A_{k+1}, [A_k^0,A^0_{i_1}], A^1_k, A_{k-1},\ldots,A_1)$ and

$$\tilde{x}_j = \rho^{A^1_{i_1}}_{A_1,\ldots,A_{k-1}}(\rho^{A^1_{i_1},A_{i_1-1},\ldots,A_{k+1},[A_k^0,A^0_{i_1}]}_{A_1,\ldots,A_{k-1},A_k^1})^{j-1}\ \rho^{A^0_{i_1}}_{A_1,\ldots,A_{k-1},A_k^1}(\tilde{w})$$

for $1 < j < \alpha_{\lambda,0}+1$, we see by Lemma 8.2.3 that $\tilde{x}_j$ has full MDC form

$(A_{i_1-1},\ldots,A_{k+1}, [A_k^0,A^0_{i_1}], A^1_k, A^1_{i_1}, A_{k-1},\ldots,A_1)$ for $1 < j < \alpha_{\lambda,0}$, and $\tilde{x}_{\alpha_{\lambda,0}+1}$ has full MDC form $(A_{i_1-1},\ldots,A_{k+1}, [A_k^0,A^0_{i_1}], [A^1_k,A^1_{i_1}], A_{k-1},\ldots,A_1)$.

On the other hand, since $\tilde{x} = \rho^{A_r,\dots,A_{k+1}}_{A_1,\dots,A_{k-1},A_k^1}(\tilde{w}) \in \tilde{H}_{\lambda,k}$ has full MDC form $([A_k^0,A_r], A_{r-1},\dots,A_{k+1},A_k^1, A_{k-1},\dots,A_1)$ which is normal for the first λ_r layers and so has full local MDC form $(A_r,\dots,A_{k+1},A_k^1,A_{k-1},\dots,A_1,A_k^0)$ which is quasi-normal for the first layer, we see by Lemma 11.5.12 that $\tilde{z} = \rho^{A_k^0}_{A_1,\dots,A_{k-1},A_k^1}(\tilde{x})$ has full MDC form $(A_r,\dots,A_{k+1},A_k^0,A_k^1,A_{k-1},\dots,A_1)$ which is quasi-normal for the first layer. So we have

<u>Lemma 12.5.3</u> <u>$\pi(\mathcal{L}(\tilde{x}_{\alpha_{\lambda,0}+1})) > \pi(\mathcal{L}(\tilde{x}_j)) = \pi(\mathcal{L}(\tilde{x}_{j'}))$ for any $1 \leq j, j' \leq \alpha_{\lambda,0}$.</u>

<u>$\pi(\mathcal{L}(\tilde{z})) = \{\lambda_1 \geq \dots \geq \lambda_{k-1} \geq \lambda_k - 1 \geq \lambda_{k+1} \geq \dots \geq \lambda_r \geq 1\}$.</u> □

§12.6 <u>THE SEQUENCE $\xi(w,k)$ BEGINNING WITH $w \in N_\lambda$.</u>

Assume that $w \in N_\lambda$ has the standard MDC form $(A_r,\dots,A_1)$ at $i \in \mathbb{Z}$. A_k^0, A_k^1, $A_{i_1}^0$, $A_{i_1}^1$ are defined in the same way as in §12.4. Then by Lemma 11.9.1, and §12.5, there exists

$$y = \rho^{A_{i_1}^0}_{A_k^1}\rho^{A_{i_1}}_{A_1,\dots,A_{k-1}}\rho^{A_k^0,A_r,\dots,A_{i_1+1}}_{A_1,\dots,A_{k-1},A_k^1}\left(\rho^{A_k^0,A_r,\dots,A_{k+1}}_{A_1,\dots,A_{k-1},A_k^1}\right)^{\alpha_{\lambda,0}-1}\rho^{A_r,\dots,A_{k+1}}_{A_1,\dots,A_{k-1},A_k^1}(w)$$

$$= \rho^{A_{i_1}^0}_{A_k^1}\rho^{A_{i_1}}_{A_1,\dots,A_{k-1}}\rho^{A_r,\dots,A_{i_1+1}}_{A_1,\dots,A_{k-1},A_k^1}\left(\rho^{A_r,\dots,A_{k+1},A_k^0}_{A_1,\dots,A_{k-1},A_k^1}\right)^{\alpha_{\lambda,0}}(w)$$

but $\rho^{A_{i_1}^1}_{A_k^1}(y)$ doesn't exist. Then we can define the sequence $\xi(w,k)$ beginning with w in the same way as that in §12.5 and conclude that the length of $\xi(w,k)$ is independent of the choice of w in N_λ. We can also define $x_0 = w, x_1,\dots,x_{\alpha_{\lambda,0}+1}$, z be such that for $1 \leq j \leq \alpha_{\lambda,0} + 1$,

$$x_j = \rho^{A_{i_1}^0}_{A_k^1}\rho^{A_{i_1}}_{A_1,\dots,A_{k-1}}\rho^{A_r,\dots,A_{i_1+1}}_{A_1,\dots,A_{k-1},A_k^1}\left(\rho^{A_r,\dots,A_{k+1},A_k^0}_{A_1,\dots,A_{k-1},A_k^1}\right)^{j-1}(w)$$

$$z = \rho^{A_r,\dots,A_{k+1},A_k^0}_{A_1,\dots,A_{k-1},A_k^1}(w)$$

and assert that all x_j, $0 \leq j \leq \alpha_{\lambda,0} + 1$, z lie in $\xi(w,k)$.

We also have

Lemma 12.6.1 $\pi(\mathcal{L}(x_{\alpha_{\lambda,0}+1})) > \pi(\mathcal{L}(x_j)) = \pi(\mathcal{L}(x_{j'}))$ for any $1 < j, j' < \alpha_{\lambda,0}$, $\pi(\mathcal{L}(z)) = \{\lambda_1 > \ldots > \lambda_{k-1} > \lambda_k - 1 > \lambda_{k+1} > \ldots > \lambda_r > 1\}$ □

§12.7 ANTICHAINS

For $w \in N_\lambda$, assume that w has the standard MDC form $(A_r,\ldots,A_1)$ at $i \in \mathbb{Z}$. Let $E_u = \{j_h^u(w) \mid 1 < h < \mu_u\} \subseteq \underline{n}$, $1 < u < \lambda_1$. Let $\zeta(E_u)$ be the set of all columns g of an affine matrix such that g are labelled by integers which, when taken mod n, lie in E_u. Let $\bar{e}((x),E_u)$ be the set of entries of x lying in $\zeta(E_u)$. Then for any $e_1, e_2 \in \bar{e}((w),E_u)$, we have $r(e_1,e_2)\cdot c(e_1,e_2) > 0$.

Definition 12.7.1 For any $w \in A_n$, we call a set $\bar{e}$ of entries of w an antichain if $\bar{e}$ is a union of some entry classes and for $e_1, e_2 \in \bar{e}$, $r(e_1.e_2).c(e_1,e_2) > 0$ holds.

Note that this definition is essnetially the same as that in §4.4 (viii).

By this definition, the sets $\bar{e}((w),E_u)$, $1 < u < \lambda_1$ are all antichains of w. Now we shall consider $\bar{e}((x),E_u)$, $1 < u < \lambda_1$, where x is some special element lying in the sequence $\xi(w,k)$ for $1 < k < r$ with $\lambda_k > \lambda_{k+1}$. We want to prove that, for this special element, the sets $\bar{e}((x),E_u)$ are all antichains of x. This property will be essential in §12.8 for us to make the D-function tables of pairs of special elements.

(1) The special elements in the sequence $\xi(w,k)$ will be introduced first in the case k = r, when they will be denoted by $x_1,\ldots,x_{\lambda_r}, x'_1,\ldots,x'_{\lambda_r}$. These elements are defined as follows.

$$x_h = (\rho^{A_r^0}_{A_1,\ldots,A_{r-1},A_r^1})^{h-1}(w), \quad 1 < h < \lambda_r$$

$$x'_{h'} = \rho^{A_r^0}_{A_1,\ldots,A_{r-1}} (\rho^{A_r^0}_{A_1,\ldots,A_{r-1},A_r^1})^{h'-1}(w), \quad 1 < h' < \lambda_r$$

where $A_r^0(w)$ (resp. $A_r^1(w)$) is the block consisting of the first row (resp. the last (λ_r-1) rows) of $A_r(w)$. Then by the result in §12.2, all x_h, $1 < h < \lambda_r$, exist. Let us denote the u-th entry of $A_r^1(x_h)$ (resp. $A_r^0(x_h)$, $A_t(x_h)$, $1 < t < r$) by $e((x_h), j^u_{A_r^1}(x_h))$ (resp. $e((x_h), j^u_{A_r^0}(x_h))$, $e((x_h), j^u_t(x_h))$). Then we have: for $1 < h < \lambda_r$,

$$\begin{cases} j_1^u(x_h),\dots,j_{\mu_u}^u(x_h)) = (j_1^u(w),\dots,j_{\mu_u}^u(w)), & \lambda_r < u \leq \lambda_1. \\ j_1^u(x_h),\dots,j_{r-1}^u(x_h),j_{A_r^1}^{u-1}(x_h)) = (j_1^u(w),\dots,j_r^u(w)), & h+1 \leq u \leq \lambda_r. \\ j_1^u(x_h),\dots,j_{r-1}^u(x_h),j_{A_r^1}^u(x_h)) = (j_r^u(w)+n,j_1^u(w),\dots,j_{r-1}^u(w)), & 1 \leq u \leq h-1. \\ j_1^h(x_h),\dots,j_{r-1}^h(x_h),j_{A_r^o}^1(x_h)) = (j_1^h(w),\dots,j_r^h(w)) \end{cases} \quad (5)$$

and for $1 \leq h' \leq \lambda_r$,

$$\begin{cases} j_1^u(x'_{h'}),\dots,j_{\mu_u}^u(x'_{h'})) = (j_1^u(w),\dots,j_{\mu_u}^u(w)), & \lambda_r < u \leq \lambda_1. \\ j_1^u(x'_{h'}),\dots,j_{r-1}^u(x'_{h'}),j_{A_r^1}^{u-1}(x'_{h'})) = (j_1^u(w),\dots,j_r^u(w)), & h'+1 \leq u \leq \lambda_r. \\ j_1^u(x'_{h'}),\dots,j_{r-1}^u(x'_{h'}),j_{A_r^1}^u(x'_{h'})) = (j_r^u(w)+n,j_1^u(w),\dots,j_{r-1}^u(w)), & 1\leq u\leq h'-1. \\ j_1^{h'}(x'_{h'}),\dots,j_{r-1}^{h'}(x'_{h'}),j_{A_r^o}^1(x'_{h'})) = (j_r^{h'}(w)+n,j_1^{h'}(w),\dots,j_{r-1}^{h'}(w)). \end{cases} \quad (6)$$

Hence, from the fact that $\bar{e}((w),E_u)$, $1 \leq u \leq \lambda_1$, are all antichains of w, we see that $\bar{e}((x_h),E_u)$ (resp. $\bar{e}(x'_{h'}), E_u)$), $1 \leq u \leq \lambda_1$, are all antichains of x_h (resp. $x'_{h'}$) for any $1 \leq h, h' \leq \lambda_r$.

(2) We now define our special elements in the case when k satisfies $1 \leq k < r$ and $\alpha_{\lambda,0} = \lambda_k - \lambda_{k+1} > 0$. The elements in this case will be denoted by $y_1,y_2,\dots,y_{\alpha_{\lambda,0}}$, $y'_0,y'_1,\dots,y'_{\alpha_{\lambda,0}},z$. They are defined as follows.

$$\begin{cases} y_h = (\rho^{A_k^o,A_r,\dots,A_{k+1}}_{A_1,\dots,A_{k-1},A_k^1})^{h-1} \rho^{A_r,\dots,A_{k+1}}_{A_1,\dots,A_{k-1},A_k^1}(w), & 1 \leq h \leq \alpha_{\lambda,0} \\ y'_{h'} = \rho^{A_{i_1}^o}_{A_k^1} \rho^{A_{i_1}}_{A_1,\dots,A_{k-1}} \rho^{A_r,\dots,A_{i_1+1}}_{A_1,\dots,A_{k-1},A_k^1} (\rho^{A_r,\dots,A_{k+1},A_k^o}_{A_1,\dots,A_{k-1},A_k^1})^{h'}(w), & 0\leq h' \leq \alpha_{\lambda,0} \\ z = \rho^{A_r,\dots,A_{k+1},A_k^o}_{A_1,\dots,A_{k-1},A_k^1}(w) \end{cases}$$

where $1 \leq k < r$ with $\lambda_k > \lambda_{k+1}$ and A_k^o, A_k^1, $A_{i_1}^o$, $A_{i_1}^1$, $\alpha_{\lambda,0}$, $i_1,\dots,i_t$ are defined as before. Then it is clear that all y_h, $y'_{h'}$, $0 \leq h, h' \leq \alpha_{\lambda,0}$, z lie in $\xi(w,k)$.

(2a) We first investigate the detailed properties of y_1. $y_1 \in H_{\lambda,k}$ has the semi-standard MDC form

$$([A_k^0,A_r],A_{r-1},\ldots,A_{k+1},A_k^1,A_{k-1},\ldots,A_1) \text{ at } i', \ i' = i + \sum_{v=k+1}^{r} \lambda_v, \text{ where}$$

$H_{\lambda,k} = \eta^{-1}(\tilde{H}_{\lambda,k})$. Since w has local MDC form $(A_k^1,A_{k-1},\ldots,A_1, A_r,\ldots,A_{k+1})$ at $i' + 1$ which is quasi-normal, by Lemma 8.4.5, for $1 \leq u \leq \alpha$, we have

$$(j_r^u(y_1),\ldots,j_{k+1}^u(y_1), j^u_{A_k^1}(y_1), j_{k-1}^u(y_1),\ldots,j_1^u(y_1))^{(om)}$$

$$= (j^u_{A_k^1}(w), j_{k-1}^u(w),\ldots,j_1^u(w), j_r^u(w) + n,\ldots,j_{k+1}^u(w) + n)^{(om)}$$

where

$$\alpha = \begin{cases} \lambda_1 & \text{if } k \neq 1 \\ \lambda_1 - 1 & \text{if } k = 1 \end{cases}$$

Let

$$\tilde{j}_h^u(y_1) = \begin{cases} j_h^{u-1}(y_1) \text{ for } & \text{(i) } h = i_v,\ 1 \leq v \leq t,\ \lambda_{i_{v+1}}+2 \leq u \leq \lambda_{i_v}+1 \\ & \text{(ii) } h = A_k^1,\ \lambda_{i_1} + 2 \leq u \leq \lambda_k. \\ j_h^u(y_1) & \text{otherwise.} \end{cases}$$

Then

$$(\tilde{j}^u_{A_k^0}(y_1), \tilde{j}_r^u(y_1),\ldots,\tilde{j}_{k+1}^u(y_1), \tilde{j}^u_{A_k^1}(y_1), \tilde{j}_{k-1}^u(y_1),\ldots,\tilde{j}_1^u(y_1))^{(om)}$$

$$= (j_k^u(w),\ldots,j_1^u(w), j_r^u(w) + n,\ldots,j_{k+1}^u(w) + n)^{(om)} \text{ for } 1 \leq u \leq \lambda_1.$$

It follows that the sets $\bar{e}((y_1),E_u)$, $1 \leq u \leq \lambda_1$, are antichains of y_1.

If we observe y_1 more closely, we find that for any $1 \leq u \leq \lambda_1$, when $\lambda_{i_{m+1}} + 1 \leq u \leq \lambda_{i_m}$, $1 \leq m \leq t$, $\bar{e}((y_1),E_u)$ corresponds to the set of the $(u-\lambda_{i_{m+1}})$-th entries of all MDC blocks of $\tilde{y}_1^m$ under the map $d(\tau^{m+1},\tau^{m+1}_{i_{m+1}})\ldots d(\tau^t,\tau^t_{i_t}).\eta$, where $\tilde{y}^m = d(\tau^{m+1},\tau^{m+1}_{i_{m+1}})\ldots d(\tau^t,\tau^t_{i_t})\cdot(\eta(y_1))$. When $\lambda_{i_1} < u \leq \lambda_1$, $\bar{e}((y_1),E_u)$ corresponds to the set of the $(u-\lambda_{i_1})$-th entries of all MDC blocks A_v, $1 \leq v \leq \mu_u$, of $\tilde{y}_1^0$, under the map $d(\tau^1,\tau^1_{i_1})\ldots d(\tau^t,\tau^t_{i_t})\cdot\eta$. This fact will be useful for the investigation in (2b), (2c) and (2d).

(2b) We now investigate the properties of y_h for $2 \leq h \leq \alpha_{\lambda,0}$. We have

$$y_h = (\rho^{[A_k^0,A_r],A_{r-1},\ldots,A_{k+1}}_{A_1,\ldots,A_{k-1},A_k^1})^{h-1}(y_1).$$

For $1 \leq m \leq t$, $\tilde{y}_1^m \in \tilde{H}_{\tau^m,k} \subseteq \tilde{F}(\tau^m,\tau^m_{i_m})$ has the semi-standard MDC form $([A_k^0,A_{i_m}],A_{i_m-1},\ldots,A_{k+1},A_k^1,A_{k-1},\ldots,A_1)$ which is normal for the first $\tau^m_{i_m}$ layers and $\tilde{y}_1^o \in \tilde{N}_{\tau^o}$ has the standard MDC form $([A_k^0,A_k^1],A_{k-1},\ldots,A_1)$, where for $0 \leq m \leq t$, $A_k^0(\tilde{y}_1^m)$ comes from $[A_k^0(\tilde{y}_1^{m+1}), A_{i_{m+1}}(\tilde{y}_1^{m+1}),\ldots,A_{i_m+1}(\tilde{y}_1^{m+1})]$ and $A(\tilde{y}_1^m)$ comes from $A(\tilde{y}_1^{m+1})$ for $A \in \{A_{i_m},\ldots,A_{k+1},A_k^1,A_{k-1},\ldots,A_1\}$.

By Lemma 11.5.13, for $2 \leq h \leq \alpha_{\lambda,0}$, $\tilde{y}_h^t = \eta(y_h)$ has full MDC form $([A_k^0,A_r]$, $A_{r-1},\ldots,A_{k+1}$, A_k^1, $A_{k-1},\ldots,A_1)$ which is normal for the first $\tau^t_{i_t}$ layers. So there exists $\tilde{y}_h^{t-1} = d(\tau^t,\tau^t_{i_t})\cdot(\eta(y_h))$. By Corollary 11.8.2 and Lemma 11.9.1,

$$\tilde{y}_h^{t-1} = (\rho^{[A_k^0,A_{i_{t-1}}],A_{i_{t-1}-1},\ldots,A_{k+1}}_{A_1,\ldots,A_{k-1},A_k^1})^{h-1}(\tilde{y}_1^{t-1}).$$

If $t > 1$, then by Lemma 11.5.13, $\tilde{y}_h^{t-1}$ has full MDC form $([A_k^0,A_{i_{t-1}}]$, $A_{i_{t-1}-1},\ldots,A_{k+1}$, $A_k^1,A_{k-1},\ldots,A_1)$ which is normal for the first $\tau^{t-1}_{i_{t-1}}$ layers. By applying induction on $t-m \geq 0$, this implies that for $2 \leq h \leq \alpha_{\lambda,0}$, there exists

$$\tilde{y}_h^m = d(\tau^{m+1},\tau^{m+1}_{i_{m+1}})\ldots d(\tau^t,\tau^t_{i_t})(\eta(y_h)) = (\rho^{[A_k^0,A_{i_m}],A_{i_m-1},\ldots,A_{k+1}}_{A_1,\ldots,A_{k-1},A_k^1})^{h-1}(\tilde{y}_1^m)$$

having full MDC form $([A_k^0,A_{i_m}]$, $A_{i_m-1},\ldots,A_{k+1},A_k^1,A_{k-1},\ldots,A_1)$ which is normal for the first $\tau^m_{i_m}$ layers for $1 \leq m \leq t$, and

$$\tilde{y}_h^o = d(\tau^1,\tau^1_{i_1})\ldots d(\tau^t,\tau^t_{i_t})(\eta(y_h)) = (\rho^{A_k^0}_{A_1,\ldots,A_{k-1},A_k^1})^{h-1}(\tilde{y}_1^o) \in \tilde{H}^h_{\tau^o,k}$$

having full MDC form $(A_k^0,A_k^1,A_{k-1},\ldots,A_1)$, where $\tilde{H}^h_{\tau^o,k} = (H^h_{\tau^o,k})$.

So for any u with $\lambda_{i_{m+1}} + 1 \leq u \leq \lambda_{i_m}$, $1 \leq m \leq t$, the set $\bar{e}((\tilde{y}_h^m),E_{u-\lambda_{i_{m+1}}})$ of the $(u-\lambda_{i_{m+1}})$-th entries of all MDC blocks of $\tilde{y}_h^m$ comes from the set $\bar{e}((\tilde{y}_1^m),E_{u-\lambda_{i_{m+1}}})$ of the $(u-\lambda_{i_{m+1}})$-th entries of all MDC blocks of $\tilde{y}_1^m$ under the operation

$(\rho^{[A_k^0,A_{i_m}],A_{i_m-1},\ldots,A_{k+1}}_{A_1,\ldots,A_{k-1},A_k^1})^{h-1}$, and $\bar{e}((\tilde{y}_1^m),E_{u-\lambda_{i_{m+1}}})$ comes from $\bar{e}((\tilde{y}_1),E_u)$ under the map $d(\tau^{m+1},\tau^{m+1}_{i_{m+1}})\ldots d(\tau^t,\tau^t_{i_t})\cdot\eta$. This implies by Corollary 11.8.2 and Lemma 11.9.1 that $\bar{e}((\tilde{y}_h^m),E_{u-\lambda_{i_{m+1}}})$ comes from $\bar{e}((y_h),E_u)$. Since $e((\tilde{y}_h^m),E_{u-\lambda_{i_{m+1}}})$ is an antichain of $\tilde{y}_h^m$, it follows that $\bar{e}((y_h),E_u)$ is an antichain of y_h, too.

For any u with $\lambda_{i_1}+1 < u < \lambda_1$, the set $\bar{e}((\tilde{y}_1^0),E_{u-\lambda_{i_1}})$ of the $(u-\lambda_{i_1})$-th entries of all MDC blocks of $\tilde{y}_1^0$ comes from $\bar{e}((y_1),E_u)$ under the map $d(\tau^1,\tau^1_{i_1})\ldots d(\tau^t,\tau^t_{i_t})\cdot\eta$. Assume that $\bar{e}(\tilde{y}_h^0)$ is the entry set of $\tilde{y}_h^0$ coming from $\bar{e}((\tilde{y}_1^0), E_{u-\lambda_{i_1}})$ under the operation $(\rho^{A_k^0}_{A_1,\ldots,A_{k-1},A_k^1})^{h-1}$. Then by Corollary 11.8.2 and Lemma 11.9.1, $\bar{e}(\tilde{y}_h^0)$ comes from $\bar{e}((y_h),E_u)$. By equation (4), $\bar{e}(\tilde{y}_h^0)$ is an antichain of $\tilde{y}_h^0$. So $\bar{e}((y_h),E_u)$ is an antichain of y_h, too. Therefore for all u,h with $1 < u < \lambda_1$ and $1 < h < \alpha_{\lambda,0}$, $\bar{e}((y_h),E_u)$ is an antichain of y_h.

(2c) We next investigate the properties of y_h' for all h. Let us rewrite

$$y_0' = \rho^{A_{i_1}^0}_{A_k^1}\,\rho^{A_{i_1}}_{A_1,\ldots,A_{k-1}}\,\rho^{A_r,\ldots,A_{i_1+1}}_{A_1,\ldots,A_{k-1},A_k^1}(w)$$

$$y_h' = \rho^{A_{i_1}^0}_{A_k^1}\,\rho^{A_{i_1}}_{A_1,\ldots,A_{k-1}}\,\rho^{[A_k^0,A_r],A_{r-1},\ldots,A_{i_1+1}}_{A_1,\ldots,A_{k-1},A_k^1}(y_h),\ 1 < h < \alpha_{\lambda,0}$$

Clearly, y_h' belongs to the sequence $\xi(w,k)$ beginning with w and has full MDC form

$$(A_{i_1-1},\ldots,A_{k+1},[A_k^0,A_r],A_{r-1},\ldots,A_{i_1+1}A_{i_1}^0,A_k^1,A_{i_1}^1,A_{k-1},\ldots,A_1)$$

at $i + \sum_{v=i_1}^{r}\lambda_v + h\cdot(\sum_{v=h+1}^{r}\lambda_v + 1)$ for $0 < h < \alpha_{\lambda,0}$, and $y'_{\alpha_{\lambda,0}}$ has full MDC form $(A_{i_1-1},\ldots,A_{k+1},[A_k^0,A_r],A_{r-1},\ldots,A_{i_1+1},A_{i_1}^0,[A_k^1,A_{i_1}^1], A_{k-1},\ldots,A_1)$ at $i + \sum_{v=i_1}^{r}\lambda_v + \alpha_{\lambda,0}\cdot(\sum_{v=k+1}^{r}\lambda_v + 1)$ since $\rho^{A_{i_1}^1}_{A_k^1}(y_h')$ exists for any $0 < h < \alpha_{\lambda,0}$ but $\rho^{A_{i_1}^1}_{A_k^1}(y'_{\alpha_{\lambda,0}})$ doesn't.

Let $\tilde{y}_h'^{,t} = \eta(y_h')$, $0 \leq h \leq \alpha_{\lambda,0}$. Let $\tilde{y}_h'^{,m} \in \tilde{A}_{\beta_{\lambda,m}}$ be obtained from $\tilde{y}_h'^{,m+1}$ by deleting the first $\tau_{i_{m+1}}^{m+1}$ entries of the MDC blocks $A_{i_1-1}(\tilde{y}_h'^{,m+1}),\ldots A_{k+1}(\tilde{y}_h'^{,m+1})$, $[A_k^o(\tilde{y}_h'^{,m+1}),\ A_{i_{m+1}}(\tilde{y}_h'^{,m+1})]$, $A_{i_{m+1}-1}(\tilde{y}_h'^{,m+1}),\ldots,A_{i_1+1}(\tilde{y}_h'^{,m+1}),A_{i_1}^o(\tilde{y}_h'^{,m+1}),A_k^1(\tilde{y}_h'^{,m+1})$, $A_{k-1}(\tilde{y}_h'^{,m+1}),\ldots,A_1(\tilde{y}_h'^{,m+1})$, $0 \leq m < t$. Then $\tilde{y}_h'^{,m}$ has full MDC form

$$\begin{cases} (A_{i_1-1},\ldots,A_{k+1},[A_k^o,A_{i_m}],A_{i_m-1},\ldots,A_{i_1+1},A_{i_1}^o\ A_k^1,A_{i_1}^1,A_{k-1},\ldots,A_1), & m > 1. \\ (A_{i_1-1},\ldots,A_{k+1},\ [A_k^o,A_{i_1}^o],\ A_k^1,\ A_{i_1}^1,\ A_{k-1},\ldots,A_1), & m = 1. \\ (A_k^1,\ A_{i_1}^1,A_{k-1},\ldots,A_1), & m = 0 \end{cases}$$

$\tilde{y}_{\alpha_{\lambda,0}}'^{,m}$ has full MDC form

$$\begin{cases} (A_{i_1-1},\ldots,A_{k+1},[A_k^o,A_{i_m}],A_{i_m-1},\ldots,A_{i_1+1},A_{i_1}^o,[A_k^1,A_{i_1}^1],A_{k-1},\ldots,A_1), & m > 1. \\ (A_{i_1-1},\ldots,A_{k+1},[A_k^o,A_{i_1}^o],\ [A_k^1,A_{i_1}^1],A_{k-1},\ldots,A_1), & m = 1. \\ ([A_k^1,A_{i_1}^1],A_{k-1},\ldots,A_1), & m = 0 \end{cases}$$

where $A_k^o(\tilde{y}_h'^{,m})$ comes from $[A_k^o(\tilde{y}_h'^{,m+1}),A_{i_{m+1}}(\tilde{y}_h'^{,m+1}),\ldots,A_{i_m+1}(\tilde{y}_h'^{,m+1})]$ for $m \geq 1$ and $A(\tilde{y}_h'^{,m})$ comes from $A(\tilde{y}_h'^{,m+1})$ for $A \neq A_k^o$ under the deletion operation. Clearly, for any MDC blocks $A(\tilde{y}_h'^{,m})$ of $\tilde{y}_h'^{,m}$ with $A \neq A_{i_1}^1$, we have $|A(y_h'^{,m})| \geq \lambda_{i_m} - \lambda_{i_{m+1}}$ when $1 \leq m \leq t$.

By Corollary 11.8.2, and Lemma 11.9.1, we can easily show that

$$\begin{cases} \tilde{y}_0'^{,m} = \rho_{A_k^1}^{A_{i_1}^o}\ \rho_{A_1,\ldots,A_{k-1}}^{A_{i_1}}\ \rho_{A_1,\ldots,A_{k-1},A_k^1}^{A_{i_m},\ldots,A_{i_1+1}}(\tilde{y}_0^m) \\ \tilde{y}_h'^{,m} = \rho_{A_k^1}^{A_{i_1}^o}\ \rho_{A_1,\ldots,A_{k-1}}^{A_{i_1}}\ \rho_{A_1,\ldots,A_{k-1},A_k^1}^{[A_k^o,A_{i_m}],A_{i_m-1},\ldots,A_{i_1+1}}(\tilde{y}_h^m), & 1 \leq h \leq \alpha_{\lambda,0} \end{cases}$$

and $2 \leq m \leq t$, and

$$\begin{cases} \tilde{y}_o'^{\,1} = \rho_{A_k^1}^{A_{i_1}^o} \rho_{A_1,\ldots,A_{k-1}}^{A_{i_1}} (\tilde{y}_o^1) \\ \tilde{y}_h'^{\,1} = \rho_{A_k^1}^{A_{i_1}^o} \rho_{A_1,\ldots,A_{k-1}}^{A_{i_1}} \rho_{A_1,\ldots,A_{k-1},A_k^1}^{A_k^o} (\tilde{y}_h^1),\ 1 < h < \alpha_{\lambda,0} \end{cases}$$

$$\tilde{y}_h'^{\,o} = \rho_{A_1,\ldots,A_{k-1}}^{A_{i_1}^1} (\tilde{y}_h^o),\quad 0 < h < \alpha_{\lambda,0}.$$

First, let us consider y_h', $1 < h < \alpha_{\lambda,0}$.

For any $1 < u < \lambda_{i_1}$, there exists m, $1 < m < t$, such that $\lambda_{i_{m+1}} + 1 < u < \lambda_{i_m}$. Let $\bar{e}(\tilde{y}_h'^{\,m})$ be the entry set of $\tilde{y}_h'^{\,m}$ consisting of the $(u-\lambda_{i_{m+1}})$-th entries of all MDC blocks except $A_{i_1}^1$. Let $\bar{e}(\tilde{y}_h^m)$ be the entry set of $\tilde{y}_h^m$ consisting of the $(u-\lambda_{i_{m+1}})$-th entries of all MDC blocks. Then since $\tilde{y}_h^m$ has full MDC form $([A_k^o,A_{i_m}],A_{i_m-1},\ldots,A_{k+1},A_k^1,A_{k-1},\ldots,A_1)$ which is normal for the first $\lambda_{i_m} - \lambda_{i_{m+1}}$ layers, by Lemma 11.5.12, $\bar{e}(\tilde{y}_h'^{\,m})$ comes from $\bar{e}(\tilde{y}_h^m)$ under the ρ operations from $\tilde{y}_h^m$ to $\tilde{y}_h'^{\,m}$ and so is an antichain of $\tilde{y}_h'^{\,m}$. Since $\bar{e}(\tilde{y}_h^m)$ comes from $\bar{e}(y_h),E_u)$ under the map $d(\tau^{m+1},\tau_{i_{m+1}}^{m+1})\ldots d(\tau^t,\tau_{i_t}^t)\cdot\eta$, by Corollary 11.8.2 and Lemma 11.9.1, $e(\tilde{y}_h'^{\,m})$ comes from $\bar{e}((\tilde{y}_h'),E_u)$ under the map η and the deleting operation. So $\bar{e}((y_h'),E_u)$ is an antichain of y_h'.

For any u with $\lambda_{i_1} + 1 < u < \lambda_1$, let $\bar{e}(\tilde{y}_h'^{\,o})$, $0 < h < \alpha_{\lambda,0}$, be the entry set of $\tilde{y}_h'^{\,o}$ coming from the entry set $\bar{e}(\tilde{y}_1^o)$ under the operation

$$\rho_{A_1,\ldots,A_{k-1}}^{A_{i_1}^1} \left(\rho_{A_1,\ldots,A_{k-1},A_k^1}^{A_{i_1}^1}\right)^{h-1},$$

where $\bar{e}(\tilde{y}_1^o)$ is the entry set of $\tilde{y}_1^o$ consisting of the $(u-\lambda_{i_1})$-th entries of all MDC blocks. Then by Equation (6) and noting that $A_{i_1}^1$ plays the role of A_k^o in this equation, we see that $\bar{e}(\tilde{y}_h'^{\,o})$ is an antichain of $\tilde{y}_h'^{\,o}$. Since $\bar{e}(\tilde{y}_1^o)$ comes from $\bar{e}((y_1),E_u)$ under the map $d(\tau^1,\tau_{i_1}^1)\ldots d(\tau^t,\tau_{i_t}^t)\cdot\eta$, by Corollary 11.8.2 and Lemma 11.9.1, $\bar{e}(\tilde{y}_h'^{\,o})$ comes from $\bar{e}((y_h'),E)$ under the deleting operation. So $\bar{e}((y_h'),E_u)$ is also an antichain of y_h'.

Secondly, we consider y_o' which has full MDC form

$$(A_{i_1-1},\dots,A_{k+1},[A_k^o,A_r],A_{r-1},\dots,A_{i_1+1},A_{i_1}^o,A_k^1,A_{i_1}^1,A_{k-1},\dots,A_1)$$

at $i + \sum_{v=i_1}^{r} \lambda_v$. Now w has full local MDC form $(A_k^1,A_{k-1},\dots,A_1,A_r,\dots,A_{i_1})$ at $i + 1 + \sum_{v=k+1}^{r} \lambda_v$ which is quasi-normal. By Lemma 8.4.5, we have

$$(j_r^u(y_o'),\dots,j_{i_1}^u(y_o'),j_{A_k^1}^u(y_o'),\ j_{k-1}^u(y_o'),\dots,j_1^u(y_o'))^{(om)}$$

$$= (j_{A_k^1}^u(w),j_{k-1}^u(w),\dots,j_1^u(w),j_r^u(w)+n,\dots,j_{i_1}^u(w)+n)^{(om)} \text{ for } 1 < u < \lambda_1.$$

So if we let

$$\tilde{j}_v^u(y_o') = \begin{cases} j_v^{u-1}(y_o') \text{ for (i) } v = i_m,\ 2 < m < t,\ \lambda_{i_{m+1}} + 2 < u < \lambda_{i_m} + 1 \\ \qquad\qquad \text{(ii) } v = A_{i_1}^o,\ \lambda_{i_2} + 2 < u < \lambda_{i_1} \\ \qquad\qquad \text{(iii) } v = A_k^1,\ \ \lambda_{i_1} + 1 < u < \lambda_k \\ j_{A_{i_1}^1}^{u-\lambda_{i_1+1}}(y_o') \text{ for } v = A_{i_1}^1 \\ j_v^u(y_o') \qquad \text{otherwise} \end{cases}$$

then

$$(\tilde{j}_{i_1-1}^u(y_o'),\dots,\tilde{j}_{k+1}^u(y_o'),\tilde{j}_{A_k^o}^u(y_o'),\tilde{j}_r^u(y_o'),\dots,\tilde{j}_{i_1+1}^u(y_o'),\tilde{j}_{A_{i_1}^o}^u(y_o'),\tilde{j}_{A_k^1}^u(y_o'),\tilde{j}_{A_{i_1}^1}^u(y_o'),$$

$$\tilde{j}_{k-1}^u(y_o'),\dots,\tilde{j}_1^u(y_o'))^{(om)} = (j_{i_1-1}^u(w),\dots,j_1^u(w),j_r^u(w)+n,\dots,j_{i_1}^u(w)+n)^{(om)} \text{ for}$$

$1 < u < \lambda_1$.

This implies that $\bar{e}((y_o'),E_u)$, $1 < u < \lambda_1$, are all antichains of y_o'.

(2d) Finally we investigate the element z. Let us rewrite $z = \rho_{A_1,\dots,A_{k-1},A_k^1}^{A_k^o}(y_1)$ which has full MDC form $(A_r,\dots,A_{k+1},A_k^o,A_k^1,A_{k-1},\dots,A_1)$ at $i + \sum_{v=k+1}^{r} \lambda_v + 1$. Since

y_1 has MDC form $(A_k^1, A_{k-1}, \ldots, A_1, A_k^0)$ at $i + 2 \sum_{v=k+1}^{r} \lambda_v + 1$ which is normal for the first layer, it follows that

$$(j^1_{r-1}(z), \ldots, j^1_{k+1}(z), j^1_{A_k^0}(z), j^1_{A_k^1}(z), j^1_{k-1}(z), \ldots, j^1_1(z))$$

$$= (j^1_{r-1}(y_1), \ldots, j^1_{k+1}(y_1), j^1_{A_k^1}(y_1), j^1_{k-1}(y_1), \ldots, j^1_1(y_1), j^1_{A_k^0}(y_1) + n).$$

So $\bar{e}((z),E_1)$ is an antichain of z since $\bar{e}((y_1),E_1)$ is an antichain of y_1. For $2 \leq u \leq \lambda_1$, it is easily seen from the corresponding assertion on $\bar{e}((y_1),E_u)$ that $\bar{e}((z),E_u)$ is an antichain of z.

By combining (1), (2a), (2b), (2c) and (2d), we have now proved the main result in this section, which is as follows.

Proposition 12.7.2 *Let $w \in N_\lambda$ and x_h, x'_h, $y_{h'}$, $y'_{h'}$, z in $\xi(w,k)$ and $E_u \subseteq \underline{n}$ be defined as above with $1 \leq k \leq r$, $0 \leq h \leq \lambda_r$, $0 \leq h' \leq \alpha_{\lambda,0}$ and $1 \leq u \leq \lambda_1$. Then for any $1 \leq u \leq \lambda_1$, $\bar{e}((x),E_u)$ is an antichain of x for $x = x_h, x'_h, y_{h'}, y'_{h'}, z$.* □

As a by-product of the above discussion, we also have:

Lemma 12.7.3 *Let $w \in N_\lambda$ and x_h, x'_h, $y_{h'}$, $y'_{h'}$, z in $\xi(w,k)$ and $E_u \subseteq \underline{n}$ be defined as in Lemma 12.7.2 with $1 \leq k \leq r$, $0 \leq h \leq \lambda_r$, $0 \leq h' \leq \alpha_{\lambda,0}$ and $1 \leq u \leq \lambda_1$. For any MDC block A(x) of x with $x = x_h, x'_h, y_{h'}, y'_{h'}, z$, let $e((x),j^\alpha_A(x))$, $e((x),j^\beta_A(x))$ be the α-th and the β-th entry of A(x), respectively. Suppose $e((x), j^\alpha_A(x)) \in \bar{e}((x),E_u)$ and $e((x), j^\beta_A(x)) \in \bar{e}((x),E_v)$. Then $\alpha > \beta$ implies $u > v$.* □

§12.8 THE D-FUNCTION

We shall in this section define a D-function which will be used in Chapter 13 to show that each raising operator on layers transforms an element w into an element w' in the same left cell as w , provided that both w and w' lie in N_λ.

Definition 12.8.1 For any $E \subseteq \underline{n}$, we define a map $d_E: \mathbb{Z} \times \mathbb{Z} \to \frac{1}{2}\mathbb{Z}$ as follows. For $\beta_1, \beta_2 \in \mathbb{Z}$ with $\beta_1 \leq \beta_2$, let $B_{\beta_1,\beta_2} = \{a \in \mathbb{Z} | \beta_1 \leq a \leq \beta_2, \bar{a} \in E\}$. Then we define

$$d_E(\beta_1,\beta_2) = -d_E(\beta_2,\beta_1) = |B_{\beta_1,\beta_2}| - \frac{1}{2}\ |\{\bar\beta_1,\ \bar\beta_2\} \cap E|.$$

Since $d_E(\beta,\beta) = 0$ for any $\beta \in \mathbf{Z}$, the map d_E is well defined. It is easily seen that for any $\beta_1,\beta_2,\beta_3 \in \mathbb{Z}$, we have

$$d_E(\beta_1,\beta_3) = d_E(\beta_1,\beta_2) + d_E(\beta_2,\beta_3) \tag{7}$$

For any $\emptyset \neq E \subseteq \underline{n}$, let $z_h \in A_n$, $h = 1,2$, have a DC form (A_h) at α_h. Assume that $\bar e((z_h),E)$ is an antichain of z_h, then there exists at most one entry of A_h lying in $\bar e((z_h),E)$ by Lemma 5.11.

Assume that $\alpha_1 < \alpha_2$ and $|A_h| = m_h$, $h = 1,2$. Let $e_{z_1}(\beta_1,\gamma_1)$ be the entry of z_1 lying in $\bar e((z_1),E)$ with $\beta_1 > \alpha_1$ and β_1 as small as possible. Let $e_{z_2}(\beta_2,\gamma_2)$ be the entry of z_2 lying in $\bar e((z_2),E)$ with $\beta_2 < \alpha_2 + m_2$ and β_2 as large as possible. Let

$$E^{z_2,A_2}_{z_1,A_1}(E) = \begin{cases} 0 \text{ if } e_{z_1}(\beta_1,\gamma_1) \text{ lies in } A_1 \text{ and } e_{z_2}(\beta_2,\gamma_2) \text{ in } A_2. \\ 2 \text{ if neither } e_{z_1}(\beta_1,\gamma_1) \text{ lies in } A_1 \text{ nor } e_{z_2}(\beta_2,\gamma_2) \text{ in } A_2. \\ 1 \text{ otherwise.} \end{cases}$$

We define

$$D^{z_2,A_2}_{z_1,A_1}(E) = -D^{z_1,A_1}_{z_2,A_2}(E) = d_E(\gamma_1,\gamma_2) + \frac{1}{2}\, E^{z_2,A_2}_{z_1,A_1}(E)$$

and say that the layer $\bar e(E)$ is raised in $D^{z_2,A_2}_{z_1,A_1}(E)$ terms from (z_1,A_1) to (z_2,A_2). (Note that $D^{z_2,A_2}_{z_1,A_1}(E)$ is a half-integer - it need not be an integer in general).

Assume that z_3 has a DC form (A_3) at α_3 with $\bar e((z_3),E)$ being an antichain of z_3.

<u>Lemma 12.8.2. Let the elements $z_h \in A_n$, the blocks A_h of z_h, $1 < h < 3$, and the set of residue classes $E \subseteq \underline{n}$ be as above. Then</u>

$$\underline{D^{z_3,A_3}_{z_1,A_1}(E) = D^{z_2,A_2}_{z_1,A_1}(E) + D^{z_3,A_3}_{z_2,A_2}(E)}$$

<u>Proof</u>: By definition of D-function and formula (7). □

<u>Lemma 12.8.3</u> <u>Assume that w, w' have full DC forms $(A_\ell,\ldots,A_1)$, $(A'_\ell,\ldots,A'_1)$ at i,</u>

i', respectively, satisfying the following conditions.

(i) There exists a decomposition $\underline{n} = E_1 \cup \ldots \cup E_t$ such that $\bar{e}((x),E_u)$ is an antichain of x for all $x = w, w'$, $1 \leq u \leq t$.

(ii) $|A_j| = |A'_j|$, for $1 \leq j \leq \ell$.

(iii) For any $1 \leq j \leq \ell$, $1 \leq u \leq t$, $D_{w,A_j}^{w',A'_j}(E_u) = 0$.

Then $w = w'$.

Proof By (ii), (iii), the set of integers labelling the columns of w containing some entry of A_t is the same as the set of integers labelling the columns of w' containing some entry of A'_t for any $1 \leq t \leq \ell$.

Let $e((w),j_t^u(w))$ (resp. $e((w'),j_t^u(w'))$ be the u-th entry of $A_t(w)$ (resp. $A'_t(w')$). Since A_t, A'_t are both DC blocks, this implies $j_t^u(w) = j_t^u(w')$ for $1 \leq u \leq |A_t|$. So

$$\sum_{\substack{1 \leq t \leq \ell \\ 1 \leq u \leq |A_t|}} j_t^u(w) = \sum_{\substack{1 \leq t \leq \ell \\ 1 \leq u \leq |A'_t|}} j_t^u(w').$$

Since $(A_\ell,\ldots,A_1)$, $(A'_\ell,\ldots,A'_1)$ are full DC forms of w, w', respectively, by definition of affine matrix, we have $i = i'$. Hence $w = w'$. □

Corollary 12.8.4 Assume that w, w', w" have full DC forms $(A_\ell,\ldots,A_1)$, $(A'_{i_\ell},\ldots,A'_{i_1})$, $(A''_{i_\ell},\ldots,A''_{i_1})$ at i, i', i", respectively. Suppose they also satisfy the following conditions.

(i) There exists a decomposition $\underline{n} = E_1 \cup \ldots \cup E_t$ such that $\tilde{e}((x),E_u)$ is an antichain of x for any $x \in \{w,w',w''\}$, $1 \leq u \leq t$.

(ii) There exists a permutation $s \in S_\ell$ such that

$(s(\ell),\ldots,s(1)) = (i_\ell,\ldots,i_1)$ and $|A_j| = |A'_{s(j)}| = |A''_{s(j)}|$, $1 \leq j \leq \ell$.

(iii) For any $1 \leq j \leq \ell$, $1 \leq u \leq t$, we have

$$D_{w,A_j}^{w',A'_{s(j)}}(E_u) = D_{w,A_j}^{w'',A''_{s(j)}}(E_u).$$ Then $w' = w''$.

Proof By Lemma 12.8.2, we see that $D_{w',A'_{s(j)}}^{w'',A''_{s(j)}}(E_u) = 0$ for any $1 \leq j \leq \ell$, $1 \leq u \leq t$.

So our result follows from Lemma 12.8.3. □

Example 12.8.5

(i) Let $z_1, z_2 \in A_n$ be such that z_1 has full MDC form $(A_\ell, \ldots, A_1)$ at $i \in \mathbb{Z}$ which is normal for the first m layers with $|A_1| \geqslant \ldots \geqslant |A_\ell|$ and $z_2 = \rho_{A_1,\ldots,A_v}^{A_\ell,\ldots,A_{v+1}}(z_1)$ for $1 \leqslant v < \ell$. Let $E_u = \overline{\{j_h^u(z_1) \mid 1 \leqslant h \leqslant \alpha_u\}}$, $1 \leqslant u \leqslant m$, $\alpha_u = \max\{h \mid 1 \leqslant h \leqslant \ell, |A_h| \geqslant u\}$. Then $\bar{e}(E_u)$ is raised in $D_{z_1,A_t(z_1)}^{z_2,A_t(z_2)}(E_u)$ terms from $(z_1, A_t(z_1))$ to $(z_2, A_t(z_2))$, for any t, u with $1 \leqslant t \leqslant \ell$ and $1 \leqslant u \leqslant m$, where

$$D_{z_1,A_t(z_1)}^{z_2,A_t(z_2)}(E_u) = \begin{cases} \alpha_u - v & \text{if } \alpha_u \geqslant v \\ 0 & \text{if } \alpha_u \leqslant v \end{cases}$$

(ii) Let $w \in N_\lambda$ have a standard MDC form $(A_r, \ldots, A_1)$ at $i \in \mathbb{Z}$. Let $x_1 = w$, $x_2, \ldots, x_{\lambda_r}$, $x'_1, \ldots, x'_{\lambda_r}$ be such that for every $1 < j \leqslant \lambda_r$, $1 \leqslant j' \leqslant \lambda_r$, $x_j = \rho_{A_1,\ldots,A_{r-1},A_r^1}^{A_r^0}(x_{j-1})$, $x'_{j'} = \rho_{A_1,\ldots,A_{r-1}}^{A_r^0}(x_{j'})$, where A_r^0 (resp. A_r^1) is the block consisting of the first row (resp. the last $\lambda_r - 1$ rows) of A_r. Let $E_u = \overline{\{j_h^u(w) \mid 1 \leqslant h \leqslant \mu_u\}}$, $1 \leqslant u \leqslant \lambda_1$. Then by Proposition 12.7.2 and the formulae of ρ-operations, we can calculate the D-functions as follows:

Table (I)

u \ A : $D^{x_j, A(x_j)}_{x_{j-1}, A(x_{j-1})}(E_u)$	$A_1, \ldots, A_{r-1}$	A_r^1	A_r^o
$1 < u < \lambda_1$, $u \neq j-1, j$	o	o	o
$j-1$	1	$\frac{1}{2}$	$\frac{1}{2}$
j	o	$\frac{1}{2}$	$\frac{1}{2}$

$1 < j < \lambda_r$

u \ A : $D^{x'_j, A(x'_j)}_{x_j, A(x_j)}(E_u)$	$A_1, \ldots, A_{r-1}$	A_r^1	A_r^o
$1 < u < \lambda_r$, $u \neq j$	o	o	1
j	1	o	1
$\lambda_r < u < \lambda_1$	o	o	o

$1 < j < \lambda_r$

Table (II)

u \ A : $D^{x_j, A(x_j)}_{w, A(w)}(E_u)$	$A_1, \ldots, A_{r-1}$	A_r^1	A_r^o
1	1	$\frac{1}{2}$	$\frac{1}{2}$
$2 < u < j-1$	1	1	1
j	o	$\frac{1}{2}$	$\frac{1}{2}$
$j < u < \lambda_r$	o	o	o
$\lambda_r < u < \lambda_1$	o	o	o

$1 < j < \lambda_r$

u \ A : $D^{x'_j, A(x'_j)}_{w, A(w)}(E_u)$	$A_1, \ldots, A_{r-1}$	A_r^1	A_r^o
1	1	$\frac{1}{2}$	$\frac{3}{2}$
$2 < u < j-1$	1	1	2
j	1	$\frac{1}{2}$	$\frac{3}{2}$
$j < u < \lambda_r$	o	o	1
$\lambda_r < u < \lambda_1$	o	o	o

$1 < j < \lambda_r$

(iii) Let $w \in N$ have the standard MDC form $(A_r,\ldots,A_1)$ at $i \in \mathbf{Z}$. Let $y_0 = w$, $y_1,\ldots,y_{\alpha_{\lambda,0}}$, $y'_1,\ldots,y'_{\alpha_{\lambda,0}}$ with $\alpha_{\lambda,0} = \lambda_k-\lambda_{k+1} > 0$ and $k < r$ such that

$$y_h = \Big(\rho^{A_k^0,A_r,\ldots,A_{k+1}}_{A_1,\ldots,A_{k-1},A_k^1}\Big)^{h-1}\rho^{A_r,\ldots,A_{k+1}}_{A_1,\ldots,A_{k-1},A_k^1}(w),\ 1 \leq h \leq \alpha_{\lambda,0}$$

$$y'_{h'} = \rho^{A^0_{i_1}}_{A_k^1}\ \rho^{A_{i_1}}_{A_1,\ldots,A_{k-1}}\ \rho^{A_r,\ldots,A_{i_1+1}}_{A_1,\ldots,A_{k-1},A_k^1}\Big(\rho^{A_r,\ldots,A_{k+1},A_k^0}_{A_1,\ldots,A_{k-1},A_k^i}\Big)^{h'}(w),\ 0 \leq h' \leq \alpha_{\lambda,0}$$

where A_k^0(resp. A_k^1) is the block consisting of the first row (resp. the last λ_k-1 rows) of A_k, (resp. $A^0_{i_1}$) is the block consisting of the first λ_{i_1}-1 rows (resp. the last row) of A_{i_1}.

Let $E_v = \{j_h^v(w) \mid 1 \leq h \leq \mu_v\}$, $1 \leq v \leq \lambda_1$. Then by Proposition 12.7.2 and the formulae of ρ-operations, we can calculate the D-functions as follows, where the notation δ_{ij} in these tables is a Kronecker delta.

TABLE (III)

$D^{y_i, A(y_1)}_{w, A(w)}(E_v)$ — v \ A	A_k^o	A_{i_g} $2 \le g \le t$	$A_{i_{g-1}+1}, \ldots, A_{i_g-1}$ $2 \le g \le t$	$A^o_{i_1}$	$A^1_{i_1}$	$A_{i_o+1}, \ldots, A_{i_1-1}$	A_k^1	A_a $1 \le a < k$
1	0	$(i_t - i_o) + \frac{1}{2}\delta_{gt}$	$i_t - i_o$	$(i_t - i_o) + \frac{1}{2}\delta_{t,1}$	$(i_t - i_o) + \frac{1}{2}\delta_{t1} \cdot \delta_{\alpha_{\lambda,1} 1}$	$i_t - i_o$	$i_t - i_o - \frac{1}{2}$	$i_t - i_o$
$\lambda_{i_h} + 1$ $3 \le h \le t$	0	$(i_{h-1} - i_o) + \frac{1}{2}(\delta_{g,h} + \delta_{g,h-1})$	$(i_{h-1} - i_o) + \delta_{gh}$	$i_{h-1} - i_o$				
$\lambda_{i_h} + 1 < v \le \lambda_{i_{h-1}}$ $1 \le h \le t+1$ $v \ne \lambda_{i_1}$	0	$i_{h-1} - i_o$						
$\lambda_{i_2} + 1$ $v \ne 1$	0	$(i_1 - i_o) + \frac{1}{2}\delta_{g,2}$	$(i_1 - i_o) + \delta_{g,2}$	$(i_1 - i_o) + \frac{1}{2}(1 + \delta_{\alpha_{\lambda,1},1})$	$(i_1 - i_o) + \frac{1}{2}\delta_{\alpha_{\lambda,1},1}$	$i_1 - i_o$		
λ_{i_1} $v \ne \lambda_{i_2} + 1$	0	$i_1 - i_o$		$(i_1 - i_o) + \frac{1}{2}$	$(i_1 - i_o) + \frac{1}{2}$	$i_1 - i_o$		
$\lambda_{i_1} + 1$	0				$\frac{1}{2}$	1	$\frac{1}{2}$	0
$\lambda_{i_o} < v \le \lambda_1$	0							

TABLE (III)

$1 < j \le \alpha_{\lambda,0}$

$D^{Y_j, A(Y_j)}_{Y_{j-1}, A(Y_{j-1})}(E_v)$ — A / v	A^o_k, A_a, $1 \le a < k$ or $i_1 < a \le r$	$A^o_{i_1}$	$A^1_{i_1}$	A^1_k	A_b, $k < b < i_1$
$\lambda_{i_h} < v \le \lambda_{i_{h-1}}$, $1 \le h \le t+1$, $v \ne \lambda_{i_1} + j-1, \lambda_{i_1}+j$		i_{h-1} — i_o			
$\lambda_{i_1} + j - 1$	1			$\frac{1}{2}$	0
$\lambda_{i_1} + j$	0			$\frac{1}{2}$	1
$\lambda_{i_o} < v \le \lambda_1$		0			

TABLE (IV)

D $Y'_o, A(Y'_o)$ (E_v) w, A(w) / v \ A	A^o_k	A_{i_g} $2 \le g \le t$	$A_{i_{g-1}+1}, \ldots, A_{i_g-1}$ $2 \le g \le t$	$A^o_{i_1}$	$A^1_{i_1}$	A_a $k < a < i_1$	A^1_k	A_b $1 \le b < k$
1	0	$(i_t - i_o) + \frac{1}{2}\delta_{gt}$	$i_t - i_o$	$(i_t - i_o) + \frac{1}{2}\delta_{t,1}$	$(i_t - i_o) + 1$	0	$(i_t - i_1) + \frac{1}{2}(1 - \delta_{t1} \cdot \delta_{\alpha_{\lambda,1},1})$	$(i_t - i_1) + 1$
$\lambda_{i_h} + 1$ $3 \le h \le t$	0	$(i_{h-1} - i_o) + \frac{1}{2}(\delta_{gh} + \delta_{g,h-1})$	$(i_{h-1} - i_o) + \delta_{hg}$	$i_{h-1} - i_o$	$(i_{h-1} - i_o) + 1$	0	$(i_{h-1} - i_1) + 1$	
$\lambda_{i_h} + 1 < v \le \lambda_{i_{h-1}}$ $2 \le h \le t+1$ $v \ne \lambda_{i_1}$	0	$i_{h-1} - i_o$			$(i_{h-1} - i_o) + 1$	0	$(i_{h-1} - i_1) + 1$	
$\lambda_{i_2} + 1$ $v \ne 1$	0	$(i_1 - i_o) + \frac{1}{2}\delta_{g,2}$	$(i_1 - i_o) + \delta_{g,2}$	$(i_1 - i_o) + \frac{1}{2}(1 + \delta_{\alpha_{\lambda,1},1})$	$(i_1 - i_o) + 1$	0	$1 - \frac{1}{2}\delta_{\alpha_{\lambda,1},1}$	1
λ_{i_1} $v \ne \lambda_{i_2} + 1$	0	$i_1 - i_o$		$(i_1 - i_o) + \frac{1}{2}$	$(i_1 - i_o) + 1$	0	$\frac{1}{2}$	1
$\lambda_{i_1} + 1$	0				1	0		
$\lambda_{i_1} + 1 < v \le \lambda_{i_o}$	0				1	0		
$\lambda_{i_o} < v \le \lambda_1$	0							

TABLE (IV)

$1 \le j \le \alpha_{\lambda,0}$

$D_{Y_j, A(Y_j)}^{Y'_j, A(Y'_j)}(E_v)$ — A / v	$A^o_{k'}$, A_a ($i_1 < a \le r$), $A^o_{i_1}$	$A^1_{i_1}$	A_b ($k < b < i_1$)	A^1_k	A_c ($1 \le c < k$)
$\lambda_{i_h} < v \le \lambda_{i_{h-1}}$, $1 < h \le t+1$	$i_{h-1} - i_o$	$i_{h-1} - i_o + 1$	0	$i_{h-1} - i_1 + 1$	
$\lambda_{i_1} < v \le \lambda_{i_o}$, $v \ne \lambda_{i_1} + j$	0	1	0		
$\lambda_{i_1} + j$	1		0		1
$\lambda_{i_o} < v \le \lambda_1$	0				

TABLE (V)

$1 \le j \le \alpha_{\lambda,0}$

$D^{y_j, A(y_j)}_{w, A(w)}(E_v)$ — v \ A	A^o_k	A_{i_g} $2 \le g \le t$	$A_{i_{g-1}+1}, \ldots, A_{i_g-1}$ $2 \le g \le t$	$A^o_{i_1}$	$A^1_{i_1}$	$A_{i_o+1}, \ldots, A_{i_1-1}$	A^1_k	A_a $1 \le a < k$
1	$(j-1)(i_t-i_o)$	$j(i_t-i_o) + \frac{1}{2}\delta_{gt}$	$j(i_t - i_o)$	$j(i_t-i_o) + \frac{1}{2}\delta_{t1}$	$j(i_t-i_o)+\frac{1}{2}\delta_{t1}\cdot\delta_{\alpha_{\lambda,1},1}$	$j(i_t-i_o)$	$j(i_t-i_o)-\frac{1}{2}$	$j(i_t-i_o)$
$\lambda_{i_h}+1$ $3 \le h \le t$	$(j-1)(i_{h-1}-i_o)$	$j(i_{h-1}-i_o)+\frac{1}{2}(\delta_{gh}+\delta_{g,h-1})$	$j(i_{h-1}-i_o)+\delta_{gh}$	$j(i_{h-1} - i_o)$				
$\lambda_{i_h}+1 < v \le \lambda_{i_{h-1}}$ $2 \le h \le t+1$ $v \ne \lambda_{i_1}$	$(j-1)(i_{h-1}-i_o)$	$j(i_{h-1} - i_o)$						
$\lambda_{i_2}+1$ $v \ne 1$	$(j-1)(i_1-i_o)$	$j(i_1-i_o)+\frac{1}{2}\delta_{g,2}$	$j(i_1-i_o)+\delta_{g,2}$	$j(i_1-i_o)+\frac{1}{2}(1+\delta_{\alpha_{\lambda,1},1})$	$j(i_1-i_o)+\frac{1}{2}\delta_{\alpha_{\lambda,1},1}$	$j(i_1 - i_o)$		
λ_{i_1} $v \ne \lambda_{i_2}+1$	$(j-1)(i_1-i_o)$	$j(i_1 - i_o)$		$j(i_1-i_o) + \frac{1}{2}$	$j(i_1-i_o) + \frac{1}{2}$	$j(i_1 - i_o)$		
$\lambda_{i_1}+\ell$ $1 \le \ell < j-1$	1							
$\lambda_{i_1}+j-1$ $v \ne \lambda_{i_1}$	1							
$\lambda_{i_1}+j$	0				$\frac{1}{2}$	1	$\frac{1}{2}$	0
$\lambda_{i_1}+\ell'$ $j < \ell' \le \alpha_{\lambda,0}$	0							
$\lambda_{i_o} < v \le \lambda_1$	0							

TABLE (VI)

$1 \le j \le \alpha_{\lambda,0}$

$Y'_j, A(y'_j)$ / $D_{w,A(w)}$ (E_v) / v \ A	A_k^o	A_{i_g} $2 \le g \le t$	$A_{i_{g-1}+1},\dots,A_{i_g-1}$ $2 \le g \le t$	$A_{i_1}^o$	$A_{i_1}^1$	$A_{i_o+1},\dots A_{i_1-1}$	A_k^1	A_a $1 \le a < k$
1	$j(i_t-i_o)$	$(j+1)(i_t-i_o)+\frac{1}{2}\delta_{gt}$	$(j+1)(i_t-i_o)$	$(j+1)(i_t-i_o)+\frac{1}{2}\delta_{t1}$	$(j+1)(i_t-i_o)+1+\frac{1}{2}\delta_{t1}\cdot\delta_{\alpha_{\lambda,1},1}$	$j(i_t-i_o)$	$(j+1)(i_t-i_o)+i_o-i_1+\frac{1}{2}$	$(j+1)(i_t-i_o)+i_o-i_1+1$
$\lambda_{i_h}+1$ $3 \le h \le t$	$j(i_{h-1}-i_o)$	$(j+1)(i_{h-1}-i_o)+\frac{1}{2}(\delta_{gh}+\delta_{g,h-1})$	$(j+1)(i_{h-1}-i_o)+\delta_{gh}$	$(j+1)(i_{h-1}-i_o)$	$(j+1)(i_{h-1}-i_o)+1$	$j(i_{h-1}-i_o)$	$(j+1)(i_{h-1}-i_o)+i_o-i_1+1$	
$\lambda_{i_h}+1<v\le\lambda_{i_{h-1}}$ $2 \le h \le t+1$ $v \ne \lambda_{i_1}$	$j(i_{h-1}-i_o)$	$(j+1)(i_{h-1}-i_o)$			$(j+1)(i_{h-1}-i_o)+1$	$j(i_{h-1}-i_o)$	$(j+1)(i_{h-1}-i_o)+i_o-i_1+1$	
$\lambda_{i_2}+1$ $v \ne 1$	$j(i_1-i_o)$	$(j+1)(i_1-i_o)+\frac{1}{2}\delta_{g,2}$	$(j+1)(i_1-i_o)+\delta_{g,2}$	$(j+1)(i_1-i_o)+\frac{1}{2}(1+\delta_{\alpha_{\lambda,1},1})$	$(j+1)(i_1-i_o)+1+\frac{1}{2}\delta_{\alpha_{\lambda,1},1}$	$j(i_1-i_o)$	$j(i_1-i_o)+1$	
λ_{i_1} $v \ne \lambda_{i_2}+1$	$j(i_1-i_o)$	$(j+1)(i_1-i_o)$		$(j+1)(i_1-i_o)+\frac{1}{2}$	$(j+1)(i_1-i_o)+\frac{3}{2}$	$j(i_1-i_o)$	$j(i_1-i_o)+1$	
$\lambda_{i_1}+\ell$ $1 \le \ell < j-1$	1				2	1		
$\lambda_{i_1}+j-1$ $v \ne \lambda_{i_1}$	1				2	1		
$\lambda_{i_1}+j$	1				$\frac{3}{2}$	1	$\frac{1}{2}$	1
$\lambda_{i_1}+\ell'$ $j < \ell' \le \alpha_{\lambda,0}$	0				1	0		
$\lambda_{i_o} < v \le \lambda_{i_1}$	0							

CHAPTER 13 : RAISING OPERATIONS ON LAYERS

So far, we have considered various operations on the affine Weyl group A_n which are successions of left star operations and which therefore, when applied to an element w, give us another element in the same left cell as w. In the present chapter, we introduce a new type of operation which, when applied to an element w in N_λ, gives us another element of N_λ in the same left cell as w but which is not in general a succession of star operations. This operation is called a raising operation on a layer. It is defined in §13.2 and is proved to give an element in the same left cell in §13.3 and §13.4. By means of a succession of these operations and left star operations, every element of A_n can be transformed into a principal normalized element.

§13.1 REFLECTIVE PAIRS

In this section, we introduce a criterion for elements y, w $\in A_n$ to satisfy $y < w$ and $\ell(w) = \ell(y) + 1$. This criterion will be used subsequently in this chapter.

For $x \in A_n$, we call an entry pair $\{e(i_1,j_1), e(i_2,j_2)\}$ of x reflective if the following conditions are satisfied:

(i) either $|i_1-i_2| < n$ or $|j_1-j_2| < n$

(ii) For any entry e(i,j) of x, $(i-i_1)(i-i_2) < 0$ implies $(j-j_1)(j-j_2) > 0$.

In that case, when $(i_1-i_2)(j_1-j_2) > 0$ (resp. $(i_1-i_2)(j_1-j_2) < 0$), $\{e(i_1,j_1), e(i_2,j_2)\}$ is called a positive (resp. negative) reflective pair.

For example, suppose $n > 4$. Then in the following elements $w, y \in A_n$.

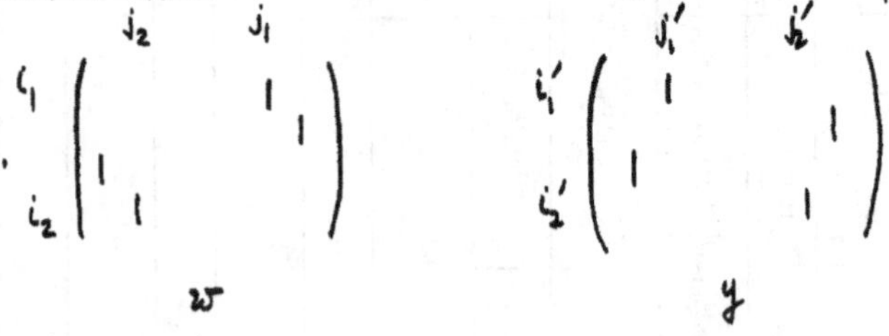

$$E_w = \{e_w(i_1,j_1), e_w(i_2,j_2)\}, \quad E_y = \{e_y(i_1',j_1'), \quad e_y(i_2',j_2')\}$$

are reflective pairs. In particular, E_w is negative and E_y positive.

<u>Lemma 13.1.1</u> *Let $y, w \in A_n$. Then $y < w$ and $\ell(w) = \ell(y) + 1$ if and only if w is obtained from y by transposing the i_1-th and the i_2-th rows of y for some integers i_1, i_2 with $i_1 \neq i_2$ and $\{e_y(i_1,j_1), e_y(i_2,j_2)\}$ a positive reflective pair.*

In the above example, suppose that $(i_1',j_1',i_2',j_2') = (i_1,j_2,i_2,j_1)$ and suppose that w is obtained from y by transposing the i_1-th and the i_2-th rows. Then we can see that $w = s_{i_1+1}s_{i_1}s_{i_1+2}s_{i_1+1}s_{i_1+2}s_{i_1}s_{i_1+1}y$ with $\ell(w)-4 = \ell(s_{i_1+2}s_{i_1}s_{i_1+1}y) = \ell(y) - 3$. So we have $y < w$ and $\ell(w) = \ell(y) + 1$.

<u>Proof of Lemma 13.1.1</u>

($\Rightarrow$) We can write $y = xz$, $w = xsz$ with $s \in \Delta$ and $x,z \in A_n$ such that $\ell(y) = \ell(x) + \ell(z)$ and $\ell(w) = \ell(x) + \ell(z) + 1$. If $x = \mathbf{1}$, then the result is obvious. Now assume that $x \neq \mathbf{1}$. We write $x = s_v x'$ for some $s_v \in \mathcal{L}(x)$. Then $s_v \in \mathcal{L}(y) \cap \mathcal{L}(w)$. Let $y' = s_v y$ and $w' = s_v w$. We have $y' = x'z$ and $w' = x'sz$ which satisfy $\ell(y') = \ell(x') + \ell(z)$ and $\ell(w') = \ell(x') + \ell(z) + 1 = \ell(y') + 1$. Since $y' < w'$, by applying induction on $\ell(x)$, we see that there exists i_1', i_2' with $i_1' \neq i_2'$ such that w' is obtained from y' by transposing the i_1'-th and the i_2'-th rows of y' with $\{e_{y'}(i_1',j_1'), e_{y'}(i_2',j_2')\}$ a positive reflective pair. Without loss of generality, we may assume $i_1' < i_2'$. If $\{\bar{v}, \overline{v+1}\} \cap \{\bar{i}_1',\bar{i}_2'\} = \emptyset$, then w is obtained from y by transposing the i_1'-th and the i_2'-th rows of y with $\{e_y(i_1',j_1'), e_y(i_2',j_2')\}$ a positive reflective pair. If $|\{\bar{v}, \overline{v+1}\} \cap \{\overline{i_1'}, \overline{i_2'}\}| = 1$, then since $s_v \in \mathcal{L}(y) \cap \mathcal{L}(w)$, one of the following four cases must occur.

(i)

$$
y' = \begin{pmatrix} \overset{j}{1} & & & \\ & \overset{j_1'}{1} & & \\ & & & \\ & & \overset{j_2'}{1} & \end{pmatrix} \begin{matrix} \\ -i_1'- \\ \\ -i_2'- \end{matrix}
\qquad
w' = \begin{pmatrix} \overset{j}{1} & & \\ & & \overset{j_2'}{1} \\ & & \\ & \overset{j_1'}{1} & \end{pmatrix}
$$

where $\overline{v+1} = \bar{i}_1'$, $\bar{v} \neq \bar{i}_2'$ and $j < j_1' < j_2'$. Let $(i_1,j_1,i_2,j_2) = (i_1'-1, j_1', i_2', j_2')$.

(ii)

$$y' = \begin{array}{c} \begin{array}{ccc} j_1' & j_2' & j \end{array} \\ \left(\begin{array}{ccc} 1 & & \\ & & 1 \\ & & \\ & 1 & \end{array}\right) \end{array} \begin{array}{c} -i_1'- \\ \\ \\ -i_2'- \end{array} \qquad w' = \begin{array}{c} \begin{array}{ccc} j_1' & j_2' & j \end{array} \\ \left(\begin{array}{ccc} & 1 & \\ & & 1 \\ & & \\ 1 & & \end{array}\right) \end{array}$$

where $\bar{v} = \bar{i}_1'$, $\overline{v+1} \neq \bar{i}_2'$ and $j_1' < j_2' < j$. Let $(i_1,j_1,i_2,j_2) = (i_1'+1,\ j_1',\ i_2',\ j_2')$.

(iii)

$$y' = \begin{array}{c} \begin{array}{ccc} j & j_1' & j_2' \end{array} \\ \left(\begin{array}{ccc} & 1 & \\ & & \\ 1 & & \\ & & 1 \end{array}\right) \end{array} \begin{array}{c} -i_1'- \\ \\ \\ -i_2'- \end{array} \qquad w' = \begin{array}{c} \begin{array}{ccc} j & j_1' & j_2' \end{array} \\ \left(\begin{array}{ccc} & & 1 \\ & & \\ 1 & & \\ & 1 & \end{array}\right) \end{array}$$

where $\overline{v+1} = \bar{i}_2'$, $\bar{v} \neq \bar{i}_1'$ and $j < j_1' < j_2'$. Let $(i_1,j_1,i_2,j_2) = (i_1',\ j_1',\ i_2'-1,\ j_2')$.

(iv)

$$y' = \begin{array}{c} \begin{array}{ccc} j_1' & j_2' & j \end{array} \\ \left(\begin{array}{ccc} 1 & & \\ & & \\ & 1 & \\ & & 1 \end{array}\right) \end{array} \begin{array}{c} -i_1'- \\ \\ -i_2'- \\ \\ \end{array} \qquad w' = \begin{array}{c} \begin{array}{ccc} j_1' & j_2' & j \end{array} \\ \left(\begin{array}{ccc} & 1 & \\ & & \\ 1 & & \\ & & 1 \end{array}\right) \end{array}$$

where $\bar{v} = \bar{i}_2'$, $\overline{v+1} \neq \bar{i}_1'$ and $j_1' < j_2' < j$. Let $(i_1,j_1,i_2,j_2) = (i_1',j_1',\ i_2'+1,\ j_2')$. Clearly, in any of the above cases, we see that $|i_1-i_2| < n$ if and only if $|i_1' - i_2'| < n$ and also that $|j_1-j_2| < n$ if and only if $|j_1'-j_2'| < n$. So $w = s_v w'$ is obtained from $y = s_v y'$ by transposing the i_1-th and the i_2-th rows of y with $\{e_y(i_1,j_1),\ e_y(i_2,j_2)\}$ a positive reflective pair. If $\{\bar{v},\ \overline{v+1}\} = \{\bar{i}_1',\bar{i}_2'\}$, then since $s_v \in \mathcal{L}(y) \cap \mathcal{L}(w)$, y', w' have to be of the form:

$$\underset{y'}{\begin{array}{c} \begin{array}{cccc} j_1' & j_2' & j_1'+qn & j_2'+qn \end{array} \\ \left(\begin{array}{cccc} 1 & & & \\ & & & \\ & 1 & & \\ & & 1 & \\ & & & \\ & & & 1 \end{array}\right) \end{array}} \begin{array}{c} \\ - i_1' - \\ \\ - i_2' - \\ - i_1'+qn - \\ \\ - i_2'+qn - \end{array} \underset{w'}{\begin{array}{c} \begin{array}{cccc} j_1' & j_2' & j_1'+qn & j_2'+qn \end{array} \\ \left(\begin{array}{cccc} & 1 & & \\ & & & \\ 1 & & & \\ & & & 1 \\ & & & \\ & & 1 & \end{array}\right) \end{array}}$$

where $\bar{v} = \bar{i}_2'$, $\overline{v+1} = \bar{i}_1'$ and $i_1' + qn = i_2' + 1$, $j_1' + qn > j_2'$ for some $q > 0$. Clearly, by the hypothesis that either $|i_1' - i_2'| < n$ or $|j_1' - j_2'| < n$, we have $|j_1' - j_2'| < n$. Let $(i_1,j_1,i_2,j_2) = (i_1' - 1,\ j_1',\ i_2' + 1,\ j_2')$. Then $|j_1-j_2| < n$. So $w = s_v w'$ is obtained from $y = s_v y'$ by transposing the i_1-th and the i_2-th rows of y with $\{e_y(i_1,j_1),\ e_y(i_2,j_2)\}$ a positive reflective pair.

($\Leftarrow$) Since w is also obtained from y by transposing the j_1-th and the j_2-th columns of y, we may assume $0 < i_2 - i_1 < n$ by symmetry. When $i_2 - i_1 = 1$, we have $w = s_{i_1} y$ with $\ell(w) = \ell(y) + 1$ and so our result is true. Now assume $i_2 - i_1 > 1$. If, for any i, $i_1 < i < i_2$, the entry $e_y(i,j)$ satisfies $j > j_2$, then $s_{i_2-1} \in \mathcal{L}(y) \cap \mathcal{L}(w)$. Let $y' = s_{i_2-1}y$, $w' = s_{i_2-1}w$. Hence we see that $y' \prec w'$ if and only if $y \prec w$ and also that $\ell(w') = \ell(y') + 1$ if and only if $\ell(w) = \ell(y) + 1$. But w' is obtained from y' by transposing the i_1-th and the (i_2-1)-th rows of y' with $\{e_{y'}(i_1,j_1),\ e_{y'}(i_2-1,j_2)\}$ a positive reflective pair. By applying induction on $i_2 - i_1 > 1$, we get $y' \prec w'$ and $\ell(w') = \ell(y') + 1$. So $y \prec w$ and $\ell(w) = \ell(y) + 1$. If there exists an i with $i_1 < i < i_2$ such that the entry $e_y(i,j)$ satisfies $j < j_2$, then by the conditions that $i_2 - i_1 < n$ and $\{e_y(i_1,j_1),\ e_y(i_2,j_2)\}$ is a positive reflective pair, we have $j < j_1$. Let i be as small as possible. Then for any i_0 with $i_1 < i_0 < i$, the entry $e(i_0,j_0)$ satisfies $j_0 > j$. Let $y' = s_{i_1} s_{i_1+1} \cdots s_{i-1} y$ and $w' = s_{i_1} s_{i_1+1} \cdots s_{i-1} w$. We have $\ell(y') = \ell(y) - (i-i_1)$ and $\ell(w') = \ell(w) - (i-i_1)$. So $y' \prec w'$ if and only if $y \prec w$, and also $\ell(w') = \ell(y') + 1$ if and only if $\ell(w) = \ell(y) + 1$. By the same argument as the above, we get $y \prec w$ and $\ell(w) = \ell(y)+1$. By induction, our conclusion is proved. □

§13.2 RAISING OPERATIONS ON LAYERS

Let $w \in A_n$ and $i_1,\ldots,i_t$ be integers satisfying $i_1 < \ldots < i_t < i_1 + n$. Let

$f(i_h)$ be the i_h-th row of w for $1 \leq h \leq t$.

Procedure (i): Let $B_1,\ldots,B_t$ be the blocks of w such that for every h, $1 \leq h \leq t$, B_h consists of rows of w from the $(i_{h-1}+1)$-th to the i_h-th and B_1 from the (i_t+1-n)-th to the i_1-th.

Let $w' \in A_n$ be obtained from w by permuting $f(i_h)$ from the bottom to the top in B_h for all h, $1 \leq h \leq t$.

Procedure (ii): Let $B'_1,\ldots,B'_t$ be the blocks of w such that for every h, $1 \leq h \leq t$ B'_h consists of rows of w from the i_h-th to the $(i_{h+1}-1)$-th and B'_t from the i_t-th to the (i_1-1+n)-th.

Let $w'' \in A_n$ be obtained from w by permuting $f(i_h)$ from the top to the bottom in B'_h for all h, $1 \leq h \leq t$.

Let F be the set of row classes of w consisting of $\{f(i_1),\ldots,f(i_t)\}$ and their congruent rows. Then procedure (i) is called "raising F by one term" and (ii) called "lowering F by one term". Let $\bar{e}(w)$ be the set of entry classes of w consisting of all entries lying in F. Then we can also say that procedure (i) raises $\bar{e}(w)$ by one term and (ii) lowers $\bar{e}(w)$ by one term. So w' is obtained from w by raising $\bar{e}(w)$ by one term and w" from w by lowering $\bar{e}(w)$ by one term. Let $E \subseteq \underline{n}$ be such that $E = \{\bar{h} \mid$ there exists some $e_w(i,h) \in \bar{e}(w)\}$, then we write $w' = L_E(w)$, $w'' = L^E(w)$ and write $\bar{e}(w) = \bar{e}((w),E)$. Clearly, $L_E.L^E(w) = L^E.L_E(w) = w$.

For example, assume n = 7. Let $w,w' \in A_7$ be as follows.

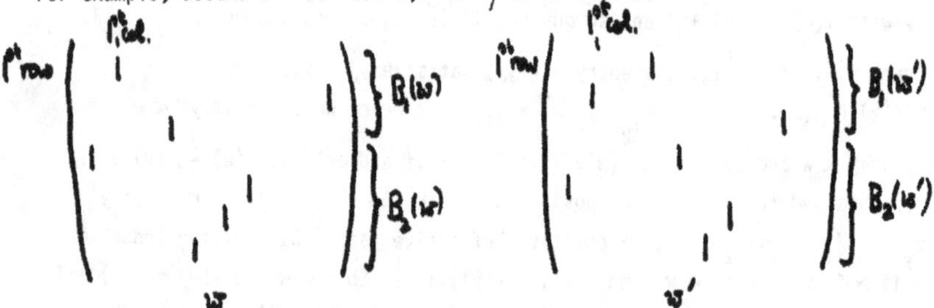

Let $F_w = \{f_w(3), f_w(7)\}$, $F_{w'} = \{f_{w'}(1), f_{w'}(4)\}$ be row sets of w, w', respectively. Then w' (resp. w) is obtained from w (resp. w') by raising F_w (resp. lowering $F_{w'}$) by one term.

Assume that $w \in N$ has a standard MDC form $(A_r,\ldots,A_1)$ at $i \in \mathbf{Z}$. Let $E_u \subseteq \underline{n}$ for $1 \leq u \leq \lambda_1$ such that $E_u = \{\overline{j^u_t(w)} \mid 1 \leq u \leq \mu_u\}$, where $\lambda = \{\lambda_1 \geq \ldots \geq \lambda_r\} \in \Lambda_n$ and $\mu = \{\mu_1 \geq \ldots \geq \mu_m\}$ is the dual partition of λ. Recall that in §8.3 $\bar{e}((w),E_u)$ was called the u-th layer of w.

Definition 13.2.1 For any u, $1 \leq u \leq \lambda_1$, the u-th layer $\bar{e}((w),E_u)$ of w is movable if $L_{E_u}(w) \in N_\lambda$ and immovable otherwise.

Clearly, the layer $\bar{e}((w),E_1)$ is always movable. $w \in \bar{N}_\lambda$ if and only if the layers $\bar{e}((w),E_u)$, $1 < u \le \lambda_1$, are all immovable.

When $w' = L_{E_u}(w) \in N_\lambda$ (resp. $w'' = L^{E_u}(w) \in N_\lambda$) for some u, $1 \le u \le \lambda_1$, we let $(A_r,\ldots,A_1)$ be the standard MDC form of w' at i+1 (resp. of w" at i-1).

Clearly, for any v, $1 \le v \le \lambda_1$, the layer $\bar{e}((w'),E_v)$ is an antichain of w' in that case. So we can work out the following $D_w^{w'}$-table.

<u>Table (VII)</u>

$D^{w',A(w')}_{w,A(w)}(E_u)$ (rows: u; columns: A)	A_t, $A_{t,h}$, $A'_{t,h}$; $1 \le t \le r$, $1 \le h < \lambda_t$
$1 \le u \le \lambda_1$, $u \neq v$	0
v	1

where $A_{t,h}$ is the block consisting of the first h rows of A_t, $A'_{t,h}$ is the block consisting of the last λ_t-h rows of A_t.

<u>Lemma 13.2.2</u> <u>Let $w \in N_\lambda$ and $E_u \subseteq n$ be defined as above. Then there exists a sequence of elements $w_0 = w, w_1,\ldots,w_\ell$ in N_λ such that for each i, $1 \le i \le \ell$, $w_i = L_{E_{u_i}}(w_{i-1})$ for some u_i, $1 < u_i \le \lambda_1$, and $w_\ell \in \bar{N}_\lambda$.</u>

<u>Proof</u> If the layers $\bar{e}((w),E_u)$ are immovable for all u with $1 < u \le \lambda_1$, then $w \in \bar{N}_\lambda$ and the result is trivial. Now assume that we are not in such a case. Then there exists a smallest number k with $1 < k \le \lambda_1$ such that the layer $\bar{e}((w),E_k)$ is movable. By definition, we have $w_1 = L_{E_k}(w) \in N_\lambda$. w_1 has a standard MDC form $(A_r,\ldots,A_1)$ at i+1 with

$$\begin{cases} j_t^u(w_1) = j_t^u(w) \text{ for any } u,t \text{ with } 1 \le t \le r \text{ and } 1 \le u \le \min\{k-1,\lambda_t\} \\ j_t^{k-1}(w) > j_t^k(w_1) > j_t^k(w) \text{ for any } t \text{ with } 1 \le t \le r \text{ and } k \le \lambda_t. \end{cases}$$

Clearly, the layers $\bar{e}((w_1),E_h)$, $1 < h < k$, are immovable. If $\bar{e}((w_1),E_k)$ is still movable, then the same procedure could be carried on. We then get $w_0 = w$. $w_1,\ldots,$

in N_λ such that w_h has a standard MDC form $(A_r,\dots,A_1)$ at $i+h$ for $h = 0,1,2,\dots$, satisfying

$$\begin{cases} j_t^u(w) = j_t^u(w_1) = j_t^u(w_2) = \dots \text{ for any } t,u \text{ with } 1 \le t \le r \text{ and } 1 \le u \le \min\{k-1,\lambda_t\} \\ j_t^k(w) < j_t^k(w_1) < j_t^k(w_2) < \dots \\ j_t^k(w_h) < j_t^{k-1}(w) \text{ for any } h > 0 \text{ and } t \text{ with } 1 \le t \le r \text{ and } k \le \lambda_t. \end{cases}$$

So after a finite number of steps, such a procedure has to stop. i.e. there exists $h > 0$ such that $w_h \in N_\lambda$ has a standard MDC form $(A_r,\dots,A_1)$ at i' and the layers $\bar{e}((w_h),E_u)$, $1 < u \le k$, are all immovable. By applying induction on $\lambda_1 - k \ge 0$, our result follows. □

In the following two sections, we shall show:

Proposition 13.2.3 Let $w \in N_\lambda$ and $E_u \subseteq \underline{n}$, $1 \le u \le \lambda_1$, be defined as in Lemma 13.2.2. If $w' = L_{E_u}(w)$ lies in N_λ, then we have $w' \underset{L}{\sim} w$.

By Lemma 13.2.2, if Proposition 13.2.3 is true, then it is immediate that:

Proposition 13.2.4 For any $w \in N_\lambda$, there exists $y \in \bar{N}_\lambda$ such that $y \underset{L}{\sim} w$. □

Furthermore, by combining Proposition 13.2.4 with Propositions 9.3.7 and 10.4, we see that:

Proposition 13.2.5 For any $w \in \sigma^{-1}(\lambda)$, there exists $y \in \bar{N}_\lambda$ such that $y \underset{L}{\sim} w$. □

§13.3 PROOF OF PROPOSITION 13.2.3 WHEN $1 \le u \le \lambda_r$.

In this section, we wish to show that if $w, w' \in N_\lambda$ and w' is obtained from w by raising the u-th layer where $1 \le u \le \lambda_r$, then w' lies in the same left cell as w. The idea for proving this will be as follows.

We consider the sequences $\xi(w,r)$, $\xi(w',r)$ beginning with w, w' respectively. We pick a certain element y_a in the sequence $\xi(w,r)$ and define $y'_a = w'$. We define y_c to be the last term in the sequence $\xi(w,r)$ and define y'_c to be the element in the sequence $\xi(w',r)$ obtained by applying the same sequence of left star operations to w' as are needed to obtain y_c from y_a. We then consider the cycle of elements shown in the figure

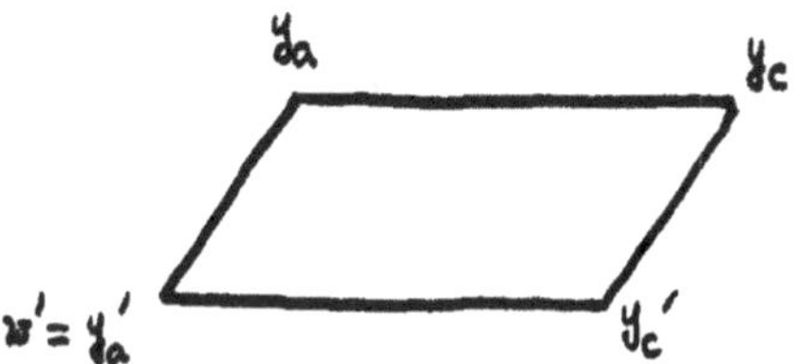

We show that $y_a' \underset{L}{<} y_a$ and $y_c \underset{L}{<} y_c'$. It follows that all the elements in this cycle lie in the same left cell. Since y_a lies in the same left cell as w and $w' = y_a'$, we see that w, w' are in the same left cell.

In order to show $y_a' \underset{L}{<} y_a$ and $y_c \underset{L}{<} y_c'$, we argue as follows. We find a certain element y_b in the sequence $\xi(w,r)$ which lies between y_a and y_c, and define y_b' to be the element in $\xi(w',r)$ obtained by applying the same succession of left star operations to y_a' as are needed to obtain y_b from y_a in $\xi(w,r)$. We thus have a figure

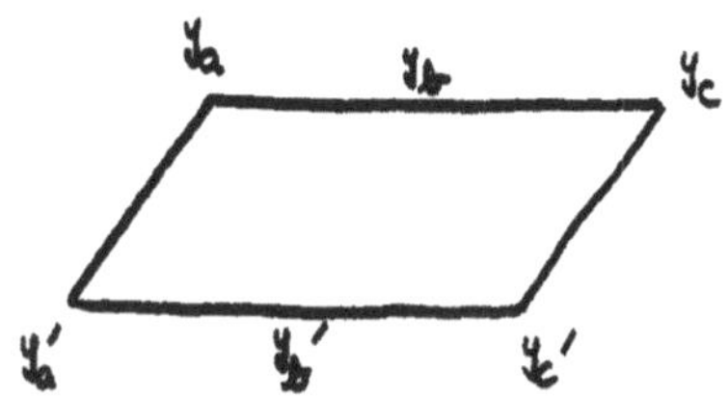

To show $y_a' \underset{L}{<} y_a$, it is enough to show the two statements. (i) $\mathcal{L}(y_a') \not\subseteq \mathcal{L}(y_a)$. (ii) Either $y_a' \prec y_a$ or $y_a \prec y_a'$. Statement (i) will follow readily by observing the full MDC form of y_a' and y_a. In order to prove statement (ii) it will be sufficient to find y_b, y_b' such that either $y_b \prec y_b'$ or $y_b' \prec y_b$ by Theorem 1.6.1. The proof of $y_c \underset{L}{<} y_c'$ is similar. We shall show in fact that y_b, y_b' can be chosen so that $y_b > y_b'$ with $\ell(y_b) = \ell(y_b') + 1$ or $y_b' > y_b$ with $\ell(y_b') = \ell(y_b) + 1$. This will certainly imply $y_b \succ y_b'$ or $y_b' \succ y_b$.

Now let us start our proof. Assume that $w \in N_\lambda$ has a standard MDC form $(A_r,\ldots,A_1)$ at $i \in \mathbf{Z}$. Let A_r^o (resp. A_r^1) be the block consisting of the first row (resp. the last λ_r-1 rows) of A_r.

If $\lambda_r = 1$, then $w' = \rho_{A_1,\ldots,A_{r-1}}^{A_r}(w)$ and the result is true. So we may assume $\lambda_r > 1$.

Let $y_a = \rho^{A_r^o}_{A_1,\dots,A_{r-1},A_r^1}(w)$, $y_a' = w'$. Then y_a lies in the sequence $\xi(w,r)$ beginning with w and the $D_w^{y_a}$-table is obtained from Table (I) by substituting y_a, w for x_2, x_1. Clearly, y_a (resp. y_a') has the full MDC form $(A_r^o, A_r^1, A_{r-1},\dots,A_1)$ (resp. $(A_r, A_{r-1},\dots,A_1)$) at i+1. This implies $\mathcal{L}(y_a') \not\subseteq \mathcal{L}(y_a)$.

When u = 1, we have the following D-tables.

$D^{y_a,A(y_a)}_{w,A(w)}(E_v)$ — v \ A	$A_1,\dots,A_{r-1}$	A_r^1	A_r^o
$3 < v < \lambda_1$	0	0	0
1	1	$\frac{1}{2}$	$\frac{1}{2}$
2	0	$\frac{1}{2}$	$\frac{1}{2}$

$D_w^{y_a}$-Table

$D^{y_a',A(y_a')}_{w,A(w)}(E_v)$ — v \ A	$A_1,\dots,A_{r-1}$	A_r^1	A_r^o
$3 < v < \lambda_1$	0	0	0
1	1	1	1
2	0	0	0

$D_w^{y_a'}$-Table

Then by Lemma 12.8.2, we get

$D_{y_a,A(y_a)}^{y'_a,A(y'_a)}(E_v)$ (v \ A)	$A_1,\ldots,A_{r-1}$	A_r^1	A_r^o
$3 < v < \lambda_1$	0	0	0
1	0	$\frac{1}{2}$	$\frac{1}{2}$
2	0	$-\frac{1}{2}$	$-\frac{1}{2}$

$D_{y_a}^{y'_a}$-Table

By Tables $D_w^{y_a}$, $D_w^{y'_a}$, we see that $e((y_a), j^1_{A_r^o}(y_a)) \in \bar{e}((y_a), E_2)$ and $e((y'_a), j^1_{A_r^o}(y'_a)) \in \bar{e}((y'_a),E_1)$. Then by Table $D_{y_a}^{y'_a}$ and the full DC forms of y_a, y'_a at i+1, we see that y'_a is obtained from y_a by transposing the (i+2)-th and the α-th rows, where the α-th row of y_a contains an entry belonging to $\bar{e}((y_a),E_1)$ and also to $A_r^1(y_a)$. So by Lemma 12.7.3, y_a, y'_a must have the following forms

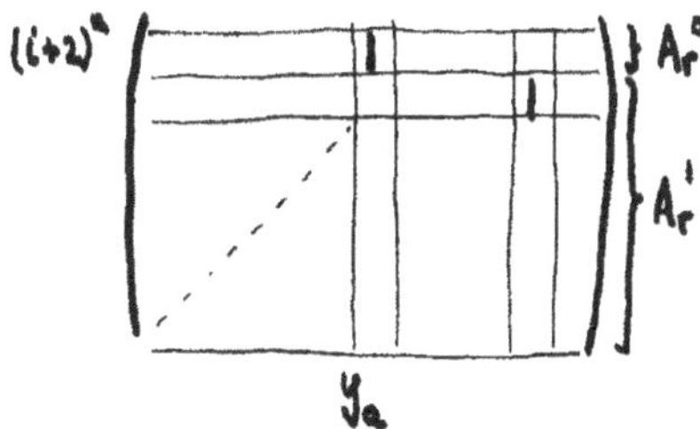

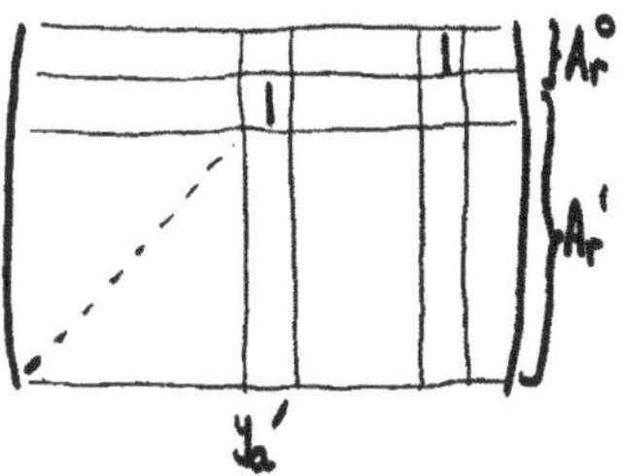

Thus by Lemma 13.1.1, we have $y'_a > y_a$ with $\ell(y'_a) = \ell(y_a) + 1$ and hence $y'_a \succ y_a$. So $y'_a \underset{L}{<} y_a$.

When u > 1, let

$$y_b = \rho^{A_r^o}_{A_1,\dots,A_{r-1}} (\rho^{A_r^o}_{A_1,\dots,A_{r-1},A_r^1})^{u-1} (w) = \rho^{A_r^o}_{A_1,\dots,A_{r-1}} (\rho^{A_r^o}_{A_1,\dots,A_{r-1},A_r^1})^{u-2} (y_a)$$

$$y_b' = \rho^{A_r^o}_{A_1,\dots,A_{r-1}} (\rho^{A_r^o}_{A_1,\dots,A_{r-1},A_r^1})^{u-2} (y_a').$$

Then y_b (resp. y_b') has the full DC form $(A_r^1, A_r^o, A_{r-1},\dots,A_1)$ at $i+u$ which is an element of the sequence $\xi(w,r)$ (resp. $\xi(y_a',r)$) beginning with w (resp. y_a'). y_b' is obtained from y_a' by applying the same succession of left star operations as are needed to obtain y_b from y_a in $\xi(w,r)$. From Tables (II), (III) and Lemma 12.8.2, we get

$D_w^{y_b}$-Table

v \ A : $D^{y_b,A(y_b)}_{w,A(w)}(E_v)$	$A_1,\dots,A_{r-1}$	A_r^1	A_r^o
1	1	$\frac{1}{2}$	$\frac{3}{2}$
$2 < v < u-2$	1	1	2
$u - 1$	1	1	2
u	1	$\frac{1}{2}$	$\frac{3}{2}$
$u < v < \lambda_r$	0	0	1
$\lambda_r < v < \lambda_1$	0	0	0

$D_w^{y'_b}$-Table

$D_{w,A(w)}^{y'_b,A(y'_b)}(E_v)$ (A / v)	$A_1,\ldots,A_{r-1}$	A_r^1	A_r^0
1	1	$\frac{1}{2}$	$\frac{3}{2}$
$2 \leq v \leq u-2$	1	1	2
$u - 1$	1	$\frac{1}{2}$	$\frac{3}{2}$
u	1	1	2
$u < v \leq \lambda_r$	0	0	1
$\lambda_r < v \leq \lambda_1$	0	0	0

Hence by Lemma 12.8.2, we have

$D_{y_b,A(y_b)}^{y'_b,A(y'_b)}(E_v)$ (A / v)	$A_1,\ldots,A_{r-1}$	A_r^1	A_r^0
$1 \leq v \leq \lambda_1$, $v \neq u-1,u$	0	0	0
$u-1$	0	$-\frac{1}{2}$	$-\frac{1}{2}$
u	0	$\frac{1}{2}$	$\frac{1}{2}$

$D_{y_b}^{y'_b}$-Table

By tables $D_W^{y_b}$, $D_W^{y_b'}$, we see that $e((y_b), j^1_{A_r^o}(y_b)) \in \bar{e}((y_b),E_u)$ and $e((y_b'), j^1_{A_r^o}(y_b')) \in \bar{e}((y_b'),E_{u-1})$. Then by Table $D_{y_b}^{y_b'}$ and the full DC forms of y_b, y_b' at i+u, we see that y_b' is obtained from y_b by transposing the α-th and the $(i+u+\lambda_r)$-th rows, where the α-th row of y_b contains an entry belonging to $\bar{e}((y_b),E_{u-1})$ and also to $A_r^1(y_b)$. So by Lemma 12.7.3, y_b, y_b' must have the following forms

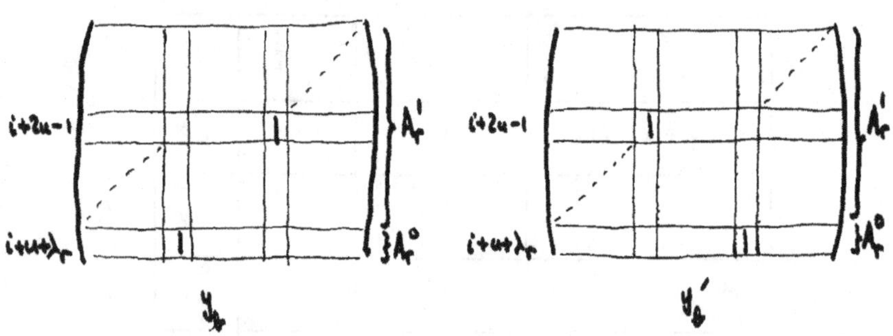

Then by Lemma 13.1.1, we have $y_b > y_b'$ and $\ell(y_b) = \ell(y_b') + 1$ and hence $y_b \succ y_b'$. Therefore, either $y_a' \succ y_a$ or $y_a' \prec y_a$ holds by Theorem 1.6.1. This implies $y_a' \underset{L}{<} y_a$.

On the other hand, let

$$y_c = \rho^{A_r^o}_{A_1,\ldots,A_{r-1}}(\rho^{A_r^o}_{A_1,\ldots,A_{r-1},A_r^1})^{\lambda_r-1}(w) = \rho^{A_r^o}_{A_1,\ldots,A_{r-1}}(\rho^{A_r^o}_{A_1,\ldots,A_{r-1},A_r^1})^{\lambda_r-2}(y_a)$$

$$y_c' = \rho^{A_r^o}_{A_1,\ldots,A_{r-1}}(\rho^{A_r^o}_{A_1,\ldots,A_{r-1},A_r^1})^{\lambda_r-2}(y_a').$$

Then y_c is the last term of the sequence $\xi(w,r)$ but y_c' is not the last term of $\xi(y_a',r)$. So y_c has the full MDC form $([A_r^1,A_r^o],A_{r-1},\ldots,A_1)$ at $i+\lambda_r$ and y_c' has the full MDC form $(A_r^1,A_r^o,A_{r-1},\ldots,A_1)$ at $i+\lambda_r$. This implies $\mathcal{L}(y_c) \not\subseteq \mathcal{L}(y_c')$. Now y_c' is obtained from y_a' by the same left star operations as are needed to obtain y_c from y_a in $\xi(w,r)$. According to Theorem 1.6.1 either $y_c \succ y_c'$ or $y_c \prec y_c'$ holds and so $y_c \underset{L}{<} y_c'$.

Therefore, we have $y_a \underset{P_L}{\sim} y_c \underset{L}{<} y_c' \underset{P_L}{\sim} y_a' \underset{L}{<} y_a$ and this implies $y_a \underset{L}{\sim} y_a'$. So $w \underset{P_L}{\sim} y_a$ and $w' = y_a'$ imply $w \underset{L}{\sim} w'$.

§13.4 PROOF OF PROPOSITION 13.2.3 WHEN $\lambda_{k+1} < u < \lambda_k$ AND $1 < k < r$.

We now consider the situation when u satisfies $\lambda_{k+1} < u < \lambda_k$ for some k with $1 < k < r$.

The idea of the proof is generally similar to §13.3 but is a little more complicated. As before, we pick a certain element y_a in the sequence $\xi(w,k)$. We also consider a certain element $y_a' \in N_\lambda$ obtained from w' by certain ρ-operations. We compare the sequences $\xi(w,k)$ and $\xi(y_a',k)$. Let y_c be the last term of the sequence $\xi(w,k)$ and y_c' be the term in $\xi(y_a',k)$ obtained by applying the same sequence of left star operations to y_a' as are needed to obtain y_c from y_a in $\xi(w,k)$. We then have a cycle of elements as in the figure

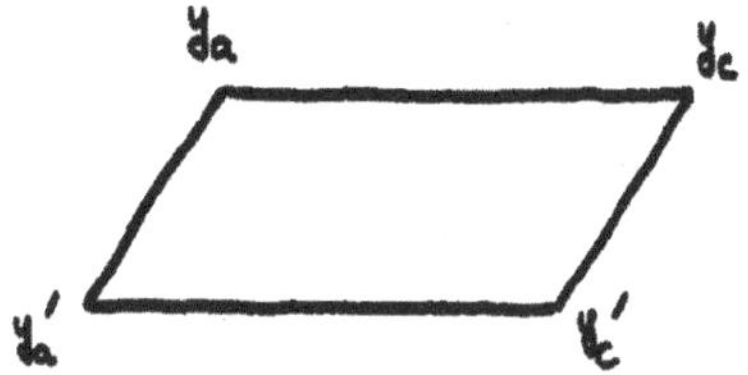

We show that all the elements in this cycle lie in the same left cell as before by finding corresponding elements y_b, y_b' in the two sequences satisfying $y_b > y_b'$ and $\ell(y_b) = \ell(y_b') + 1$. Since $y_a \underset{L}{\sim} w$ and $y_a' \underset{L}{\sim} w'$, we obtain $w \underset{L}{\sim} w'$ as required.

Now we start to prove our result.

Assume that w has a standard MDC form $(A_r,\dots,A_1)$ at $i \in \mathbb{Z}$ and let blocks A_k^0, A_k^1, $A_{i_1}^0$ and $A_{i_1}^1$ be defined as in §12.6.

Let $y_a = \rho^{A_r,\dots,A_{k+1},A_k^0}_{A_1,\dots,A_{k-1},A_k^1}(w)$, $y_a' = \rho^{A_r,\dots,A_{k+1}}_{A_1,\dots,A_k}(w')$. Then y_a' lies in N_λ and y_a is in the sequence $\xi(w,k)$ beginning with w. Both y_a and y_a' have the full DC form

$$(A_r,\dots,A_{k+1},\ A_k^0,A_k^1,A_{k-1},\dots,A_1) \text{ at } i',\ i' = i + \sum_{j=k+1}^{r} \lambda_j + 1.$$

Since, by Lemma 12.6.1, $[A_k^0(y_a'),\ A_k^1(y_a')]$ is a DC block but $[A_k^0(y_a), A_k^1(y_a)]$ is not, this implies that $\mathcal{L}(y_a') \not\subseteq \mathcal{L}(y_a)$.

For $u' = u - \lambda_{i_1}$, let

$$y_b = \rho_{A_k^1}^{A_{i_1}^0} \rho_{A_1,\ldots,A_{k-1}}^{A_{i_1}} \rho_{A_1,\ldots,A_{k-1},A_k^1}^{A_k^0,A_r,\ldots,A_{i_1+1}} \left(\rho_{A_1,\ldots,A_{k-1},A_k^1}^{A_k^0,A_r,\ldots,A_{k+1}}\right)^{u'-1} \rho_{A_1,\ldots,A_{k-1},A_k^1}^{A_r,\ldots,A_{k+1}}(w)$$

$$= \rho_{A_k^1}^{A_{i_1}^0} \rho_{A_1,\ldots,A_{k-1}}^{A_{i_1}} \rho_{A_1,\ldots,A_{k-1},A_k^1}^{A_r,\ldots,A_{i_1+1}} \left(\rho_{A_1,\ldots,A_{k-1},A_k^1}^{A_r,\ldots,A_{k+1},A_k^0}\right)^{u'-1} (y_a)$$

$$y_b' = \rho_{A_k^1}^{A_{i_1}^0} \rho_{A_1,\ldots,A_{k-1}}^{A_{i_1}} \rho_{A_1,\ldots,A_{k-1},A_k^1}^{A_r,\ldots,A_{i_1+1}} \left(\rho_{A_1,\ldots,A_{k-1},A_k^1}^{A_r,\ldots,A_{k+1},A_k^0}\right)^{u'-1} (y_a')$$

Then y_b (resp. y_b') is an element of the sequence $\xi(w,k)$ (resp. $\xi(y_a',k)$) beginning with w (resp. y_a'). y_b' is obtained by applying the same sequence of left star operations to y_a' as are needed to obtain y_b from y_a in $\xi(w,k)$. The tables for functions $D_w^{y_b}$ and $D_w^{y_b'}$ can be worked out by Example 12.8.5(i), Tables (IV), (VI), (VII) and Lemma 12.8.2 as follows.

$D_w^{Y_b}$ - Table

$D_{w,\,A(w)}^{Y_b;\,A(Y_b)}(E_v)$ (v \ λ)	A_k^0	A_{i_g}, $2 \le g \le t$	$A_{i_{g-1}+1}, \dots, A_{i_g-1}$, $2 \le g \le t$	$A_{i_1}^0$	$A_{i_1}^1$	$A_{i_o+1}, \dots, A_{i_1-1}$	A_k^1	A_a, $1 \le a < k$
1	$u'(i_t-i_o)$	$(u'+1)(i_t-i_o)+\frac{1}{2}\delta_{gt}$	$(u'+1)(i_t-i_o)$	$(u'+1)(i_t-i_o)+\frac{1}{2}\delta_{1t}$	$(u'+1)(i_t-i_o)+1+\frac{1}{2}\delta_{1t}\cdot\delta_{\alpha_{\lambda,1},1}$	$u'(i_t-i_o)$	$(u'+1)(i_t-i_o)+i_o-i_1+\frac{1}{2}$	$(u'+1)(i_t-i_o)+i_o-i_1+1$
$\lambda_{i_h}+1$, $3 \le h \le t$	$u'(i_{h-1}-i_o)$	$(u'+1)(i_{h-1}-i_o)+\frac{1}{2}(\delta_{hg}+\delta_{h-1,g})$	$(u'+1)(i_{h-1}-i_o)+\delta_{hg}$	$(u'+1)(i_{h-1}-i_o)$	$(u'+1)(i_{h-1}-i_o)+1$	$u'(i_{h-1}-i_o)$	$(u'+1)(i_{h-1}-i_o)+i_o-i_1+1$	
$\lambda_{i_h}+1<v\le\lambda_{i_{h-1}}$, $2 \le h \le t+1$, $v \ne \lambda_{i_1}$	$u'(i_{h-1}-i_o)$	$(u'+1)(i_{h-1}-i_o)$			$(u'+1)(i_{h-1}-i_o)+1$	$u'(i_{h-1}-i_o)$	$(u'+1)(i_{h-1}-i_o)+i_o-i_1+1$	
$\lambda_{i_2}+1$, $v \ne 1$	$u'(i_1-i_o)$	$(u'+1)(i_1-i_o)+\frac{1}{2}\delta_{2;g}$	$(u'+1)(i_1-i_o)+\delta_{2,g}$	$(u'+1)(i_1-i_o)+\frac{1}{2}(1+\delta_{\alpha_{\lambda,1},1})$	$(u'+1)(i_1-i_o)+1+\frac{1}{2}\delta_{\alpha_{\lambda,1},1}$	$u'(i_1-i_o)$	$u'(i_1-i_o)+1$	
λ_{i_1}, $v \ne \lambda_{i_2}+1$	$u'(i_1-i_o)$	$(u'+1)(i_1-i_o)$		$(u'+1)(i_1-i_o)+\frac{1}{2}$	$(u'+1)(i_1-i_o)+\frac{3}{2}$	$u'(i_1-i_o)$	$u'(i_1-i_o)+1$	
$\lambda_{i_1}+\ell$, $1 \le \ell < u'-1$	1				2	1		
$\lambda_{i_1}+u'-1$, $v \ne \lambda_{i_1}$	1				2	1		
$\lambda_{i_1}+u'$	1				$\frac{3}{2}$	1	$\frac{1}{2}$	1
$\lambda_{i_1}+\ell'$, $u'<\ell'\le\alpha_{\lambda,0}$	0				1	0		
$\lambda_{i_o}<v\le\lambda_1$	0							

$D_w^{y_b'}$ - table (when u' = 1)

$D_{w,\,A(w)}^{y_b',\,A(y_b')}$ (E_v) A / v	A_k^o	A_{i_g} $2 \le g \le t$	$A_{i_{g-1}+1},\ldots,A_{i_g-1}$ $2 \le g \le t$	$A_{i_1}^o$	$A_{i_1}^1$	A_a $k<a<i_1$	A_k^1	A_b $1 \le b < k$
1	i_t-i_o	$2(i_t-i_o)+\frac{1}{2}\delta_{gt}$	$2(i_t-i_o)$	$2(i_t-i_o)+\frac{1}{2}\delta_{1,t}$	$2(i_t-i_o)+1$	i_t-i_o	$2(i_t-i_o)+i_o-i_1+\frac{1}{2}(1-\delta_{1t}\cdot\delta_{a_{\lambda,1},1})$	$2(i_t-i_o)+i_o-i_1+1$
$\lambda_{i_h}+1$ $3 \le h \le t$	$i_{h-1}-i_o$	$2(i_{h-1}-i_o)+\frac{1}{2}(\delta_{gh}+\delta_{g,h-1})$	$2(i_{h-1}-i_o)+\delta_{hg}$	$2(i_{h-1}-i_o)$	$2(i_{h-1}-i_o)+1$	$i_{h-1}-i_o$	$2(i_{h-1}-i_o)+i_o-i_1+1$	
$\lambda_{i_h}+1<v\le\lambda_{i_{h-1}}$ $2 \le h \le t+1$ $v \ne \lambda_{i_1}$	$i_{h-1}-i_o$	$2(i_{h-1}-i_o)$			$2(i_{h-1}-i_o)+1$	$i_{h-1}-i_o$	$2(i_{h-1}-i_o)+i_o-i_1+1$	
$\lambda_{i_2}+1$ $v \ne 1$	i_1-i_o	$2(i_1-i_o)+\frac{1}{2}\delta_{g,2}$	$2(i_1-i_o)+\delta_{g,2}$	$2(i_1-i_o)+\frac{1}{2}(1+\delta_{a_{\lambda,1},1})$	$2(i_1-i_o)+1$	i_1-i_o	$(i_1-i_o)+(1-\frac{1}{2}\delta_{a_{\lambda,1},1})$	$(i_1-i_o)+1$
λ_{i_1} $v \ne \lambda_{i_2}+1$	i_1-i_o	$2(i_1-i_o)$		$2(i_1-i_o)+\frac{1}{2}$	$2(i_1-i_o)+1$	i_1-i_o	$(i_1-i_o)+\frac{1}{2}$	$(i_1-i_o)+1$
$\lambda_{i_1}+1$	1				2	1		
$\lambda_{i_1}+1<v\le\lambda_{i_o}$	0				1	0		
$\lambda_{i_o}<v\le\lambda_1$	0							

$D_w^{y_b'}$ – TABLE (when $\alpha_{\lambda,0} \geq u' > 1$)

<table>
<tr><td>$D_{y_1,A(w)}^{y_b',A(y_b')}$ (E_v) / A / v</td><td>A_k^0</td><td>A_{i_g}
$2 \leq g \leq t$</td><td>$A_{i_{g-1}+1},\dots,A_{i_g-1}$
$2 \leq g \leq t$</td><td>$A_{i_1}^0$</td><td>$A_{i_1}^1$</td><td>$A_{i_0+1},\dots,A_{i_1-1}$</td><td>$A_k^1$</td><td>$A_a$
$1 \leq a < k$</td></tr>
<tr><td>1</td><td>$u'(i_t-i_o)$</td><td>$(u'+1)(i_t-i_o)+\frac{1}{2}\delta_{gt}$</td><td>$(u'+1)(i_t-i_o)$</td><td>$(u'+1)(i_t-i_o)+\frac{1}{2}\delta_{1t}$</td><td>$(u'+1)(i_t-i_o)+1+\frac{1}{2}\delta_{1t}\cdot\delta_{\alpha_{\lambda,1},1}$</td><td>$u'(i_t-i_o)$</td><td>$(u'+1)(i_t-i_o)+i_o-i_1+\frac{1}{2}$</td><td>$(u'+1)(i_t-i_o)+i_o-i_1+1$</td></tr>
<tr><td>$\lambda_{i_h}+1$
$3 \leq h \leq t$</td><td>$u'(i_{h-1}-i_o)$</td><td>$(u'+1)(i_{h-1}-i_o)+\frac{1}{2}(\delta_{gh}+\delta_{g,h-1})$</td><td>$(u'+1)(i_{h-1}-i_o)+\delta_{gh}$</td><td>$(u'+1)(i_{h-1}-i_o)$</td><td>$(u'+1)(i_{h-1}-i_o)+1$</td><td>$u'(i_{h-1}-i_o)$</td><td colspan="2">$(u'+1)(i_{h-1}-i_o)+i_o-i_1+1$</td></tr>
<tr><td>$\lambda_{i_h}+1<v\leq\lambda_{i_{h-1}}$
$2 \leq h \leq t+1$
$v \neq \lambda_{i_1}$</td><td>$u'(i_{h-1}-i_o)$</td><td colspan="3">$(u'+1)(i_{h-1}-i_o)$</td><td>$(u'+1)(i_{h-1}-i_o)+1$</td><td>$u'(i_{h-1}-i_o)$</td><td colspan="2">$(u'+1)(i_{h-1}-i_o)+i_o-i_1+1$</td></tr>
<tr><td>$\lambda_{i_2}+1$
$v \neq 1$</td><td>$u'(i_1-i_o)$</td><td>$(u'+1)(i_1-i_o)+\frac{1}{2}\delta_{g,2}$</td><td>$(u'+1)(i_1-i_o)+\delta_{g,2}$</td><td>$(u'+1)(i_1-i_o)+\frac{1}{2}(1+\delta_{\alpha_{\lambda,1},1})$</td><td>$(u'+1)(i_1-i_o)+1+\frac{1}{2}\delta_{\alpha_{\lambda,1},1}$</td><td>$u'(i_1-i_o)$</td><td colspan="2">$u'(i_1-i_o)+1$</td></tr>
<tr><td>λ_{i_1}
$v \neq \lambda_{i_2}+1$</td><td>$u'(i_1-i_o)$</td><td colspan="2">$(u'+1)(i_1-i_o)$</td><td>$(u'+1)(i_1-i_o)+\frac{1}{2}$</td><td>$(u'+1)(i_1-i_o)+\frac{3}{2}$</td><td>$u'(i_1-i_o)$</td><td colspan="2">$u'(i_1-i_o)+1$</td></tr>
<tr><td>$\lambda_{i_1}+\ell$
$1 \leq \ell < u'-1$</td><td colspan="4">1</td><td>2</td><td colspan="3">1</td></tr>
<tr><td>$\lambda_{i_1}+u'-1$
$v \neq \lambda_{i_1}$</td><td colspan="4">1</td><td>$\frac{3}{2}$</td><td>1</td><td>$\frac{1}{2}$</td><td>1</td></tr>
<tr><td>$\lambda_{i_1}+u'$</td><td colspan="4">1</td><td>2</td><td colspan="3">1</td></tr>
<tr><td>$\lambda_{i_1}+\ell'$
$u' < \ell' \leq \alpha_{\lambda,0}$</td><td colspan="4">0</td><td>1</td><td colspan="3">0</td></tr>
<tr><td>$\lambda_{i_o} < v \leq \lambda_1$</td><td colspan="8">0</td></tr>
</table>

Then also by Lemma 12.8.2, we get the following table.

$v \backslash A$; $D^{y_b',A(y_b')}_{y_b,A(y_b)}(E_v)$	A_h, $1<h<r$, $A^o_{i_1}$, $h \neq k$, A^o_k i_1	$A^1_{i_1}$	A^1_k
$1 < v < \lambda_1$ $v \neq u-1,u$	0	0	0
$u - 1$	0	$-\frac{1}{2}$	$-\frac{1}{2}$
u	0	$\frac{1}{2}$	$\frac{1}{2}$

$D^{y_b'}_{y_b}$-Table

Since both y_b and y_b' have the full DC forms

$$(A_{i_1-1},\ldots,A_{k+1}, [A^o_k,A_r], A_{r-1},\ldots,A_{i_1+1},A^o_{i_1},A^1_k,A^1_{i_1},A_{k-1},\ldots,A_1)$$

at i", i" = u' ($\sum_{h=k+1}^{r} \lambda_h + 1) + \sum_{h=i_1}^{r} \lambda_h$, we see from tables $D^{y_b}_w$, $D^{y_b'}_w$, $D^{y_b'}_{y_b}$ that $e((y_b), j_{A^1_{i_1}}^1(y_b)) \in \bar{e}((y_b),E_u)$, $e((y_b'), j_{A^1_{i_1}}^1(y_b')) \in \bar{e}((y_b'),E_{u-1})$ and y_b' is obtained from y_b by transposing the α-th and the β-th rows, where $\alpha = i" + \sum_{h=k}^{r} \lambda_h$, the β-th row of y_b contains an entry belonging to $\bar{e}((y_b),E_{u-1})$ and also to $A^1_k(y_b)$. So by Lemma 12.7.3, y_b, y_b' must have the forms as below:

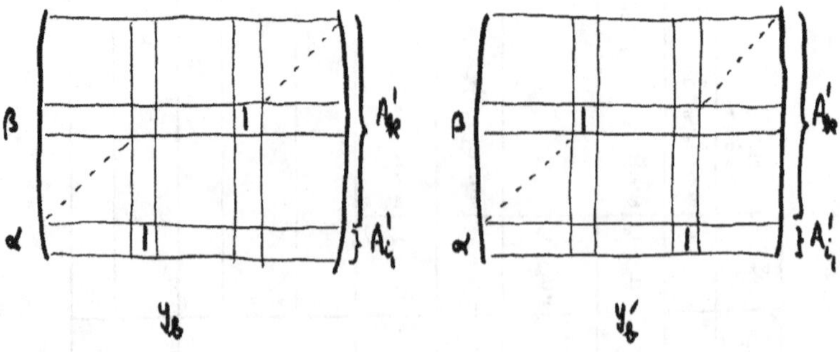

So by Lemma 13.1.1, we have $y_b > y'_b$ with $\ell(y_b) = \ell(y'_b) + 1$ and hence $y_b \succ y'_b$. According to Theorem 1.6.1, either $y_a \succ y'_a$ or $y_a \prec y'_a$ holds. Therefore $y'_a \underset{L}{<} y_a$.

On the other hand, let

$$y_c = \rho^{A^0_{i_1}}_{A^1_k} \rho^{A_{i_1}}_{A_1,\dots,A_{k-1}} \rho^{A^0_k,A_r,\dots,A_{i_1+1}}_{A_1,\dots,A_{k-1},A^1_k} \left(\rho^{A^0_k,A_r,\dots,A_{k+1}}_{A_1,\dots,A_{k-1},A^1_k}\right)^{\alpha_{\lambda,0}-1} \rho^{A_r,\dots,A_{k+1}}_{A_1,\dots,A_{k-1},A^1_k}(w)$$

$$= \rho^{A^0_{i_1}}_{A^1_k} \rho^{A_{i_1}}_{A_1,\dots,A_{k-1}} \rho^{A_r,\dots,A_{i_1+1}}_{A_1,\dots,A_{k-1},A^1_k} \left(\rho^{A_r,\dots,A_{k+1},A^0_k}_{A_1,\dots,A_{k-1},A^1_k}\right)^{\alpha_{\lambda,0}-1} (y_a)$$

$$y'_c = \rho^{A^0_{i_1}}_{A^1_k} \rho^{A_{i_1}}_{A_1,\dots,A_{k-1}} \rho^{A_r,\dots,A_{i_1+1}}_{A_1,\dots,A_{k-1},A^1_k} \left(\rho^{A_r,\dots,A_{k+1},A^0_k}_{A_1,\dots,A_{k-1},A^1_k}\right)^{\alpha_{\lambda,0}-1} (y'_a)$$

Then y_c is the last term of $\xi(w,k)$ but y'_c is not the last term of $\xi(y'_a,k)$. By Lemma 12.6.1, we see that $\mathcal{L}(y_c) \not\subseteq \mathcal{L}(y'_c)$. We know that y'_c is obtained from y'_a by applying the same sequence of left star operations as are needed to obtain y_c from y_a in $\xi(w,k)$. According to Theorem 1.6.1, we have either $y_c \prec y'_c$ or $y_c \succ y'_c$. So $y_c \underset{L}{<} y'_c$. Hence $y_a \underset{P_L}{\sim} y_c \underset{L}{<} y'_c \underset{P_L}{\sim} y'_a \underset{L}{<} y_a$. This implies $y_a \underset{L}{\sim} y'_a$. Finally, it follows from $w \underset{P_L}{\sim} y_a$ and $w' \underset{P_L}{\sim} y'_a$ that $w \underset{L}{\sim} w'$.

CHAPTER 14 : THE LEFT AND RIGHT CELLS IN $\sigma^{-1}(\lambda)$

In the present chapter, we are able to determine the left cells of A_n which lie in $\sigma^{-1}(\lambda)$. We recall from Chapter 9 that $\sigma^{-1}(\lambda)$ is a union of left cells. We define in §14.1 a map $T: N_\lambda \to \check{C}_\lambda$ from the set of normalized elements of type λ into the set of λ-tabloids, and we shall show that two elements of N_λ lie in the same left cell if and only if they have the same image under T.

In proving this result, it is of crucial importance to identify certain particular left cells in $\sigma^{-1}(\lambda)$. These are obtained from one fixed left cell by applying powers of the automorphism ϕ. This fixed left cell is obtained as follows: We show in §14.3 that there exists a subset J of Δ such that $\{w \in \sigma^{-1}(\lambda) \mid R(w) = J\}$ is a left cell.

The result that any two elements x, y lying in the same left cell can be transformed from one to another by a succession of raising operations in N_λ and left star operations will be applied in Chapter 18 to prove the connectedness of cells of A_n, where $\lambda = \sigma(x)$.

§14.1 THE MAP T FROM N_λ TO THE SET OF λ-TABLOIDS

Fix $\lambda = \{\lambda_1 \geq \ldots \geq \lambda_r\} \in \Lambda_n$. Let $\mu = \{\mu_1 \geq \ldots \geq \mu_m\}$ be the dual partition of λ.

Recall the definition of a λ-tabloid of $\check{C}$ in §4.4 (xiii). Let $\check{C}_\lambda$ be the set of all λ-tabloids. Then we have

Lemma 14.1.1

$$|\check{C}_\lambda| = \frac{n!}{\prod_{j=1}^{m} \mu_j!} \quad . \qquad \square$$

For any $w \in N_\lambda$, w has a standard MDC form $(A_r, \ldots, A_1)$ at $i \in \mathbf{Z}$. Let $E_u = \{\overline{j_t^u(w)} \mid 1 \leq t \leq \mu_u\} \subseteq \bar{n}$. Clearly, E_u is independent of the choice of the standard MDC form of w. Let T(w) be the element of $\check{C}_\lambda$ whose u-th column contains all numbers α with $1 \leq \alpha \leq n$ and $\bar{\alpha} \in E_u$ for any u, $1 \leq u \leq \lambda_1$. Then T(w) is uniquely determined by w. So we can define a map $T: N_\lambda \to \check{C}_\lambda$ by sending w to T(w).

Example Let $w \in A_9$ have the form

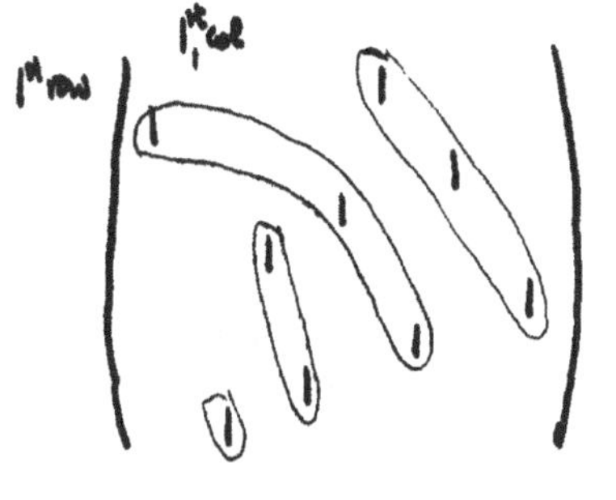

Then $w \in N_\lambda$ with $\lambda = \{4 > 3 > 2\}$ and $T(w) = \begin{array}{|c|c|c|c|} \hline & & 5 & 1 \\ & 3 & 7 & 6 \\ 2 & 4 & 9 & 8 \\ \hline \end{array}$

Lemma 14.1.2 The map T is surjective.

Proof For any $X \in \check{C}_\lambda$, assume that the set of numbers in the t-th column of X is

$$E_t = \{i_{th} \mid 1 \le h \le \mu_t,\ 1 \le i_{t\mu_t} < i_{t,\mu_t-1} < \cdots < i_{t1} \le n\}$$

for $1 \le t \le \lambda_1$. Let $M = (M_{\lambda_1}, M_{\lambda_1-1}, \ldots, M_1)$ be an $n \times \lambda_1 n$ matrix with M_t being an $n \times n$ matrix, $1 \le t \le \lambda_1$, such that

(i) for any t, $1 \le t \le \lambda_1$, $M_t = \begin{matrix} M_{tr} \\ M_{t,r-1} \\ \vdots \\ M_{t1} \end{matrix}$ with M_{tj} being a $\lambda_j \times n$ matrix, $1 \le j \le r$.

(ii) the entries of M_{tj} are all zero if $\mu_t < j \le r$; and are all zero except the (t, i_{tj})-entry which is 1 if $1 \le j \le \mu_t$.

It is clear that there exists a unique element $\tilde{X}(M)$ of $\tilde{A}_n$ with M as a submatrix.

Suppose $i_{\alpha\beta} = 1$ for some α, β with $1 \le \alpha \le \lambda_1$ and $1 \le \beta \le \mu_\alpha$. Then there also exists a unique element X(M) of A_n with M as a submatrix and with $(\alpha, i_{\alpha\beta})$-entry of $M_{\alpha\beta}$ lying in its first column, where $M_{\alpha\beta}$ is a submatrix of X(M) in M which, regarded as a submatrix of M, is just defined as above.

Clearly, X(M) has the full MDC form $(A_r, \ldots, A_1)$ at i which is normal for some $i \in \mathbf{Z}$, with $|A_h| = \lambda_h$, $1 \le h \le r$. By Lemma 12.1.1, X(M) is in N_λ. Also clearly, we have $T(X(M)) = X$. This implies that T is surjective. □

If $w, w' \in N_\lambda$ with $w' = L_{E_u}(w)$ for some u, $1 \le u \le \lambda_1$, where E_u is the subset of $\underline{n}$ such that $\bar{e}((w), E_u)$ is the u-th layer of w, then we have $T(w) = T(w')$. So by Lemma 14.1.2, it is immediate that

Corollary 14.1.3

The map $\bar{T} = T|_{\bar{N}_\lambda} : \bar{N}_\lambda \to \check{C}$ is surjective. □

§14.2 THE SET $\bar{N}_\lambda$ OF PRINCIPAL NORMALIZED ELEMENTS

We shall show in this section that for any given $X \in \check{C}_\lambda$ and any number α in the first column of X, there exists a unique element w of $\bar{N}_\lambda$ such that w has a standard MDC form $(A_r,\ldots,A_1)$ at i for some $i \in \mathbf{Z}$ with $j_1^1(w) = \bar{\alpha}$ and $T(w) = X$. By using this result, we can precisely calculate the cardinal of $\bar{T}^{-1}(X)$ for any $X \in \check{C}_\lambda$ and then the cardinal of $\bar{N}_\lambda$.

The main purpose of this section is to show that for any $X \in \check{C}_\lambda$, $\bar{T}^{-1}(X)$ lies in some left cell of A_n.

Suppose that $E = \{\bar{i}_1,\ldots,\bar{i}_t\} \subseteq \underline{n}$ is a non-empty set with $1 \leq i_1 < \ldots < i_t \leq n$. Let $\bar{e}((w),E)$ be the set of entry classes of $w \in A_n$ such that $\bar{e}((w),E)$ contains all entries $e(i,j)$ with $\bar{j} \in E$. Let $\bar{e}((w),j_1)$, $e((w),j_2)$ be two entries of w lying in $\bar{e}((w),E)$ with $j_1 < j_2$ such that there exists no entry $e((w),j)$ of w lying in $\bar{e}((w),E)$ with $j_1 < j < j_2$. Suppose $\bar{j}_1 = \bar{i}_\alpha$ (resp. $\bar{j}_2 = \bar{i}_\alpha$) for some α, $1 \leq \alpha \leq t$. This implies $\bar{j}_2 = \bar{i}_{\alpha+1}$ (resp. $\bar{j}_1 = \bar{i}_{\alpha-1}$) and $0 < j_2-j_1 \leq n$ with the convention that $i_\beta = i_{\beta+qt}$ for any β, q with $1 \leq \beta \leq t$ and $q \in \mathbf{Z}$.

Now suppose $\lambda = \{\lambda_1 \geq \ldots \geq \lambda_r\} \in \Lambda_n$ and $w \in N_\lambda$ has a standard MDC form $(A_r,\ldots,A_1)$ at i for some $i \in \mathbf{Z}$ with $T(w) = X \in \check{C}_\lambda$. Let X_k be the set of all numbers in the k-th column of X, $1 \leq k \leq \lambda_1$. Then $\bar{X}_k = \{j_t^k(w) | 1 \leq t \leq \mu_k\}$. Since $\bar{e}((w),E_k)$, $1 \leq k \leq \lambda_1$, are antichains of w, we see by the above statement that for any k, $1 \leq k \leq \lambda_1$, once one number $j_t^k(w)$ is known for some t, $1 \leq t \leq \mu_k$, all the numbers $j_v^k(w)$, $1 \leq v \leq \mu_k$, will be determined.

We say an increasing sequence $E:i_1,\ldots,i_t$ of $\mathbf{Z}$ is compact with respect to n, $n \geq 1$, if inequality $i_t - n < i_1 < \ldots < i_t$ holds. (Subsequently, the number n is always fixed. So we may just say E is compact without danger of confusion.)

Clearly, if $E:i_1,\ldots,i_t$ and $F:j_1,\ldots,j_t$ are two compact increasing sequences of $\mathbf{Z}$ such that $\bar{E}$, $\bar{F} \subseteq \underline{n}$ with $\bar{E} = \bar{F}$ and $i_1 = j_1$, then we have $i_\alpha = j_\alpha$ for any α, $1 \leq \alpha \leq t$.

Suppose $E:i_1,\ldots,i_t$ and $F:j_1,\ldots,j_t$ are two compact increasing sequences. We say E dominates F if the following conditions are satisfied:

(i) $\bar{E}$, $\bar{F} \subset \underline{n}$ with $\bar{E} \cap \bar{F} = \emptyset$.

(ii) $i_\alpha > j_\alpha$ for any α, $1 \leq \alpha \leq t$.

Suppose E dominates F. We say E, F are interlocking if the following additional condition is also satisfied:

(iii) There exists β, $1 \leq \beta \leq t$, such that $i_{\beta-1} < j_\beta$ with the convention that $i_a = i_{a+qt} - qn$ and $j_b = j_{b+qt} - qn$ for any $a,b,q \in \mathbf{Z}$ with $1 \leq a, b \leq t$.

Example: Suppose $w \in N_\lambda$ has a standard MDC form $(A_r,\ldots,A_1)$ at $i \in \mathbf{Z}$. For any k, h with $1 \le k \le \lambda_1$ and $1 \le h < \lambda_1$, let $E_k : j^k_{\mu_k}(w), j^k_{\mu_k-1}(w),\ldots,j^k_1(w)$ and let $E'_h : j^h_{\mu_{h+1}}(w), j^h_{\mu_{h+1}-1}(w), \ldots\ldots\ldots, j^h_1(w)$. Then by definition of N_λ, we see that E_k, E'_h are all compact increasing sequences of $\mathbf{Z}$ and that E'_h dominates E_{h+1} for any h, $1 \le h < \lambda_1$. Moreover, w is in $\bar{N}_\lambda$ if and only if E'_h, E_{h+1} are interlocking for any h, $1 \le h < \lambda_1$.

<u>Lemma 14.2.1</u> <u>Suppose $E : i_1,\ldots,i_t$ is a compact increasing sequence of $\mathbf{Z}$. Suppose there exists $D \subset \underline{n}$ satisfying $|D| = t$ and $D \cap \bar{E} = \emptyset$ in $\underline{n}$. Then there exists a unique compact increasing sequence $F : j_1,\ldots,j_t$ such that $\bar{F} = D$, E dominates F, and and E,F are interlocking.</u>

<u>Proof</u> Obviously, there exists a compact increasing sequence $F_1 : j_{11},\ldots,j_{1t}$ such that $\bar{F}_1 = D$ and E dominates F_1. If E, F_1 are not interlocking, then for any β, $1 \le \beta \le t$, we have $i_{\beta-1} > j_{1\beta}$. Let $F_2 : j_{21},\ldots,j_{2t}$ such that $j_{2\alpha} = j_{1,\alpha+1}$ for $1 \le \alpha \le t$ with the convention that $j_{1\beta} = j_{1,\beta+qt} - qn$ for any β, $q \in \mathbf{Z}$ with $1 \le \beta \le t$. Then $\bar{F}_2 = D$ and E dominates F_2. If E, F_2 are still not interlocking, then the same procedure can be carried on and we get $F_1, F_2, \ldots$ such that for every $c \ge 1$, the compact increasing sequence $F_c : j_{c1},\ldots,j_{ct}$ satisfies $j_{c\alpha} = j_{1,\alpha+c-1}$ for all α, $1 \le \alpha \le t$. It is clear that $\bar{F}_c = D$ for $c \ge 1$. If E, F_{b-1} are not interlocking, then E dominates F_b for any $b > 1$. We claim that there exists some $c \ge 1$ such that E, F_c are interlocking. Otherwise, E dominates F_c for any $c \ge 1$. This implies $i_\alpha > j_{c\alpha} = j_{1,\alpha+c-1}$ for any $c \ge 1$ and any α with $1 \le \alpha \le t$. In particular, $i_1 > j_{1,\alpha+qt}$ for any $q > 0$. But i_1 is finite and $\lim_{q\to\infty} j_{1,\alpha+qt} = \lim_{q\to\infty} (j_{1\alpha} + qn) = \infty$, this gives a contradiction. Thus the existence of F is proved.

Now we shall show that such a sequence F is unique. Suppose $F : j_1,\ldots,j_t$ and $F' : j'_1,\ldots,j'_t$ are two compact increasing sequences of $\mathbf{Z}$ such that $\bar{F} = \bar{F}' = D$ and E is dominantly interlocking with both F and F'. We may assume $j_1 \le j'_1$. If $j_1 < j'_1$, then there exists some $c \ge 1$ such that $j'_1 = j_{1+c}$ and hence $j'_\alpha = j_{\alpha+c}$ for all α, $1 \le \alpha \le t$ with the convention that $j_\alpha = j_{\alpha+qt} - qn$ for any α, q with $1 \le \alpha \le t$ and $q \in \mathbf{Z}$. Since E dominates F', we have $i_{\beta-1} > j'_{\beta-1} = j_{\beta+c-1} > j_\beta$ for any β, $1 \le \beta \le t$. This contradicts the fact that E, F are interlocking. Hence we must have $j_1 = j'_1$ and then $j_\alpha = j'_\alpha$ for all α, $1 \le \alpha \le t$. This proves the uniqueness of F. □

Given $\lambda = \{\lambda_1 \ge \ldots \ge \lambda_r\} \in \Lambda_n$ and $X \in \check{C}_\lambda$. Let X_j be the set of all numbers in the j-th column of X, $1 \le j \le \lambda_1$. Then for any $\alpha \in X_1$, we denote the following conditions on an element w of N_λ by $A(X,\alpha)$:

(i) $T(w) = X$

(ii) w has a standard MDC form $(A_r,\ldots,A_1)$ at i for some $i \in \mathbf{Z}$ with $\overline{j^1_1(w)} = \bar{\alpha}$.

Lemma 14.2.2 Let $\lambda \in \Lambda_n$, $X \in \check{C}_\lambda$ and $\alpha \in X_1$ be defined as above. Then there exists an element $w \in N_\lambda$ which satisfies $A(X,\alpha)$.

Proof We know from Lemma 14.1.2 that there exists an element $y \in N_\lambda$ which has a standard MDC form $(A_r,\ldots,A_1)$ at i for some $i \in \mathbb{Z}$ with $T(y) = X$. Suppose that E is the sequence $j^1_r(y), j^1_{r-1}(y),\ldots,j^1_1(y)$. Then E is a compact increasing sequence of $\mathbb{Z}$ with $\bar{E} = \bar{X}_1$. Suppose $\overline{j^1_k(y)} = \bar{\alpha}$ for some k, $1 \leqslant k \leqslant r$. Then there exists an element $w = (\rho^{A_r}_{A_1,\ldots,A_{r-1}})^{r+1-k}(y)$ of A_n by Lemma 8.3.5. Hence we see from Corollary 8.4.7 that w is in N_λ and has a standard MDC form $(A_r,\ldots A_1)$ at $i + (r+1-k)\lambda_r$ with $T(w) = X$ and $\overline{j^1_1(w)} = \overline{j^1_k(y)} = \bar{\alpha}$. So w is as required. □

Lemma 14.2.3 Assume that $w \in N_\lambda$ has a standard MDC form $(A_r,\ldots,A_1)$ at i for some $i \in \mathbb{Z}$. Assume that $w' \in N_\lambda$ satisfies $w' = L_{E_u}(w)$ for some u, $1 < u \leqslant \lambda_1$. Then w satisfies $A(X,\alpha)$ if and only if w' satisfies $A(X,\alpha)$, where $\lambda \in \Lambda_n$, $X \in \check{C}_\lambda$ and $\alpha \in X_1$ are as in Lemma 14.2.2.

Proof This is obvious. □

Proposition 14.2.4 Let $\lambda \in \Lambda_n$, $X \in \check{C}_\lambda$ and $\alpha \in X_1$ be as in Lemma 14.2.2 Then there exists a unique element w in $\bar{N}_\lambda$ which satisfies $A(X,\alpha)$.

Proof The existence of w follows from Lemmas 14.2.2, 14.2.3 and Proposition 13.2.4.

Now we shall show the uniqueness of w. Assume that $w \in \bar{N}_\lambda$ satisfies $A(X,\alpha)$ and has a standard MDC form $(A_r,\ldots,A_1)$ at i for some $i \in \mathbb{Z}$. We may assume $\underline{1 \leqslant j^1_1}(w) \leqslant n$ in this form. Then the number $j^1_1(w)$ is uniquely determined by equation $\overline{j^1_1(w)} = \bar{\alpha}$.

Let $E_t : j^t_{\mu_t}(w),\ldots,j^t_1(w)$ for any t, $1 \leqslant t \leqslant \lambda_1$. Then E_t are all compact increasing sequences of $\mathbb{Z}$ with $\bar{E}_t = \bar{X}_t$. Since $\overline{j^1_1(w)} = \bar{\alpha}$ and $\bar{E}_1 = \bar{X}_1$, the sequence $j^1_r(w),\ldots,j^1_1(w)$ is uniquely determined by $j^1_1(w)$. In particular, the sequence $E'_1 : j^1_{\mu_2}(w),\ldots,j^1_1(w)$ is uniquely determined. It is obvious that E'_1 is also a compact increasing sequence of $\mathbb{Z}$ such that E'_1, E_2 are interlocking. Since $\bar{E}_2 = \bar{X}_2 \subset \underline{n}$ satisfies $|\bar{E}_2| = |E'_1|$ and $\bar{E}_2 \cap \bar{E}'_1 = \emptyset$ in $\underline{n}$, we see by Lemma 14.2.1 that the sequence E_2 is uniquely determined by E'_1. In general, assume that for some k, $1 \leqslant k < \lambda_1$, the sequences E_h, $1 \leqslant h \leqslant k$, are all uniquely determined. We know that $E'_k : j^k_{\mu_{k+1}}(w),\ldots,j^k_1(w)$ and E_{k+1} are two compact increasing sequences of $\mathbb{Z}$ with E'_k, E_{k+1} interlocking. We also know that the sequence E'_k and the set $\bar{E}_{k+1} = \bar{X}_{k+1}$ have been determined. This implies by Lemma 14.2.1 that the sequence E_{k+1} must be uniquely determined. By applying induction on k, we see that for any k, h with $1 < k < \lambda_1$ and $1 < h < \mu_k$, $j_h^k(w)$ is uniquely determined. Then by equation

$$\sum_{\substack{1<k<\lambda_1 \\ 1<h<\mu_k}} j_h^k(w) = in + \sum_{h=1}^{n} h$$

we can get a unique integer solution for i. This means that w is uniquely determined by condition $A(X,\alpha)$. Our proof is complete. □

Now we can prove the main result in this section.

Proposition 14.2.5 For any $X \in \check{\mathcal{C}}_\lambda$, the fibre $T^{-1}(X)$ belongs to some left cell of A_n.

Proof Take any number α in the first column of X. By Proposition 14.2.4, there exists a unique element w in $\bar{N}_\lambda$ which satisfies $A(X,\alpha)$.

Now let y be any element of $T^{-1}(X)$. By the proof of Lemma 14.2.2, we see that there exists $x \in N_\lambda$ satisfying $x \underset{P_L}{\sim} y$ and $A(X,\alpha)$. Then by Lemmas 14.2.3, 13.2.2 and Proposition 13.2.3, we can transform x by a succession of raising operations on the u-th layers of w with $1 < u < \lambda_1$ and get an element of $\bar{N}_\lambda$ which is in the same left cell as x and satisfies $A(X,\alpha)$. Thus by Proposition 14.2.4, this element must be w. Hence we have $y \underset{L}{\sim} w$. Thus any element of $T^{-1}(X)$ lies in the same left cell as w. Our conclusion follows. □

For any $X \in \check{\mathcal{C}}_\lambda$, let $y, y' \in \bar{T}^{-1}(X)$ be such that y (resp. y') has a standard MDC form $(A_r,\ldots,A_1)$ at i (resp. i'). Let X_1 be the set of numbers in the first column of X. Then by Proposition 14.2.4, we see that $\overline{j_1^1(y)} = \overline{j_1^1(y')}$ implies $y = y'$. If $\lambda_1 = \lambda_r$, then by proper choice of standard MDC forms of y, y', we can always make $\overline{j_1^1(y)} = \overline{j_1^1(y')}$. So in that case, the cardinal of $\bar{T}^{-1}(X)$ is 1. But if $\lambda_1 > \lambda_r$, then the residue class $\overline{j_1^1(y)}$ (resp. $\overline{j_1^1(y')}$) is independent of the choice of the standard MDC form of y (resp. y'). Clearly, when $\overline{j_1^1(y)} \neq \overline{j_1^1(y')}$, we have $y \neq y'$. So the cardinal of $\bar{T}^{-1}(X)$ must be r.

Proposition 14.2.6 Assume $\lambda = \{\lambda_1 > \ldots > \lambda_r\} \in \Lambda_n$. Then for any $X \in \check{\mathcal{C}}_\lambda$ $|\bar{T}^{-1}(X)|$ is a constant: it is equal to 1 if $\lambda_1 = \lambda_r$, or equal to r if $\lambda_1 > \lambda_r$. So $|\bar{N}_\lambda|$ is equal to $\dfrac{n!}{\prod_{j=1}^{m} \mu_j!}$ if $\lambda_1 = \lambda_r$ or equal to $\dfrac{r(n!)}{\prod_{j=1}^{m} \mu_j!}$ if $\lambda_1 > \lambda_r$, where $\mu = \{\mu_1 > \ldots > \mu_m\}$ is the dual partition of λ.

Proof By the above statement and Lemmas 14.1.1, 14.1.2. □

§14.3 THE SUBSET X_λ OF N_λ

For any v, t with $0 \leq v < n$ and $1 \leq t \leq \lambda_1$, let $E_{vt} = \{v + \sum_{h=j}^{r} \lambda_h + 1 - t \mid 1 \leq j \leq \mu_t\}$. Let X^v be a tabloid of $\check{C}_\lambda$ with the set of numbers modulo n in its t-th column to be E_{vt}. Let $X_\lambda = \bigcup_{v=0}^{n-1} T^{-1}(X^v)$.

Example Let $w \in A_7$ have the form

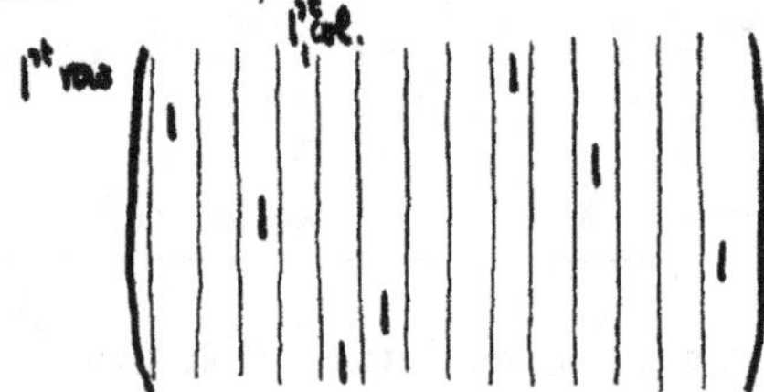

Then $w \in T^{-1}(X^4) \subset X_\lambda$ with $\lambda = \{3 > 2 > 2\}$ and $X^0 =$

	3	4
2	5	6
	7	1

In this section, we shall show $X_\lambda = \{w \in N_\lambda | w^{-1} \in H_\lambda\}$ and then show that for any $x,y \in \sigma^{-1}(\lambda)$ with $x^{-1} \in H_\lambda$, $x \underset{L}{\sim} y$ if and only if $R(x) = R(y)$.

For any v with $0 \leq v < n$, let $J_\lambda(v) = J_{r,v} \cup \ldots \cup J_{1,v}$, where $J_{h,v} = \{s_{\alpha_h+1}, s_{\alpha_h+2}, \ldots, s_{\alpha_h+\lambda_h-1}\}$, $\alpha_h = v + \sum_{\ell=h+1}^{r} \lambda_\ell$. Then by the latter part of Chapter 5, we see that $w_0^{J_\lambda(v)}$ is in $T^{-1}(X^v)$. So by Proposition 14.2.5, it is immediate that

Lemma 14.3.1 Any $w \in T^{-1}(X^v)$ is in the same left cell as $w_0^{J_\lambda(v)}$ for any v, $0 \leq v < n$. □

Let $w \in N_\lambda$ have a standard MDC form $(A_r,\ldots,A_1)$ at $i \in \mathbb{Z}$. Let $\bar{e}((w),E_u)$ be the u-th layer of w for any u, $1 \leq u \leq \lambda_1$.

Lemma 14.3.2 Assume that $e(i_t,j_t)$ are in $\bar{e}((w),E_{u_t})$, $t = 1,2$, with $i_1 < i_2$ and $j_1 > j_2$. Then $u_1 < u_2$.

Proof By Lemma 5.11, we know $u_1 \neq u_2$. Suppose $u_1 > u_2$. We may assume without loss of generality that $j_1 = j_h^{u_1}(w)$ for some h, $1 \leq h \leq \mu_1$. Let $e(i_3,j_3)$ be $e((w),j_h^{u_2}(w))$. Then $i_3 < i_1$ and $j_3 = j_h^{u_2}(w) > j_h^{u_1}(w) = j$. So $e(i_3,j_3)$ is not $e(i_2,j_2)$. Since $\bar{e}((w),E_{u_2})$ is an antichain of w, we have $(i_2-i_3)(j_2-j_3) > 0$. But on the other hand,

$i_2 - i_3 = (i_2-i_1) + (i_1-i_3) > 0$ and $j_2-j_3 = (j_2-j_1) + (j_1-j_3) < 0$. This implies $(i_2-i_3)(j_2-j_3) < 0$, a contradiction. Therefore $u_1 < u_2$. □

If $w \in N_\lambda$ with $w^{-1} \in H_\lambda$, then there exists $i' \in \mathbf{Z}$ such that $\{e(i_t^v(w), i' + \sum_{h=t}^{r} \lambda_h + 1-v) | 1 \leq v \leq \lambda_t\}$ is a descending chain of w for all t, $1 \leq t \leq r$. Assume that the entry $e(i_t^v(w), i' + \sum_{h=t}^{r} \lambda_h + 1-v)$ is in $\tilde{e}((w), E_{\alpha_{tv}})$. Then by Lemma 14.3.2, we have $\alpha_{t1} < \alpha_{t2} \cdots < \alpha_{t\lambda_t}$ for $1 \leq t \leq r$. Since $\overline{\{i' + \sum_{h=t}^{r} \lambda_h + 1-v | 1 \leq t \leq r,\ 1 \leq v \leq \lambda_t\}} = \underline{n}$ by successively considering the sets $\{\alpha_{tv} | 1 \leq v \leq \lambda_t\}$ where t runs from 1 to r, we can see that $\alpha_{tv} = v$ for any t, $1 \leq t \leq \mu_v$. Therefore $E_v = \{i' + \sum_{h=t}^{r} \lambda_h + 1-v | 1 \leq t \leq \mu_v\}$ for all v, $1 \leq v \leq \lambda_1$, and we have $w \in X_\lambda$.

Proposition 14.3.3 $X_\lambda = \{w \in N_\lambda | w^{-1} \in H_\lambda\}$.

Proof We have shown $\{w \in N_\lambda | w^{-1} \in H_\lambda\} \subseteq X_\lambda$. By Theorem 1.5.2(i), we know $R(x) = R(y)$ for any x,y with $x \underset{L}{\sim} y$. So Lemma 14.3.1 tells us that for any $w \in X_\lambda$, $R(w) = J_\lambda(v)$ with some v, $0 \leq v < n$. i.e. $w^{-1} \in H_\lambda$. This implies $X_\lambda \subseteq \{w \in N_\lambda | w^{-1} \in H_\lambda\}$. Our result follows. □

Proposition 14.3.4 Assume $x,y \in \sigma^{-1}(\lambda)$ with $x^{-1} \in H_\lambda$. Then $x \underset{L}{\sim} y$ if and only if $R(x) = R(y)$.

Proof (⇒) By Theorem 1.5.2(i).

(⇐) By Proposition 13.2.5, there exist x', $y' \in N_\lambda$ such that $x' \underset{L}{\sim} x$ and $y' \underset{L}{\sim} y$. The condition $R(x) = R(y)$ implies $R(x') = R(y') = R(x)$ by Theorem 1.5.2(i). Since $x^{-1} \in H_\lambda$, it follows that $x'^{-1}, y'^{-1} \in H_\lambda$. Hence Proposition 14.3.3 together with $R(x') = R(y')$ implies that $T(x') = T(y') = X^v$ for some v, $0 \leq v < n$. So by Proposition 14.2.5, we get $x' \underset{L}{\sim} y'$ and then $x \underset{L}{\sim} y$. □

§14.4 THE NUMBER OF LEFT CELLS IN $\sigma^{-1}(\lambda)$

So far, we have shown that for any $X \in \check{C}_\lambda$ the fibre $T^{-1}(X)$ lies in some left cell by using the properties of a principal normalized element. In any principal normalized element, we see that the distance between any two adjacent layers is very short. Now we shall define a new kind of set which consists of all elements w of N_λ such that the distance between any two adjacent layers of w goes beyond some fixed bound. By using the properties of these elements, we shall find all the left cells of $\sigma^{-1}(\lambda)$.

Assume that $w \in N_\lambda$ has a standard MDC form $(A_r,\dots,A_1)$ at $i \in \mathbf{Z}$ and $E_u = \{\overline{j_h^u(w)} \mid 1 \le h \le \mu_u\}$, $1 \le u \le \lambda_1$. For $\ell > 0$, we say that $x = (L_{E_u})^\ell(w)$ is available from w in N_λ if $(L_{E_u})^j(w) \in N_\lambda$ for any j, $1 \le j \le \ell$. We say that

$x = (L_{E_{u_1}})^{\ell_1}(L_{E_{u_2}})^{\ell_2} \dots (L_{E_{u_t}})^{\ell_t}(w)$ is available from w in N_λ if

$x_h = (L_{E_{u_h}})^{\ell_h}((L_{E_{u_{h+1}}})^{\ell_{h+1}} \dots (L_{E_{u_t}})^{\ell_t}(w))$ is available from

$(L_{E_{u_{h+1}}})^{\ell_{h+1}} \dots (L_{E_{u_t}})^{\ell_t}(w)$ in N_λ for any h, $1 \le h \le t$.

Suppose that $w_1 = (L_{E_v})^{\mu_v}(w)$ is available from w in N_λ, then w_1 has a standard MDC form $(A_r,\dots,A_1)$ at i', $i' = i + \mu_v$, and for any t, $1 \le t \le r$,

$$j_t^u(w_1) = \begin{cases} j_t^u(w) & \text{if } 1 \le u \le \lambda_t,\ u \ne v \\ j_t^v(w)+n & \text{if } u = v. \end{cases}$$

By this formula, it follows that if $w_1 = (L_{E_v})^{m\mu_v}(w)$ is available from w in N_λ, $1 \le v < \lambda_1$, $m > 0$, then $w_2 = (L_{E_{v+1}})^\ell(w_1)$ is available from w_1 in N_λ for any ℓ, $1 \le \ell \le m\mu_{v+1}$. In particular, $(L_{E_1})^m(w)$ is always available from w in N_λ for any $m > 0$. Given any integer $N > 0$, let m_α, $1 \le \alpha \le \lambda_1$, be a set of integers such that $m_1 > \dots > m_{\lambda_1} > 0$ with $m_j - m_{j+1} \ge N + 2$, $1 \le j < \lambda_1$ then

$w' = (L_{E_{\lambda_1}})^{m_{\lambda_1}\mu_{\lambda_1}}(L_{E_{\lambda_1-1}})^{m_{\lambda_1-1}\mu_{\lambda_1-1}} \dots (L_{E_1})^{m_1\mu_1}(w)$ is available from w in N_λ such

that

$$\begin{cases} j_t^u(w') - j_{t'}^{u'}(w') > Nn, \text{ for any } t, t', u, u' \text{ with } u < u', \\ \qquad\qquad 1 \le t \le \mu_u \text{ and } 1 \le t' \le \mu_{u'} \\ j_1^u(w')-n < j_{\mu_u}^u(w') < j_{\mu_u-1}^u(w') < \dots < j_1^u(w'), \text{ for } 1 \le u \le \lambda_1. \end{cases}$$

By Proposition 13.2.3, we see that $w' \in N_\lambda$ satisfies $w' \underset{L}{\sim} w$.

For $a \in \mathbf{Z}$, we define $N_\lambda(a)$ to be the set of all elements w of N_λ such that w has a standard MDC form $(A_r,\dots,A_1)$ at $i \in \mathbf{Z}$ satisfying $j_t^u(w) - j_{t'}^{u'}(w) > a$ for any t,t',u,u' with $u < u'$, $1 \le t \le \mu_u$ and $1 \le t' \le \mu_{u'}$.

For example, suppose $n = 7$, $\lambda = \{3 \ge 3 \ge 1\}$. Then the following element w is in $N_\lambda(1)$ but not in $N_\lambda(2)$.

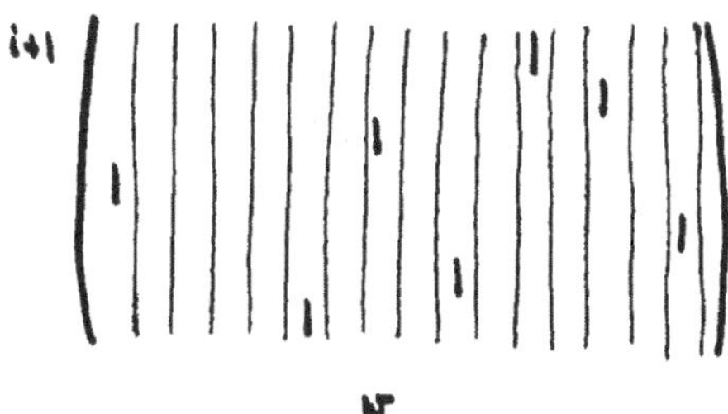

It is clear that $N_\lambda(-n) = N_\lambda$ and $N_\lambda(a) \subseteq N_\lambda(a')$ for any a, $a' \in \mathbf{Z}$ with $a \geq a'$.

By the above discussion, for any $a \in \mathbf{Z}$, $w \in N_\lambda$, there always exists an element w' in $N_\lambda(a)$ such that $w' \underset{L}{\sim} w$.

<u>Lemma 14.4.1</u> <u>For $a \geq n$, let $w \in N_\lambda(a)$ have a standard MDC form $(A_r,\ldots,A_1)$ at $i \in \mathbf{Z}$. Let $e(i_t,t)$ be in the layer $\bar{e}((w),E_{\alpha_t})$ and $e(i_{t+\varepsilon},t+\varepsilon)$ in the layer $\bar{e}((w),E_{\alpha_{t+\varepsilon}})$, $\varepsilon = 1,2$. Then</u>

(i) $\alpha_t = \alpha_{t+\varepsilon} \Leftrightarrow 0 < i_{t+\varepsilon} - i_t < n$

(ii) $\alpha_t < \alpha_{t+\varepsilon} \Leftrightarrow i_{t+\varepsilon} - i_t > n$

(iii) $\alpha_t > \alpha_{t+\varepsilon} \Leftrightarrow i_{t+\varepsilon} - i_t < 0$

<u>Proof</u> Since $\bar{t} \neq \overline{t+\varepsilon}$, we have $\bar{i}_t \neq \overline{i_{t+\varepsilon}}$. So $i_{t+\varepsilon} - i_t \neq 0,n$. We may assume that $e(i_t,t)$ is $e((w),j_h^{\alpha_t}(w))$ for some h, $1 \leq h \leq \mu_{\alpha_t}$. We need only show the implication in the direction "$\Leftarrow$" for all these three cases.

(i) Either $e(i_{t+\varepsilon}, t+\varepsilon)$ or $e(i_{t+\varepsilon}-n, t+\varepsilon-n)$ is $e((w),j_{h'}^{\alpha_{t+\varepsilon}}(w))$ for some h', $1 \leq h' \leq \mu_{\alpha_{t+\varepsilon}}$. If $\alpha_t \neq \alpha_{t+\varepsilon}$, then $|j_h^{\alpha_t}(w) - j_{h'}^{\alpha_{t+\varepsilon}}(w)| > a \geq n$. But in fact, $|j_h^{\alpha_t}(w) - j_{h'}^{\alpha_{t+\varepsilon}}(w)| = |t-(t+\varepsilon)|$ or $|t-(t+\varepsilon-n)|$ which is less than n. This is a contradiction. So $\alpha_t = \alpha_{t+\varepsilon}$.

(ii) Let $i_{t+\varepsilon} - i_t = qn + p$ with $q, p \in \mathbf{Z}$ and $0 < p < n$.

Then $q > 0$. Let $e(i_0,j_0)$ be $e(i_{t+\varepsilon} - qn, t+\varepsilon-qn)$. Then $e(i_0,j_0)$ is in $\bar{e}((w),E_{\alpha_{t+\varepsilon}})$. Since $i_t < i_0$ and $t > j_0$, this implies $\alpha_t < \alpha_{t+\varepsilon}$ by Lemma 14.3.2.

(iii) By Lemma 14.3.2. □

Let w be as in Lemma 14.4.1. Let $e(i_\beta(w),\beta)$ be in $\bar{e}((w), E_{\alpha_\beta(w)})$: for $\beta = t$, $t+1$, $t+2$ with $t \in \mathbf{Z}$. Then $w \in \mathcal{D}_R(s_t)$ if and only if one of the following cases occurs by §4.3.

(i) $i_{t+1}(w) < i_t(w) < i_{t+2}(w)$, (ii) $i_{t+1}(w) < i_{t+2}(w) < i_t(w)$;

(iii) $i_t(w) < i_{t+2}(w) < i_{t+1}(w)$: (iv) $i_{t+2}(w) < i_t(w) < i_{t+1}(w)$.

By Lemma 14.4.1, we have

inequality (i) holds $\Longleftrightarrow \alpha_{t+1}(w) < \alpha_t(w) < \alpha_{t+2}(w)$

inequality (ii) holds $\Longleftrightarrow \alpha_{t+1}(w) < \alpha_{t+2}(w) < \alpha_t(w)$

inequality (iii) holds $\Longleftrightarrow \alpha_t(w) < \alpha_{t+2}(w) < \alpha_{t+1}(w)$

inequality (iv) holds $\Longleftrightarrow \alpha_{t+2}(w) < \alpha_t(w) < \alpha_{t+1}(w)$.

When $w \in \mathcal{D}_R(s_t)$, let $w' = w^*$ in $\mathcal{D}_R(s_t)$. Let $e(i_\beta(w'),\beta)$ be the entry of w', $\beta = t, t+1, t+2$. Then we have

$$(i_{t+1}(w'), i_t(w'), i_{t+2}(w')) = (i_{t+2}(w), i_t(w), i_{t+1}(w)) \text{ in case (i)}$$

$$(i_{t+1}(w'), i_{t+2}(w'), i_t(w')) = (i_t(w), i_{t+2}(w), i_{t+1}(w)) \text{ in case (ii)}$$

$$(i_t(w'), i_{t+2}(w'), i_{t+1}(w')) = (i_{t+1}(w), i_{t+2}(w), i_t(w)) \text{ in case (iii)}$$

$$(i_{t+2}(w'), i_t(w'), i_{t+1}(w')) = (i_{t+1}(w), i_t(w), i_{t+2}(w)) \text{ in case (iv).}$$

Since w' is obtained from w by transposing two consecutive columns of w which contain the entries lying in distinct layers of w, we see that w' lies in $N_\lambda(a-2)$ which also has a standard MDC form $(A_r,\dots,A_1)$ at i.

Assume that $e(i_\beta(w'),\beta)$ is in $\bar{e}((w'), E_{\alpha_\beta(w')})$. Then we have

$$(\alpha_{t+1}(w'),\alpha_t(w'),\alpha_{t+2}(w')) = (\alpha_{t+2}(w),\alpha_t(w),\alpha_{t+1}(w) \text{ in case (i)}$$

$$(\alpha_{t+1}(w'), \alpha_{t+2}(w'),\alpha_t(w')) = (\alpha_t(w),\alpha_{t+2}(w),\alpha_{t+1}(w)) \text{ in case (ii)}$$

$$(\alpha_t(w'),\alpha_{t+2}(w'),\alpha_{t+1}(w')) = (\alpha_{t+1}(w),\alpha_{t+2}(w),\alpha_t(w)) \text{ in case (iii)}$$

$$(\alpha_{t+2}(w'),\alpha_t(w'),\alpha_{t+1}(w')) = (\alpha_{t+1}(w),\alpha_t(w),\alpha_{t+2}(w)) \text{ in case (iv).}$$

So in other words, we can say that $w \in \mathcal{D}_R(s_t)$ if and only if one of the following cases occurs:

(i') $\alpha_t(w)$, $\alpha_{t+1}(w)$ and $\alpha_{t+2}(w)$ are distinct, one of $\alpha_t(w)$ and $\alpha_{t+2}(w)$, say $\alpha(w)$, lies between the others.

(ii') Two of $\alpha_t(w)$, $\alpha_{t+1}(w)$, $\alpha_{t+2}(w)$ are equal, but the remaining one, say $\alpha(w)$, is distinct. When $\alpha(w) \neq \alpha_{t+1}(w)$, then either $\alpha(w) = \alpha_t(w)$ is maximum or $\alpha(w) = \alpha_{t+2}(w)$ is minimum among $\{\alpha_\beta(w) | \beta = t,t+1,t+2\}$.

In case (i'), let $\alpha'(x) = \{\alpha_t(x), \alpha_{t+2}(x)\} - \{\alpha(x)\}$ for $x = w, w'$. Then $\{\alpha_{t+1}(w'), \alpha(w'), \alpha'(w')) = (\alpha'(w), \alpha(w), \alpha_{t+1}(w))$. In case (ii'), when $\alpha(w) \neq \alpha_{t+1}(w)$, let $\alpha'(x) = \{\alpha_t(x), \alpha_{t+2}(x)\} - \{\alpha(x)\}$. Then $(\alpha_{t+1}(w'), \alpha'(w'), \alpha(w')) = (\alpha(w), \alpha'(w), \alpha_{t+1}(w))$.

When $\alpha(w) = \alpha_{t+1}(w)$, then

$$(\alpha_{t+1}(w'),\alpha_t(w'),\alpha_{t+2}(w')) = (\alpha_t(w),\alpha_{t+1}(w),\alpha_{t+2}(w)) \text{ if } \alpha_{t+1}(w) \text{ is maximum.}$$

$$(\alpha_{t+1}(w'),\alpha_t(w'),\alpha_{t+2}(w')) = (\alpha_{t+2}(w), \alpha_t(w), \alpha_{t+1}(w)) \text{ if } \alpha_{t+1}(w) \text{ is minimum.}$$

Therefore we have

Lemma 14.4.2 For $w \in N_\lambda(a)$, $a > n$ and $t \in \mathbb{Z}$, whether w is in $\mathcal{D}_R(s_t)$ or not is entirely determined by $T(w)$. That is, for any $X \in \mathcal{C}_\lambda$ and $w, y \in N_\lambda(a) \cap T^{-1}(X)$, we have that w is in $\mathcal{D}_R(s_t)$ if and only if y is in $\mathcal{D}_R(s_t)$. When they are in $\mathcal{D}_R(s,t)$, let $w' = w^*$, $y' = y^*$ in $\mathcal{D}_R(s_t)$. Then $w', y' \in N_\lambda(a-2)$ and $T(w') = T(y')$. □

Now we can prove the following result.

Proposition 14.4.3 For $x,y \in N_\lambda$, we have $x \underset{L}{\sim} y \Leftrightarrow T(x) = T(y)$

Proof ($\Leftarrow$) By Proposition 14.2.5

($\Rightarrow$) Otherwise, for any $a,b \in \mathbb{Z}$ with $a > b > 0$, there exists x,y with $x \underset{L}{\sim} y$ such that for some $X, Y \in \mathcal{C}_\lambda$ with $X \neq Y$, we have $x \in N_\lambda(a) \cap T^{-1}(X)$ and $y \in N_\lambda(a) \cap T^{-1}(Y)$. By Proposition 9.3.7, there exists a sequence $x_0 = x, x_1,\dots,x_\ell$ such that for every j, $1 \leq j \leq \ell$, $x_j^{-1} = {}^*(x_{j-1}^{-1})$ in $\mathcal{D}_L(s_{t_j})$ with some $t_j \in \mathbb{Z}$, i.e. $x_j = x_{j-1}^*$ in $\mathcal{D}_R(s_{t_j})$, and $x_\ell^{-1} \in H_\lambda$. Clearly, the number ℓ is independent of the choice of a. Now we take $a,b \in \mathbb{Z}$ with $b > n$ and $a > b + 2\ell$. Then we have $x_j \in N_\lambda(b)$ for any j, $0 \leq j \leq \ell$. Since $x \underset{L}{\sim} y$, we see by Theorems 1.5.2(i) and 1.6.2(ii) that there also exists a sequence $y_0 = y, y_1,\dots,y_\ell$ such that for every j, $1 \leq j \leq \ell$, $y_j = y_{j-1}^*$ in $\mathcal{D}_R(s_{t_j}) \cap N_\lambda(b)$, $y_j \underset{L}{\sim} x_j$ and then $R(y_j) = R(x_j)$. In particular, $y_\ell \underset{L}{\sim} x_\ell$ and $R(y_\ell) = R(x_\ell)$. So by Proposition 14.3.3, we have $x_\ell, y_\ell \in T^{-1}(X^v)$ for some v, $0 \leq v < n$. Now we start with the pair $\{x_\ell, y_\ell\}$ and consider the sequence of pairs $\{x_\ell,y_\ell\}, \{x_{\ell-1},y_{\ell-1}\} \dots, \{x_0,y_0\} = \{x,y\}$. Then by repeatedly applying Lemma 14.4.2, we have $T(y_j) = T(x_j)$ for all j, $0 \leq j \leq \ell$. In particular, $T(y) = T(x)$. This gives a contradiction. So our conclusion is shown. □

Proposition 14.4.4 Any two elements x,y in the same left cell of A_n can be transformed from one to another by a succession of raising operations in N_λ and left star operations, where $\lambda = \sigma(x)$.

Proof By Proposition 9.4.1, we have $y \in \sigma^{-1}(\lambda)$ by $x \underset{L}{\sim} y$ and $x \in \sigma^{-1}(\lambda)$. It follows from Propositions 9.3.7 and 10.4 that there exists $x', y' \in N_\lambda$ such that $x' \underset{P_L}{\sim} x$ and $y' \underset{P_L}{\sim} y$. Proposition 14.4.3 tells us that $T(x') = T(y')$ since $x' \underset{L}{\sim} y'$. Let α be any number in the first column of $T(x')$. Then we see in the proof of Proposition 14.2.5 that both x' and y' can be transformed to the unique element of $\bar{N}_\lambda$ which satisfies $A(T(x'),\alpha)$ by a succession of raising operations in N_λ and left star operations. This implies that x and y can be transformed from one to another by a succession of raising operations in N_λ and left star operations. □

Theorem 14.4.5 For any $\lambda \in \Lambda_n$, $\sigma^{-1}(\lambda)$ is a disjoint union of $\dfrac{n!}{\prod_{j=1}^{m} \mu_j!}$ left (resp. right) cells of A_n, where $\mu = \{\mu_1 \geq \ldots \geq \mu_m\}$ is the dual partition of λ.

Proof By Propositions 13.2.5, 14.4.3 and Lemma 14.1.1, we see that $\sigma^{-1}(\lambda)$ is a disjoint union of $\dfrac{n!}{\prod_{j=1}^{m} \mu_j!}$ left cells of A_n. Since the map $w \to w^{-1}$ induces a bijection from the set of left cells of A_n to the set of right cells of A_n and since $\sigma^{-1}(\lambda)$ is invariant under this map by Lemma 5.4 the remaining result of this theorem follows. □

CHAPTER 15 : $\sigma^{-1}(\lambda)$ IS AN RL-EQUIVALENCE CLASS OF A_n

Fix $\lambda = \{\lambda_1 \geq \ldots \geq \lambda_r\} \in \Lambda_n$. We have shown in §9.4 that $\sigma^{-1}(\lambda)$ is a union of some RL-equivalence classes of A_n. Now we shall prove that $\sigma^{-1}(\lambda)$ is an RL-equivalence class of A_n. We may assume $\lambda \neq \{1 \geq \ldots \geq 1\}$ and $n \geq 3$ by §1.7(v) and Chapter 9. We know from Proposition 9.3.7 that any element of $\sigma^{-1}(\lambda)$ is in the same left cell as some element of H_λ. But by Lemma 5.4 this is equivalent to saying that any element of $\sigma^{-1}(\lambda)$ is in the same right cell as some element of H_λ^{-1}, where $H_\lambda^{-1} = \{x \in \sigma^{-1}(\lambda) \mid x^{-1} \in H_\lambda\}$. By Proposition 14.3.4, we see that for any $x,y \in H_\lambda^{-1}$, if $R(x) = R(y)$ then $x \underset{L}{\sim} y$. So to reach our goal, it is sufficient to show that for any $x,y \in H_\lambda^{-1}$, there exists an element z of H_λ^{-1} such that $z \underset{R}{\sim} x$ and $R(z) = R(y)$. But this is equivalent to show that for any $x,y \in H_\lambda$, there exists an element z of H_λ such that $z \underset{L}{\sim} x$ and $\mathcal{L}(z) = \mathcal{L}(y)$. By Proposition 10.4 and the fact that $N_\lambda \subseteq H_\lambda$, we need only show that for any $x \in N_\lambda$, $y \in H_\lambda$, there exists an element z of N_λ such that $z \underset{L}{\sim} x$ and $\mathcal{L}(z) = \mathcal{L}(y)$. Note that for any $u,v \in H_\lambda$, $\mathcal{L}(u) = \mathcal{L}(v)$ if and only if there exists an integer i such that both u and v have a standard MDC form at i. Now assume that $x \in N_\lambda$ has a standard MDC form at h and $y \in H_\lambda$ has a standard MDC form at k. We may assume $h \leq k < h + n$. Let $E = \{j_t(x) \mid 1 \leq t \leq r\} \subseteq \underline{n}$. Then by Proposition 13.2.3, we see that the element $z = (L_E)^k(x)$ has all required properties. Therefore, we get

<u>Theorem 15.1</u> For any $n \geq 2$ and $\lambda \in \Lambda_n$, the set $\sigma^{-1}(\lambda)$ is an RL-equivalence class of A_n. □

CHAPTER 16 : LEFT CELLS ARE CHARACTERIZED BY THE GENERALIZED RIGHT τ-INVARIANT

In Chapter 15, we have characterized any left cell of A_n by a tabloid of size n by means of the map T which maps all the normalized elements of this left cell to the corresponding tabloid. In the present chapter, we shall give another characterization of any left cell of A_n, i.e. the generalized right τ-invariant which has been defined on an element of any Coxeter group in §1.6.

Let P_n be the standard parabolic subgroup of A_n generated by $\{s_1,\ldots,s_{n-1}\}$ which is isomorphic to the symmetric group S_n. In §16.2, we shall apply the above result to discuss the relations between the RL-equivalence classes (resp. left, right cells) in A_n and in P_n. Our main results in §16.2 are as follows: The intersection of any RL-equivalence class of A_n with P_n is non-empty and is just an RL-equivalence class of P_n. The intersection of any left (resp. right) cell of A_n with P_n is either empty or a left (resp. right) cell of P_n.

In the proof of these results, we actually give a proof for determination of all left, right cells and all RL-equivalence classes of the symmetric group S_n in our own way which differs from Kazhdan, Lusztig [KL1] and Vogan [Vo].

§16.1 LEFT CELLS ARE CHARACTERIZED BY THE GENERALIZED RIGHT τ-INVARIANT

Lemma 16.1.1 If $w,y \in A_n$ have the same generalized right τ-invariant, then $w \underset{RL}{\sim} y$.

Proof By Theorem 15.1, it is sufficient to show that $\sigma(w) = \sigma(y)$. Assume $\lambda = \sigma(w)$ and $\lambda' = \sigma(y)$. Then $w^{-1} \in \sigma^{-1}(\lambda)$ and $y^{-1} \in \sigma^{-1}(\lambda')$ by Lemma 5.4. Hence by Proposition 9.3.7, there exists a sequence $w_0 = w^{-1}, w_1,\ldots,w_\ell$ such that for every j, $1 < j < \ell$, $w_j = {}^*w_{j-1}$ in $\mathcal{D}_L(s_{i_j})$ for some $s_{i_j} \in \Delta$ and $w_\ell \in H_\lambda$. Since y has the same generalized right τ-invariant as w, this implies that there also exists a sequence $y_0 = y^{-1}, y_1,\ldots,y_\ell$ such that for every j, t with $1 < j < \ell$ and $0 < t < \ell$, $y_j = {}^*y_{j-1}$ in $\mathcal{D}_L(s_{i_j})$ and $\mathcal{L}(y_t) = \mathcal{L}(w_t)$. In particular, $\mathcal{L}(y_\ell) = \mathcal{L}(w_\ell)$. So

$$\lambda' = \sigma(y) = \sigma(y^{-1}) = \sigma(y_\ell) \geq \pi(\mathcal{L}(y_\ell)) = \pi(\mathcal{L}(w_\ell)) = \sigma(w_\ell) = \sigma(w^{-1}) = \sigma(w) = \lambda.$$

i.e. $\lambda' \geq \lambda$. By symmetry, we also have $\lambda \geq \lambda'$. So $\lambda = \lambda'$. □

Theorem 16.1.2 For $w,y \in A_n$, $w \underset{L}{\sim} y \Leftrightarrow w,y$ have the same generalized right τ-invariant.

Proof ($\Rightarrow$) By Theorems 1.5.2 and 1.6.2.

($\Leftarrow$) By Lemma 16.1.1, we have $w,y \in \sigma^{-1}(\lambda)$ for some $\lambda \in \Lambda_n$. Then $w^{-1}, y^{-1} \in \sigma^{-1}(\lambda)$. Since w^{-1}, y^{-1} have the same generalized left τ-invariant, there exist two sequences

$w_0 = w^{-1}, w_1,\dots,w_\ell$ and $y_0 = y^{-1}, y_1,\dots,y_\ell$ such that for every j, $1 \le j \le \ell$, $w_j = {}^*w_{j-1}$ and $y_j = {}^*y_{j-1}$ in $\mathcal{D}_L(s_{i_j})$ with some $s_{i_j} \in \Delta$, and $\mathcal{L}(w_h) = \mathcal{L}(y_h)$, $0 \le h \le \ell$, $w_\ell, y_\ell \in H_\lambda$. By Proposition 14.3.4, we have $w_\ell^{-1} \underset{L}{\sim} y_\ell^{-1}$, i.e. $w_\ell \underset{R}{\sim} y_\ell$. Owing to Theorem 1.6.2, we get $w_j \underset{R}{\sim} y_j$ for all j. In particular, $w^{-1} \underset{R}{\sim} y^{-1}$. So $w \underset{L}{\sim} y$. □

§16.2 THE STANDARD PARABOLIC SUBGROUP P_n

For any $w \in A_n$, let $\{e_w(u,j_u) | u \in \mathbf{Z}\}$ be the entry set of w. Then we see that $w \in P_n$ if and only if j_u satisfies $1 \le j_u \le n$ for any u, $1 \le u \le n$.

Now assume that w is in P_n. By the property that $j_{u+n} = j_u+n$ for any $u \in \mathbf{Z}$, we see that if $u,v \in \mathbf{Z}$ satisfy $u-v \ge n$ then $j_u > j_v$. So for any descending chain $\{e_w(u_i,j_{u_i}) | 1 \le i \le t,\ u_1 < u_2 < \dots < u_t\}$ of entries of w, there must exist some $q \in \mathbf{Z}$ such that $qn + 1 \le u_1 < u_2 < \dots < u_t \le (q+1)n$.

Bearing these facts in mind, we shall first show that any element x of P_n can be transformed to some element of $P_n \cap H_\lambda$ by a succession of left star operations in P_n, where $\lambda = \sigma(x)$.

Lemma 16.2.1 For $w \in A_n$, let A(w) be the block of w consisting of rows from the (i+1)-th to the (i+j)-th with $1 \le j \le n$.

Let $S = \{e_w(u_\alpha, j_{u_\alpha}) | {}^{1 \le \alpha \le t}_{u_1 < u_2 < \dots < u_t}\}$ be the longest descending chain of w in A(w). Then we can transform w to w' by a succession of left star operations on the block A(w) such that w' has a descending chain $\{e_{w'}(u,j'_u) | i+j+1-t \le u \le i+j\}$.

Remark: We say that a left star operation on w acts on the block A(w) it it is defined in $\mathcal{D}_L(s_u)$ with $i+1 \le u \le i+j-2$, where j are defined as above.

Proof Clearly, t satisfies $1 \le t \le j$. Fix $t \ge 1$, let us apply induction on $j-t \ge 0$. It is trivial for the case $j-t = 0$, since in that case, we just take $w' = w$. Now assume $j-t > 0$.

(i) If $u_1 \ne i+1$, let A'(w) be the block of w consisting of rows from the (i+2)-th to the (i+j)-th. Then S is a longest descending chain of w in A'(w) and so our result follows from the inductive hypothesis.

(ii) If $u_1 = i+1$ and $j_{i+1} < j_{i+2}$, where $e_w(i+2,j_{i+2})$ is an entry of w, then $e_w(i+2,j_{i+2})$ is not in S. Let $S' = S \cup \{e_w(i+2,j_{i+2})\} - \{e_w(i+1,j_{i+1})\}$. Then S' is also a longest descending chain of w in A(w). This reduces to case (i).

(iii) Let $u_1 = i+1$ and $j_{i+1} > j_{i+2} > \dots > j_{i+h}$ and $j_{i+h} < j_{i+h+1}$ with $h \ge 2$, where $e_w(i+k,j_{i+k})$ are the entries of w for all k, $1 \le k \le h+1$. By our assumption, we see that h is less than j. Then there exists $x \in A_n$ with $w \xrightarrow{*(i+1,h,1)} x$. Clearly, x is obtained from w by a succession of left star operations in the block A(w) and the maximal length of the descending chains of x in A(x) is also equal to t by Lemma 5.8 and its proof, where A(x) is the block of x consisting of rows from the (i+1)-th to the (i+j)-th. Let $e_x(i+1,j'_{i+1})$ and $e_x(i+2, j'_{i+2})$ be two entries of x. Then we have $j'_{i+1} < j'_{i+2}$. This reduces to case (ii).

Therefore our conclusion follows by induction. □

The following is an example for Lemma 16.2.1, where $w,x, w' \in A_n$ with $n \geq 6$, $j = 6$, $t = 4$, $w \xrightarrow{*(i+1,2,1)} x$ and $x \xrightarrow{*(i+2,3,1)} w'$

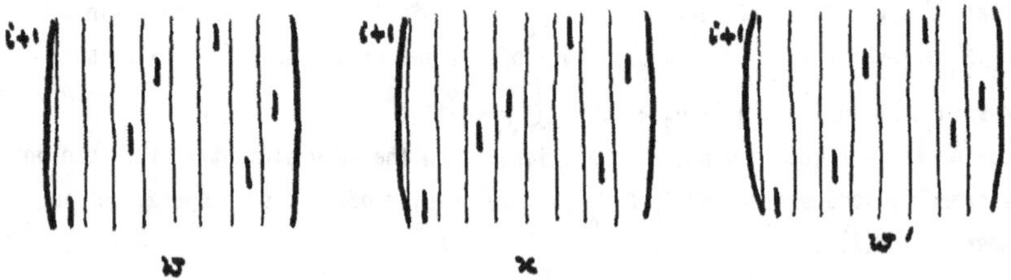

<u>Lemma 16.2.2</u> <u>For $w \in A_n$, let A be the block of w consisting of rows from the (i+1)-th to the (i+j)-th with $1 \leq j \leq n$. Assume that $D = \{e_w(u,j_u) \mid i+j+1-t \leq u \leq i+j\}$ is a longest descending chain of w in A. Assume that $S = S_1 \cup \ldots \cup S_h$ is any disjoint union of h descending chains of w in A. Then there exists a disjoint union $S' = S'_1 \cup \ldots \cup S'_h$ of h descending chains of w in A such that $|S'| \geq |S|$ and $S'_1 = D$.</u>

<u>Proof</u> We may assume $S \cap D \neq \emptyset$, as otherwise, $S' = D \cup S_2 \cup \ldots \cup S_h$ would be as required.

Let k be the smallest integer with $i+j+1-t \leq k \leq i+j$ such that $e_w(k,j_k) \in S$. We may assume $e_w(k,j_k) \in S_1$. Let $S' = S'_1 \cup \ldots \cup S'_h$ be as follows:

$$S'_1 = S_1 \cup \{e_w(u,j_u) \mid k < u \leq i+j\}$$

$$S'_\ell = S_\ell - \{e_w(u,j_u) \mid k < u \leq i+j\}, \qquad 2 \leq \ell \leq h.$$

Then $S' = S'_1 \cup \ldots \cup S'_h$ is a disjoint union of h descending chains of w in A satisfying $|S'| \geq |S|$. If $k = i+j+1-t$, then S' is as required. If not, then $S'' = D \cup S'_2 \cup \ldots \cup S'_h$ is as required since $|D| \geq |S'_1|$. □

<u>Lemma 16.2.3</u> <u>For $\lambda = \{\lambda_1 \geq \ldots \geq \lambda_r\} \in \Lambda_n$, assume that $w \in P_n \cap \sigma^{-1}(\lambda)$ has an MDC form $(A_k, A_{k-1}, \ldots, A_1)$ at $\sum_{h=k+1}^{r} \lambda_h$ with $0 \leq k < r$ and $|A_j(w)| = \lambda_j$ for every j,</u>

$1 \le j \le k$. Then the maximal length ℓ of the descending chains of w in B is λ_{k+1}, where B is the block of w consisting of rows from the 1-st to the $(\sum_{h=k+1}^{r} \lambda_h)$-th.

Proof By our assumption, it is easily seen that $\ell \le \lambda_{k+1}$. Let B_j be the block of w consisting of rows from the 1-st to the $(\sum_{h=j}^{r} \lambda_h)$-th for any j, $1 \le j \le k+1$. Then $B_{k+1} = B$. By the remark at the beginning of this section, we see that for any h with $1 \le h \le r$, there exists a disjoint union $S = S_1 \cup \ldots \cup S_h$ of h descending chains of w in B_1 satisfying $C_n(w,h)$ and $|S| = \sum_{j=1}^{h} \lambda_j$. Now we take $h = k+1$. Since A_1 is a longest descending chain of w in B_1, it follows from Lemma 16.2.2 that there exists a disjoint union $S^{(1)} = S_1^{(1)} \cup S_2^{(1)} \cup \ldots \cup S_{k+1}^{(1)}$ of $k+1$ descending chains of w in B_1 satisfying $C_n(w,k+1)$, $|S^{(1)}| \ge |S|$ and $S_1^{(1)} = A_1$. Since $w \in \sigma^{-1}(\lambda)$, we must have $|S^{(1)}| = |S|$. Now $\tilde{S}^{(1)} = S_2^{(1)} \cup \ldots \cup S_{k+1}^{(1)}$ is a disjoint union of k descending chains of w in B_2 satisfying $C_n(w,k)$ and $|\tilde{S}^{(1)}| = \sum_{h=2}^{k+1} \lambda_h$. Since A_2 is a longest descending chain of w in B_2, it follows from Lemma 16.2.2 that there exists a disjoint union $S^{(2)} = S_2^{(2)} \cup \ldots \cup S_{k+1}^{(2)}$ of k descending chains of w in B_2 satisfying $C_n(w,k)$, $|S^{(2)}| \ge |\tilde{S}^{(1)}|$ and $S_2^{(2)} = A_2$. Again by the fact $w \in \sigma^{-1}(\lambda)$, we see $|S^{(2)}| = |\tilde{S}^{(1)}| = \sum_{h=2}^{k+1} \lambda_h$. In such a way, for any j with $1 \le j \le k$, we see that there exists a disjoint union $S^{(j)} = S_j^{(j)} \cup \ldots \cup S_{k+1}^{(j)}$ of $k+2-j$ descending chains of w in B_j satisfying $C_n(w,k+2-j)$, $|S^{(j)}| = \sum_{h=j}^{k+1} \lambda_h$ and $S_j^{(j)} = A_j$. In particular, when $j = k$, we have a disjoint union $S^{(k)} = S_k^{(k)} \cup S_{k+1}^{(k)}$ of 2 descending chains of w in B_k satisfying $C_n(w,2)$, $S_k^{(k)} = A_k$ and $|S^{(k)}| = \sum_{h=k}^{k+1} \lambda_h$. So $S_{k+1}^{(k)}$ is a descending chain of w in B with $|S_{k+1}^{(k)}| = \lambda_{k+1}$. This implies $\ell = \lambda_{k+1}$. □

Proposition 16.2.4 For any $\lambda = \{\lambda_1 \ge \ldots \ge \lambda_r\} \in \Lambda_n$, if $w \in \sigma^{-1}(\lambda) \cap P_n$, then there exists a sequence of elements $w_0 = w, w_1, \ldots, w_t$ in P_n such that for every j, $1 \le j \le t$, $w_j = {}^*w_{j-1}$ in $\mathcal{D}_L(s_{i_j})$ with some i_j, $1 \le i_j \le n-2$ and $w_t \in H_\lambda \cap P_n$.

Proof Let B_j, $1 \le j \le r$, be defined as in the proof of Lemma 16.2.3. By Lemma 16.2.3, if w has an MDC form $(A_k, \ldots, A_1)$ at $\sum_{h=k+1}^{r} \lambda_h$ with $0 \le k < r$ and $|A_j(w)| = \lambda_j$ for every j, $1 \le j \le k$, then the maximal length of the descending chains of w in B_{k+1} is λ_{k+1}. So by Lemma 16.2.1, we can transform w to x by a succession of left star operations in the block B_{k+1} such that x has an MDC form $(A_{k+1}, \ldots, A_1)$ at $\sum_{h=k+2}^{r} \lambda_h$ with $|A_j(x)| = \lambda_j$ for every j, $1 \le j \le k+1$. Now we go through the same process by taking $k = 0,1,\dots,r-1$ in turn. Finally we get an element w' of P_n which has an MDC form $(A_r,\dots,A_1)$ at 0 with $|A_j(w')| = \lambda_j$ for every j, $1 \leq j \leq r$, i.e. $w' \in H_\lambda$. Since the whole process from w to w' is a succession of left star operations in P_n, this implies our result. □

Corollary 16.2.5 For any $\lambda \in \Lambda_n$, if $w \in \sigma^{-1}(\lambda) \cap P_n$, then there exists a sequence of elements $w_0 = w, w_1,\dots,w_t$ in P_n such that for every j, $1 \leq j \leq t$, either $w_j = {}^*w_{j-1}$ in $\mathcal{D}_L(s_{i_j})$ or $w_j = w_{j-1}^*$ in $\mathcal{D}_R(s_{i_j})$ with some i_j, $1 \leq i_j \leq n-2$, and w_t, w_t^{-1} are both in $P_n \cap H_\lambda$.

Proof By Proposition 16.2.4, we may assume $w \in P_n \cap H_\lambda$. Clearly w^{-1} is also in the set $\sigma^{-1}(\lambda) \cap P_n$. Also by Proposition 16.2.4, there exists a sequence of elements $w_0 = w^{-1}, w_1,\dots,w_t$ in P_n such that for every j, $1 \leq j \leq t$, $w_j = {}^*w_{j-1}$ in $\mathcal{D}_L(s_{i_j})$ with some i_j, $1 \leq i_j \leq n-2$, and $w_t \in H_\lambda$. i.e. there exists a sequence of elements $w_0^{-1} = w, w_1^{-1},\dots,w_t^{-1}$ in P_n such that $w_j^{-1} = (w_{j-1}^{-1})^*$ in $\mathcal{D}_R(s_{i_j})$ for every j, $1 \leq j \leq t$ and $w_t \in H_\lambda$. Since this sequence is in $\sigma^{-1}(\lambda)$ and also in some left cell of P_n, we have $\mathcal{L}(w_t^{-1}) = \mathcal{L}(w_{t-1}^{-1}) = \dots = \mathcal{L}(w)$ by Theorem 1.5.2 and hence $w_t^{-1} \in H_\lambda$. Our assertion follows. □

Now we can consider the relations between the RL-equivalence classes (resp. left, right cells) in A_n and in P_n.

For any $\lambda = \{\lambda_1 \geq \dots \geq \lambda_r\} \in \Lambda_n$, let $J_\lambda(0) = J_r \cup \dots \cup J_1 \subset \Delta$ with $\{s_{\alpha_h+1}, s_{\alpha_h+2},\dots,s_{\alpha_h+\lambda_h-1}\}$, $\alpha_h = \sum_{\ell=h+1}^{r} \lambda_\ell$. Then it is clear that $w_0^{J_\lambda(0)} \in P_n \cap \sigma^{-1}(\lambda)$. So we get the following result.

Proposition 16.2.6 For any $\lambda \in \Lambda_n$, the intersection of any RL-equivalence class of A_n with P_n is non-empty.

Proof By Theorem 15.1, any RL-equivalence class of A_n has the form $\sigma^{-1}(\lambda)$ for some $\lambda \in \Lambda_n$. Thus our result follows immediately. □

By Proposition 16.2.6, we see that the intersection of any RL-equivalence class of A_n with P_n is non-empty and hence it is a union of some RL-equivalence classes of P_n. Now we shall show that such an intersection is actually an RL-equivalence class of P_n.

Lemma 16.2.7 For $\lambda = \{\lambda_1 \geq \dots \geq \lambda_r\} \in \Lambda_n$, assume that $w, w^{-1} \in H_\lambda \cap P_n$. Then $w = w_0^{J_\lambda(0)}$.

Proof By the assumption that $w \in H_\lambda \cap P_n$, w has a standard MDC form $_r(A_r,\ldots,A_1)$ at 0. Let $e(i_t^u(w), j_t^u(w))$ be the u-th entry of $A_t(w)$. Then $i_t^u(w) = \sum_{h=t+1}^{r} \lambda_h + u$. We claim $j_1^1(w) = n$. For suppose that $j_1^1(w) < n$. Then there exists an entry $e_w(i,n)$ of w with $1 \leq i \leq n$ and $i \neq i_1^1(w)$. We have $i \not> i_1^1(w)$ by the fact that $A_1(w)$ is a DC block. So we must have $i < i_1^1(w)$. But then $S = \{e_w(i,n)\} \cup A_1$ is a descending chain of w with $|S| = \lambda_1+1$. This contradicts $w \in \sigma^{-1}(\lambda)$. So w has the entry $e_w(n+1-\lambda_1,n)$. Also by the assumption that $w^{-1} \in H_\lambda \cap P_n$ and the fact that the matrix w^{-1} is the transpose of w, we see that $e_{w^{-1}}(n+1-\lambda_1,n)$ is an entry of w^{-1} and so $e_w(n,n+1-\lambda_1)$ is an entry of w, i.e. $j_1^{\lambda_1}(w) = n+1-\lambda_1$. Now $n = j_1^1(w) > j_1^2(w) > \ldots > j_1^{\lambda_1}(w) = n+1-\lambda_1$. So we get $j_1^h(w) = n+1-h$ for any h, $1 \leq h \leq \lambda_1$.

Now suppose that we have shown $j_t^u(w) = \sum_{h=t}^{r} \lambda_h + 1 - u$ for some k, $1 \leq k < r$ and all t, u with $1 \leq t \leq k$ and $1 \leq u \leq \lambda_t$. Then we have $1 \leq j_{h'}^{u'}(w) \leq \sum_{h=k+1}^{r} \lambda_h$ for any h', u' with $k < h' \leq r$ and $1 \leq u' \leq \lambda_{h'}$. Note that the maximal length of the descending chains of w in its submatrix D is λ_{k+1}, where D is the $(\sum_{h=k+1}^{r} \lambda_h) \times (\sum_{h=k+1}^{r} \lambda_h)$ matrix lying between the 1st row (resp. column) and the $(\sum_{h=k+1}^{r} \lambda_h)$-th row (resp. column) of w. By the same argument as above with the number $\sum_{h=k+1}^{r} \lambda_h$ in place of n, we can show that $j_{k+1}^u(w) = \sum_{h=k+1}^{r} \lambda_h + 1 - u$ for any u, $1 \leq u \leq \lambda_{k+1}$. So by applying induction on $k \geq 1$, finally we see that $j_t^u(w) = \sum_{h=t}^{r} \lambda_h + 1-u$ for any t, u with $1 \leq t \leq r$ and $1 \leq u \leq \lambda_t$. This implies $w = w^{J_\lambda(0)}$. □

The following result gives a one-to-one correspondence between the set of RL-equivalence classes of A_n and those of P_n.

Theorem 16.2.8 For any $\lambda \in \Lambda_n$, the set $P_n \cap \sigma^{-1}(\lambda)$ is an RL-equivalence class of P_n. Conversely, any RL-equivalence class of P_n has such a form.

Proof By Theorem 15.1 and Proposition 16.2.6, it is sufficient to show that any two elements of $P_n \cap \sigma^{-1}(\lambda)$ are in the same RL-equivalence class of P_n. But by Corollary 16.2.5 and Lemma 16.2.7, they are all P-equivalent to $w_0^{J_\lambda(0)}$. So our result follows by Theorem 1.6.3(iii). □

Now we consider the relations between the left (resp. right) cells in A_n and in P_n.

In general, the intersection of any left (resp. right) cell of A_n with P_n is not necessarily non-empty. But what will happen when such an intersection is non-empty? Clearly we can at least say that it is a union of some left (resp. right) cells of P_n. The following result gives us a more satisfactory answer.

Theorem 16.2.9 The intersection of any left (resp. right) cell of A_n with P_n is either empty or a left (resp. right) cell of P_n. Conversely, any left (resp. right) cell of P_n can be expressed as an intersection of some left (resp. right) cell of A_n with P_n.

Proof Let $x,y \in P_n$. It is sufficient to show that if x and y are in the same left cell of A_n then they are in the same left cell of P_n.

Now assume that x and y are in the same left cell of A_n. By Proposition 9.4, we have $\sigma(x) = \sigma(y) = \lambda$ for some $\lambda \in \Lambda_n$. By Proposition 16.2.4, we may assume $x,y \in H_\lambda \cap P_n$. Lemma 5.4 and Theorem 16.1.2 tell us that x^{-1}, y^{-1} are in $\sigma^{-1}(\lambda)$ and, they have the same generalized left τ-invariant in A_n and hence in P_n. So by Proposition 16.2.4, there exist two sequences of elements $x_0 = x^{-1}, x_1,\ldots,x_t$ and $y_0 = y^{-1}, y_1,\ldots,y_t$ in P_n such that for every j, k with $1 \le j \le t$ and $0 \le k \le t$, $x_j = {}^*x_{j-1}$ and $y_j = {}^*y_{j-1}$ in $\mathcal{D}_L(s_{i_j})$ with some i_j, $1 \le i_j \le n-2$, $\mathcal{L}(x_k) = \mathcal{L}(y_k)$ and $x_t, y_t \in H_\lambda \cap P_n$. Since $x_0 = x^{-1}, x_1,\ldots,x_t$ are all in the same left cell of P_n, we have $\mathcal{R}(x_t) = \mathcal{R}(x^{-1}) = \mathcal{L}(x)$ and thus $x_t^{-1} \in H_\lambda \cap P_n$. By Lemma 16.2.7, we must have $x_t = w_0^{J_\lambda(0)}$. Similarly we can show $y_t = w_0^{J_\lambda(0)}$ and so $x_t = y_t$. Thus $x_j = y_j$ for all j. In particular, $x^{-1} = y^{-1}$, i.e. $x = y$. So x and y are in the same left cell of P_n. Our proof is complete. □

We know that any standard parabolic subgroup of A_n which is isomorphic to the symmetric group S_n has the form $P_t = \langle s_i | 1 \le i \le n, i \ne t\rangle$ for some t, $1 \le t \le n$. Clearly, the automorphism Φ^t of A_n fixes all $\sigma^{-1}(\lambda)$, $\lambda \in \Lambda_n$ and induces an isomorphism from P_n to P_t which preserves the left, right and two-sided cells as well as RL-equivalence classes for them. So, if we replace P_n by P, where $P \in \{P_t | 1 \le t \le n\}$, then all results in this section still hold.

We have shown that for any $\lambda \in \Lambda_n$, $\sigma^{-1}(\lambda)$ is a RL-equivalence class of A_n which consists of $\frac{n!}{\prod_{j=1}^{m} \mu_j!}$ left (resp. right) cells of A_n, where $\mu = \{\mu_1 \geq \dots \geq \mu_m\}$ is the dual partition of λ. From this, we know that the total number of left (resp. right) cells of A_n is

$$\sum_{\{\lambda_1 \geq \dots \geq \lambda_r\} \in \Lambda_n} \frac{n!}{\prod_{j=1}^{r} \lambda_j!} .$$

Since any two-sided cell of A_n is a union of some RL-equivalence classes of A_n, this implies that the number of two-sided cells of A_n is finite and is less than or equal to the number of partitions of n. So far, all the results here have been obtained by our own elementary methods and do not rely on the knowledge of the intersection cohomology theory. But now we shall use a recent result of Lusztig, i.e. Theorem 1.5.3 [L12] in Chapter 1 to prove that any RL-equivalence class of A_n is actually a two-sided cell of A_n. Then we can determine the exact number of two-sided cells of A_n which is equal to the number of partitions of n.

This result of Lusztig comes from the deep theory of intersection cohomology.

We understand that Lusztig [L13] showed the main result of this chapter in the similar way as we do. But we do this independently.

To show our result, we shall first give some lemmas. The notation $y \underset{\Gamma}{\sim} w$ means that y, w are in the same two-sided cell of A_n.

Lemma 17.1 If $y, w \in A_n$ with $y \underset{\Gamma}{\sim} w$, then

(i) $y \underset{L}{\leq} w$ implies $R(y) = R(w)$.

(ii) $y \underset{R}{\leq} w$ implies $\mathcal{L}(y) = \mathcal{L}(w)$.

Proof It is enough to show (i). Since $y \underset{L}{\leq} w$, there exists a sequence $y_0 = y$, $y_1, \dots, y_t = w$ such that either $y_{i-1} \prec y_i$ or $y_i \prec y_{i-1}$, and $\mathcal{L}(y_{i-1}) \not\subset \mathcal{L}(y_i)$ for every i, $1 \leq i \leq t$. Clearly, $y = y_0 \underset{L}{\leq} y_1 \underset{L}{\leq} \dots \underset{L}{\leq} y_t = w$. Since $y \underset{\Gamma}{\sim} w$, this implies that all y_i, $0 \leq i \leq t$, are in the same two-sided cell of A_n. By Theorem 1.5.2(i), we have

$$R(y) = R(y_0) \supseteq R(y_1) \supseteq \dots \supseteq R(y_t) = R(w).$$

In particular, $R(y) \supseteq R(w)$. If $R(y) \supsetneq R(w)$, then there exists at least one j, $1 < j < r$, such that $R(y_{j-1}) \supsetneq R(y_j)$, i.e. $R(y_{j-1}) \not\subset R(y_j)$. So by Theorem 1.5.3, y_{j-1} and y_j are not in the same two-sided cell. This gives a contradiction. Hence we must have $R(y) = R(w)$. □

Lemma 17.2 Let $y,w \in A_n$ with $y \underset{\Gamma}{\sim} w$.

(i) Let y, w be two elements in $\mathcal{D}_L(s_t)$. If $y \underset{R}{<} w$, then $*y \underset{R}{<} *w$ and $*y \underset{\Gamma}{\sim} *w$.

(ii) Let y,w be two elements in $\mathcal{D}_R(s_t)$. If $y \underset{L}{<} w$, then $y^* \underset{L}{<} w^*$ and $y^* \underset{\Gamma}{\sim} w^*$.

Proof We first note that, if $x \in \mathcal{D}_L(s_u)$, then $*x \underset{L}{\sim} x$, hence, by Theorem 1.5.2(i), we have $R(*x) = R(x)$. Now let y, w be two elements in $\mathcal{D}_L(s_t)$ such that $y \underset{R}{\leq} w$ and $y \underset{\Gamma}{\sim} w$. Then there exists a sequence $y_0 = y, y_1, \ldots, y_r = w$ such that either $y_{i-1} \prec y_i$ or $y_{i-1} \succ y_i$, and $R(y_{i-1}) \not\subset R(y_i)$ for every i, $1 < i < r$. Clearly,

$$y = y_0 \underset{R}{<} y_1 \underset{R}{<} \cdots \underset{R}{<} y_r = w.$$

Since $y \underset{\Gamma}{\sim} w$, this implies that all y_i, $0 < i < r$, are in the same two-sided cell. So by Lemma 17.1(ii), we have

$$\mathcal{L}(y) = \mathcal{L}(y_0) = \mathcal{L}(y_1) = \cdots = \mathcal{L}(y_r) = \mathcal{L}(w).$$

Hence, $*y_i$, $0 < i < r$, are well defined. Theorem 1.6.1 shows that either $*y_{i-1} \prec *y_i$ or $*y_i \prec *y_{i-1}$ for every i, $1 < i < r$. By the remark at the beginning of the proof, we have $R(y_i) = R(*y_i)$ for all i. It follows that $R(*y_{i-1}) \not\subset R(*y_i)$ for i, $1 < i < r$ and hence $*y = *y_0 \underset{R}{<} *y_1 \underset{R}{<} \cdots \underset{R}{<} *y_r = *w$. In particular, $*y \underset{R}{<} *w$. On the other hand, we have $*y \underset{L}{\sim} y \underset{\Gamma}{\sim} w \underset{L}{\sim} *w$ and hence $*y \underset{\Gamma}{\sim} *w$. (i) is proved. The proof of (ii) is entirely similar. □

Lemma 17.3 Let y, $w \in A_n$ with $y \underset{\Gamma}{\sim} w$. Then

(i) $y \underset{L}{<} w$ implies $y \underset{L}{\sim} w$.

(ii) $y \underset{R}{<} w$ implies $y \underset{R}{\sim} w$.

Proof It is enough to show (i). Assume $y \in \sigma^{-1}(\lambda)$ and $w \in \sigma^{-1}(\lambda')$ for some λ, $\lambda' \in \Lambda_n$. Then $y^{-1} \in \sigma^{-1}(\lambda)$ by Lemma 5.4. So by Proposition 9.3.7, there exists a sequence $y_0 = y^{-1}, y_1, \ldots, y_r$ in $\sigma^{-1}(\lambda)$ such that for every j, $1 < j < r$, $y_j = *y_{j-1}$ in $\mathcal{D}_L(s_{\alpha_j})$ with some $s_{\alpha_j} \in \Delta$, and $y_r \in H_\lambda$. Now we define a sequence $w_0, w_1, \ldots, w_r$

as follows: Let $w_0 = w^{-1}$. Then by $y \underset{\Gamma}{\sim} w$ and $y \underset{L}{\leq} w$, we have $y_0 \underset{\Gamma}{\sim} w_0$ and $y_0 \underset{R}{\leq} w_0$. Thus by Lemma 17.1(ii), we have $\mathcal{L}(y_0) = \mathcal{L}(w_0)$ and then $w_0 \in \mathcal{D}_L(s_{\bar{\alpha}_1})$. Let $w_1 = {}^*w_0$ in $\mathcal{D}_L(s_{\alpha_1})$. By Lemma 17.2(i), we see that $y_1 \underset{R}{\leq} w_1$ and $y_1 \underset{\Gamma}{\sim} w_1$. Again by Lemma 17.1(ii), we have $\mathcal{L}(y_1) = \mathcal{L}(w_1)$ and then $w_1 \in \mathcal{D}_L(s_{\alpha_2})$. Let $w_2 = {}^*w_1$ in $\mathcal{D}_L(s_{\alpha_2})$. In such a way, we get a sequence $w_0 = w^{-1}, w_1, \ldots, w_r$ such that $w_j = {}^*w_{j-1}$ in $\mathcal{D}_L(s_{\alpha_j})$, $y_h \underset{R}{\leq} w_h$, $y_h \underset{\Gamma}{\sim} w_h$ and $\mathcal{L}(y_h) = \mathcal{L}(w_h)$ for every j, h with $1 \leq j \leq r$ and $0 \leq h \leq r$. In particular, $\mathcal{L}(w_r) = \mathcal{L}(y_r)$. So

$$\lambda' = \sigma(w) = \sigma(w^{-1}) = \sigma(w_r) \geq \pi(\mathcal{L}(w_r)) = \pi(\mathcal{L}(y_r)) = \sigma(y_r) = \lambda.$$

On the other hand, by Lemma 5.4, $w^{-1} \in \sigma^{-1}(\lambda')$. So by Proposition 9.3.7, there exists a sequence $x_0 = w^{-1}, x_1, \ldots, x_t$ in $\sigma^{-1}(\lambda')$ such that for every j, $1 \leq j \leq t$, $x_j = {}^*x_{j-1}$ in $\mathcal{D}_L(s_{\beta_j})$ with some $s_{\beta_j} \in \Delta$ and $x_t \in H_{\lambda'}$. Then we can define a sequence $z_0 = y^{-1}, z_1, \ldots, z_t$ from the sequence $x_0, x_1, \ldots, x_t$ in the similar way as $w_0, w_1, \ldots, w_r$ from $y_0, y_1, \ldots, y_r$ such that $z_j = {}^*z_{j-1}$ in $\mathcal{D}_L(s_{\beta_j})$, $z_h \underset{R}{\leq} x_h$, $z_h \underset{\Gamma}{\sim} x_h$ and $\mathcal{L}(z_h) = \mathcal{L}(x_h)$ for every j, h with $1 \leq j \leq t$ and $0 \leq h \leq t$. In particular, $\mathcal{L}(z_t) = \mathcal{L}(x_t)$. So

$$\lambda = \sigma(y) = \sigma(y^{-1}) = \sigma(z_t) \geq \pi(\mathcal{L}(z_t)) = \pi(\mathcal{L}(x_t)) = \sigma(x_t) = \lambda'.$$

This implies $\lambda' = \lambda$ and then $y_r, w_r \in H_\lambda$ with $\mathcal{L}(y_r) = \mathcal{L}(w_r)$. So $(y_r^{-1})^{-1}, (w_r^{-1})^{-1} \in H_\lambda$ with $R(y_r^{-1}) = R(w_r^{-1})$. By Proposition 14.3.4, we have $y_r^{-1} \underset{L}{\sim} w_r^{-1}$. Therefore, by repeatedly applying Theorem 1.6.2(ii) to the sequence $y_0^{-1} = y, y_1^{-1}, \ldots, y_r^{-1}$ and $w_0^{-1} = w, w_1^{-1}, \ldots, w_r^{-1}$, we get $y_j^{-1} \underset{L}{\sim} w_j^{-1}$ for all j, $0 \leq j \leq r$. In particular, $y \underset{L}{\sim} w$. □

We now can determine all the two-sided cells of A_n.

<u>Theorem 17.4</u> <u>The set $\sigma^{-1}(\lambda)$ is a two-sided cell of A_n for any $\lambda \in \Lambda_n$. Conversely, any two-sided cell of A_n has such a form.</u>

<u>Proof</u> We have already shown that $\sigma^{-1}(\lambda)$ is in some two-sided cell of A_n. So it is sufficient to show that for any y, $w \in A_n$ with $y \underset{\Gamma}{\sim} w$, we have $\sigma(y) = \sigma(w)$. Now assume $y \underset{\Gamma}{\sim} w$. Then there exists a sequence $y_0 = y, y_1, \ldots, y_r = w$ in some two-sided cell of A_n such that for every j, $1 \leq j \leq r$, with $y_{j-1} \underset{L}{\leq} y_j$ or $y_{j-1} \underset{R}{\leq} y_j$. So by Lemma 17.3, we have either $y_{j-1} \underset{L}{\sim} y_j$ or $y_{j-1} \underset{R}{\sim} y_j$ and then $\sigma(y_{j-1}) = \sigma(y_j)$ for all j, $1 \leq j \leq r$. This implies $\sigma(y) = \sigma(w)$. □

Corollary 17.5 *The number of two-sided cells of A_n is equal to the number of partitions of n.*

Proof This follows immediately from Theorem 17.4. □

Note that if $x,y \in P$ (for P, see the end of Chapter 16) are in the same two-sided cell of P then they are automatically in the same two-sided cell of A_n. So by Theorems 16.2.8 and 17.4, we can describe all the two-sided cells of P as well.

Theorem 17.6 *For any $\lambda \in \Lambda_n$, the set $P \cap \sigma^{-1}(\lambda)$ is a two-sided cell of P. Any two-sided cell of P has such a form.* □

CHAPTER 18 : SOME PROPERTIES OF CELLS AND OTHER EQUIVALENCE CLASSES OF A_n

In this chapter, we shall discuss some properties of the equivalence classes of A_n. Three such properties are particularly interesting: the first one is the commutativity between a left star operation and a right star operation: the second one is the connectedness of these equivalence classes, and the last one is for the intersection of any left cell with any right cell in the same two-sided cell.

§18.1 THE COMMUTATIVITY BETWEEN A LEFT STAR OPERATION AND A RIGHT STAR OPERATION

Given two integers, t,u, we say $w \in A_n$ has a special (t,u) form if w has one of the following forms.

(i) $\begin{array}{c} \\ t \\ \\ \\ \end{array}\begin{array}{c} u+qn \\ \left(\begin{array}{ccc} & 1 & \\ 1 & & \\ & & 1 \end{array}\right) \end{array}$

(ii) $\begin{array}{c} \\ t \\ \\ \\ \end{array}\begin{array}{c} u+qn \\ \left(\begin{array}{ccc} & 1 & \\ & & 1 \\ 1 & & \end{array}\right) \end{array}$

(iii) $\begin{array}{c} \\ t \\ \\ \\ \end{array}\begin{array}{c} u+qn \\ \left(\begin{array}{ccc} & & 1 \\ 1 & & \\ & 1 & \end{array}\right) \end{array}$

(iv) $\begin{array}{c} \\ t \\ \\ \\ \end{array}\begin{array}{c} u+qn \\ \left(\begin{array}{ccc} 1 & & \\ & & 1 \\ & 1 & \end{array}\right) \end{array}$

for some $q \in \mathbf{Z}$.

Lemma 18.1.1 Assume $w, w' \in A_n$, $s_t \in \Delta$.

(i) If $w,w' \in \mathcal{D}_R(s_t)$ with $w' = w^*$ in $\mathcal{D}_R(s_t)$, then $\mathcal{L}(w') = \mathcal{L}(w)$.

(ii) If $w,w' \in \mathcal{D}_L(s_t)$ with $w' = {}^*w$ in $\mathcal{D}_L(s_t)$, then $\mathcal{R}(w') = \mathcal{R}(w)$.

Proof (i) $w' = w^*$ in $\mathcal{D}_R(s_t)$ implies $w' \underset{R}{\sim} w$.

By Theorem 1.5.2, we have $\mathcal{L}(w') = \mathcal{L}(w)$.

(ii) Similarly. □

Assume $w \in A_n$, s_t, $s_u \in \Delta$ such that $w \in \mathcal{D}_L(s_t) \cap \mathcal{D}_R(s_u)$. Let $x = w^*$ in $\mathcal{D}_R(s_u)$ and let $y = {}^*w$ in $\mathcal{D}_L(s_t)$. Then by Lemma 18.1.1, we have $x,y \in \mathcal{D}_L(s_t) \cap \mathcal{D}_R(s_u)$. Let $z = {}^*x$ in $\mathcal{D}_L(s_t)$.

<u>Lemma 18.1.2</u> <u>Let $w, x, y, z \in A_n$ be as above. Then $yw^{-1} \neq zx^{-1}$ if and only if w has a special (t,u) form.</u>

<u>Proof</u> ($\Leftarrow$) Suppose w has a special (t,u) form. We can see that if w has special (t,u) form (i), then $x = w^* = ws_{u+1}$ in $\mathcal{D}_R(s_u)$ has special (t,u) form (iii). So $y = {}^*w = s_{t+1}w$, $z = {}^*x = s_t w$ in $\mathcal{D}_L(s_t)$ have special (t,u) forms (ii), (iv), respectively. Thus $yw^{-1} = s_{t+1} \neq s_t = zx^{-1}$. That is, $yw^{-1} \neq zx^{-1}$. If w has special (t,u) form (ii), (iii) or (iv), then the similar argument can be used.

($\Rightarrow$) Now suppose $yw^{-1} \neq zx^{-1}$. Since $yw^{-1}, zx^{-1} \in \{s_t, s_{t+1}\}$, by symmetry, we may assume $yw^{-1} = s_t$ and $zx^{-1} = s_{t+1}$. That is, $y = s_t w$ and $z = s_{t+1}x$. Since $y = {}^*w$ and $z = {}^*x$ in $\mathcal{D}_L(s_t)$, this implies that w has one of the following forms:

(i') $\begin{array}{c} \\ t \\ \\ \\ \end{array}\begin{array}{ccc} j_1 & j_2 & j_3 \\ \end{array}$
$$t\begin{pmatrix} 1 & & \\ & & 1 \\ & 1 & \end{pmatrix}$$

(ii') columns $j_1 \; j_2 \; j_3$
$$t\begin{pmatrix} & & 1 \\ 1 & & \\ & 1 & \end{pmatrix}$$

and x has one of the following forms:

(i")
$$t\begin{pmatrix} & 1 & \\ 1 & & \\ & & 1 \end{pmatrix}$$

(ii")
$$t\begin{pmatrix} & 1 & \\ & & 1 \\ 1 & & \end{pmatrix}$$

But $x = w^*$ in $\mathcal{D}_R(s_u)$. So x is obtained from w by transposing two adjacent columns which contain their entries lying in the block $A(w)$, where $A(w)$ consists of three rows of w from the t-th to the $(t+2)$-th. In case (i'), these two adjacent columns of w must be the j_1-th and the j_2-th columns and then by the fact that $x = {}^*w$ in $\mathcal{D}_R(s_u)$, w must have special (t,u) form (iv). In case (ii'), these two adjacent columns must be the j_2-th and the j_3-th columns and then by the fact that $x = w^*$ in $\mathcal{D}_R(s_u)$, w must have special (t,u) form (iii). Our proof is complete. □

<u>Proposition 18.1.3</u> <u>Assume $w \in \mathcal{D}_L(s_t) \cap \mathcal{D}_R(s_u)$ for some $s_t, s_u \in \Delta$. Let $x = w^*$ in $\mathcal{D}_R(s_u)$, $x' = {}^*x$ in $\mathcal{D}_L(s_t)$, $y = {}^*w$ in $\mathcal{D}_L(s_t)$ and $y' = y^*$ in $\mathcal{D}_R(s_u)$. Then $x' = y'$.</u>

<u>Proof</u> We see in the proof of Lemma 18.1.2 that if any element $z \in \mathcal{D}_L(s_t) \cap \mathcal{D}_R(s_u)$ has a special (t,u) form, then z^* in $\mathcal{D}_R(s_u)$ and ${}^*z$ in $\mathcal{D}_L(s_t)$ also have special (t,u)

forms.

First assume that w has special (t,u) form (i), then $x = w^* = ws_{u+1}$ in $\mathcal{D}_R(s_u)$ has special (t,u) form (iii) and so $x' = {}^*x = s_tx = s_tws_{u+1}$ in $\mathcal{D}_L(s_t)$ has special (t,u) form (iv). On the other hand, $y = {}^*w = s_{t+1}w$ in $\mathcal{D}_L(s_t)$ has special (t,u) form (ii) and so $y' = y^* = ys_u = s_{t+1}ws_u$ in $\mathcal{D}_R(s_u)$ has special (t,u) form (iv). Since the process of passing from w to x' or from w to y' only involves the transposes among the rows from the t-th to the (t+2)-th and among the columns from the u-th to the (u+2)-th, we have $x' = y'$. Similarly for the cases when w has special (t,u) form (ii), (iii) or (iv).

Next assume that w has no special (t,u) form. By Lemma 18.1.2, we have $yw^{-1} = x'x^{-1}$ (1). Also since w^{-1} has no special (u,t) form, we see by Lemma 18.1.2 that $x^{-1}(w^{-1})^{-1} = y'^{-1}(y^{-1})^{-1}$. (2) From (1) and (2), we get $y' = yw^{-1}x = x'$.

So in any case, we always have $x' = y'$. Our result is proved. □

Proposition 18.1.3 tells us that left star operations commute with right star operations on any element of A_n. This implies the following result.

Corollary 18.1.4 Assume that y, w are in A_n with $y \underset{P_L}{\sim} w$. Then $R(w) = R(y)$. If y, w are also in $\mathcal{D}_R(s_t)$ for some $s_t \in \Delta$, let $y' = y^*$, $w' = w^*$ in $\mathcal{D}_R(s_t)$. Then $y' \underset{P_L}{\sim} w'$.

Proof By Theorems 1.6.3 and 1.5.2, we see that $y \underset{P_L}{\sim} w$ implies $y \underset{L}{\sim} w$ and then $R(w) = R(y)$. For the remainder of this corollary, we may assume that y, w are in $\mathcal{D}_L(s_u)$ with $y = {}^*w$ in $\mathcal{D}_L(s_u)$ for some $s_u \in \Delta$. By Proposition 18.1.3, we see that $y' = ({}^*w)^* = {}^*(w^*) = {}^*w'$ in $\mathcal{D}_R(s_t) \cap \mathcal{D}_L(s_u)$. So $y' \underset{P_L}{\sim} w'$. □

§18.2 CONNECTEDNESS OF CELLS AND OTHER EQUIVALENCE CLASSES OF A_n

Recall that in §7.2 we defined a connected (resp. left connected; right connected) set of A_n. We proved there that any $X \in S$, when regarded as a set of elements of A_n, is left connected. Now we shall discuss the connectedness of cells of A_n. This is one of the important properties concerning the structure of cells of A_n.

Our main result in this section is as follows.

Theorem 18.2.1

(i) Any left cell of A_n is left connected and is also a maximal left connected component in the two-sided cell containing it.

(ii) Any right cell of A_n is right connected and is also a maximal right connected component in the two-sided cell containing it.

(iii) Any two-sided cell of A_n is connected.

Recall that in §7.2 we defined a left (resp. two-sided) cell on S. We also defined a map $\hat{\sigma}: S \to \Lambda_n$ which coincides the map $\sigma: A_n \to \Lambda_n$ in the sense that the diagram

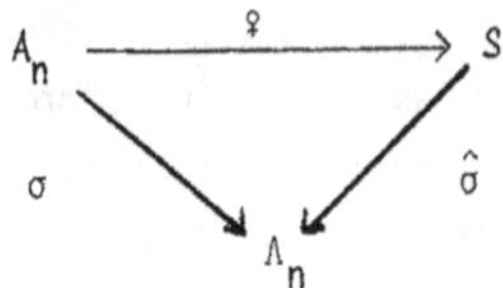

is commutative.

Proposition 18.2.2 Suppose that Theorem 18.2.1 is true. Then the map $\varsigma: A_n \to S$ induces a bijection between the set of two-sided (resp. left) cells of A_n and that of S (for the definition of these cells of S, see 7.2).

Proof The result for the two-sided cells follows from the above diagram and Theorem 17.4. Recall that any left cell L of S is a maximal connected component of $\hat{\sigma}^{-1}(\hat{\sigma}(L))$ in S (see Lemma 7.2.3). So by Theorem 18.2.1, the result for the left cells also follows. □

To show Theorem 18.2.1, we need some lemmas.

Lemma 18.2.3

(i) Every P_L-equivalence class of A_n is left connected.

(ii) Every P_R-equivalence class of A_n is right connected.

(iii) Every P-equivalence class of A_n is connected.

Proof The P_L-equivalence relation is generated by $w \underset{P_L}{\sim} {}^*w$ in $\mathcal{D}_L(s_u)$ with $s_u \in \Delta$. If $w \in \mathcal{D}_L(s_u)$, then ${}^*w$ in $\mathcal{D}_L(s_u)$ is either $s_u w$ or $s_{u+1}w$. This implies that every P_L-equivalence class is left connected. The remaining results can be proved similarly. □

Lemma 18.2.4 Assume $\lambda = \{\lambda_1 > \ldots > \lambda_r\} \in \Lambda_n$. Let $w, w' \in N_\lambda$ be such that $w' = L_{E_u}(w)$ with $1 < u < \lambda_1$. Then there exists a sequence $x_0 = w, x_1, \ldots, x_t = w'$ in $\sigma^{-1}(\lambda)$ such that for every j, $1 < j < t$, $x_j = s_{i_j} x_{j-1}$ with some $s_{i_j} \in \Delta$.

Proof Assume that $\mu = \{\mu_1 > \ldots > \mu_m\}$ is the dual partition of λ. Assume $\lambda_{k+1} < u < \lambda_k$ for some k, $1 < k < r$, with the convention that $\lambda_{r+1} = 0$. Assume that w has a standard

MDC form $(A_r,\dots,A_1)$ at $i \in \mathbf{Z}$. Then w' has a standard MDC form $(A_r,\dots,A_1)$ at i+1 and the u-th layer $\bar{e}((w),E_u)$ of w is movable. So the following properties are obvious:

(i) $j_t^u(w) < j_{t+1}^{u-1}(w)$ for any t, $1 \le t < k$.

(ii) $j_k^u(w) < j_1^{u-1}(w)-n$ and so $j_k^u(w) < j_h^v(w)$ for all h, v with $k < h \le r$ and $1 \le v \le \lambda_h$.

(iii) $j_t^v(w') = j_t^v(w)$ for any t, v with $1 \le t \le r$, $1 \le v \le \lambda_t$ and $v \neq u$.

(iv) $j_t^u(w') = j_{t-1}^u(w)$ for any t, $1 < t \le k$, and $j_1^u(w') = j_k^u(w) + n$.

For any h, $1 \le h \le k$, let $f_{hu}(w)$ be the row of w containing the u-th entry of $A_h(w)$ and let $A_h^0(w)$ (resp. $A_h^1(w)$) be the block consisting of the first u-1 rows (resp. the last λ_h-u rows) of $A_h(w)$.

Now we define an element y of A_n which is obtained from w by permuting the row $f_{hu}(w)$ from the bottom to the top in the block $[A_h^0(w), f_{hu}(w)]$ for all h, $1 \le h < k$ and also permuting the row $f_{ku}(w)$ from the bottom to the top in the block $[A_r,\dots,A_{k+1},A_k^0,f_{ku}(w)]$. Then by properties (i), (ii), we see that there exists a sequence $w_0 = w, w_1,\dots,w_\alpha = y$ such that for every j, $1 \le j \le \alpha$, $w_j = s_{i_j} w_{j-1}$ with $s_{i_j} \in \Delta$ and $\ell(w_j) = \ell(w_{j-1})-1$. By Lemma 5.5 this implies

$\lambda = \sigma(w) = \sigma(w_0) \geq \sigma(w_1) \geq \dots \geq \sigma(w_\ell) = \sigma(y)$.

y has the DC form

$$(f_{ku},A_r,\dots,A_{k+1},A_k^0,A_k^1,f_{k-1,u},A_{k-1}^0,A_k^1,\dots,f_{1u},A_1^0,A_1^1) \text{ at } i,$$

where each block of y occurring in this form comes from the corresponding block of w under the above transformation from w to y.

It is clear that for every DC block A in this form of y, we have $j_A^h(y) = j_A^h(w)$, $1 \le h \le |A|$.

By definition of the raising operation from w to w', we see that w' can be obtained from y by permuting the row $f_{h,u}(y)$ from the bottom to the top in the block $[A_{h+1}^1(y), f_{hu}(y)]$ for any h, $1 \le h < k$ and permuting the row $f_{ku}(y)$ from the bottom to the top in the block $[\tilde{A}_1^1(y), f_{ku}(y)]$, where $\tilde{A}_1^1(y)$ is the congruent block of $A_1^1(y)$ between the (i+1-n)-th row and the i-th row of y.

Since $j_t^u(w) > j_{t+1}^{u+1}(w)$ for any t with $1 \le t < k$ and $1 \le t+1 \le \mu_{u+1}$, and since $j_k^u(w) > j_1^{u+1}(w)-n$ when $u < \lambda_1$ we see by properties (iii), (iv) that there exists a sequence $y_0 = y, y_1,\dots,y_\beta = w'$ such that for every j, $1 \le j \le \beta$, $y_j = s_{h_j} y_{j-1}$ with $s_{h_j} \in \Delta$ and $\ell(y_j) = \ell(y_{j-1}) + 1$. By Lemma 5.5, this implies

$\sigma(y) = \sigma(y_0) \le \sigma(y_1) \le \dots \le \sigma(y_\beta) = \sigma(w') = \lambda$.

Now we define a sequence $\xi: x_0 = w, x_1, \ldots, x_{\alpha+\beta} = w'$ such that for any j, $0 \leq j \leq \alpha$, we have $x_j = w_j$ and for any h, $\alpha \leq h \leq \alpha + \beta$, we have $x_h = y_{h-\alpha}$. Then it is clear that for every j, $1 \leq j \leq \alpha + \beta$, we have $x_j = s_{\ell_j} x_{j-1}$ with $s_{\ell_j} \in \Delta$. To show the elements in the sequence ξ are in $\sigma^{-1}(\lambda)$, it is enough to show that y is in $\sigma^{-1}(\lambda)$.

Assume $y \in \sigma^{-1}(\lambda')$ for some $\lambda' \in \Lambda_n$. We already know $\lambda' \leq \lambda$. For the DC form $(f_{ku}, A_r, \ldots, A_{k+1}, A_k^0, A_k^1, f_{k-1,u}, A_{k-1}^0, A_{k-1}^1, \ldots, f_{1u}, A_1^0, A_1^1)$ at i of y, let $S = S_1 \cup \ldots \cup S_r$, where S_h = {all the entries of $A_h(y)$} for $k+1 \leq h \leq r$.

S_ℓ = {all the entries of $A_{\ell+1}^0(y)$} ∪ {the entry of $f_{\ell u}(y)$} ∪ {all the entries of $A^1(y)$} for $1 \leq \ell < k$.

S_k = {all the entries of $\tilde{A}_1^0(y)$} ∪ {the entry of $f_{ku}(y)$} ∪ {all the entries of $A_k^1(y)$}

where $\tilde{A}_1^0(y)$ is the congruent block of $A_1^0(y)$ between the (i+1-n)-th row and the i-th row of y. Then we see that $S = S_1 \cup \ldots \cup S_r$ satisfies $C_n(y,r)$ with $|S_h| = \lambda_h$ for any h, $1 \leq h \leq r$. This implies $\lambda' \geq \lambda$. So $\lambda' = \lambda$ and ξ is the required sequence. □

An example for the elements w, w', y in Lemma 18.2.4 is as follows.

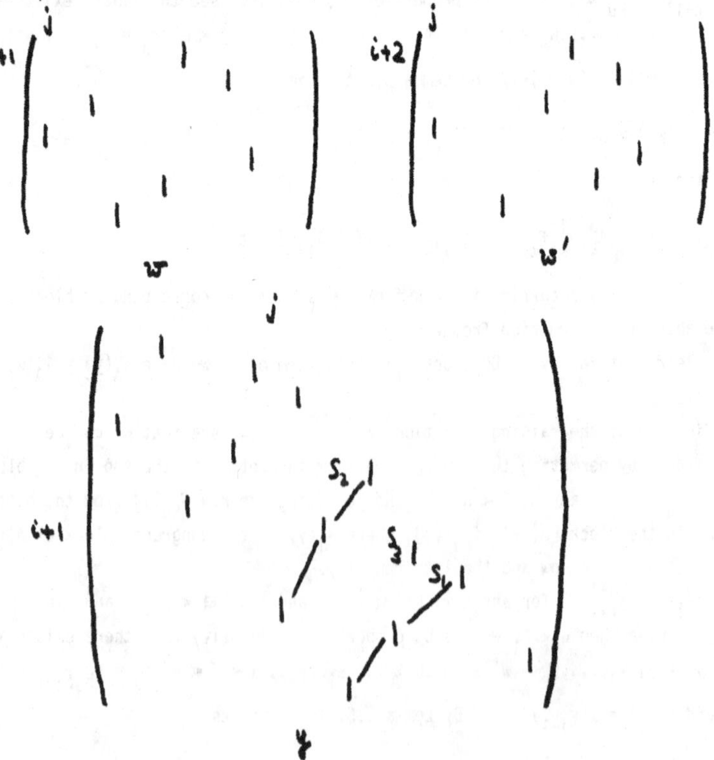

where $n = 7$, $\lambda = \{3 > 3 > 1\}$, $k = 2$ and $u = 2$. $S = S_1 \cup S_2 \cup S_3$ is as defined in the proof of Lemma 18.2.4, where for any h, $1 \le h \le 3$, S_h consists of all entries of y occurring as the vertices of the corresponding broken line.

Lemma 18.2.5 Let K be any two-sided cell of A_n.

(i) Any left connected set of K is in some left cell of A_n.

(ii) Any right connected set of K is in some right cell of A_n.

Proof Assume that $M \subseteq K$ is a left connected set. Let $x, y \in M$. Then there exists a sequence $x_0 = x, x_1, \ldots, x_t = y$ in M such that for every j, $1 \le j \le t$, $x_j = s_{i_j} x_{j-1}$ with some $s_{i_j} \in \Delta$ and so either $x_j \underset{L}{\le} x_{j-1}$ or $x_{j-1} \underset{L}{\le} x_j$. By Lemma 17.3, we must have $x_{j-1} \underset{L}{\sim} x_j$ for all j, $1 \le j \le t$. This implies $x \underset{L}{\sim} y$ and hence M is in some left cell of A_n. (i) follows.

The proof of (ii) is entirely similar. □

Lemma 18.2.6 Let $\lambda \in \Lambda_n$, $w, w' \in N_\lambda$ and the sequence $\xi : x_0 = w, x_1, \ldots, x_t = w'$ be defined as in Lemma 18.2.5. Then the elements of ξ lie in some left cell of A_n.

Proof We see that ξ is in some left connected set of some two-sided cell of A_n. So our result follows from Lemma 18.2.5. □

Now we can prove our main result.

Proof of Theorem 18.2.1

By Proposition 14.4.4, we see that any two elements x, y in the same left cell can be transformed from one to another by a succession of raising operations in N_λ and left star operations, where $\lambda = \sigma(x)$. Thus it follows from Lemmas 18.2.3 and 18.2.6 that any left cell of A_n is left connected. Then the latter part of (i) follows by Lemma 18.2.5.

The proof of (ii) is entirely similar.

We know that any two-sided cell of A_n is a RL-equivalence class. For any x, y with $x \underset{\Gamma}{\sim} y$, there exists a sequence $x_0 = x, x_1, \ldots, x_t = y$ such that for every j, $1 \le j \le t$, either $x_j \underset{L}{\sim} x_{j-1}$ or $x_j \underset{R}{\sim} x_{j-1}$. So by (i), (ii), we get (iii) easily. □

§18.3 THE INTERSECTION OF A LEFT CELL WITH A RIGHT CELL IN A_n

Let L (resp. R) be a left (resp. right) cell of A_n. Then $L \cap R \neq \emptyset$ only if L and R are in the same two-sided cell of A_n.

We know from Theorems 17.4 and 15.1 that any two-sided cell of A_n has the form

$\sigma^{-1}(\lambda)$ for some $\lambda \in \Lambda_n$, and that this is also an RL-equivalence class of A_n.

We shall show the following result.

Theorem 18.3.1 Let L (resp. R) be a left (resp. right) cell of A_n. Then $L \cap R \neq \emptyset$ if and only if L and R are in the same two-sided cell of A_n. When $L \cap R \neq \emptyset$, the cardinality of $L \cap R$ is infinite provided that $L \neq \{1\}$ (or $R \neq \{1\}$).

By the above statement, to show this theorem, we need only show that if $L, R \subset \sigma^{-1}(\lambda)$ for some $\lambda \in \Lambda_n$ with $\lambda \neq \{1 > \dots > 1\}$ then $|L \cap R| = \infty$. So from now on we assume $L, R \subset \sigma^{-1}(\lambda)$ with $\lambda = \{\lambda_1 > \dots > \lambda_r\} \neq \{1 > \dots > 1\}$.

By Theorem 1.5.2, we see that the restriction on L (resp. on R) of the function $\mathcal{R}(w)$ (resp. $\mathcal{L}(w)$) is constant. Then we can define

$$\mathcal{R}(L) = \mathcal{R}(x) \text{ for any } x \in L.$$

and $$\mathcal{L}(R) = \mathcal{L}(y) \text{ for any } y \in R.$$

Hence for any $s \in \Delta$, we have either $L \subset \mathcal{D}_R(s)$ or $L \cap \mathcal{D}_R(s) = \emptyset$. If $L \subset \mathcal{D}_R(s)$, let $L^* = \{w' \in \mathcal{D}_R(s) | w' = w^* \text{ in } \mathcal{D}_R(s) \text{ with } w \in L\}$. Thus by Theorem 1.6.2, L^* is also a left cell of A_n which is in $\sigma^{-1}(\lambda)$. We say that L^* is obtained from L by a right star operation (in $\mathcal{D}_R(s)$). Clearly, in that case, $L^* \cap R = \{w' \in \mathcal{D}_R(s) | w' = w^* \text{ in } \mathcal{D}_R(s)$ with $w \in L \cap R\}$. So the right star operation in $\mathcal{D}_R(s)$ restricts to a bijection between $L \cap R$ and $L^* \cap R$. Similarly, if $R \subset \mathcal{D}_L(s)$, let ${}^*R = \{y' \in \mathcal{D}_L(s) | y' = {}^*y \text{ in } \mathcal{D}_L(s)$ with $y \in R\}$. Then ${}^*R$ is a right cell of A_n which is in $\sigma^{-1}(\lambda)$. We say that ${}^*R$ is obtained from R by a left star operation (in $\mathcal{D}_L(s)$). The left star operation in $\mathcal{D}_L(s)$ restricts to a bijection between $L \cap R$ and $L \cap {}^*R$.

By repeatedly using the above argument and by Proposition 18.1.3, we get the following result immediately.

Lemma 18.3.2 Let L, R be as above. Let L' (resp. R') be a left (resp. right) cell of $\sigma^{-1}(\lambda)$ obtained from L (resp. R) by a succession of right (resp. left) star operations. Then there exists a bijection between $L \cap R$ and $L' \cap R'$. This bijective map is defined by a succession of left star operations, which corresponds to passing from L to L', followed by a succession of right star operations, which corresponds to passing from R to R'. □

By Proposition 9.3.7, we may choose L' and R' in the above lemma such that $L' \subset H_\lambda^{-1}$ and $R' \subset H_\lambda$, where $H_\lambda^{-1} = \{w | w^{-1} \in H_\lambda\}$.

So to show Theorem 18.3.1, we may assume $L \subset H_\lambda^{-1}$ and $R \subset H_\lambda$ without loss of generality by Lemma 18.3.2.

Let $J_\lambda = J_{\lambda 1} \cup \dots \cup J_{\lambda r}$ with $J_{\lambda h} = \{s_t \in \Delta | \sum_{\alpha=h+1}^{r} \lambda_\alpha < t < \sum_{\alpha=h}^{r} \lambda_\alpha\}$, $1 < h < r$. Let Δ_λ^ℓ (resp. Δ_λ^r) be the set of all subsets X of Δ so that there exists some $x \in H_\lambda$ (resp. $x \in H_\lambda^{-1}$) with $\mathcal{L}(x) = X$ (resp. $\mathcal{R}(x) = X$). Recall that ϕ is an automorphism of

A_n which sends s_t to s_{t+1} for any $s_t \in \Delta$. Then

$$\begin{cases} \Delta_\lambda^\ell = \Delta_\lambda^r = \{\phi^h(J_\lambda) \mid 1 \leq h \leq n\} & \text{if } \lambda_1 \neq \lambda_r. \\ \Delta_\lambda^\ell = \Delta_\lambda^r = \{\phi^h(J_\lambda) \mid 1 \leq h \leq \lambda_1\} & \text{if } \lambda_1 = \lambda_r. \end{cases}$$

In both cases, we can write $\Delta_\lambda = \Delta_\lambda^\ell = \Delta_\lambda^r$.

Assume $R(L) = J$ and $\mathcal{L}(R) = K$. By Proposition 14.3.4, we have $L = \{w \in H_\lambda^{-1} \mid R(w) = J\}$ and $R = \{y \in H_\lambda \mid \mathcal{L}(y) = K\}$. So $L \cap R = \{w \in H_\lambda \cap H_\lambda^{-1} \mid R(w) = J \text{ and } \mathcal{L}(w) = K\}$.

By Proposition 9.3.7 and 10.4, we can find $w \in L$ with $w \in N_\lambda$. Assume that w has a standard MDC form $(A_r,\ldots,A_1)$ at $i \in \mathbf{Z}$. Let $e((w), j_t^u(w))$ be the u-th entry of $A_t(w)$. Let $E_1 = \{j_t^1(w) \mid 1 \leq t \leq r\}$. Then we know that $(L_{E_1})^h(w) \in N_\lambda$ for any $h \geq 0$ and so by Proposition 13.2.3 we have $(L_{E_1})^h(w) \in L$. We see that $\mathcal{L}((L_{E_1})^h(w)) = \phi^h(\mathcal{L}(w))$. Since $\mathcal{L}(w)$, $K \in \Delta_\lambda^\ell$, there exists some α, $0 \leq \alpha \leq n$, satisfying $K = \phi^\alpha(\mathcal{L}(w))$. Then $\{(L_{E_1})^{\alpha+qn}(w) \mid q \geq 0\} \subset L \cap R$. Obviously, by the assumption that $\lambda \neq \{1 \geq \ldots \geq 1\}$, we have $(L_{E_1})^{\alpha+qn}(w) \neq (L_{E_1})^{\alpha+q'n}(w)$ for any q, $q' \geq 0$ with $q \neq q'$. This implies that $L \cap R$ is an infinite set and Theorem 18.3.1 is proved.

In contrast to cells of A_n, the intersection of a left cell with a right cell in A_n has an entirely different structure.

Let us consider the more general case where $W = \langle W,S \rangle$ is any Coxeter group. We say a non-empty set $M \subset W$ is discrete if for any $x,y \in M$, neither xy^{-1} nor $x^{-1}y$ is in S.

<u>Proposition 18.3.3</u> <u>Let L and R be a left cell and a right cell of W, respectively. Then $L \cap R$ is either empty or a discrete set. In particular, this conclusion is true when $W = A_n$.</u>

<u>Proof</u> It is enough to show that if $x \in L \cap R$ and $t \in S$ then neither tx nor xt is in $L \cap R$. We see that t is either in $\mathcal{L}(x)$ or in $\mathcal{L}(tx)$ but not in both $\mathcal{L}(x)$ and $\mathcal{L}(tx)$. So by Theorem 1.5.2, $tx \not\sim_R x$. Thus $tx \notin L \cap R$. Similarly, we can show $xt \notin L \cap R$. □

CHAPTER 19 : SOME SPECIAL KINDS OF SIGN TYPES

Recall that in §4.4(xiii) we defined C to be the set of generalized tabloids of rank n. We also defined $\hat{C}$ (resp. $\check{C}$) to be the set of all proper (resp. opposed) tabloids of C.

Let $\mathcal{N}$ be the set of all normalized elements of A_n. Recall that in Chapter 14 we defined a map $T:\mathcal{N} \to \check{C}$. We showed that T is surjective. We know from Chapters 9 and 10 that for any $x \in A_n$ there exists some $x' \in \mathcal{N}$ with $x \underset{L}{\sim} x'$. Although such an element x' may not be unique, we know from Chapter 14 that if x'' is another element of $\mathcal{N}$ satisfying $x \underset{L}{\sim} x''$ then $T(x') = T(x'')$. This means that x determines a unique element of $\check{C}$. So there is a well defined map from A_n to $\check{C}$ by sending x to $T(x')$ for any $x' \in \mathcal{N}$ with $x \underset{L}{\sim} x'$. This map extends the map $T:\mathcal{N} \to \check{C}$ and is still denoted by T. Obviously, it is surjective. We showed in Chapter 14 that T induces a bijection from the set of left cells of A_n to $\check{C}$.

A Young tableau of shape λ is by definition a labelling of the cells of the Young diagram F_λ with integers $1,2,\ldots,n$ for $\lambda \in \Lambda_n$. Given $X = (X_1,\ldots,X_t) \in \hat{C}$, we define a Young tableau $\tilde{X}$ of size n such that the set of numbers in the i-th row of $\tilde{X}$ is X_i for any i and such that the numbers in each row of $\tilde{X}$ increase from left to right. Clearly, $\tilde{X}$ is uniquely determined by X. We call $\tilde{X}$ the Young tableau associated to X.

For example, the Young tableau associated to

$$X = \begin{array}{|c|c|c|} \hline 4 & 9 & 8 \\ 2 & 3 & 1 \\ 7 & 5 & \\ 6 & & \\ \hline \end{array}$$

$$\tilde{X} = \begin{array}{|c|c|c|c|} \hline 2 & 4 & 6 & 7 \\ \hline 3 & 5 & 9 \\ \cline{1-3} 1 & 8 \\ \cline{1-2} \end{array}$$

Note that $\tilde{X}$ is not necessarily a standard Young tableau.

In the last three chapters of our book, we shall define a map $\hat{T}:A_n \to \hat{C}$ which satisfies the following three properties.

(i) $\hat{T}$ induces a bijection from the set of left cells of A_n to $\hat{C}$.

(ii) For any $x \in A_n$, $\hat{T}(x) \underset{L}{\sim} T(x)$, regarded as elements of $\mathcal{S}$ (note that in Chapter 19 we shall identify C with a subset of $\mathcal{S}$).

(iii) When restricted to the subgroup P_n (see Chapter 16 for the definition of P_n and note that $P_n \cong S_n$), $\hat{T}$ coincides with the Robinson-Schensted map $w \to (P(w), P(w^{-1}))$ in the sense that, for any $w \in P_n$, let $X = \hat{T}(w)$ and $\tilde{X}$ the Young tableau associated to X then $\tilde{X} = P(w)$. In particular, $\tilde{X}$ is a standard Young tableau.

In the present chapter, we shall develop the results on sign types, which were first discussed in Chapter 7, by introducing some kinds of sign types. The most important one is called a Coxeter sign type which can be identified with a generalized tabloid. We shall define two kinds of actions on the set C of generalized tabloids which will be used in Chapter 20 to construct the inserting algorithm on C. Another kind of sign type is called a dominant sign type. The left cells of S consisting of dominant sign types have a relatively simple structure. We shall also introduce a so called special sign type. The consideration of these kinds of sign types will help us to understand the structure of general left cells of A_n.

§19.1 COXETER SIGN TYPES AND GENERALIZED TABLOIDS

A sign type $X = (X_{ij})_{1\leq i<j\leq n} \in S$ is called a Coxeter sign type if there exists a decomposition of the set $[n]$ into a disjoint union of subsets, say $[n] = \bigcup_{\alpha=1}^{t} \{h_{\alpha 1}, h_{\alpha 2},\dots,h_{\alpha k_\alpha}\}$, such that $X_{ij} = 0$ if and only if $\{i,j\}\subset \{h_{\alpha 1},h_{\alpha 2},\dots,h_{\alpha k_\alpha}\}$ for some α, $1 \leq \alpha \leq t$, with the convention that the set $\{X_{pq},X_{qp}\}$ is either $\{+,-\}$ or $\{0,0\}$ for any $p,q \in [n]$. Clearly, such a decomposition of $[n]$, when it exists, is unique up to the numbering of these subsets. We can call it the decomposition of $[n]$ associated to a Coxeter sign type X. Note that two different Coxeter sign types may give rise to the same associated decomposition of $[n]$.

<u>Lemma 19.1.1</u> <u>Let $\bigcup_{\alpha=1}^{t} \{h_{\alpha 1},h_{\alpha 2},\dots,h_{\alpha k_\alpha}\}$ be the decomposition of $[n]$ associated to a Coxeter sign type $X = (X_{ij})_{1\leq i<j\leq n}$. Let $\alpha,\beta \in \mathbf{Z}$ with $\alpha \neq \beta$ and $1 \leq \alpha, \beta \leq t$. Suppose $X_{ij} = +$ (resp. $X_{ij} = -$) for $i = h_{\alpha 1}$, $j = h_{\beta 1}$. Then $X_{i'j'} = +$ (resp. $X_{i'j'} = -$) for all $i' \in \{h_{\alpha 1},\dots,h_{\alpha k_\alpha}\}$, $j' \in \{h_{\beta 1},\dots,h_{\beta k_\beta}\}$.</u>

<u>Proof</u> Note that for any $k,\ell, m \in [n]$, where k, ℓ, m is a cyclic permutation of their natural order, the triple $X_{k\ell}\overset{X_{km}}{\quad}X_{\ell m}$ must be admissible.

First assume $X_{ij} = +$ for $i = h_{\alpha 1}$, $j = h_{\beta 1}$. Let us consider $X_{i'j'}$ for any $i' \in \{h_{\alpha 1},\dots,h_{\alpha k_\alpha}\}$, $j' \in \{h_{\beta 1},\dots,h_{\beta k_\beta}\}$. We see that either

$X_{ij}\overset{X_{ij'}}{\quad}X_{jj'} = +\overset{X_{ij'}}{\quad}o$ or $X_{ij'}\overset{X_{ij}}{\quad}X_{j'j} = X_{ij'}\overset{+}{\quad}o$ is admissible. This implies $X_{ij'} = +$. Also, we have that either $X_{i'i}\overset{X_{i'j'}}{\quad}X_{ij'} = o\overset{X_{i'j'}}{\quad}+$ or $X_{ii'}\overset{X_{ij'}}{\quad}X_{i'j'} = o\overset{+}{\quad}X_{i'j'}$ is admissible. So $X_{i'j'} = +$.

The proof is entirely similar for the case $X_{ij} = -$, where $i = h_{\alpha 1}$ and $j = h_{\beta 1}$. □

Lemma 19.1.1 enables us to define a total order on the set $\{X_\alpha = \{h_{\alpha 1},\dots,h_{\alpha k_\alpha}\} \mid 1 \leq \alpha \leq t\}$ with respect to a given Coxeter sign type $X = (X_{ij})_{1\leq i<j\leq n}$, where $[n] = \bigcup_{\alpha=1}^{t} X_\alpha$ is the decomposition associated to X. For $1 \leq \alpha, \beta \leq t$, $i = h_{\alpha 1}$, $j = h_{\beta 1}$, we define

$$X_\alpha = X_\beta \Leftrightarrow \alpha = \beta$$
$$X_\alpha < X_\beta \Leftrightarrow X_{ij} = -$$
$$X_\alpha > X_\beta \Leftrightarrow X_{ij} = +$$

Such an order is well defined by Lemma 19.1.1. By properly numbering X_α, $1 \leq \alpha \leq t$, we may assume $X_1 < X_2 < \dots < X_t$.

Recall the definition of a generalized tabloid in §4.4 (xiii). We now define a generalized tabloid $\xi(X)$ whose ℓ-th column is X_ℓ for $1 \leq \ell \leq t$, written $\xi(X) = (X_1, X_2, \dots, X_t)$. Thus $X \to \xi(X)$ is a map from the set Ω of Coxeter sign types of S to the set C of generalized tabloids. The following result enables us to identify Ω with C.

Proposition 19.1.2 $\xi : \Omega \to C$ is bijective.

Proof It is clear that ξ is injective. Now let $T = (T_1, \dots, T_t)$ with $T_\ell = \{\alpha_{\ell 1}, \dots, \alpha_{\ell k_\ell} \mid 1 \leq \alpha_{\ell 1} < \dots < \alpha_{\ell h_\ell} \leq n\}$, $1 \leq \ell \leq t$. Let $M = (M_1, N, M_2, N, \dots, N, M_r)$ be an $n \times ((t-1)mn+1)n$ matrix for any $m > 0$ such that N is an $n \times n(mn-1)$ matrix whose entries are all zero and M_j, $1 \leq j \leq t$, is an $n \times n$ matrix whose entries are all zero except for those in the $(\sum_{u=1}^{j-1} k_u + h, \alpha_{jh})$-positions which is 1. Then there exists a unique element w of A_n which has M as its submatrix lying in the section of w consisting of rows from the 1-st to the n-th. We can see that the α-th column of M is contained in the 1-st column of w with $\alpha = mn(\sum_{h=1}^{t-1} hk_{h+1}) + 1 \equiv 1 \pmod{n}$. So for any non-zero entry of M, if it is contained in the a-th column of M and also in the b-th column of w then $\bar{a} = \bar{b}$. Then by Proposition 6.2.3, we have $X = \zeta(w) \in \Omega$ and hence $\xi(X) = T$. So ξ is surjective and is thus bijective. □

From now on, we shall identify Ω with C. Then C will be regarded as a subset of S.

Examples 19.1.3

(i) Recall in §6.1 that A_n can be written as the form $N \rtimes P_n$, where P_n is the standard parabolic subgroup of A_n generated by $\{s_1, \dots, s_{n-1}\}$. N is the maximal normal subgroup of A_n which is a free abelian group of rank $n-1$. In terms of a matrix, N can be described as the set of all elements w of A_n, where w has an entry set $\{e_w(t, j_t) \mid t \in [n]\}$ with $\bar{j}_t = \bar{t}$ for all t. In that case, let $j_t = q_t n + t$. Then there exists a unique decomposition of $[n]$, say $[n] = \bigcup_{i=1}^{m} X_i$ for some $m \geq 1$, satisfying: (a) $t, u \in [n]$ lie in the same X_j if and only if $q_t = q_u$. (b) For $t \in X_i$ and $u \in X_j$ with $i \neq j$, $q_t < q_u$ implies $i < j$.

Then by Proposition 6.2.3, we can easily see that w determines the Coxeter sign type $X = (X_1, \dots, X_m)$.

(ii) Recall that in Chapter 14 we defined a subset $N_\lambda(a)$ of N_λ for $\lambda = \{\lambda_1 \geq \ldots \geq \lambda_r\} \in \Lambda_n$ with $\lambda \neq \{1 \geq \ldots \geq 1\}$ and $a \in \mathbf{Z}$. Assume that $w \in N_\lambda(n)$ has a standard MDC form $(A_r, \ldots, A_1)$ at i. Let $e((w), j_t^u)$ be the u-th entry of A_t. Let $a_{tu} \in [n]$ satisfy $\overline{a_{tu}} = j_t^u$. Let $X_u = \{a_{tu} | 1 \leq t \leq \mu_u\}$ for $1 \leq u \leq \lambda_1$, where $\mu = \{\mu_1 \geq \ldots \geq \mu_{\lambda_1}\}$ is the dual partition of λ. Then by Proposition 6.2.3, we see that w determines the Coxeter sign type $X = (X_{\lambda_1}, X_{\lambda_1 - 1}, \ldots, X_1)$.

For any $T = (T_1, \ldots, T_t) \in C$, $w \in \zeta^{-1}(T)$ and $1 \leq \ell \leq t$, we define $d_\ell(w) = \min\{((\beta)\bar{w}^{-1})w - ((\alpha)\bar{w}^{-1})w | \beta \in T_{\ell+1}, \alpha \in T_\ell\}$ and $d(w) = \min\{d_\ell(w) | 1 \leq \ell < t\}$, with the convention that $d(w) = 0$ when $t = 1$. Then by Proposition 6.2.3, we have $d(w) \geq 0$ in general. $d(w) = 0$ if and only if $w = 1$. In that case, T consists of a single column. We call it the trivial tabloid.

<u>Corollary 19.1.4</u> <u>Assume that $T = (T_1, \ldots, T_t) \in C$ is not trivial and $c > 0$. Then there exists an element $w \in \zeta^{-1}(T)$ with $d(w) > c$. In particular, $\zeta^{-1}(T)$ is an infinite set of A_n.</u>

<u>Proof</u> Assume that the columns T_ℓ, $1 \leq \ell \leq t$, of T, the matrix M with the number $m > c$, and the element $w \in A_n$ are all as in the proof of Proposition 19.1.2. By proper choice of m, we can make $n(mn-1) \geq c$. Since $d(w) > n(mn-1)$, w is as required. The infinitivity of the set $\zeta^{-1}(T)$ is an immediate consequence of this result. □

Let $\lambda \in \Lambda_n$ with $\lambda \neq \{1 \geq \ldots \geq 1\}$ and let $w \in \sigma^{-1}(\lambda)$. From the statement preceding Lemma 14.4.1 we know that for any $a \in \mathbf{Z}$ there exists some $y \in N_\lambda(a)$ such that $w \underset{L}{\sim} y$. So by Example 19.1.3(ii) and Proposition 18.2.2 we see that each left cell of S contains at least one Coxeter sign type. This together with Proposition 18.2.2 and Corollary 19.1.4 imply the following result immediately.

<u>Proposition 19.1.5</u> <u>Each non-trivial left cell of A_n is an infinite set of elements of A_n.</u> □

§19.2 <u>THREE MAPS $\hat{\sigma}$, f, ϕ AND THEIR RELATIONS</u>

Let $\tilde{\Lambda}_n = \{(\alpha_1, \ldots, \alpha_t) \in \bigcup_{m=1}^{n} \mathbf{Z}^m | \sum_{h=1}^{t} \alpha_h = n, \alpha_h > 0\}$ be the set of generalized partitions of n. For any $T = (T_1, \ldots, T_t) \in C$, we define $f(T) = (|T_1|, \ldots, |T_t|)$. Then $f(T) \in \tilde{\Lambda}_n$. So we have a well defined map $f: C \to \tilde{\Lambda}_n$. Obviously, f is surjective. For any $\alpha = (\alpha_1, \ldots, \alpha_t) \in \tilde{\Lambda}_n$, the cardinal of the fibre $f^{-1}(\alpha)$ is $\frac{n!}{\prod_{j=1}^{t} \alpha_j!}$. We say that T, T' $\in$ C have the same shape if $f(T) = f(T')$.

By the definition of the map $\hat{\sigma}: S \to \Lambda_n$ and the identification of C with $\Omega \subseteq S$, we see that for any $\alpha \in \tilde{\Lambda}_n$ $f^{-1}(\alpha)$ is in $\hat{\sigma}^{-1}(\lambda) \cap C$ for some $\lambda \in \Lambda_n$. More precisely, let $\alpha = (\alpha_1,\ldots,\alpha_t)$ and let $i_1,\ldots,i_t$ be a permutation of $1,2,\ldots,t$ with $\alpha_{i_1} > \ldots > \alpha_{i_t}$. Then λ is the dual partition of $\{\alpha_{i_1} > \ldots > \alpha_{i_t}\}$. So there exists a unique map $\phi: \tilde{\Lambda}_n \to \Lambda_n$ such that the following diagram is commutative.

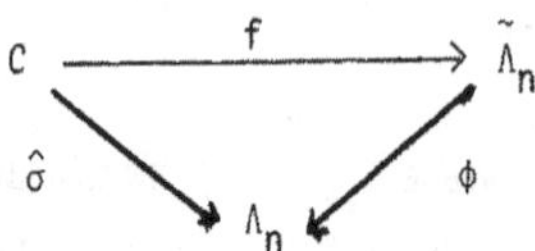

Clearly, for any $\lambda \in \Lambda_n$ and $\alpha = (\alpha_1,\ldots,\alpha_t)$, $\beta = (\beta_1,\ldots,\beta_u) \in \phi^{-1}(\lambda)$, we have $t = u$ and that $\alpha_1,\ldots,\alpha_t$ is a permutation of $\beta_1,\ldots,\beta_t$.

§19.3 TWO KINDS OF ACTIONS ON THE SET C

Now we shall define two kinds of actions on C.

(i) The right action of A_n on C.

Assume $T \in C$, $x,y \in A_n$ and $u \in [n]$. We define $T' = (T)s_u$ in C which is obtained from T by exchanging the positions of u and $u+1$ (here and later, the numbers in a generalized tabloid are always regarded as in $\underline{n} = \mathbb{Z}/n\mathbb{Z}$). We also define $(T)xy = ((T)x)y$.

Clearly, such an action of A_n on C is well defined. This action preserves the shape of a tabloid. Moreover, $f^{-1}(\alpha)$ is a single A_n-orbit for any $\alpha \in \tilde{\Lambda}_n$.

Say $X = (X_{ij})_{1 \leq i<j \leq n} \in S$ is dominant if $X_{ij} \in \{+,o\}$ for any i,j, $1 \leq i < j \leq n$. In particular, $T = (T_1,\ldots,T_t) \in C$ is dominant if and only if $i \in T_\alpha$, $j \in T_\beta$ and $\alpha < \beta$ together imply $i > j$ for any $i,j \in [n]$.

It is well known that each A_n-orbit of C contains a unique dominant tabloid.

(ii) The right star operations on C.

For $T = (T_1,\ldots,T_r) \in C$, we define a subset of Δ: $R(T) = \{s_t \in \Delta \mid \alpha < \beta$, where $t \in T_\alpha$ and $t+1 \in T_\beta\}$. Then by the definition of ξ and ζ, $R(T) = R(w)$ for any $w \in \zeta^{-1}(T)$. Thus for any t, $1 \leq t < n$, we define $\bar{\mathcal{D}}_R(s_t) = \{T \in C \mid R(T) \cap \{s_t, s_{t+1}\}$ contains exactly one element$\}$. When $T \in \bar{\mathcal{D}}_R(s_t)$, we see that there exists exactly one element in $\bar{\mathcal{D}}_R(s_t)$ between $(T)s_t$ and $(T)s_{t+1}$ which is not T. Let T^* be this element. The transformation from T to T^* is called a right star operation.

<u>Remark 19.3.1</u> For any $w \in A_n$ with $\zeta(w) = T \in C$ and any $s_t \in \Delta$, it does not always hold that $\zeta(ws_t) = (T)s_t$. We can't even always expect $\zeta(ws_t) \in C$. The reason for this is that the number of zeros in $\{k_{ij}^{w}\}_{1 \leq i<j \leq n}$ is not always the same as that in $\{k_{ij}^{ws_t}\}_{1 \leq i<j \leq n}$. But the number of zeros in $X = \{X_{ij}\}_{1 \leq i<j \leq n}$ is always the same as

that in $X' = \{X'_{ij}\}_{1\le i<j\le n}$, where $X = \xi^{-1}(T)$ and $X' = \xi^{-1}((T)s_t)$. For the same reason, although $w \in \mathcal{D}_R(s_t)$ is equivalent to $T \in \bar{\mathcal{D}}_R(s_t)$, it is not always true that $w^* \in \zeta^{-1}(T^*)$ when $w \in \mathcal{D}_R(s_t)$ (and so $T \in \bar{\mathcal{D}}_R(s_t)$). But nevertheless, in certain circumstances, we may expect $\zeta(ws_t) = (T)s_t$ and also $w^* \in \zeta^{-1}(T^*)$ when $w \in \mathcal{D}_R(s_t)$.

<u>Lemma 19.3.2 Let $w \in A_n$ satisfy $\zeta(w) = T \in C$ and let $s_t \in \Delta$.</u>

<u>Then (i) $\zeta(ws_t) = (T)s_t$ if and only if $k^w_{t,t+1} \neq 0, -1$.</u>

<u>(ii) In case where $w \in \mathcal{D}_R(s_t)$, we have $w^* \in \zeta^{-1}(T^*)$ if and only if</u>

$$\begin{matrix} & k^w_{t,t+2} & \\ k^w_{t,t+1} & & k^w_{t+1,t+2}\end{matrix} \neq \begin{matrix} & -1 & \\ 0 & & -1\end{matrix}, \quad \begin{matrix} & -1 & \\ -1 & & 0\end{matrix}$$

<u>Proof</u> (i) We know that $\zeta(ws_t) = (T)s_t$ if and only if $\{k^w_{ij}\}_{1\le i<j\le n}$ and $\{k^{ws_t}_{ij}\}_{1\le i<j\le n}$ contain the same number of zeros. But this is just equivalent to the condition $k^w_{t,t+1} \neq 0, -1$.

(ii) We see that when $w \in \mathcal{D}_R(s_t)$, $w^* \in \zeta^{-1}(T^*)$ if and only if $\{k^w_{ij}\}_{1\le i<j\le n}$ and $\{k^{w^*}_{ij}\}_{1\le i<j\le n}$ contains the same number of zeros. But this is equivalent to say that

$$\begin{matrix} & k^w_{t,t+2} & \\ k^w_{t,t+1} & & k^w_{t+1,t+2}\end{matrix} \neq \begin{matrix} & -1 & \\ 0 & & -1\end{matrix}, \begin{matrix} & -1 & \\ -1 & & 0\end{matrix}, \begin{matrix} & 0 & \\ -1 & & 0\end{matrix}, \begin{matrix} & 0 & \\ 0 & & -1\end{matrix}.$$

Since $\zeta(w) = T$, the cases

$$\begin{matrix} & k^w_{t,t+2} & \\ k^w_{t,t+1} & & k^w_{t+1,t+2}\end{matrix} = \begin{matrix} & 0 & \\ -1 & & 0\end{matrix} \text{ or } \begin{matrix} & 0 & \\ 0 & & -1\end{matrix}$$

will never happen. This implies our conclusion. □

Recall the definition of $d(w)$ in §19.1 for $w \in A_n$ with $\zeta(w) \in C$.

<u>Lemma 19.3.3 Assume $T = (T_1,\ldots,T_r) \in C$. Let $w \in \zeta^{-1}(T)$ and $s_t \in \Delta$. Suppose $\zeta(ws_t) = (T)s_t$. Then $d(ws_t) \ge d(w)-2$.</u>

<u>Proof</u> It suffices to show $d_\ell(ws_t) \ge d_\ell(w)-2$ for any ℓ, $1 \le \ell < r$.

Suppose $t \in T_\alpha$, $t+1 \in T_\beta$. Then by the condition $\zeta(ws_t) = (T)s_t$ and Lemma 19.3.2, we have $\alpha \neq \beta$. Let $T' = (T)s_t = (T'_1,\ldots,T'_r)$. Then $|T'_j| = |T_j|$ for all j, $1 \le j \le r$. Clearly, for any u, h with $u \in [n]$ and $1 \le h \le r$, we have $u \in T_h$ if and only if $(u)\bar{s}_t \in T'_h$. So for $1 \le \ell < r$, $d_\ell(w) = \min\{((\beta)\bar{w}^{-1})w - ((\alpha)\bar{w}^{-1})w \mid \beta \in T_{\ell+1}, \alpha \in T_\ell\}$

$d_\ell(ws_t) = \min\{((\beta')\bar{s}_t\bar{w}^{-1})ws_t - ((\alpha')\bar{s}_t\bar{w}^{-1})ws_t \mid \beta' \in T'_{\ell+1}, \alpha' \in T'_\ell\}$

$\qquad = \min\{((\beta)\bar{w}^{-1})ws_t - ((\alpha)\bar{w}s_t \mid \beta \in T_{\ell+1}, \alpha \in T_\ell\}$

But $((\beta)\bar{w}^{-1})ws_t - ((\alpha)\bar{w}^{-1})ws_t \ge ((\beta)\bar{w}^{-1})w - ((\alpha)\bar{w}^{-1})w-2$ always holds for any

$\alpha,\beta \in [n]$. So $d_\ell(ws_t) \geq d_\ell(w)-2$ for any ℓ, $1 \leq \ell < r$. Our conclusion follows. □

Lemma 19.3.4 Assume $T = (T_1,\ldots,T_r) \in C$ and $w \in \zeta^{-1}(T)$ with $d(w) \geq n+2$ and $s_t \in R(w)$. Then $k^w_{t,t+1} < -1$.

Proof Let $(\ell)w^{-1} = q_\ell n + r_\ell$ for $\ell \in [n]$, where $q_\ell, r_\ell \in \mathbf{Z}$, $r_\ell \in [n]$. Then $r_\ell = (\ell)\bar{w}^{-1}$. So we have $((\ell)\bar{w}^{-1})w = \ell - q_\ell n$. Suppose $t \in T_\alpha$, $t+1 \in T_\beta$. Then $\alpha < \beta$ by the condition $s_t \in R(w)$ and Proposition 6.2.3. Hence by the condition $d(w) \geq n+2$, we have $((t+1)\bar{w}^{-1})w - ((t)\bar{w}^{-1})w = (q_t - q_{t+1})n+1 \geq n+2$ and so $q_t \geq q_{t+1} + 2$. Then $(t+1)w^{-1}-(t)w^{-1} = (q_{t+1}-q_t)n + (r_{t+1}-r_t) < -n$. But we have $(t+1)w^{-1}-(t)w^{-1} = k^w_{t,t+1}n + r^w_{ij}$. This implies $k^w_{t,t+1} < -1$. □

Lemma 19.3.5 Assume $T = (T_1,\ldots,T_r) \in C$ and $w \in \zeta^{-1}(T)$ with $d(w) \geq n+2$. Then

(i) For $s_t \in \Delta$, $T \in \bar{\mathcal{D}}_R(s_t)$ if and only if $w \in \mathcal{D}_R(s_t)$.

(ii) Assume $T \in \bar{\mathcal{D}}_R(s_t)$ (so $w \in \mathcal{D}_R(s_t)$). Let $T' = T^*$ in $\bar{\mathcal{D}}_R(s_t)$ and $w' = w^*$ in $\mathcal{D}_R(s_t)$. Then $\zeta(w') = T'$.

Proof (i) This follows from the fact that $R(w) = R(\zeta(w))$.

(ii) By Lemma 19.3.2, we must show $\begin{smallmatrix} & k^w_{t,t+2} & \\ k^w_{t,t+1} & & k^w_{t+1,t+2} \end{smallmatrix} \neq \begin{smallmatrix} & -1 & \\ 0 & & -1 \end{smallmatrix}, \begin{smallmatrix} & -1 & \\ -1 & & 0 \end{smallmatrix}$. But this is immediate by Lemma 19.3.4. □

Proposition 19.3.6 Assume $T, T' \in C$. Suppose that there exists a sequence of elements $X_0 = T, X_1, \ldots, X_\ell = T'$ in C such that for every j, $1 \leq j \leq \ell$, $X_j = X^*_{j-1}$ in $\bar{\mathcal{D}}_R(s_{\alpha_j})$ with some $s_{\alpha_j} \in \Delta$. Then there also exists a sequence of elements $x_0, x_1, \ldots, x_\ell$ in A_n such that for every j, $1 \leq j \leq \ell$, $x_j = x^*_{j-1}$ in $\mathcal{D}_R(s_{\alpha_j})$ and for every h, $0 \leq h \leq \ell$, $\zeta(x_h) = X_h$.

Proof We may assume that T, T' are both non-trivial. Assume $T = (T_1,\ldots,T_r) \in C$ and $c = n + 2\ell$. Then by Corollary 19.1.4, there exists $x_0 \in \zeta^{-1}(X_0)$ with $d(x_0) > c$. Hence by repeatedly applying Lemmas 19.3.3 and 19.3.5, we can find the required sequence $x_0, x_1, \ldots, x_\ell$ in A_n. □

§19.4 THE MAP $\hat{T}$: $A_n \to \hat{C}$

The main task of this section is to define a map $\hat{T}: A_n \to \hat{C}$. The results of this section are basic to the ideas which will be developed in Chapter 20 to construct the inserting algorithm on C.

Let N be the set of all normalized elements of A_n. Let $N(a) = \bigcup_{\lambda\in\Lambda_n} N_\lambda(a)$ (for the definition of $N_\lambda(a)$, see §14.4).

Lemma 19.4.1 Assume $w \in N(a)$ for some $a > 2n$. Then $\zeta(w) = T(w) \in \check{C}$ with $d(w) > a-n$.

Proof Assume $\sigma(w) = \lambda = \{\lambda_1 > \ldots > \lambda_r\}$. Let $\mu = \{\mu_1 > \ldots > \mu_m\}$ be the dual of λ. Let w have a standard MDC form $(A_r,\ldots,A_1)$ at some $i \in \mathbf{Z}$. Let $e\,(i_t^u, j_t^u)$ be the u-th entry of A_t. Let $E_u = \{a_{u\ell} \mid 1 \leq \ell \leq \mu_u,\ 1 \leq a_{u1} < \ldots < a_{u\mu_u} \leq n\}$ for $1 \leq u \leq \lambda_1$ with $\bar{E}_u = \{i_t^u \mid 1 \leq t \leq \mu_u\}$. Then we see that $w \in N(a)$ implies $(a_{u\ell})w-(a_{u'\ell'})w > a-n$ for all $1 \leq u < u' \leq \lambda_1$, $1 \leq \ell \leq \mu_u$, $1 \leq \ell' \leq \mu_{u'}$ and $(a_{h\mu_h})w-n < (a_{h1})w < \ldots < (a_{h\mu_h})w$ for $1 \leq h \leq \lambda_1$. By Proposition 6.2.3 and the condition that $a-n > n$, we have

$$k_{ij}^w \begin{cases} = 0 & \text{if } \{i,j\} \in (E_u)\bar{w} \text{ for some } u,\ 1 \leq u \leq \lambda_1. \\ > 0 & \text{if } i \in (E_u)\bar{w},\ j \in (E_v)\bar{w} \text{ with } u < v. \\ < 0 & \text{if } i \in (E_u)\bar{w},\ j \in (E_v)\bar{w} \text{ with } u > v. \end{cases}$$

So $\zeta(w) = T(w) \in \check{C}$ with $\zeta(w) = ((E_{\lambda_1})\bar{w},\ldots,(E_1)\bar{w})$ and $d(w) > a-n$. □

Recall that at the beginning of this part we defined a surjective map $T: A_n \to \check{C}$. We know that for any $w,y \in A_n$, the following results hold.

(i) $w \underset{L}{\sim} y \iff T(w) = T(y)$ (by Proposition 14.4.3)

(ii) $w \underset{L}{\sim} y \iff \zeta(w) \underset{L}{\sim} \zeta(y)$ (by Proposition 18.2.2)

(iii) for $w \in N(2n)$, $\zeta(w) = T(w)$ (by Lemma 19.4.1)

(iv) for any non trivial left cell L of A_n, $L \cap N(2n) \neq \emptyset$ (see §14.4)

(v) for any $T \in \check{C}$, $\zeta^{-1}(T) \cap N(2n) \neq \emptyset$ (by (i), (iv) and the fact that T is a surjective map from A_n to $\check{C}$).

Lemma 19.4.2 Let $T, T' \in \check{C}$. Then $T \underset{L}{\sim} T' \iff T = T'$.

Proof ($\Leftarrow$) Obviously.

($\Rightarrow$) We may assume that T and T' are both non-trivial. For otherwise, the result is obvious.

By (v), there exist $w, y \in A_n$ with $w \in \zeta^{-1}(T) \cap N(2n)$ and $y \in \zeta^{-1}(T') \cap N(2n)$. Then by (iii), we have $T(w) = T$ and $T(y) = T'$. Owing to (ii), $T \underset{L}{\sim} T'$ implies $w \underset{L}{\sim} y$. So by (i), we must have $T(w) = T(y)$, i.e. $T = T'$. □

Let $\check{C}_\lambda = \{T \in \check{C} \mid \hat{\sigma}(T) = \lambda\}$. We know from Proposition 18.2.2 and Theorem 14.4.5 that there are $\dfrac{n!}{\prod_{j=1}^{m} \mu_j!}$ left cells in $\hat{\sigma}^{-1}(\lambda)$, where $\mu = \{\mu_1 > \ldots > \mu_m\}$ is the dual of λ.

So Lemma 19.4.2 tells us that $\check{C}_\lambda$ is a set of representatives of $\hat{\sigma}^{-1}(\lambda)$ in S.

Lemma 19.4.3 Assume $T, T' \in C$ with $T \underset{L}{\sim} T'$. Then $R(T) = R(T')$ and so for any $s_t \in \Delta$, we have $T \in \bar{\mathcal{D}}_R(s_t) \Leftrightarrow T' \in \bar{\mathcal{D}}_R(s_t)$.

Proof It is enough to show that $R(T) = R(T')$. Let $w \in \zeta^{-1}(T)$ and $w' \in \zeta^{-1}(T')$. Then $T \underset{L}{\sim} T \Rightarrow w \underset{L}{\sim} w' \Rightarrow R(w) = R(w')$. But $R(T) = R(w)$ and $R(T') = R(w')$. So $R(T) = R(T')$. □

Lemma 19.4.4 Assume $T, T' \in C$ with $T \underset{L}{\sim} T'$ and $T, T' \in \bar{\mathcal{D}}_R(s_t)$ for some $s_t \in \Delta$. Then $T^* \underset{L}{\sim} T'^*$.

Proof By Corollary 19.1.4, there exist $w \in \zeta^{-1}(T)$ and $w' \in \zeta^{-1}(T')$ with $d(w)$, $d(w') > n+2$. Then by Lemma 19.3.5, we have $w, w' \in \mathcal{D}_R(s_t)$ and $\zeta(w^*) = T^*$, $\zeta(w'^*) = T'^*$. So $T \underset{L}{\sim} T' \Leftrightarrow w \underset{L}{\sim} w' \overset{(1)}{\Leftrightarrow} w^* \underset{L}{\sim} w'^* \Leftrightarrow T^* \underset{L}{\sim} T'^*$, where the implication (1) follows by Theorem 1.6.2. □

Let e_0 be the element of A_n corresponding to the cyclic permutation $(12\ldots n)$.

Lemma 19.4.5 Assume $T, T' \in C$ with $T \underset{L}{\sim} T'$. Then $(T)e_0 \underset{L}{\sim} (T')e_0$.

Proof Let $w \in \zeta^{-1}(T)$ and $w' \in \zeta^{-1}(T')$. Let ϕ be the automorphism of A_n with $\phi(s_t) = s_{t+1}$ for $t \in [n]$. Then $T \underset{L}{\sim} T' \Leftrightarrow w \underset{L}{\sim} w' \Leftrightarrow \phi(w) \underset{L}{\sim} \phi(w')$. Since $(T)e_0 = \zeta(\phi(w))$ and $(T')e_0 = \zeta(\phi(w'))$, this implies $(T)e_0 \underset{L}{\sim} (T')e_0$. □

To show the next lemma we need a fact which will follow from Lemma 19.6.2. This fact says that for any $\alpha \in \tilde{\Lambda}_n$, the fibre $f^{-1}(\alpha) \subseteq C$ is invariant and also transitive under the right star operations together with the right action of e_0.

Lemma 19.4.6 For $\alpha \in \tilde{\Lambda}_n$ and $T, T' \in f^{-1}(\alpha)$, we have $T \underset{L}{\sim} T' \Leftrightarrow T = T'$.

Proof Clearly, there exists some $\lambda \in \Lambda_n$ with $f^{-1}(\alpha) \subseteq \hat{\sigma}^{-1}(\lambda)$.

Since $\check{C}_\lambda$ is a set of representatives of $\hat{\sigma}^{-1}(\lambda)$ in S, we see that for any $X \in f^{-1}(\alpha)$ there exists $Y \in \check{C}_\lambda$ with $X \underset{L}{\sim} Y$. We know that the sets $\check{C}_\lambda$ and $f^{-1}(\alpha)$ are all invariant and transitive under the right star operations and the right actions of e_0. Also, by Lemma 19.4.3, $X \underset{L}{\sim} Y$ implies "$X \in \bar{\mathcal{D}}_R(s_t) \Leftrightarrow Y \in \bar{\mathcal{D}}_R(s_t)$", for any $s_t \in \Delta$. By Lemmas 19.4.4 and 19.4.5, we see that when Y is transformed to any element Y' of $\check{C}_\lambda$ by a succession of right star operations together with the right actions of e_0, X is transformed to some element X' of $f^{-1}(\alpha)$ by the same operations with $X' \underset{L}{\sim} Y'$. This implies that for any element $Y' \in \check{C}_\lambda$, there exists $X' \in f^{-1}(\alpha)$ satisfying $X' \underset{L}{\sim} Y'$. Since $|\check{C}_\lambda| = |f^{-1}(\alpha)|$, we see from Lemma 19.4.2 that for any $Y \in \check{C}_\lambda$ there can't exist more than one element of $f^{-1}(\alpha)$ which is in the same left cell as Y. So there just exists a unique element X of $f^{-1}(\alpha)$ with $X \underset{L}{\sim} Y$ for any $Y \in \check{C}_\lambda$. Our result follows by Lemma 19.4.2. □

From the proof of Lemma 19.4.6, we can deduce the following result.

Proposition 19.4.7 Let $\alpha, \beta \in \tilde{\Lambda}_n$ with $\phi(\alpha) = \phi(\beta) = \lambda$. Then for any $X \in f^{-1}(\alpha)$, there exists a unique $Y \in f^{-1}(\beta)$ satisfying $X \underset{L}{\sim} Y$.

Proof We know that $|f^{-1}(\alpha)| = |f^{-1}(\beta)| = |\check{C}_\lambda|$. Now for any $Z \in \check{C}_\lambda$, there exists a unique element $X \in f^{-1}(\alpha)$ (resp. $Y \in f^{-1}(\beta)$) satisfying $X \underset{L}{\sim} Z$ (resp. $Y \underset{L}{\sim} Z$). So our assertion follows form Lemma 19.4.2. □

By Proposition 19.4.7 we can define a map $\hat{T}: A_n \to \hat{C}$ such that for any $w \in A_n$, $\hat{T}(w)$ is the unique element of $\hat{C}$ satisfying the condition that $\hat{T}(w) \underset{L}{\sim} T(w)$.

Proposition 19.4.8 (i) $\hat{T}$ is surjective

(ii) $\hat{T}$ induces a bijection between the set of left cells of A_n and $\hat{C}$.

(iii) For any $w \in A_n$, we have $\zeta(w) \underset{L}{\sim} \hat{T}(w)$.

Proof All these results follow from the corresponding properties of the map T and Proposition 19.4.7. □

§19.5 DOMINANT SIGN TYPES AND DOMINANT TABLOIDS

Recall that in §19.3 we defined a dominant sign type and in particular a dominant generalized tabloid. We know that for any $\alpha \in \tilde{\Lambda}_n$ there exists a unique dominant tabloid lying in $f^{-1}(\alpha)$. In particular, for any $\lambda \in \Lambda_n$, there exists a unique dominant proper tabloid $X \in \hat{C}$ with $\hat{\sigma}(X) = \lambda$.

Examples (i) Let $\alpha = (2,3,3,2)$. Then the dominant tabloid in $f^{-1}(\alpha)$ is

	6	3	
9			1
	7	4	
10			2
	8	5	

(ii) The dominant proper tabloid X with $\hat{\sigma}(X) = \{4 > 3 > 2\}$ is

7	4	2	
8	5		1
9	6	3	

Lemma 19.5.1 Assume that $T, T' \in C$ are two dominant tabloids. Then $T \underset{L}{\sim} T' \Leftrightarrow \hat{\sigma}(T) = \hat{\sigma}(T')$.

Proof ($\Rightarrow$) Obviously.

($\Leftarrow$) Assume $f(T) = \alpha$ and $f(T') = \beta$. Then by Proposition 19.4.7, there exists a unique element $X \in f^{-1}(\beta)$ satisfying $T \underset{L}{\sim} X$. We need only show that X is dominant. Now we regard T, X as two sign types. Then $T \underset{L}{\sim} X$ means that there exists a sequence $X_0 = T, X_1, \ldots, X_u = X$ in $\hat{\sigma}^{-1}(\hat{\sigma}(T))$ such that for every j, $1 \leq j \leq u$, X_j is obtained from X_{j-1} by replacing a certain zero sign by a non-zero sign or vice versa. So our result can be implied from the following lemma.

Lemma 19.5.2 Assume that $X \in S$ is dominant and assume that Y is obtained from X by replacing a certain zero sign by a non-zero sign with $\hat{\sigma}(Y) = \hat{\sigma}(X)$. Then Y is also dominant.

Proof Let $w \in \zeta^{-1}(X)$ and $y \in \zeta^{-1}(Y)$. Then our condition implies $w \underset{L}{\sim} y$. If Y is not dominant then Y is obtained from X by replacing a certain zero sign by a negative sign. We claim that this negative sign in Y must be $Y_{t,t+1}$ for some $1 \leq t < n$, where we assume $Y = (Y_{ij})_{1 \leq i < j \leq n}$. For otherwise, we have $Y_{ij} = -$ for some $1 \leq i < j \leq n$ with $j - i > 1$. Then for $i < t < j$, the admissibility of the triple $\begin{smallmatrix} & Y_{ij} & \\ Y_{it} & & Y_{tj} \end{smallmatrix}$ implies either Y_{it} or Y_{tj} must be a negative sign. But Y has at most one negative sign. This leads to a contradiction. However, $Y_{t,t+1} = -$ implies $k^y_{t,t+1} < 0$ and then $s_t \in R(y)$. Since $s_t \notin R(w)$ and $R(w) = R(y)$, this is also impossible. Our proof is complete. □

Corollary 19.5.3 Assume $X, Y \in S$ are both dominant. Then $X \underset{L}{\sim} Y \Leftrightarrow \hat{\sigma}(X) = \hat{\sigma}(Y)$.

Proof ($\Rightarrow$) Obviously.

($\Leftarrow$) Let $x \in \zeta^{-1}(X)$, $y \in \zeta^{-1}(Y)$ and let $\lambda = \hat{\sigma}(X) = \hat{\sigma}(Y)$.

We may assume that $\lambda \neq \{1 > \ldots > 1\}$, for otherwise the result is obvious. By Theorem 18.2.1 and the fact that the intersection of $N(2n)$ with any non-identity left cell of A_n is non-empty, there exists a sequence of elements $x_0 = x, x_1, \ldots, x_\alpha$ in $\sigma^{-1}(\lambda)$ such that for every j, $1 \leq j \leq \alpha$, $x_j = s_{i_j} x_{j-1}$ with some $s_{i_j} \in \Delta$ and $x_\alpha \in N(2n)$. Let $X_h = \zeta(x_h)$ for $0 \leq h \leq \alpha$. Then $X_0 = X, X_1, \ldots, X_\alpha$ is a sequence in $\hat{\sigma}^{-1}(\lambda)$ such that for every j, $1 \leq j \leq \alpha$, either $X_j = X_{j-1}$ or X_j is obtained from X_{j-1} by replacing a certain zero sign by a non-zero sign or vice versa, and $X_\alpha \in C$. That is, $X \underset{L}{\sim} X_\alpha$ for some $X_\alpha \in C$. By Lemma 19.5.2, X dominant implies X_α dominant. Similarly, there exists a dominant tabloid $Y_\beta \in C$ satisfying $Y \underset{L}{\sim} Y_\beta$. We get $X_\alpha \underset{L}{\sim} X_\beta$ from $\hat{\sigma}(X_\alpha) = \hat{\sigma}(Y_\beta) = \lambda$ by Lemma 19.5.1. So $X \underset{L}{\sim} Y$. □

Proposition 19.5.4 Let $\mu = \{\mu_1 > \ldots > \mu_m\} \in \Lambda_n$. Then for any $w \in \delta^{-1}(\mu)$ (see Definition 5.1.4 for the map δ) with $\zeta(w)$ dominant, we have $\hat{T}(w) = (X_1, \ldots, X_m)$, where

$X_\alpha = \{ \sum_{i=\alpha+1}^{m} \mu_i + k | 1 \leq k \leq \mu_\alpha \}$ $1 \leq \alpha \leq m$.

Proof We have $\zeta(w) \underset{L}{\sim} \hat{T}(w)$. So by the condition that $\zeta(w)$ is dominanet and Lemma 19.5.2, we see that $\hat{T}(w)$ is also dominant. Thus our result follows by Corollary 19.5.3 and Proposition 5.15(iv). □

Let us consider the set $D = \{w \in A_n \big| |R(w)| \leq 1\}$. D can be described as the set of all elements w of A_n such that $\zeta(\Phi^m(w))$ are dominant for some $m \geq 0$. D can also be regarded as the set of all elements w of A_n such that $\pi(R(w))$ take the minimum values among the set $\{\pi(R(y)) | y \in \sigma^{-1}(\sigma(w))\}$ with respect to the partial order on Λ_n. Let $S^{min} = \{X \in S \big| |R(w)| \leq 1\}$. Then we have

Proposition 19.5.5

(i) For any $x,y \in D$, $x \underset{L}{\sim} y \Leftrightarrow \sigma(x) = \sigma(y)$ and $R(x) = R(y)$.

(ii) For any $X,Y \in S^{min}$, $X \underset{L}{\sim} Y \Leftrightarrow \hat{\sigma}(X) = \hat{\sigma}(Y)$ and $R(X) = R(Y)$.

(iii) Suppose $x \in D$ with $\delta(x) = \{\mu_1 \geq \ldots \geq \mu_m\}$ and $R(x) = s_t$. Then $\hat{T}(x) = (X_1,\ldots,X_m)$ with $X_\alpha = \{t + \sum_{i=\alpha+1}^{m} \mu_i + k | 1 \leq k \leq \mu_\alpha\}$, $1 \leq \alpha \leq m$.

Proof For (i), the implication "$\Rightarrow$" follows from Theorem 1.5.2 and Proposition 9.4.1. Now assume $\sigma(x) = \sigma(y)$ and $R(x) = R(y)$. Then there exists some $m \geq 0$ such that $\zeta(\Phi^m(x))$, $\zeta(\Phi^m(y))$ are both dominant. Hence by Corollary 19.5.3 and Proposition 18.2.2 we have $\Phi^m(x) \underset{L}{\sim} \Phi^m(y)$ and so $x \underset{L}{\sim} y$.

(ii) follows from (i) by Proposition 18.2.2.

For (iii), we see by Proposition 19.5.4 that $\hat{T}(\Phi^{n-t}(x)) = (Y_1,\ldots,Y_m)$ with $Y_\alpha = \{ \sum_{i=\alpha+1}^{m} \mu_i + k | 1 \leq k \leq \mu_\alpha\}$, $1 \leq \alpha \leq m$. So (iii) follows by the equation $\hat{T}(x) = (\hat{T}(\Phi^{n-t}(x)))e^t$. □

§19.6 SPECIAL SIGN TYPES AND SPECIAL TABLOIDS

Recall that in Chapter 9 for any $\lambda = \{\lambda_1 \geq \ldots \geq \lambda_r\} \in \Lambda_n$, H_λ is the set of all elements w of $\sigma^{-1}(\lambda)$ which have an MDC form $(A_r,\ldots,A_1)$ at i for some $i \in \mathbf{Z}$ with $|A_j| = \lambda_j$, $1 \leq j \leq r$. Let $H_\lambda^{-1} = \{w | w^{-1} \in H_\lambda\}$. Then Proposition 14.3.4 asserts that for any $x,y \in \sigma^{-1}(\lambda)$ with $x \in H_\lambda^{-1}$, we have $x \underset{L}{\sim} y$ if and only if $R(x) = R(y)$.

Recall the map $\pi:\Delta \to \Lambda_n$ in Definition 5.1. Lemma 5.7 asserts that $\pi(\mathcal{L}(w)) \leq \sigma(w)$ for any $w \in A_n$. We see that $\pi(\mathcal{L}(w)) = \sigma(w)$ for any $w \in H_\lambda$. Now we define $\hat{H}_\lambda$ to be the set of all elements w of $\sigma^{-1}(\lambda)$ satisfying $\pi(\mathcal{L}(w)) = \sigma(w)$. i.e. the set of all elements w of $\sigma^{-1}(\lambda)$ such that $\pi(\mathcal{L}(w))$ take the maximum values among the set

$\{\pi(\mathcal{L}(y))|y \in \sigma^{-1}(\lambda)\}$ with respect to the partial order on Λ_n. Then any element y of $\hat{H}_\lambda$ has an MDC form $(A_r,\ldots,A_1)$ at i for some $i \in \mathbf{Z}$ with $|A_j| = \lambda_{i_j}$, $1 \leq j \leq r$, where $i_1,\ldots,i_r$ is a permutation of $1,2,\ldots,r$. Let S_r be the permutation group on the letters $1,2,\ldots,r$. For any $\tau \in S_r$ and $k \in [n]$, let $H_{\lambda,\tau,k}$ be the set of all elements y of $\hat{H}_\lambda$ which have an MDC form $(A_r,\ldots,A_1)$ at k with $|A_j| = \lambda_{(j)_\tau}$, $1 \leq j \leq r$. Then we see that the function $\mathcal{L}(y)$ on $H_{\lambda,\tau,k}$ is constant and that for $\tau,\tau' \in S_r$ and $k,k' \in [n]$, either $H_{\lambda,\tau,k} = H_{\lambda,\tau',k'}$ or $H_{\lambda,\tau,k} \cap H_{\lambda,\tau',k'} = \emptyset$. We also see that $\hat{H}_\lambda = \bigcup_{\substack{k\in[n]\\ \tau\in S_r}} H_{\lambda,\tau,k}$ and $H_\lambda = \bigcup_{k\in[n]} H_{\lambda,1,k}$, where 1 is the identity of S_r.

For any subset H of A_n, let $H^{-1} = \{w|w^{-1} \in H\}$.

<u>Proposition 19.6.1</u> <u>For any $\tau \in S_r$ and $k \in [n]$, $H^{-1}_{\lambda,\tau,k}$ is a left cell of A_n.</u>

<u>Proof</u> S_r can be regarded as a Coxeter group generated by $s_1,\ldots,s_{r-1}$, where $s_t = (t,t+1)$ is the permutation which transposes t and t+1. So we can define a length function $\ell(\tau)$ on S_r in the usual way. Now we argue by induction on $\ell(\tau) \geq 0$. When $\ell(\tau) = 0$, this is just the result of Proposition 14.3.4. Now assume $\ell(\tau) > 0$. Then there exists some t, $1 \leq t < r$, such that $\tau' = s_t\tau$ and $\ell(\tau') = \ell(\tau)-1$. By inductive hypothesis, $H^{-1}_{\lambda,\tau',k}$ is a left cell of A_n. We know that for any $y \in H^{-1}_{\lambda,\tau,k}$, y^{-1} has an MDC form $(A_r,\ldots,A_1)$ at k with $|A_j| = \lambda_{(j)\tau}$. If $\lambda_{(t)_\tau} > \lambda_{(t+1)_\tau}$ then for any $y \in H^{-1}_{\lambda,\tau,k}$, the element w with $w^{-1} = \theta^{A_{t+1}}_{A_t}(y^{-1})$ must exist by Lemma 8.3.7 and this element is in $H^{-1}_{\lambda,\tau',k}$. If $\lambda_{(t)_\tau} < \lambda_{(t+1)_\tau}$ then for any $y \in H^{-1}_{\lambda,\tau,k}$, the element w with $w^{-1} = \rho^{A_t}_{A_{t+1}}(y^{-1})$ must exist by Lemma 8.3.5 and this element is in $H^{-1}_{\lambda,\tau',k}$. In any case, we have a bijection from $H^{-1}_{\lambda,\tau,k}$ to $H^{-1}_{\lambda,\tau',k}$. This bijection is a succession of right star operations on elements of $H^{-1}_{\lambda,\tau,k}$, and each element of $H^{-1}_{\lambda,\tau,k}$ is acted on by the same succession of right star operations under this bijection. So $H^{-1}_{\lambda,\tau,k}$ is a left cell of A_n by Theorem 1.6.2 and the inductive hypothesis that $H^{-1}_{\lambda,\tau',k}$ is a left cell of A_n. Therefore our assertion follows by induction. □

For any $\lambda = \{\lambda_1 \geq \ldots \geq \lambda_r\} \in \Lambda_n$, $\tau \in S_r$ and $k \in [n]$, the function R(w) on the left cell $H^{-1}_{\lambda,\tau,k}$ is constant. Moreover, let y be any element of $H^{-1}_{\lambda,\tau,k}$. Then $H^{-1}_{\lambda,\tau,k}$ is the set of all elements w of $\sigma^{-1}(\lambda)$ with $R(w) = R(y)$.

Let $\hat{H} = \bigcup_{\lambda\in\Lambda_n} \hat{H}_\lambda$. Then $\hat{H}$ is the set of all elements w of A_n such that $\pi(\mathcal{L}(w))$ take the maximum values among the set $\{\pi(\mathcal{L}(y))|y \in \sigma^{-1}(\sigma(w))\}$ with respect to the partial order on Λ_n.

Let $\mathcal{S}^{max} = \zeta(\hat{H}^{-1})$. We call X a special sign type for any $X \in \mathcal{S}^{max}$.

<u>Proposition 19.6.2</u>

(i) For any $x,y \in \hat{H}^{-1}$, $x \underset{L}{\sim} y \Leftrightarrow \sigma(x) = \sigma(y)$ and $R(x) = R(y)$.

(ii) <u>For any $X,Y \in S^{max}$, $X \underset{L}{\sim} Y \iff \hat{\sigma}(X) = \hat{\sigma}(Y)$ and $R(X) = R(Y)$.</u>

<u>Proof</u> For (i), the implication "$\Rightarrow$" follows from Proposition 9.4.1 and Theorem 1.5.2. Now assume $\sigma(x) = \sigma(y) = \lambda = \{\lambda_1 > \dots > \lambda_r\}$ and $R(x) = R(y)$. Then x,y are in $H^{-1}_{\lambda,\tau,k}$ for some $\tau \in S_r$ and $k \in [n]$ by the above discussion. Thus $x \underset{L}{\sim} y$ by Proposition 19.6.1. (ii) follows from (i) by Proposition 18.2.2. □

By the surjectivity of the map T and Proposition 18.2.2, 19.4.7 and 19.6.1, we see that the intersection of $\zeta(H^{-1}_{\lambda,\tau,k})$ with $f^{-1}(\alpha)$ (see §19.2) consists of exactly one element, say Z, for any $\alpha = (\alpha_1,\dots,\alpha_t) \in \phi^{-1}(\lambda)$ (see §19.2). That is, there exists a unique element Z of $f^{-1}(\alpha)$ satisfying $R(Z) = R(y)$, where y is any element of $H^{-1}_{\lambda,\tau,k}$.

Now let us describe the tabloid $Z = (Z_1,\dots,Z_t)$. Let $a_{j\ell} = k + \sum_{k=j+1}^{r} \lambda_{(h)\tau} + \ell$ for any j, ℓ with $1 \leq \ell \leq \lambda_{(j)\tau}$ and $1 \leq j \leq r$. Suppose $a_{j\ell} \in Z_{\alpha_{j\ell}}$. We have $a_{j1} < \dots < a_{j\lambda_{(j)\tau}}$ by Proposition 6.2.3. In particular, $\lambda_1 = t$ and $(a_{(1)\tau-1,1},\dots,a_{(1)\tau-1,\lambda_1}) = (1,\dots,\lambda_1)$. Now let $B_u = \{a_{(j)\tau-1,\ell} \mid 1 \leq j \leq u,\ 1 \leq \ell \leq \lambda_j\}$ for $0 \leq u \leq r$ and let $Z^{(u)} = (Z_1^{(u)},\dots,Z_{\beta_u}^{(u)}) = (Z_1 - B_{u-1},\dots,Z_t - B_{u-1})$ with $Z_j^{(u)} \neq \emptyset$ for $1 \leq j \leq \beta_u$, where we stipulate $(X_1,\dots,X_v) = (X_1,\dots,\hat{X}_p,\dots,X_v)$ whenever $X_p = \emptyset$. Then $\beta_u = \lambda_u$ for all u, $1 \leq u \leq r$, and $a_{(u)\tau-1,\ell} \in Z_\ell^{(u)}$ for all ℓ, $1 \leq \ell \leq \lambda_u$, i.e. $a_{(u)\tau-1,\ell}$ are in the ℓ-th such column of Z which contains at least u numbers. So in that case, Z is entirely determined by $\alpha \in \phi^{-1}(\lambda)$, the integer k and the permutation $\tau \in S_r$.

Let Σ be the set of all triples (α,τ,k) such that (i) $\alpha \in \tilde{\Lambda}_n$, say $\phi(\alpha) = \lambda = \{\lambda_1 > \dots > \lambda_r\} \in \Lambda_n$. (ii) $\tau \in S_r$. (iii) $k \in [n]$.

Then Z is determined by the triple $(\alpha,\tau,k) \in \Sigma$.

Call any element X of C a special tabloid if X, regarded as a subset of A_n, is in $\hat{H}^{-1}$.

Let C' be the set of all special tabloids of C. Then by the previous results, we see that any special tabloid X is determined by a triple $(\alpha,\tau,k) \in \Sigma$. More precisely, we have the following result.

<u>Proposition 19.6.3</u> <u>Let $X = (X_1,\dots,X_t) \in C$ be a special tabloid determined by $(\alpha,\tau,k) \in \Sigma$ with $\alpha = (\alpha_1,\dots,\alpha_t) \in \tilde{\Lambda}_n$ and $\phi(\alpha) = \lambda = \{\lambda_1 > \dots > \lambda_r\} \in \Lambda_n$. Then X can be obtained in the following way.</u>

(i) <u>Let F be a diagram corresponding to (α,τ). This diagram consists of n cells. These cells are arranged in t columns and r rows. The ℓ-th column of F (from left to right) contains α_ℓ cells. The m-th rows of F (from bottom to top) contains $\lambda_{(m)\tau}$ cells.</u>

(ii) The numbers $k+1, k+2, \ldots, k+n$, regarded as elements in $\mathbb{Z}/n\mathbb{Z}$, are put into the n cells of F in turn from the top row to the bottom row and from left to right in each row.

(iii) Then X is obtained from the filled diagram F in (ii) by forgetting rows. □

For examples, assume $n = 10$, $\alpha = (2,3,3,2)$, $\tau = 1$, $\tau' = (23)$ and $k = 3$. We have $\phi(\alpha) = \{4 > 4 > 2\}$. Then the special tabloid determined by (α,τ,k) is

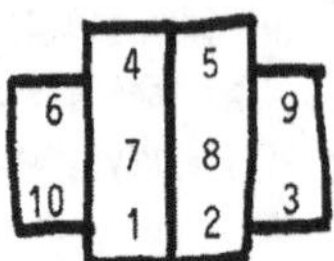

This tabloid is obtained from the diagram

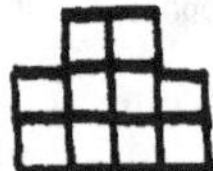

by filling the numbers of [n] into all cells in the following way

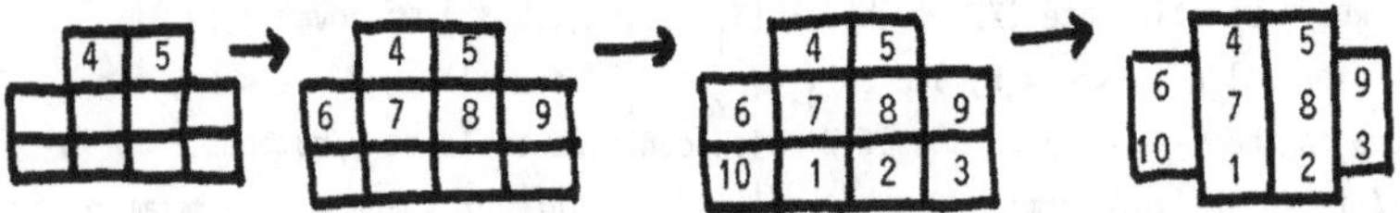

The special tabloid determined by (α,τ',k) is

This tabloid is obtained from the diagram

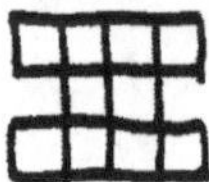

by filling the numbers of [n] into all cells in the following way:

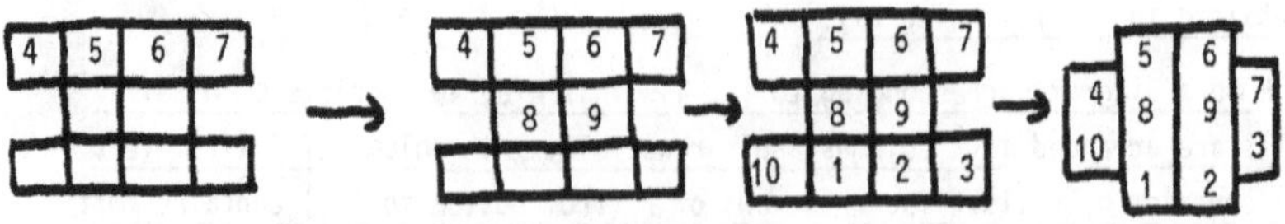

We say $(\alpha,\tau,k) \sim (\alpha',\tau',k')$ for (α,τ,k), $(\alpha',\tau',k') \in \Sigma$, if $H^{-1}_{\phi(\alpha),\tau,k} = H^{-1}_{\phi(\alpha'),\tau',k'}$. This is an equivalence relation on Σ. We see that $(\alpha,\tau,k) \sim (\alpha',\tau',k')$ if and only if the following conditions are satisfied.

(i) $\phi(\alpha) = \phi(\alpha') = \lambda = \{\lambda_1 > \dots > \lambda_r\}$ for some $\lambda \in \Lambda_n$.

(ii) If $\lambda_1 \neq \lambda_r$ then $k = k'$ and $(\lambda_{(1)\tau},\dots,\lambda_{(r)\tau}) = (\lambda_{(1)\tau'},\dots,\lambda_{(r)\tau'})$.

(iii) If $\lambda_1 = \lambda_r$ then $k \equiv k' \pmod{\lambda_1}$ and $\tau, \tau' \in S_r$ are arbitrary.

Proposition 19.6.4 Assume that $X, X' \in C'$ are determined by (α,τ,k), $(\alpha',\tau',k') \in \Sigma$, respectively. Then

$$X \underset{L}{\sim} X' \Longleftrightarrow (\alpha,\tau,k) \sim (\alpha',\tau',k')$$

In particular, when $k = k'$ and $\tau = \tau'$, we have

$$X \underset{L}{\sim} X' \Longleftrightarrow \hat{\sigma}(X) = \hat{\sigma}(X')$$

where, τ,τ' are regarded as elements in the same permutation group S_m for some sufficiently large $m > 0$. For example, we can always take $m = n$.

Proof This follows from Proposition 19.6.1 and the definition of the equivalence relation on Σ. □

By the above results, we can determine the proper tabloid $\hat{T}(x)$ for any $x \in \hat{H}^{-1}$. First we determine by the subset $R(x) \subset \Delta$ the set $H^{-1}_{\lambda,\tau,k}$ in which x lies. Then we take $\alpha = (\alpha_1,\dots,\alpha_t) \in \phi^{-1}(\lambda)$ with $|\alpha_1| > \dots > |\alpha_t|$ which is uniquely determined by the partition λ. Finally, the tabloid $\hat{T}(x)$ is determined by the trip (α,τ,k) by the procedure given in Proposition 19.6.3.

We have considered two kinds of sign types S^{min} and S^{max}, where $\pi(R(X))$ for $X \in S^{min}$ (resp. for $X \in S^{max}$) takes the minimum (resp. maximum) value among $\{\pi(R(Y)) | Y \in \hat{\sigma}^{-1}(\hat{\sigma}(X))\}$. We recall that we can think of a tabloid as a special kind of sign type, i.e. a Coxeter sign type. In a tabloid which occurs in S^{min}, the numbers $1,2,\dots,n$ are either arranged column by column in their natural order from right to left as in the example

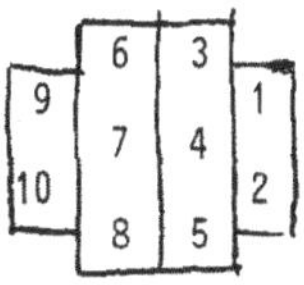

	6	3	
9	7	4	1
10	8	5	2

or arranged in a cyclic permutation of the above, as in the example

	9	6	
2	10	7	4
3	1	8	5

However, given a tabloid which occurs in S^{max}, there exists a diagram for this tabloid of the type described in Proposition 19.6.3 whose numbers are either arranged in their natural order row by row from top to bottom and from left to right in each row, as in the example

	1	2	
3	4	5	6
7	8	9	10

which gives rise to the tabloid

	1	2	
3	4	5	6
7	8	9	10

or arranged in a cyclic permutation of the above, as in the example

	4	5	
6	7	8	9
10	1	2	3

which gives rise to the tabloid

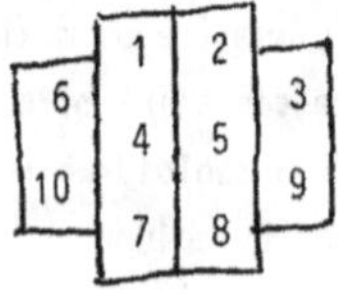

CHAPTER 20 : THE INSERTING ALGORITHM ON THE SET C

We shall first introduce an n-circle and the sets $H_Y(X)$ and $L_Y(X)$ for any $X,Y \subseteq [n]$ with $|Y| \geq |X|$ and $X \cap Y = \emptyset$. These concepts will be used in §20.2 to describe an algorithm called the inserting algorithm on C. This algorithm passes from an arbitrary generalized tabloid of C to the proper tabloid in the same left cell (note that C is regarded as a subset of S).

In the process of constructing this algorithm, we also get criteria for elements of C to belong to the same left cell of S.

§20.1 THE n-CIRCLE AND THE SETS $H_Y(X)$, $L_Y(X)$

We arrange n numbers $1,2,\ldots,n$ on a circle such that $t+1$ is the successor of t in the clockwise direction for $1 \leq t < n$ and 1 is the successor of n clockwise. We call such a circle an n-circle.

For example, the following circle is the 7-circle

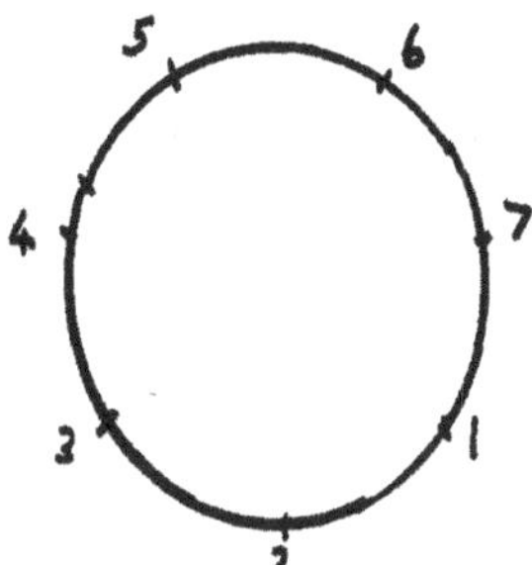

Figure 1

For $x,y \in [n]$, $x \neq y$, we denote by $\widehat{xy}$ the arc of the n-circle which, starting with x and moving clockwise, ends with y. For $Z \subseteq [n]$, let $Z|_{xy}$ be the set of all numbers on Z on xy. In Figure 1, let $Z = \{1,2,3,4,6\}$, $x = 2$, $y = 5$. Then $Z|_{xy} = \{2,3,4\}$, $Z|_{yx} = \{1,2,6\}$.

Let $X = \{\alpha_j | 1 \leq j \leq t, \alpha_1 < \ldots < \alpha_t\}$ and $Y = \{\beta_j | 1 \leq j \leq r, \beta_1 < \ldots < \beta_r\}$ be two subsets of $[n]$ with $X \cap Y = \emptyset$ and $r \geq t$.

(i) We define a subset $H_Y(X) = \{\gamma_1,\ldots,\gamma_t\}$ of Y as follows: let γ_1 be the element of Y satisfying $|(Y|_{\alpha_1\gamma_1}) \cap Y| = 1$. Let $\gamma_2 \in Y$ satisfy $|(Y|_{\alpha_2\gamma_2}) \cap (Y - \{\gamma_1\})| = 1$. In general, suppose that we have defined $\gamma_1,\ldots,\gamma_{h-1}$ for $1 \leq h \leq t$. Then we define $\gamma_h \in Y$ which satisfies $|(Y|_{\alpha_h\gamma_h}) \cap (Y - \{\gamma_1,\ldots,\gamma_{h-1}\})| = 1$. Clearly, the set $H_Y(X)$ is well defined. In particular, when $|X| = |Y|$, we have $H_Y(X) = Y$. In Figure 1, let $X = \{1,4\}$, $Y = \{2,6,7\}$. Then $H_Y(X) = \{2,6\}$.

(ii) We define a subset $L_Y(X) = \{\delta_1,\ldots,\delta_t\}$ of Y as follows: Let δ_1 be the element of Y satisfying $|(Y|_{\delta_1\alpha_t}) \cap Y| = 1$. Let $\delta_2 \in Y$ satisfy $|(Y|_{\delta_2\alpha_{t-1}}) \cap (Y - \{\delta_1\})| = 1$.

Inductively, let $\delta_h \in Y$ satisfy $|(Y|_{\delta_h\alpha_{t+1-h}}) \cap (Y - \{\delta_1,\dots,\delta_{h-1}\})| = 1$ for $1 \leq h \leq t$.
As in Figure 1, let $X = \{1,4\}$, $Y = \{2,6,7\}$. Then $L_Y(X) = \{2,7\}$.

Lemma 20.1.1. Suppose $X, Y \subset [n]$ with $X \cap Y = \emptyset$ and $|X| \leq |Y|$. Then for any $y \in Y$, $y \in H_Y(X)$ if and only if there exists some $x \in X$ satisfying $|Y|_{xy}| \leq |X|_{xy}|$.

Proof Let $X = \{\alpha_j | 1 \leq j \leq t,\ \alpha_1 < \dots < \alpha_t\}$, $Y = \{\beta_j | 1 \leq j \leq r,\ \beta_1 < \dots < \beta_r\}$ and $H_Y(X) = \{\gamma_1,\dots,\gamma_t\}$, where $\gamma_1,\dots,\gamma_t$ are defined from the sets X, Y by the above process. We see that for any v, $1 \leq v \leq t$, there exists the smallest integer h_v with $1 \leq h_v \leq v$ such that γ_k is not in $\alpha_k\ \alpha_{k+1}$ for all k, $h_v \leq k < v$. Then $Y|_{\alpha_{h_v}\gamma_v} \subset H_Y(X)$. So when $\gamma_v > \alpha_v$ or when $\gamma_v < \alpha_v$ with $\gamma_v = \min\{y | y \in H_Y(X)\}$, we have $Y|_{\alpha_{h_v}\gamma_v} = \{\gamma_{h_v}, \gamma_{h_v+1},\dots,\gamma_v\}$ and $\{\alpha_{h_v}, \alpha_{h_v+1},\dots,\alpha_v\} \subseteq X|_{\alpha_{h_v}\gamma_v}$. So $|Y|_{\alpha_{h_v}\gamma_v}| = v+1-h_v \leq |X|_{\alpha_{h_v}\gamma_v}|$. When $\gamma_v < \alpha_v$ with $\gamma_v > \min\{y | y \in H_Y(X)\}$, let $m = \max\{y | y < \gamma_v,\ y \in H_Y(X)\}$. Then $Y|_{\alpha_{h_v}\gamma_v} = \{\gamma_{h_v}, \gamma_{h_v+1},\dots,\gamma_v, \gamma_1, \gamma_2,\dots,\gamma_m\}$, and $X|_{\alpha_{h_v}\gamma_v} \supseteq \{\alpha_{h_v}, \alpha_{h_v+1},\dots,\alpha_v, \alpha_1, \alpha_2,\dots,\alpha_m\}$. So we also have $|Y|_{\alpha_{h_v}\gamma_v}| \leq |X|_{\alpha_{h_v}\gamma_v}|$. So the implication "$\Rightarrow$" has been shown.

Now assume $y \in Y$ satisfies $|Y|_{\alpha_h y}| \leq |X|_{\alpha_h y}|$ for some h, $1 \leq h \leq t$. Suppose $y \notin H_Y(X)$. We can take h such that $\alpha_h y$ contains as few numbers as possible. Suppose $\alpha_h, \alpha_{h+1},\dots,\alpha_{h+m}$ are on $\alpha_h y$ for some $m \geq 0$ with $h+m \leq t$ and α_1 is not on $\alpha_h y$ whenever $h \neq 1$. Then there exists j, $-1 \leq j < m$, such that $\gamma_h,\dots,\gamma_{h+j}$ are on $\alpha_h y$ but γ_{h+j+1} is not with the convention that $\{\gamma_h,\dots,\gamma_{h+j}\} = \emptyset$ when $j = -1$ (see Figure 2).

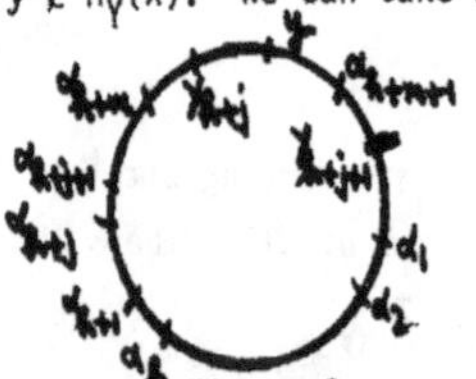

Figure 2

Since y is on $\alpha_{h+j+1}\ \gamma_{h+j+1}$ but $y \notin H_Y(X)$, this contradicts the definition of γ_{h+j+1}. Now suppose $\alpha_h, \alpha_{h+1},\dots,\alpha_t, \alpha_1, \alpha_2,\dots,\alpha_i$ are on $\alpha_h y$ but α_{i+1} is not on $\alpha_h y$ whenever $h \neq i+1$. Then by the assumption on h, we see that $\gamma_1,\dots,\gamma_i$ must be on $\alpha_1 y$. Then there must exist some j, $-1 \leq j \leq t-h$, such that $\gamma_h,\dots,\gamma_{h+j}$ are on $\alpha_h y$ but γ_{h+j+1} is not with the same convention as above. This contradicts the definition of γ_{h+j+1} for the same reason as above. Therefore, we have $y \in H_Y(X)$. □

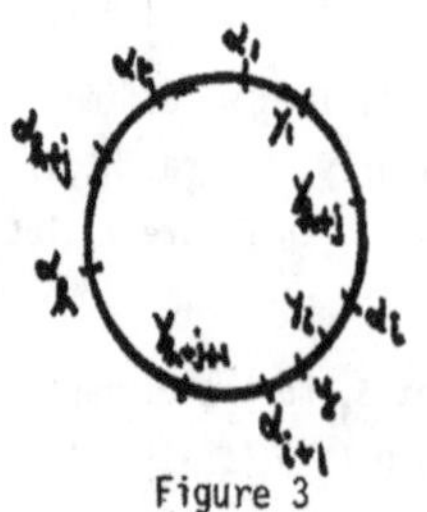

Figure 3

As in Lemma 20.1.1, if $y \in Y$ satisfies $|Y|_{xy}| < |X|_{xy}|$ for some $x \in X$ then we claim that there must exist some $z \in X$ with $|Y|_{zy}| = |X|_{zy}|$. For, let $x_1, x_2, \ldots, x_\ell = x$ be the elements of X with $|X|_{x_jy}| = j$ for every j, $1 \leq j \leq \ell$, where $\ell = |X|_{xy}|$. Then $|Y|_{x_1y}| \geq 1$ and $|Y|_{x_\ell y}| < \ell$. But we have $|Y|_{x_1y}| \leq |Y|_{x_2y}| \leq \ldots \leq |Y|_{x_\ell y}|$. So there must exist some m, $1 \leq m \leq \ell$, with $|Y|_{x_jy}| = j$. Hence $z = x_j$ is an element of X satisfying $|Y|_{zy}| = |X|_{zy}|$. So we can make the result of Lemma 20.1.1 stronger.

__Lemma 20.1.2__ __Let X,Y be as in Lemma 20.1.1. Then for any $y \in Y$, $y \in H_Y(X)$ if and only if there exists some $x \in X$ satisfying $|Y|_{xy}| = |X|_{xy}|$.__ □

__Lemma 20.1.3__ __Let X,Y be as in Lemma 20.1.1. Then for any $y \in Y$, $y \in L_Y(X)$ if and only if there exists some $x \in X$ satisfying $|Y|_{yx}| = |X|_{yx}|$.__

__Proof__ Let θ be the permutation on [n] with $\theta(u) = n+1-u$ for all $u \in [n]$. Then $\theta^2 = 1$. By the definition, we see that $L_Y(X) = \theta(H_{\theta(Y)}(\theta(X)))$. So $y \in L_Y(X) \Leftrightarrow \theta(y) \in H_{\theta(Y)}(\theta(X))$ $\Leftrightarrow$ there exists some $x \in X$ such that $|\theta(Y)|_{\theta(x)\theta(y)}| = |\theta(X)|_{\theta(x)\theta(y)}|$. i.e. $|Y|_{yx}| = |X|_{yx}|$. □

Recall the sequence $x_1, x_2, \ldots, x_\ell \in X$ for $y \in H_Y(X)$ in the proof of Lemma 20.1.2. Let us take j, $1 \leq j \leq \ell$, as small as possible which satisfies $|Y|_{x_jy}| = |X|_{x_jy}|$. Then such an $x_j \in X$ is uniquely determined by y. Let us write $\xi(y)$ for x_j. Then $\xi: H_Y(X) \to X$ is a well defined map. We claim that ξ is injective. For, suppose $y, y' \in H_Y(X)$ with $x = \xi(y) = \xi(y')$. Then either $y \in xy'$ or $y' \in xy$, where the notation $\alpha \in \beta\gamma$ means that the number α is on the arc $\beta\gamma$ of the n-circle. We may assume $y' \in xy$ without loss of generality. Let $x_1, x_2, \ldots, x_\ell = x \in X$ be such that $|X|_{x_jy}| = j$, $1 \leq j \leq \ell$. If $y' \neq y$ then there exists a smallest i, $1 \leq i \leq \ell$, with $y' \in Y|_{x_iy}$. By the condition that $|Y|_{xy}| = |X|_{xy}|$ and $|Y|_{xy'}| = |X|_{xy'}|$, we see $i > 1$. Hence $|Y|_{xy}| \geq |Y|_{x_{i-1}y}| + |Y|_{xy'}|$ and $|X|_{xy}| = |X|_{x_{i-1}y}| + |X|_{xy'}|$. This implies $|Y|_{x_{i-1}y}| \leq |X|_{x_{i-1}y}|$. By the same argument as in the proof of Lemma 20.1.2, we can find h, $1 \leq h \leq i-1$, such that $|Y|_{x_hy}| = |X|_{x_hy}|$. Since $x_h \neq x$, this contradicts the definition of $x = \xi(y)$. So we must have $y = y'$ and then ξ is injective. Thus ξ is bijective since $|H_Y(X)| = |X|$.

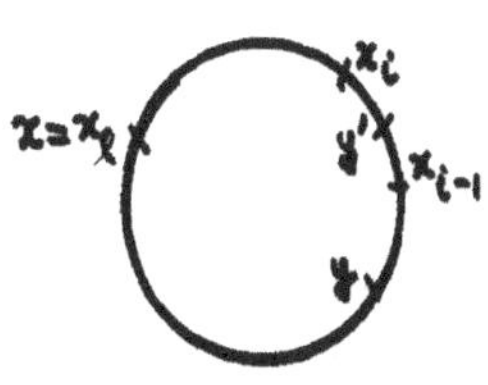

Figure 4

Now fix $x \in X$ and let $y = \xi^{-1}(x)$. Suppose $|Y|_{xy}| = \ell$. Then there exist $y_1, \ldots, y_\ell = y$ in Y such that $|Y|_{xy_j}| = j$, $1 \leq j \leq \ell$. Since $|X|_{xy_1}| \geq 1$ and

$|X|_{xy_\ell}| = |Y|_{xy_\ell}| = \ell$, we can find the smallest m, $1 \leq m \leq \ell$, satisfying $|X|_{xy_m}| = m$. Clearly $y_m \in H_Y(X)$. We claim $y = y_m$. Suppose not. Then there exist $x_1, x_2, \ldots, x_m = x$ in X such that $|X|_{x_h y_m}| = h$, $1 \leq h \leq m$ and for some α, $1 \leq \alpha < m$, $\alpha = |X|_{x_\alpha y_m}| = |Y|_{x_\alpha y_m}|$. Let β be the smallest number with $1 \leq \beta \leq m$ and $x_\alpha \notin y_\beta y_m$. Then we claim $\beta \neq 1$. For otherwise, $\alpha = |X|_{x_\alpha y_m}| = |Y|_{x_\alpha y_m}| = |Y|_{y_1 y_m}| = m$. This contradicts $\alpha < m$. Now

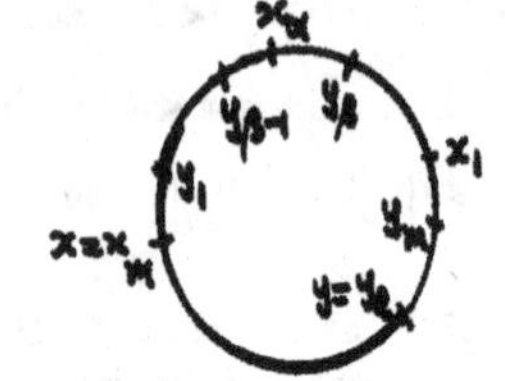

$\beta > 1$. $|X|_{xy_{\beta-1}}| \leq |Y|_{xy_{\beta-1}}| = \beta-1$ and $|X|_{xy_1}| \geq 1$ and $|X|_{xy_1}| \leq |X|_{xy_2}| \leq \ldots \leq |X|_{xy_{\beta-1}}|$. There must exist some γ with $1 \leq \gamma \leq \beta-1$ such that $|X|_{xy_\gamma}| = \gamma$. But $\gamma < m$. This contradicts the minimality of m. So $y = y_m$. The above argument tells us that for any $x \in X$, the element $y = \xi^{-1}(x) \in Y$ can be characterized by the following properties:

(i) $|Y|_{xy} = |X|_{xy}|$

(ii) If $y' \in Y$ also satisfies $|Y|_{xy'}| = |X|_{xy'}|$ then $|X|_{xy'}| \geq |X|_{xy}|$. Also, for any $y \in H_Y(X)$, the element $x = \xi(y) \in X$ can be characterized by the following properties:

(i) $|Y|_{xy}| = |X|_{xy}|$

(ii) If $x' \in X$ also satisfies $|Y|_{x'y}| = |X|_{x'y}|$ then $|X|_{x'y}| \geq |X|_{xy}|$).

Similarly, by using the map θ, we can define a bijective map $\zeta: L_Y(X) \to X$ which is uniquely determined by the following properties: for $y \in L_Y(X)$, let $x = \zeta(y) \in X$, then

(i) $|Y|_{yx}| = |X|_{yx}|$

(ii) If $y' \in Y$ satisfies $|Y|_{y'x}| = |X|_{y'x}|$ then $|X|_{y'x}| \geq |X|_{yx}|$.

Alternatively, ζ can be uniquely determined by the following properties: for $x \in X$, let $Y = \zeta^{-1}(x) \in Y$, then

(i) $|Y|_{yx}| = |X|_{yx}|$

(ii) If $x' \in X$ satisfies $|Y|_{yx'}| = |X|_{yx'}|$ then $|X|_{yx'}| \geq |X|_{yx}|$.

Sometimes, we denote ξ, ζ by $\xi_{Y,X}$, $\zeta_{X,Y}$ to indicate what sets we are concerned with. We write down a simple fact which could be checked by Lemma 20.1.1 and the definition of $\xi_{Y,X}$: For any $x \in X$, let $y = \xi^{-1}_{Y,X}(x)$. Then $Y|_{xy} \subseteq H_Y(X)$, or alternatively, $|(Y-H_Y(X))|_{xy} = 0$.

<u>Proposition 20.1.4</u> <u>Let X,Y be as in Lemma 20.1.1</u>

(i) <u>Let $Y' = H_Y(X)$ and $X' = X \cup (Y-H_Y(X))$. Then $X = L_{X'}(Y')$ and $Y = Y' \cup (X'-L_{X'}(Y'))$.</u>

(ii) Let $X'' = X \cup (Y-L_Y(X))$ and $Y'' = L_Y(X)$. Then $X = H_{X''}(Y'')$ and $Y = Y'' \cup (X''-H_{X''}(Y''))$.

Proof (i) Since $X' \cup Y' = X \cup Y$, it is enough to show $X = L_{X'}(Y')$. Since $|X| = |H_Y(X)| = |Y'| = |L_{X'}(Y')|$, we need only show $X \subseteq L_{X'}(Y')$. For any $x \in X$, let $y = \xi^{-1}_{Y,X}(x)$. Then $y \in Y'$, $x \in X'$ and $|X|_{xy}| = |Y|_{xy}|$. Now

$|X'|_{xy}| = |X|_{xy}| + |(Y-H_Y(X))|_{xy}| = |X|_{xy}| = |Y|_{xy}| = |H_Y(X)|_{xy}| = |Y'|_{xy}|$.

So by Lemma 20.1.3, $x \in L_{X'}(Y')$, i.e. $X \subseteq L_{X'}(Y')$ and then our result follows.

(ii) This follows by (i), using the map θ. □

From Lemma 20.1.2, we see that the set $H_Y(X)$ is only dependent on the relative positions of the numbers of X with those of Y on the n-circle. This implies $(H_Y(X))e_0 = H_{(Y)e_0}((X)e_0)$ when they exist.

Lemma 20.1.5 Let X,Y be as in Lemma 20.1.1. Then

$$(H_Y(X))e_0 = H_{(Y)e_0}((X)e_0). \qquad \square$$

In the remaining part of this chapter, when we talk about the numbers on the n-circle, we make a convention that any $r \in [n]$ can also be denoted by $qn + r$ for any $q \in \mathbf{Z}$.

Lemma 20.1.6 Let X,Y be as in Lemma 20.1.1 such that for some $u \in [n]$, $u \in X$ and $u+1, u+2 \in Y$. Let $X' = X \cup \{u+1\} - \{u\}$ and $Y' = Y \cup \{u\} - \{u+1\}$. Then $u+2 \notin H_Y(X) \Leftrightarrow u \notin H_{Y'}(X')$.

Proof It is equivalent to show: $u+2 \in H_Y(X) \Leftrightarrow u \in H_{Y'}(X')$. We may assume $|X| < |Y|$ since otherwise the result is trivial. By Lemma 20.1.1, (1) $u+2 \in H_Y(X) \Leftrightarrow$ (2) there exists some $x \in X$ such that $|Y|_{x\ u+2}| < |X|_{x\ u+2}|$ (3) $u \in H_{Y'}(X') \Leftrightarrow$ (4) there exists some $x' \in X'$ such that $|Y'|_{x'u}| < |X'|_{x'u}|$. We know that for any $x \in X \cap X'$,

$|Y|_{x\ u+2}| < |X|_{x\ u+2}| \Leftrightarrow |Y'|_{xu}| < |X'|_{xu}|$. We also know that $|Y|_{u\ u+2}| = 2 > 1 = |X|_{u\ u+2}|$ and $|Y'|_{u+1\ u}| = |Y| > |X| = |X'|_{u+1\ u}|$. Since the only number which is in X (resp. X') but not in X' (resp. X) is u (resp. u+1), we have (2) $\Leftrightarrow$ (4) and thus (1) $\Leftrightarrow$ (3). Our proof is complete. □

Lemma 20.1.7 Assume X,Y are as in Lemma 20.1.1. Suppose for some $u \in [n]$, $|\{u,u+1\} \cap (X \cup Y)| \leq 1$. Then $(H_Y(X))s_u = H_{(Y)s_u}((X)s_u)$.

Proof This can be checked easily by Lemma 20.1.1. □

Lemma 20.1.8 Assume X,Y are as in Lemma 20.1.1. Suppose for some $u \in [n]$, either $u \in X$, $u+1, u+2 \in H_Y(X)$ or $u+1 \in X$, $u, u+2 \in H_Y(X)$. Then $H_{(Y)s_u}((X)s_u) = (H_Y(X))s_u$.

Proof We may assume $|X| < |Y|$. For otherwise, $H_{(Y)s_u}((X)s_u) = (Y)s_u = (H_Y(X))s_u$.

Since $|H_{(Y)s_u}((X)s_u)| = |(H_Y(X))s_u|$, it suffices to show $(H_Y(X))s_u \subseteq H_{(Y)s_u}((X)s_u)$.

First assume $u \in X$, $u+1$, $u+2 \in H_Y(X)$. Then $u \in (H_Y(X))s_u$. Since $u+1, u+2 \in H_Y(X)$, there exists $x \in X$ such that $|Y|_{x\,u+2}| < |X|_{x\,u+2}|$. Clearly, $x \neq u$. Thus $|(Y)s_u|_{x\,u}| = |Y|_{x\,u+1}| = |Y|_{x\,u+2}| -1 < |X|_{x\,u+2}| -1 = |(X)s_u|_{x\,u+2}| -1 = |(X)s_u|_{x\,u}|$. This implies $u \in H_{(Y)s_u}((X)s_u)$ by Lemma 20.1.1.

Now assume $y \in (H_Y(X))s_u - \{u\}$. Then $y = (y)s_u \in H_Y(X) - \{u+1\}$. By Lemma 20.1.1, there exists $x \in X$ such that $|Y|_{x\,y}| < |X|_{xy}|$. If $x \neq u$ then one of the following cases must happen

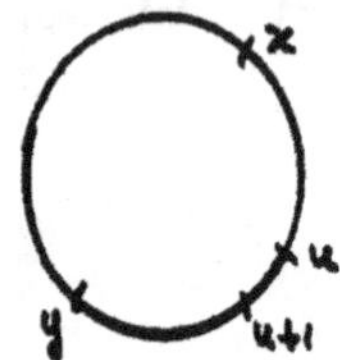

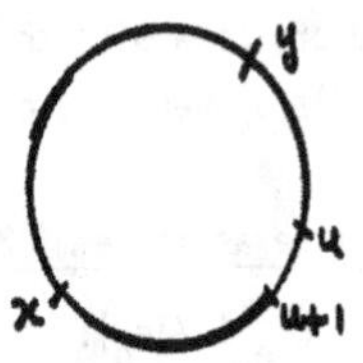

So we also have $|(Y)s_u|_{x\,y}| < |(X)s_u|_{x\,y}|$ and then $y \in H_{(Y)s_u}((X)s_u)$. If $x = u$ then $y \neq u+2$ and there must exist some $z \in X - \{x\}$ on $x\,y$ such that $|X|_{z\,y}| = |X|_{x\,y}| -1$. So $|(X)s_u|_{zy}| = |X|_{zy}| = |X|_{xy}|-1 > |Y|_{xy}| -1 > |Y|_{zy}| = |(Y)s_u|_{xy}|$. This also implies $y \in H_{(Y)s_u}((X)s_u)$. Therefore we have $(H_Y(X))s_u \subseteq H_{(Y)s_u}((X)s_u)$ and hence $(H_Y(X))s_u = H_{(Y)s_u}((X)s_u)$.

Now assume $u+1 \in X$, u, $u+2 \in H_Y(X)$. Then $u \in (X)s_u$ and $u+1$, $u+2 \in (H_Y(X))s_u \subseteq (Y)s_u$. Since $|(Y)s_u|_{u\,u+1}| = 1 = |(X)s_u|_{u,u+1}|$, we have $u+1 \in H_{(Y)s_u}((X)s_u)$. Also, since $u \in H_Y(X)$, there exists $x \in X$ such that $|Y|_{xu}| < |X|_{xu}|$. Clearly, $|X| < |Y|$ implies $x \neq u+1$. Then $|(Y)s_u|_{x\,u+2}| = |Y|_x|_{u+2}| = |Y|_{xu}| + 1 < |X|_{xu}| + 1 = |X|_{x\,u+2}| = |(X)s_u|_{x\,u+2}|$. We also have $u+2 \in H_{(Y)s_u}((X)s_u)$.

Now $u \in (X)s_u$ and $u+1$, $u+2 \in H_{(Y)s_u}((X)s_u)$. By applying the above result, we get $H_{((Y)s_u)s_u}(((X)s_u)s_u) = (H_{(Y)s_u}((X)s_u))s_u$. But this is equivalent to $H_{(Y)s_u}((X)s_u) = (H_Y(X))s_u$. □

Lemma 20.1.9 Assume X,Y are as in Lemma 20.1.1. Suppose for some $u \in [n]$, either $u \in X$, $u+1 \in H_Y(X)$, $u+2 \in Y - H_Y(X)$, or $u+1 \in X$, $u+2 \in H_Y(X)$, $u \in Y - H_Y(X)$. Then $H_{(Y)s_u}((X)s_u) = (H_Y(X))s_{u+1}$.

Proof We may assume $|X| < |Y|$. For otherwise, the result is obvious. Since $|H_{(Y)s_u}((X)s_u)| = |X| = |(H_Y(X))s_{u+1}|$, it is enough to show that $(H_Y(X))s_{u+1} \subseteq H_{(Y)s_u}((X)s_u)$.

First assume $u \in X$, $u+1 \in H_Y(X)$, $u+2 \in Y - H_Y(X)$. Then $u+2 \in (H_Y(X))s_{u+1}$ but $u+1 \notin (H_Y(X))s_{u+1}$. Since $|(Y)s_u|_{u+1\ u+2}| = 1 = |(X)s_u|_{u+1\ u+2}|$, we have $u+2 \in H_{(Y)s_u}((X)s_u)$. Now suppose $y \in (H_Y(X))s_{u+1} - \{u+1, u+2\}$. Then $y \in H_Y(X)$. We know $y \neq u$ since $u \in X$. By Lemma 20.1.1, there exists $x \in X$ with $|Y|_{xy}| < |X|_{xy}|$. If $x \neq u$ then one of the following cases must occur.

So $|(Y)s_u|_{xy}| = |Y|_{xy}| < |X|_{xy}| = |(X)s_u|_{xy}|$, $y \in H_{(Y)s_u}((X)s_u)$. If $x = u$ then there must exist $z \in X$ on $x\,y$ with $|X|_{zy}| = |X|_{uy}| -1$. So $|(Y)s_u|_{zy}| = |Y|_{zy}| < |Y|_{uy}| -2 < |X|_{uy}| -2 < |X|_{zy}| = |(X)s_u|_{zy}|$. This implies $y \in H_{(Y)s_u}((X)s_u)$ also. Therefore we have shown $(H_Y(X))s_{u+1} \subseteq H_{(Y)s_u}((X)s_u)$ and hence $(H_Y(X))s_{u+1} = H_{(Y)s_u}((X)s_u)$.

Next assume $u+1 \in X$, $u+2 \in H_Y(X)$, $u \in Y - H_Y(X)$. We shall show $u \in (X)s_u$, $u+1 \in H_{(Y)s_u}((X)s_u)$ and $u+2 \in (Y)s_u - H_{(Y)s_u}((X)s_u)$. Once we have done this, we get $H_{((Y)s_u)s_u}(((X)s_u)s_u) = (H_{(Y)s_u}((X)s_u))s_{u+1}$ by the above result. But this is just $H_{(Y)s_u}((X)s_u) = (H_Y(X))s_{u+1}$. Our proof is complete.

Now $u \in (X)s_u$ follows from $u+1 \in X$ and $u+1 \in H_{(Y)s_u}((X)s_u)$ from $|(Y)s_u|_{u\ u+1}| = 1 = |(X)s_u|_{u\ u+1}|$. We also have $u+2 \in H_Y(X) \subseteq Y$ and then $u+2 \in (Y)s_u$. Suppose $u+2 \in H_{(Y)s_u}((X)s_u)$. Then by Lemma 20.1.1, there exists $x \in (X)s_u$ satisfying $|(Y)s_u|_{x\ u+2}| < |(X)s_u|_{x\ u+2}|$. We see $x \neq u$ since $|(Y)s_u|_{u\ u+2}| = 2 > 1 = |(X)s_u|_{u\ u+2}|$. Thus $|Y|_{xu}| = |Y|_{x\ u+2}| -1 = |(Y)s_u|_{x\ u+2}| -1 < |(X)s_u|_{x\ u+2}| -1 = |X|_{x\ u+2}| -1 = |X|_{xu}|$. So $u \in H_Y(X)$. This leads to a contradiction. Then we have $u+2 \in (Y)s_u - H_{(Y)s_u}((X)s_u)$. Our goal is reached. □

§20.2 THE INSERTING ALGORITHM ON THE SET C

In this section, we shall construct the inserting algorithm on C and also give criteria for elements of C to belong to the same left cell.

Lemma 20.2.1 Assume $T = (T_1,\dots,T_r) \in C$ with $t \in T_\alpha$, $t+1 \in T_\beta$, $t+2 \in T_\gamma$ for some $t \in [n]$. Then $T \in \bar{\mathcal{D}}_R(s_t)$ if and only if one of the following cases happens:

(i) $\alpha < \gamma < \beta$. (ii) $\gamma < \alpha < \beta$. (iii) $\beta < \alpha < \gamma$. (iv) $\beta < \gamma < \alpha$.

When $T \in \bar{\mathcal{D}}_R(s_t)$, let $T' = T^*$ in $\bar{\mathcal{D}}_R(s_t)$. Then

$$T' = \begin{cases} (T)s_t & \text{in cases (i), (iv)} \\ (T)s_{t+1} & \text{in cases (ii), (iii)} \end{cases}$$

Proof This can be checked directly by definition of the set $\bar{\mathcal{D}}_R(s_t)$ and the right star operations on it. □

Proposition 20.2.2 Assume $T = (T_1,\dots,T_t) \in C$ with $|T_\ell| < |T_{\ell+1}|$ for some ℓ, $1 < \ell < t$. Let

$$T' = (T_1,\dots,T_{\ell-1},T_\ell \cup (T_{\ell+1} - H_{T_{\ell+1}}(T_\ell)),\ H_{T_{\ell+1}}(T_\ell),\ T_{\ell+2},\dots,T_t) \qquad (*)$$

Then $T \underset{L}{\sim} T'$.

Proof Let $f(T) = \alpha = (\alpha_1,\dots,\alpha_t)$. Then $f(T') = \alpha' = (\alpha_1,\dots,\alpha_{\ell-1},\alpha_{\ell+1},\alpha_\ell,\alpha_{\ell+2},\dots,\alpha_t)$. So there exists some $\lambda = \{\lambda_1 > \dots > \lambda_r\} \in \Lambda_n$ satisfying $\hat{\sigma}(T) = \hat{\sigma}(T') = \lambda$. Let $\mu = \{\mu_1 > \dots > \mu_m\}$ be the dual of λ. Then $m = t$. We know that $f^{-1}(\alpha)$ and $f^{-1}(\alpha')$ are A_n-orbits with $|f^{-1}(\alpha)| = \frac{n!}{\prod_{j=1}^{t} \mu_j!} = |f^{-1}(\alpha')|$. We also know that $f^{-1}(\alpha)$ and $f^{-1}(\alpha')$ are all invariant and transitive under the right action of e_0 together with the right star operations and that there is a unique dominant element in each A_n-orbit of C.

We shall show our result by the following process. First assume that T is dominant. Then by Lemma 19.5.1, to show $T \underset{L}{\sim} T'$, it suffices to show T' also dominant.

After doing that, we assume that we have shown our result for some pair T, T' with $T \in f^{-1}(\alpha)$, $T' \in f^{-1}(\alpha')$ and T' is related with T as in (*). Then we shall show
(i) $(T')e_0$ is also related with $(T)e_0$ as in (*) by replacing T, T' by $(T)e_0$, $(T')e_0$.
(ii) We know that $T \in \bar{\mathcal{D}}_R(s_u)$ if and only if $T' \in \bar{\mathcal{D}}_R(s_u)$ by the assumption $T \underset{L}{\sim} T'$.

When $T, T' \in \bar{\mathcal{D}}_R(s_u)$, let $Y = T^*$ and $Y' = T'^*$. Then Y' is related with Y as in (*) by replacing T, T' by Y, Y'.

Having shown these facts and noting that $(T)e_0 \underset{L}{\sim} (T')e_0$ and $Y \underset{L}{\sim} Y'$ by Lemmas 19.4.4 and 19.4.5, we have shown our result for any $T \in f^{-1}(\alpha)$ and the corresponding $T' \in f^{-1}(\alpha')$ as in (*). So our result follows.

(I) First we assume that T is dominant. We shall show that T' is also dominant. We see for any h, $1 \leq h \leq t$,

$$T_h = \{\sum_{i=h+1}^{t} \alpha_i + j \mid 1 \leq j \leq \alpha_h\}$$

In particular, $T_\ell = \{\sum_{i=\ell+1}^{t} \alpha_i + j \mid 1 \leq j \leq \alpha_\ell\}$ and $T_{\ell+1} = \{\sum_{i=\ell+2}^{t} \alpha_i + j \mid 1 \leq j \leq \alpha_{\ell+1}\}$.

So by the definition, we have $T'_{\ell+1} = H_{T_{\ell+1}}(T_\ell) = \{\sum_{i=\ell+2}^{t} \alpha_i + j \mid 1 \leq j \leq \alpha_\ell\}$ and

$T'_\ell = T_\ell \cup (T_{\ell+1} - H_{T_{\ell+1}}(T_\ell)) = \{\sum_{i=\ell+2}^{t} \alpha_i + \alpha_\ell + j \mid 1 \leq j \leq \alpha_{\ell+1}\}$.

So T' is also dominant.

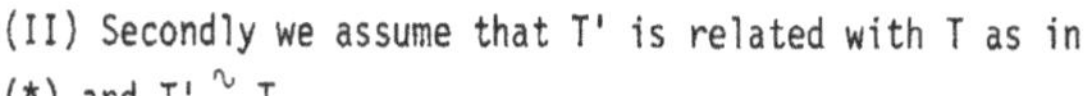

(II) Secondly we assume that T' is related with T as in (*) and $T' \underset{L}{\sim} T$.

(II(i) We have $(T)e_0 = ((T_1)e_0, \ldots, (T_t)e_0)$ and

$$(T')e_0 = ((T_1)e_0, \ldots, (T_{\ell-1})e_0, (T_\ell \cup (T_{\ell+1} - H_{T_{\ell+1}}(T_\ell)))e_0, (H_{T_{\ell+1}}(T_\ell))e_0, (T_{\ell+2})e_0, \ldots,$$

$$T_t)e_0) = ((T_1)e_0, \ldots, (T_{\ell-1})e_0, (T_\ell)e_0 \cup ((T_{\ell+1})e_0 - H_{(T_{\ell+1})e_0}((T_\ell)e_0)),$$

$$H_{(T_{\ell+1})e_0}((T_\ell)e_0), (T_{\ell+2})e_0, \ldots, (T_t)e_0)$$

by Lemma 20.1.5. So $(T')e_0$ is related with $(T)e_0$ as in (*) by substituting $(T)e_0$, $(T')e_0$ for T, T' respectively.

II(ii) Assume $T, T' \in \bar{\mathcal{D}}_R(s_u)$. Let $Y = T^*$ and $Y' = T'^*$. We shall show that Y' is related with Y as in (*) by substituting Y, Y' for T, T'. By symmetry, we may assume $Y = (T)s_u$. Then we need only show $Y'_{\ell+1} = H_{(T_{\ell+1})s_u}((T_\ell)s_u)$, where $Y' = (Y'_1, \ldots, Y'_t)$. Let $\varepsilon_{\ell,u}(T) = |\{u, u+1, u+2\} \cap (T_\ell \cup T_{\ell+1})|$. Our result is obviously true when $\varepsilon_{\ell,u}(T) \leq 1$.

Now assume $\varepsilon_{\ell,u}(T) = 2$. Then by Lemma 20.2.1, one of the following cases must happen: for some α, β, $1 \leq \alpha < \ell$, $\ell+1 < \beta \leq t$,

(1) $u \in T_\alpha$, $u+2 \in T_\ell$, $u+1 \in T_{\ell+1}$. (2) $u \in T_\alpha$, either $u+1, u+2 \in T_\ell$ or $u+1, u+2 \in T_{\ell+1}$.

(3) $u \in T_\ell$, $u+2 \in T_{\ell+1}$, $u+1 \in T_\ell$ (4) $u+1 \in T_\alpha$, $u+2 \in T_\ell$, $u \in T_{\ell+1}$

(5) $u+1 \in T_\alpha$, either $u+2, u \in T_\ell$ or $u+2, u \in T_{\ell+1}$.

We can check case by case that $Y' = (T')s_u$ and $Y'_{\ell+1} = (H_{T_{\ell+1}}(T_\ell))s_u = H_{(T_{\ell+1})s_u}((T_\ell)s_u)$

by Lemma 20.1.7.

Finally, assume $\varepsilon_{\ell,u}(T) = 3$. Then by Lemma 20.2.1 one of the following cases must occur:

(1)' $u \in T_\ell$, $u+1, u+2 \in T_{\ell+1}$ (2)' $u+1 \in T_\ell$, $u, u+2 \in T_{\ell+1}$.

In case (1)', when $u+1, u+2 \in H_{T_{\ell+1}}(T_\ell)$, we have $u \in T'_\ell$, $u+1, u+2 \in T'_{\ell+1}$. So $Y' = (T')s_u$. By Lemma 20.1.8, $Y'_{\ell+1} = (H_{T_{\ell+1}}(T_\ell))s_u = H_{(T_{\ell+1})s_u}((T_\ell)s_u)$. When $u+1 \in H_{T_{\ell+1}}(T_\ell)$, $u+2 \in T_{\ell+1} - H_{T_{\ell+1}}(T_\ell)$, we have $u, u+2 \in T'_\ell$, $u+1 \in T'_{\ell+1}$. So $Y' = (T')s_{u+1}$. By Lemma 20.1.9, $Y'_{\ell+1} = (H_{T_{\ell+1}}(T_\ell))s_{u+1} = H_{(T_{\ell+1})s_u}((T_\ell)s_u)$. Since $T' \in \bar{\mathcal{D}}_R(s_u)$, no other case could happen in case (1)'.

In case (2)', when $u, u+2 \in H_{T_{\ell+1}}(T_\ell)$, we have $u+1 \in T'_\ell$, $u,u+2 \in T'_{\ell+1}$. So $Y' = (T')s_u$. By Lemma 20.1.8, $Y'_{\ell+1} = (H_{T_{\ell+1}}(T_\ell))s_u = H_{(T_{\ell+1})s_u}((T_\ell)s_u)$. When $u+2 \in H_{T_{\ell+1}}(T_\ell)$, $u \in T_{\ell+1} - H_{T_{\ell+1}}(T_\ell)$, we have $u, u+1 \in T'_\ell$, $u+2 \in T'_{\ell+1}$. So $Y' = (T')s_{u+1}$. By Lemma 20.1.9, $Y'_{\ell+1} = (H_{T_{\ell+1}}(T_\ell))s_{u+1} = H_{(T_{\ell+1})s_u}((T_\ell)s_u)$. Since $T' \in \bar{\mathcal{D}}_R(s_u)$, the above are the only cases which could happen in case (2)'.

Combining (I), (II)(i), (II)(ii) together, our result is shown. □

<u>Corollary 20.2.3</u> <u>Assume $T = (T_1,\ldots,T_t) \in C$ with $|T_\ell| > |T_{\ell+1}|$ for some ℓ, $1 \leq \ell < t$</u>

<u>Let</u>

$$T' = (T_1,\ldots,T_{\ell-1}, L_{T_\ell}(T_{\ell+1}), T_{\ell+1} \cup (T_\ell - L_{T_\ell}(T_{\ell+1})), T_{\ell+2},\ldots,T_t)$$

<u>Then $T \underset{L}{\sim} T'$.</u>

<u>Proof</u> By Propositions 20.1.4 and 20.2.2. □

Now we can construct the inserting algorithm for passing from an arbitrary element of C to the element of $\hat{C}$ in the same left cell of S by Proposition 20.2.2.

Assume $X = (X_t, X_{t-1},\ldots,X_1) \in C$.

<u>Inserting algorithm on X</u>

(i) Let $Y^{(1)} = X_1$

(ii) Recursively, assume that we have inserted $X_1, X_2,\ldots,X_m$ into m ordered columns $Y^{(m)} = (Y_1, Y_2,\ldots,Y_m)$ such that $|Y_1| \geq \ldots \geq |Y_m|$, and $|Y_1|,\ldots,|Y_m|$ is a permutation of $|X_1|,\ldots,|X_m|$ for some m, $1 \leq m < t$. Now we shall insert the set X_{m+1} into $Y^{(m)}$ to get $Y^{(m+1)} = (Y'_1, Y'_2,\ldots,Y'_{m+1})$ as follows: (a) if $|X_{m+1}| \geq |Y_1|$ then let $Y^{(m+1)} = (X_{m+1}, Y_1,\ldots,Y_m)$; (b) if $|X_{m+1}| < |Y_1|$ then let $Y'_1 = X_{m+1} \cup (Y_1 - H_{Y_1}(X_{m+1}))$ and insert $H_{Y_1}(X_{m+1})$ into $(Y_2, Y_3,\ldots,Y_m)$ to obtain $(Y'_2, Y'_3,\ldots,Y'_{m+1})$.

(iii) Let $Y = Y^{(t)}$. Then Y is the required element of $\hat{C}$.

By the above algorithm, we get a criterion for elements of C to belong to the same left cell of S.

Proposition 20.2.4 Let $X, X' \in C$. Let $Y, Y' \in \hat{C}$ be the resulting elements under the inserting algorithm on X, X', respectively. Then $X \underset{L}{\sim} X'$ if and only if $Y = Y'$.

Proof This follows by Proposition 20.2.2 and Lemma 19.4.6. □

Sometimes the following criterion is more useful than that in the above proposition.

Let $X, Y \in C$. Assume $f(X) = \alpha = (\alpha_1, \ldots, \alpha_t)$ and $f(Y) = \beta = (\beta_1, \ldots, \beta_u)$. Then a necessary condition for $X \underset{L}{\sim} Y$ is that $t = u$ and $\beta_1, \ldots, \beta_u$ is a permutation of $\alpha_1, \ldots, \alpha_t$. So we may assume that we are in that case. We transform α to β by a succession of transpositions on pairs of adjacent terms. Suppose that $\varepsilon_0 = \alpha, \varepsilon_1, \ldots, \varepsilon_m = \beta$ is a sequence such that for every j, $1 \leqslant j \leqslant m$, ε_j is obtained from ε_{j-1} by transposing the ℓ_j-th and the (ℓ_j+1)-th terms, $1 \leqslant \ell_j < t$. Then we get the corresponding sequence $Z_0 = X, Z_1, \ldots, Z_m = Z$ with $f(Z_h) = \varepsilon_h$, $0 \leqslant h \leqslant m$, and $Z_{j-1} = (Z_{j-1,1}, \ldots, Z_{j-1,t})$ such that for every j, $1 \leqslant j \leqslant m$,

$$Z_j = \begin{cases} (Z_{j-1,1}, \ldots, Z_{j-1,\ell_j-1}, Z_{j-1,\ell_j} \cup (Z_{j-1,\ell_j+1} - {}^{H}Z_{j-1,\ell_j+1}(Z_{j-1,\ell_j})), \\ \qquad {}^{H}Z_{j-1,\ell_j+1}(Z_{j-1,\ell_j}), Z_{j-1,\ell_j+2}, \ldots, Z_{j-1,t}) \text{ if } \varepsilon_{j-1,\ell_j} < \varepsilon_{j-1,\ell_j+1} \\ (Z_{j-1,1}, \ldots, Z_{j-1,\ell_j-1}, {}^{L}Z_{j-1,\ell_j}(Z_{j-1,\ell_j+1}), Z_{j-1,\ell_j+1} \cup (Z_{j-1,\ell_j} - {}^{L}Z_{j-1,\ell_j} \\ \qquad (Z_{j-1,\ell_j+1})), Z_{j-1,\ell_j+2}, \ldots, Z_{j-1,t}) \text{ if } \varepsilon_{j-1,\ell_j} > \varepsilon_{j-1,\ell_j+1} \end{cases}$$

Proposition 20.2.5 Let $X, Y, Z \in C$ be as above. Then $X \underset{L}{\sim} Y$ if and only if $Y = Z$.

Proof By Proposition 20.2.2 and Corollary 20.2.3, we have $X \underset{L}{\sim} Z$. So by Proposition 19.4.7, we get our result. □

Example We apply the inserting algorithm on

$X = (\{1,4\}, \{2,5,8\}, \{6\}, \{3,7,9\})$ as follows.

$(\{3,7,9\}) \to (\{3,6,9\}, \{7\}) \to (\{2,5,8\}, \{3,6,9\}, \{7\})$

$\to (\{1,4,8\}, \{2,5,9\}, \{3,6\}, \{7\}) = Y$

We also apply the inserting algorithm on

$$X' = (\{1,4,8\}, \{2,5\}, \{3,7,9\}, \{6\}):$$

$(\{6\}) \rightarrow (\{3,7,9\}, \{6\}) \rightarrow (\{2,5,9\}, \{3,7\}, \{6\})$

$\rightarrow (\{1,4,8\}, \{2,5,9\}, \{3,7\}, \{6\}) = Y'$.

So this implies $X \not\sim_L X'$ from $Y \neq Y'$. The conclusion $X \not\sim_L X'$ can also follow by the second criterion since

$X' \sim_L (\{1,4\}, \{2,5,8\}, \{3,7,9\}, \{6\})$

$\sim_L (\{1,4\}, \{2,5,8\}, \{3\}, \{6,7,9\}) \neq X$.

CHAPTER 21 : THE RESTRICTION OF THE MAP $\hat{T}$ ON P_n

Recall that we defined the map $\hat{T}:A_n \to \hat{C}$ in §19.4. By Proposition 19.4.8 we see that $\hat{T}$ satisfies the first two conditions stated at the beginning of Chapter 19. In the present chapter we shall show that $\hat{T}$ satisfies the last condition as well, i.e. when restricted to P_n, $\hat{T}$ coincides with the Robinson-Schensted map defined in §1.7(iv). We shall give a criterion to tell when the intersection of a left cell of A_n with P_n is non-empty.

§21.1 SEPARATED ENTRY SETS AND THE OPERATION $w \xrightarrow[*(i+1,\alpha,m)]{} y$

To show that $\hat{T}$ satisfies the third required property, we must reformulate the Robinson-Schensted map in terms of affine matrices. First we need to introduce separated entry sets of an element of A_n.

Recall all the terminology given in §4.4(v), (viii) and (ix). We say that an entry set E of $w \in A_n$ is separated if for any $e \in E$ and $e' \notin E$, the entry set $\{e,e'\}$ is a chain.

Let $D = \{e_w(i_t,t) | t \in [n]\}$ be an entry set of w. Let $D_I = \{e_w(i_t,t) | t \in I\}$ for any $I \subseteq [n]$. Then we see from Proposition 6.2.3 that D_I is an antichain of w if and only if $k^w_{\alpha\beta} = 0$ for any $\alpha, \beta \in I$, where $A_w = \bigcap_{1\leq i<j\leq n} H^1_{ij,k^w_{ij}}$. Again by Proposition 6.2.3, we see that an entry set D_I is separated if and only if $k^w_{\alpha\beta} \neq 0$ for any $\alpha,\beta \in [n]$ with $\alpha \neq \beta$ and $|\{\alpha,\beta\} \cap I| = 1$. So by the definition of a Coxeter sign type (see §19.1), we have that for any $w \in A_n$, w determines a Coxeter sign type if and only if all maximum antichains of w are separated.

Next we wish to introduce the operation $w \xrightarrow[*(i+1,\alpha,m)]{} y$ on an element w of A_n. Assume that $w \in A_n$ has an entry set $E = \{e_w(i+t,j_t(w)) | t \in [n]\}$ for some $i \in \mathbf{Z}$. Assume that $E_1 = \{e_w(i+t,j_t(w)) | 1 \leq t \leq \alpha\}$ is an IC block of w for some α, $1 \leq \alpha < n$. If $j_{\alpha+1}(w) < j_\alpha(w)$ then there exists a sequence $w_0 = w, w_1, \ldots, w_{\alpha-1} = w'$ such that for every u, $0 \leq u < \alpha-1$, $w_{u+1} = {}^*w_u$ in $\mathcal{D}_L(s_{i+\alpha-u-1})$. If $j_{\alpha+1}(w) > j_\alpha(w)$ then let $w' = w$. In both cases, we write $w \xrightarrow[*(i+1,\alpha)]{} w'$. By this definition, we see that the operation $w \xrightarrow[*(i+1,\alpha)]{} w'$ is a succession of left star operations. In particular, in the case $j_{\alpha+1}(w) < j_\alpha(w)$, let $\{e_{w'}(i+t,j_t(w')) | t \in [n]\}$ be an entry set of w'. Then

$$j_t(w') = \begin{cases} j_t(w) & \text{if } \alpha+2 \leq t \leq n \\ j_{t-1}(w) & \text{if } 1 < t \leq \alpha+1,\ t \neq v+1 \\ j_v(w) & \text{if } t = 1 \\ j_{\alpha+1}(w) & \text{if } t = v+1 \end{cases}$$

where v is the smallest integer satisfying $1 \leq v \leq \alpha$ and $j_v(w) > j_{\alpha+1}(w)$.

For example, when $n > 5$, let $w \in A_n$ have the form

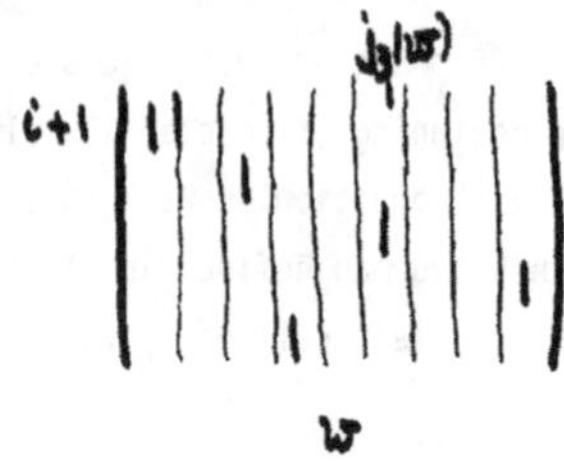

Then we have a sequence $w_0 = w$, w_1, w_2, $w_3 = w'$, where $w_1 = {}^*w_0$ in $\mathcal{D}_L(s_{i+3})$, $w_2 = {}^*w_1$ in $\mathcal{D}_L(s_{i+2})$, $w_3 = {}^*w_2$ in $\mathcal{D}_L(s_{i+1})$. Clearly, w' with $w \xrightarrow[*(i+1,4)]{} w'$ has the form

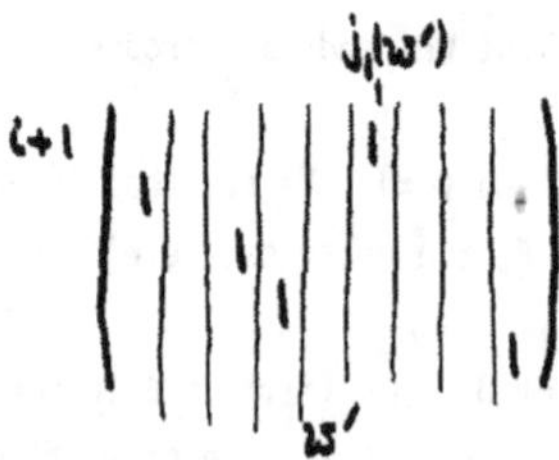

where $j_1(w') = j_3(w)$.

Let $w \in A_n$ and its entry sets E, E_1 be as above. For any integer $m > 0$, we write $w \xrightarrow[*(i+1,\alpha,m)]{} y$ for some $y \in A_n$ if w, y satisfy one of the following relations.

(i) $\alpha > 1$ and there exists a sequence $y_0 = w, y_1, \ldots, y_m = y$ such that for every u, $1 \leq u \leq m$, $y_{u-1} \xrightarrow[*(i+u-\beta_u,\alpha+\beta_u)]{} y_u$, where $\beta_u = \#\{v \mid 1 \leq v < u,\ y_{v-1} = y_v\}$.

(ii) $\alpha = 1$. Let $V_m = \{v \mid 1 \leq v < m,\ j_v(w) < j_{v+1}(w)\}$. Then when $V_m = \emptyset$, we have $y = w$. When $V_m \neq \emptyset$, let $a_m = \min\{v \mid v \in V_m\}$. Then y satisfies $w \xrightarrow[*(i+a_m,2,m-a_m)]{} y$.

Warning The reader should take care to distinguish between the operation $w \xrightarrow[*(i+1,\alpha,m)]{} y$ just defined and the operation $w \xrightarrow{*(i+1,\alpha,m)} y$ defined in Chapter 8. The former involves increasing chains whereas the latter involves descending chains.

Examples 21.1.1

(1) When $n > 6$, let $w \in A_n$ have the form

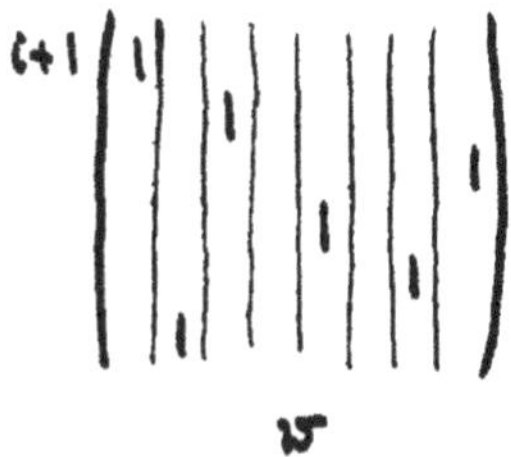

Then we have a sequence $y_0 = w, y_1, y_2, y_3 = y$, where $y_0 \xrightarrow[*(i+1,3)]{} y_1$, $y_1 \xrightarrow[*(i+2,3)]{} y_2$ $(y_1 = y_2)$, $y_2 \xrightarrow[*(i+2,4)]{} y_3$, $\beta_1 = \beta_2 = 0$ and $\beta_3 = 1$. So $w \xrightarrow[*(i+1,3,3)]{} y$ and y has the form

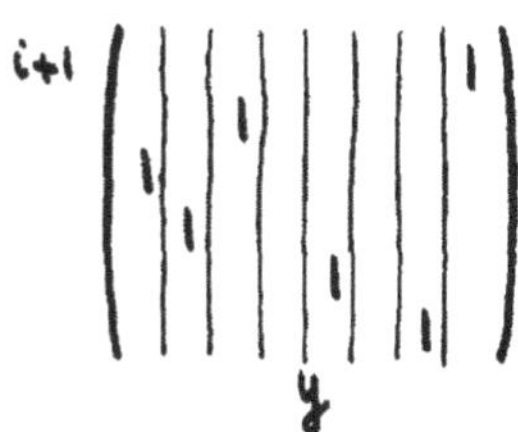

Note that in this case we can also write $w \xrightarrow[*(i+1,2,4)]{} y$. But then the corresponding sequence becomes $x_0 = w, x_1, x_2, x_3, x_4 = y$, where $x_0 \xrightarrow[*(i+1,2)]{} x_1$ $(x_0 = x_1)$, $x_1 \xrightarrow[*(i+1,3)]{} x_2$, $x_2 \xrightarrow[*(i+2,3)]{} x_3$ $(x_2 = x_3)$, $x_3 \xrightarrow[*(i+2,4)]{} x_4$, $\beta_1 = 0$, $\beta_2 = \beta_3 = 1$ and $\beta_4 = 2$.

(2) When $n > 4$, let $w \in A_n$ have the form

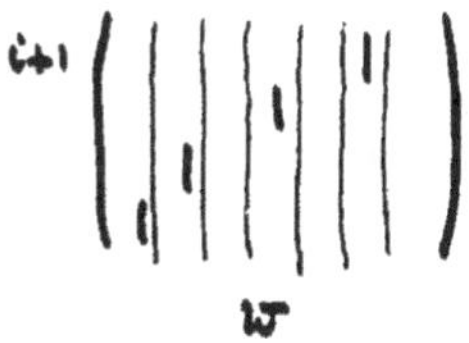

Then we can write $w \xrightarrow[*(i+1,1,3)]{} y$ $(y = w)$ since $V_4 = \emptyset$.

(3) When $n > 5$, let $w \in A_n$ have the form

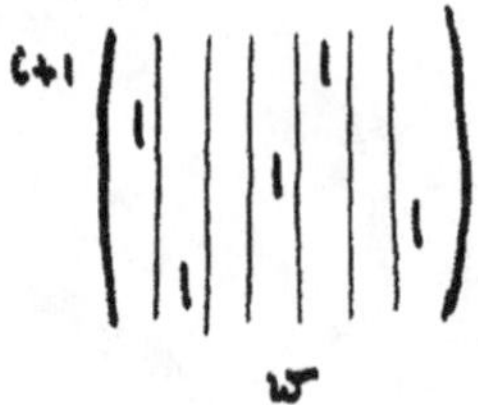

Then $V_5 = \{2,3\} \neq \emptyset$ and $a_5 = 2$. So the expression $w \xrightarrow[*(i+1,1,4)]{} y$ is equivalent to $w \xrightarrow[*(i+2,2,2)]{} y$. We see that y has the form

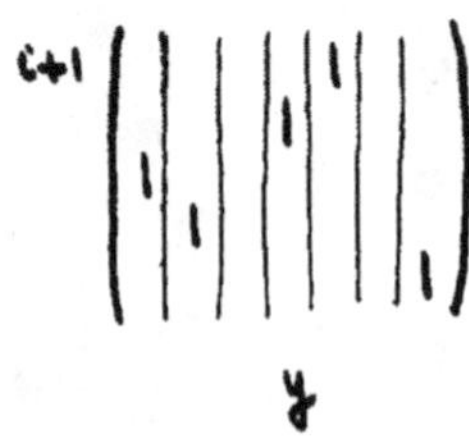

Remark 21.1.2 (i) In contrast to what we have seen for the operation $w \xrightarrow{*(i+1,\alpha,m)} y$, the element y, satisfying $w \xrightarrow[*(i+1,\alpha,m)]{} y$ for $w \in A_n$ and $\alpha < n$, always exists, provided that w has an IC block $\{e_w(i+t,j_t) \mid 1 \leq t \leq \alpha\}$.

(ii) Suppose that w has an IC block $\{e_w(i+t,j_t(w)) \mid 1 \leq t \leq k\}$ with $\alpha \leq k \leq n$ and suppose that $w \xrightarrow[*(i+1,\alpha,m)]{} y$ with $\alpha + m \geq k$. Then y has an IC block $\{e_y(i+t,j_t(w)) \mid \alpha + m-k < t \leq \alpha + m\}$. So we can say that the size of an IC block is increasing (in the weak sense) under the operation $w \xrightarrow[*(i+1,\alpha,m)]{} y$.

Lemma 21.1.3 Suppose that $w \xrightarrow[*(i+1,\alpha,m)]{} y$ for some $i,\alpha, m \in \mathbb{Z}$ with $\alpha, m > 0$ and $\alpha + m \leq n$. Let η be the largest number with $1 \leq \eta \leq \alpha + m$ so that $\{e_y(i+t,j_t(y)) \mid \alpha+m - \eta < t \leq \alpha + m\}$ is an IC block of y. Then η is the maximum size of the increasing chains of w in the block $\{e_w(i+t,j_t(w)) \mid 1 \leq t \leq \alpha + m\}$.

Proof We can argue by induction on $m \geq 1$. The result is obvious when $m = 1$. Now assume $m > 1$. When $j_{\alpha+1}(w) > j_\alpha(w)$, we can rewrite $w \xrightarrow[*(i+1,\alpha+1,m-1)]{} y$. So by inductive hypothesis, our result follows. When $j_{\alpha+1}(w) < j_\alpha(w)$, let x satisfy $w \xrightarrow[*(i+1,\alpha)]{} x$. Then $x \xrightarrow[*(i+2,\alpha,m-1)]{} y$. So by inductive hypothesis, we need only show that the maximum size of the increasing chains of w in the block $A(w) = \{e_w(i+t, j_t(w)) | 1 \leq t \leq \alpha + m\}$ is equal to that of x in the block $A'(x) = \{e_x(i+t, j_t(x)) | 2 \leq t \leq \alpha + m\}$. But this will be deduced from the following two statements.

(i) The maximum size of the increasing chains of w in A(w) is equal to that of x in the block $A(x) = \{e_x(i+t, j_t(x)) | 1 \leq t \leq \alpha + m\}$.

(ii) The maximum size of the increasing chains of x in A(x) is equal to that in A'(x).

(i) (resp. (ii)) can be shown by using the same technique as that in the proof of Lemma 5.8 (resp. Lemma 8.2.1) by replacing the term "a descending chain" by "an increasing chain". So our proof is complete. □

§21.2 REFORMULATION OF THE ROBINSON-SCHENSTED ALGORITHM

Recall that in §1.7(iv) we defined the Robinson-Schensted map $w \to (P(w), P(w^{-1}))$ from the symmetric group S_n to the set of pairs (P,Q) of standard Young tableaux of the same shape and of size n. By identifying S_n with the standard parabolic subgroup P_n of A_n, we now shall reformulate this map in terms of affine matrices.

For $w \in P_n$, let $P_\ell(w)$, $\ell \geq 1$, be the set of numbers in the ℓ-th row of P(w). Let $\delta(w) = \{\mu_1 \geq \dots \geq \mu_r\}$ (see Definition 5.14). Then it is well known that $\mu_\ell = |P_\ell(w)|$ for all ℓ, $1 \leq \ell \leq r$.

Fix $w \in P_n$. Assume that w has an entry set $\{e_w(t, j_t) | 1 \leq t \leq n\}$ with $\sigma(w) = \lambda$ and $\delta(w) = \{\mu_1 \geq \dots \geq \mu_r\}$.

(i) Let $w_1 = w$ and call $A_1(w_1) = \{e_{w_1}(1, j_1)\}$

(ii) If $j_2 > j_1$ then let $w_2 = w_1$ and call $A_1(w_2) = \{e_{w_2}(1, j_1), e_{w_2}(2, j_2)\}$.

(iii) If $j_2 < j_1$ then let $w_2 = w_1$ and call $A_1(w_2) = \{e_{w_2}(2, j_2)\}$, $A_2(w_2) = \{e_{w_2}(1, j_1)\}$.

(iv) Now assume that we have got w_ℓ from w_1 and have defined $A_1(w_\ell), \dots, A_k(w_\ell)$ with $A_\alpha(w_\ell) = \{e_{w_\ell}(t, j_t^{(\ell)}) | i_\alpha \leq t < i_{\alpha-1}\}$ all IC blocks for $1 \leq \alpha \leq k$, where $1 = i_k < i_{k-1} < \dots < i_0 = \ell+1$, and $j_1^{(\ell)}, \dots, j_\ell^{(\ell)}$ is a permutation of $j_1, \dots, j_\ell$. We also assume that $\{e_{w_\ell}(t, j_t) | \ell+1 \leq t \leq n\}$ is an entry set of w_ℓ. Then

(a) if $j_{\ell+1} > j_\ell^{(\ell)}$ then let $w_{\ell+1} = w_\ell$ and call

$$A_1(w_{\ell+1}) = \{e_{w_{\ell+1}}(t, j_t^{(\ell)}),\ i_1 \leq t \leq \ell;\ e_{w_{\ell+1}}(\ell+1, j_{\ell+1})\}$$

$$A_\alpha(w_{\ell+1}) = \{e_{w_{\ell+1}}(t, j_t^{(\ell)}),\ i_\alpha \leq t < i_{\alpha-1}\ ,\ 1 < \alpha \leq k.$$

(b) if $j_{\ell+1} < j_\ell^{(\ell)}$ then there exists a sequence $y_0 = w_\ell, y_1, \dots, y_c$ for some c, $1 \leq c \leq k$, such that for every m, $1 \leq m \leq c$, $y_{m-1} \xrightarrow{*(i_m, i_{m-1}-i_m)} y_m$ with $y_{m-1} \neq y_m$. We can take such an integer c as large as possible and then let $w_{\ell+1} = y_c$. Assume that $w_{\ell+1}$ has an entry set $\{e_{w_{\ell+1}}(t, j_t^{(\ell+1)}) | 1 \leq t \leq n\}$. Then let

$$A_\alpha(w_{\ell+1}) = \begin{cases} \{e_{w_{\ell+1}}(t,j_t^{(\ell+1)}) \mid i_\alpha < t < i_{\alpha-1}\} & \text{if } c+2 < \alpha < k \\ \{e_{w_{\ell+1}}(t,j_t^{(\ell+1)}) \mid i_\alpha < t < i_{\alpha-1}\} & \text{if } 1 < \alpha < c \\ \{e_{w_{\ell+1}}(t,j_t^{(\ell+1)}) \mid i_{c+1} < t < i_c\} & \text{if } \alpha = c+1. \end{cases}$$

where, when $c = k$, we assume $i_{c+1} = 1$.

We see that all $A_\alpha(w_{\ell+1})$'s are IC blocks, that $j_1^{(\ell+1)},\ldots,j_{\ell+1}^{(\ell+1)}$ is a permutation of $j_1,\ldots,j_{\ell+1}$, and that $j_t^{(\ell+1)} = j_t$ for any t, $\ell+2 < t < n$. So by a recursive process, we finally get $y = w_n$ which has the MIC form $(A_m, A_{m-1},\ldots,A_1)$ at o (see §4.4(xi)) for some $m > 1$. We see that $y \underset{P_L}{\sim} w$ and y has entry sets

$$A_\alpha(y) = \{e_y(t,j_t^{(n)}) \mid i'_\alpha < t < i'_{\alpha-1}, j_{i'_\alpha}^{(n)} < j_{i'_\alpha+1}^{(n)} < \ldots < j_{i'_{\alpha-1}-1}^{(n)}\},\ 1 < \alpha < m,$$

where $1 = i'_m < i'_{m-1} < \ldots < i'_0 = n+1$ and $j_1^{(n)},\ldots,j_n^{(n)}$ is a permutation of $j_1,\ldots,j_n$ and hence a permutation of $1,2,\ldots,n$. We also see that $P_\alpha(w) = \{j_{i'_\alpha}^{(n)}, j_{i'_\alpha+1}^{(n)},\ldots,j_{i'_{\alpha-1}-1}^{(n)}\}$, $1 < \alpha < m$, by our algorithm and the definition of the operation $\cdot \xrightarrow[*(-,-)]{} \cdot$ (recall that $P_\alpha(w)$ is the set of numbers in the α-th row of $P(w)$). So we must have $m = r$ and $i'_\alpha - i'_{\alpha-1} = \mu_\alpha$ for all α, $1 < \alpha < r$. Therefore, $P(w)$ can be obtained from y by filling the numbers $j_{i'_\alpha}^{(n)}, j_{i'_\alpha+1}^{(n)},\ldots,j_{i'_{\alpha}-1}^{(n)}$, $1 < \alpha < m$, into the α-th row of the Young diagram F_λ from left to right.

<u>Example 21.2.1</u> Let $w \in P_{10}$ have the form

Let y be obtained from w by the above procedure. Then y has the form

§21.3 THE IMAGE OF P_n UNDER THE MAP $\hat{T}$

We shall consider $\hat{T}(w)$ for any $w \in P_n$. Let y be obtained from w by the procedure given in the last section. Then $y \underset{L}{\sim} w$. So by Proposition 19.4.8, we have $\hat{T}(y) = \hat{T}(w)$. Thus it is enough to consider all elements y of P_n which have an MIC form $(A_m, A_{m-1}, \ldots, A_1)$ at o with $\delta(y) = \{|A_1| \geqslant |A_2| \geqslant \ldots \geqslant |A_m|\}$ for some $m \geqslant 1$.

Let us first define an operation on any element of A_n.

Assume that $w \in A_n$ has a block A which consists of rows from the $(i+1)$-th to the $(i+k+1)$-th with $1 \leqslant k < n$. Let f be the $(i+k+1)$-th row of w. We write $w \xrightarrow{(i+1,k)} w'$ for $w' \in A_n$ if w' is obtained from w by permuting the row f from the bottom to the top in A.

For example, when $n \geqslant 4$, let $w \in A_n$ have the form

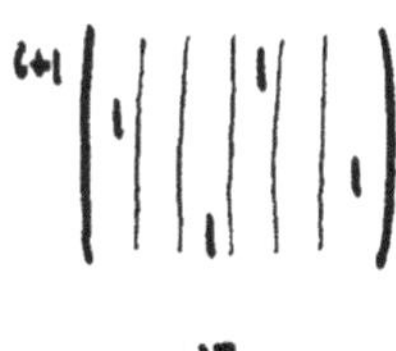

and let $y \in A_n$ satisfy $w \xrightarrow{(i+1,3)} y$. Then y has the above form.

We write $w \xrightarrow{(i+1,k,h)} y$ for $w, y \in A_n$ and $1 \leqslant k < n$, $h \geqslant 0$ if there exists a sequence of elements $w_0 = w, w_1, \ldots, w_h = y$ in A_n such that for every j, $1 \leqslant j \leqslant h$, we have $w_{j-1} \xrightarrow{(i+j,k)} w_j$.

<u>Warning</u> The reader should take care to distinguish between the operation $w \xrightarrow{(i+1,k,h)} y$ just defined and the operation $w \xrightarrow{*(i+1,k,h)} y$ defined in Chapter 8. These operations have no relation to one another.

Note that when $k+h \leqslant n$ and $k,h \geqslant 1$, the expression $w \xrightarrow{(i+1,k,h)} y$ is equivalent to

$y \xrightarrow{(i+1,h,k)} w.$

In case that $w \xrightarrow{(i+1,k)} w'$, we have $w' = s_{i+1}s_{i+2} \cdots s_{i+k}w$. So when w has an entry set $\{e_w(i+t, j_t(w)) | 1 \leq t \leq k+1\}$ with $j_t(w) < j_{k+1}(w)$ (resp. $j_t(w) > j_{k+1}(w)$) for all t, $1 \leq t \leq k$, we have $\ell(w') = \ell(w) + k$ (resp. $\ell(w') = \ell(w) - k$). In particular, by Propositions 5.15 and 6.2.3, we have $\delta(w) \geq \delta(w')$ (resp. $\delta(w) \leq \delta(w')$). When the equality holds, we have $w \underset{L}{\sim} w'$ by Proposition 5.15, Theorem 17.4 and Lemma 17.3.

Now assume that $y \in P_n$ has an MIC form $(A_m, A_{m-1}, \ldots, A_1)$ at o with $\delta(y) = \{|A_1| \geq \ldots \geq |A_m|\} = \{\mu_1 \geq \ldots \geq \mu_m\}$ for some $m \geq 1$. We define a sequence of elements $y_1 = y, y_2, \ldots, y_m = y'$ in A_n such that

$$y_{i-1} \xrightarrow{\left(n+1-\sum_{j=1}^{i-1}(i-j)\mu_j,\ \sum_{j=1}^{i-1}\mu_j,\ \sum_{j=i}^{m}\mu_j\right)} y_i \quad \text{for every } i,\ 1 < i \leq m.$$

For example, let y be the element at the end of the last section, i.e. y has an MIC form (A_3, A_2, A_1) at o.

Then the condition that $\delta(y) = \{|A_1| \geq |A_2| \geq |A_3|\} = \{5 \geq 3 \geq 2\}$ is satisfied. We have a sequence $y_1 = y$, y_2, $y_3 = y'$ such that $y_1 \xrightarrow{(6,5,5)} y_2$ and $y_2 \xrightarrow{(-2,8,2)} y_3$.

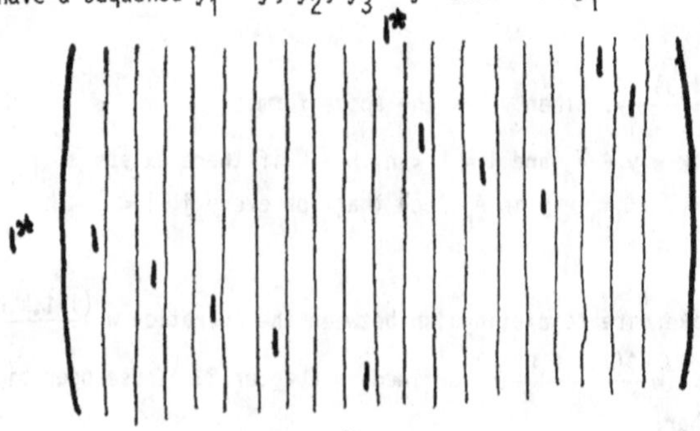

y_2

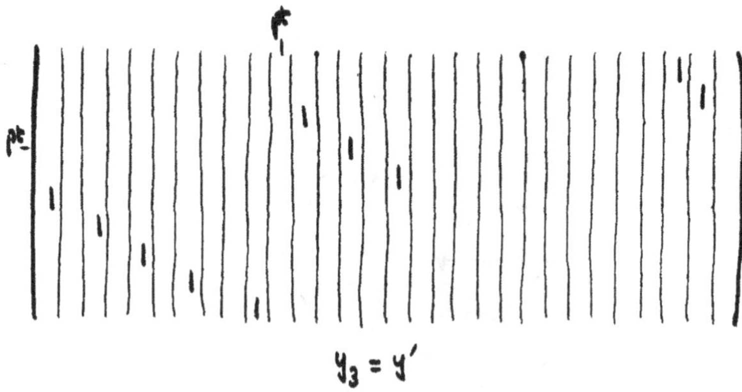

We now return to the general case. We see that y' has an entry set $\{e_{y'}(r+t, j_t(y')) \mid t \in [n]\}$ with $r = -\sum_{j=1}^{m}(m-j)\mu_j$ and $j_t(y') = j_t(y) - in$ if t satisfies $\sum_{j=m+2-i}^{m} \mu_j < t \leq \sum_{j=m+1-i}^{m} \mu_j$ for $1 \leq i \leq m$. Let $A_h = \{e_{y'}(r+t, j_t(y')) \mid$

$\sum_{j=h+1}^{m} \mu_j < t \leq \sum_{j=h}^{m} \mu_j\}$, $1 \leq h \leq m$. Then the A_h are maximum separated antichains of y' with $|A_h| = \mu_h$. So $\zeta(y')$ is a generalized tabloid. In particular, $\zeta(y') \in \hat{C}$ and so $\hat{T}(y') = \zeta(y')$. We have $\{\mu_1 \geq \ldots \geq \mu_m\} \leq \delta(y') \leq \delta(y) = \{\mu_1 \geq \ldots \geq \mu_m\}$. This implies $\delta(y') = \delta(y)$ and hence $y' \underset{L}{\sim} y$. By Proposition 19.4.8, we get $\hat{T}(y) = \hat{T}(y')$.

Now for a given $w \in P_n$, assume that y is obtained from w by the procedure given in §21.2 and assume that y' is obtained from y by the procedure given as above. Then $w \underset{L}{\sim} y'$. Let $\hat{T}(y') = X = (X_1, \ldots, X_m) \in \hat{C}$. We have $P_\ell(w) = X_\ell$ for $1 \leq \ell \leq m$. Let $\tilde{X}$ be the Young tableau associated to X. Then $\tilde{X}$ is the standard Young tableau P(w). Since $\hat{T}(w) = \hat{T}(y') = X$ by Proposition 19.4.8, we see that the map $\hat{T}: A_n \to \hat{C}$ satisfies all the required properties listed at the beginning of Chapter 19.

Now we can show the following criterion for a left cell of A_n having a non-empty intersection with P_n.

<u>Theorem 21.3.1</u> <u>For any left cell L of A_n, let $X = \hat{T}(L)$ and $\tilde{X}$ the associated Young tableau of X. Then L has a non-empty intersection with P_n if and only if $\tilde{X}$ is a standard Young tableau.</u>

<u>Proof</u> First assume $L \cap P_n \neq \emptyset$. Take $w \in L \cap P_n$. Then $\tilde{X} = P(w)$ is a standard Young tableau as showed above. Conversely, assume that X is a standard Young tableau. Then by the surjectivity of the Robinson-Schensted map, there exists an element $w' \in P_n$ satisfying $P(w') = \tilde{X}$. Let L' be the left cell of A_n containing w', $X' = \hat{T}(L')$ and $\tilde{X}'$ the associated Young tableau of X'. Then $\tilde{X}' = P(w')$, i.e. $\tilde{X}' = \tilde{X}$. This implies $X' = X$. Again by Proposition 19.4.8, we have L' = L and so L has a non-empty

intersection with P_n. □

We shall conclude this section by giving an example. Recall that in §1.1. we described an affine Weyl group as the set of alcoves of a Euclidean space. Thus the set of alcoves corresponding to A_3 is as in the following diagram.

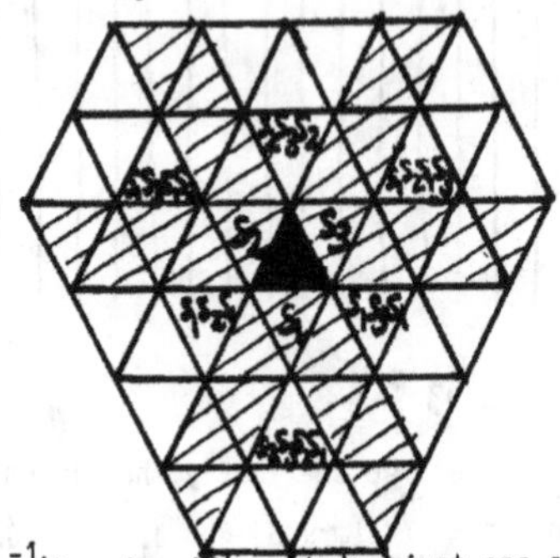

Then the two-sided cell $\sigma^{-1}(\{1 > 1 > 1\})$ contains just one solid alcove and consists of a unique left cell parametrized by the tabloid $\begin{matrix}1\\2\\3\end{matrix}$ whose associated Young tableau $1\ 2\ 3$ is a standard Young tableau. The two-sided cell $\sigma^{-1}(\{2 > 1\})$ consists of all shaded alcoves. Each connected component of $\sigma^{-1}(\{2 > 1\})$ is a left cell and so $\sigma^{-1}(\{2 > 1\})$ consists of 3 left cells. The left cells L_{s_t} containing s_t are parametrized by the tabloids $\begin{matrix}t & t+1\\ t+2 & \end{matrix}$, $1 \leq t \leq 3$, where the numbers in the tabloids are regarded as those in $\mathbb{Z}/3\mathbb{Z}$. Their associated Young tableaux are

1	3
2	

(t=1),

1	2
3	

(t=2),

2	3
1	

(t=3).

We see that the first two are standard Young tableaux but the last one is not. We also see that $L_{s_t} \cap P_3 = s_t$ for $t = 1,2$ and $L_{s_3} \cap P_3 = \emptyset$. Finally, the two-sided cell $\sigma^{-1}(\{3\})$ consists of all the remaining alcoves. Each connected component of $\sigma^{-1}(\{3\})$ is a left cell and so $\sigma^{-1}(\{3\})$ consists of 6 left cells. The left cells $L_{s_t s_{t+1} s_t}$ containing $s_t s_{t+1} s_t$ are parametrized by the tabloids $\boxed{t \mid t+1 \mid t+2}$. Their associated Young tableau are

1
2
3

(t=1),

2
3
1

(t=2),

3
1
2

(t=3)

Only the first one is a standard Young tableau. On the other hand, $L_{s_1 s_2 s_1} \cap P_3 = s_1 s_2 s_1$ and $L_{s_t s_{t+1} s_t} \cap P_3 = \emptyset$ for $t = 2,3$. The left cells $L_{s_t s_{t+1} s_t s_{t+2}}$ containing $s_t s_{t+1} s_t s_{t+2}$ are parametrized by the tabloids $t+2\ \ t+1\ \ t$. Their associated Young tableaux are $\begin{array}{|c|}\hline 3\\\hline 2\\\hline 1\\\hline\end{array}$ (t=1), $\begin{array}{|c|}\hline 1\\\hline 3\\\hline 2\\\hline\end{array}$ (t=2), $\begin{array}{|c|}\hline 2\\\hline 1\\\hline 3\\\hline\end{array}$ (t=3). None of them is a standard Young tableau. Clearly, $L_{s_t s_{t+1} s_t s_{t+2}} \cap P_3 = \emptyset$ for any t.

REFERENCES

[AL] Dean Alvis & G. Lusztig. The representations and generic degrees of the Hecke algebra of type H_4, J. Reine Angew. Math. 336 (1982), 201-212.

[A] H.H. Andersen. An inversion formula for the Kazhdan-Lusztig polynomials for affine Weyl groups, Adv. in Math.

[BV1] D. Barbasch & D. Vogan. Primitive ideals and orbital integrals in complex classical groups, Math. Ann. 259 (1982), 153-199.

[BV2] ——————————. Primitive ideals and orbital integrals in complex exceptional groups, J. of Alg. 80 (1983), 350-382.

[BB] A.A. Beilinson & J. Bernstein. Localisation des $\underline{g}$-modules, C.R. Acad. Sci. Paris t. 292 (1981), Serie I, 15-18.

[Bor] A. Borel. Properties and linear representations of Chevalley groups, Lecture Notes in Math. 131. Springer-Verlag, Berlin, 1970.

[BJ] W. Borho & Jens C. Jantzen. Über primitive Ideale in der Einhüllenden einer halbeinfachen Lie-Algebra, Invent. Math. 39 (1977), 1-53.

[Bou] N. Bourbaki. Groupes et algèbres de Lie, Ch. 4-6, Hermann, Paris, 1968.

[BK] J.L. Brylinski & M. Kashiwara, Kazhdan-Lusztig conjecture and holonomic systems, Invent. Math. 64 (1981), 387-410.

[Ca1] R.W. Carter. Simple groups of Lie type, John Wiley, London (1972).

[Ca2] —————. Finite groups of Lie type: conjugacy classes and complex characters, Wiley series in pure and applied mathematics, 1985.

[Cu] Curtis, C.W. Modular representations of finite groups with split BN-pairs, Lecture Notes in Math. 131, Springer-Verlag, Berlin, 1970.

[CIK] Curtis, C.W., N. Iwahori & R. Kilmoyer. Hecke algebras and characters of parabolic type of finite groups with BN-pairs, Publ. Math. IHES, 40 (1972), 81-116.

[De] P. Deligne. Letter to D. Kazhdan and G. Lusztig, 20 April 1979.

[Du] M. Duflo. Sur la classification des ideaux primitifs dans l'algèbre enveloppante d'une algèbre de Lie semi-simple, Ann. of Math. 105 (1977), 107-120.

[Go] R.M. Goresky, Kazhdan-Lusztig polynomials for classical groups. Northeastern Univ.

[GM] R.M. Goresky & R. Macpherson. Intersection homology theory, Topology, 19 (1980), 135-162.

[Gr1] C. Greene. An extension of Schensted's theorem, Adv. in Math., 14 (1974), 254-265.

[Gr2] —————. The structure of Sperner k-families, J. of Combinatorics Theory. (A)20 (1976), 41-68.

[Gr3] —————. Some partitions associated with a partially ordered set, J. of Combinatorics Theory (A)20 (1976), 69-79.

[HC] Harish-Chandra. Some applications of the universal enveloping algebra of a semi-simple Lie algebra, Trans. Amer. Math. Soc. 70 (1951), 28-96.

[Hi] H.L. Hiller. Geometry of Coxeter groups, Research Notes in Math. 54 (1982), Pitman.

[Hu1] J.E. Humphreys. Introduction to Lie algebras and representation theory (GTM9) (1972), Springer.

[Hu2] ——————. Modular representations of finite groups of Lie type (Finite Simple Groups II, Academic Press, London, 1980, 259-290.)

[I] N. Iwahori. On the structure of the Hecke ring of a Chevalley group over a finite field, J. Fac. Sci. Univ. Tokyo Sect. 1A. Math. 10 (part 2) (1964), 215-236.

[JK] G. James & A. Kerber. The representation theory of the symmetric group, Encyclopaedia of Mathematics and its Applications, Vol. 16, G.C. Rota ed., Addison-Wesley, Reading, 1981.

[Ja] Jens C. Jantzen. Weyl modules for groups of Lie type. (Finite Simple Groups II, Academic Press, London, 1980, 291-300).

[Jo1] A. Joseph. Goldie rank in the enveloping algebra of a semisimple Lie algebra, J. of Algebra 65 (1980), 269-283; 65 (1980), 284-306; 73 (1981), 295-326.

[Jo2] ————. On the classification of primitive ideals in the enveloping algebra of a semisimple Lie algebra, Lecture Notes in Math., No. 1024, Springer-Verlag, Berlin (1983), 30-76.

[Jo3] ————. W-module structure in the primitive spectrum of a semisimple Lie algebra. **Lecture Notes in Mathematics 728 (1979), 116-135. Springer.**

[Kac] V.G. Kac. Infinite dimensional Lie algebras. Progress in Math. Vol. 44 (1984).

[Kat] S. Kato. A realization of irreducible representations of affine Weyl groups, Proc. Kon. Nederl. Akad. A 86(2) (1983), 193-201.

[KL1] D. Kazhdan & G. Lusztig. Representations of Coxeter groups and Hecke algebras, Invent. Math. <u>53</u> (1979), 165-184.

[KL2] ——————————. Schubert varieties and Poincaré duality, Proc. Symp. Pure Math. Vol. <u>36</u>, 185-203. Amer. Math. Soc. 1980.

[L1] G. Lusztig. A class of irreducible representations of a Weyl group, Indag. Math. 41 (1979), 323-335.

[L2] ————. Some problems in the representation theory of finite Chevalley groups, Proc. Symp. Pure Math. (A.M.S), 37 (1980), 313-317.

[L3] ————. Hecke algebras and Jantzen's generic decomposition patterns, Adv. in Math. 37 (1980), 121-164.

[L4] ————. On a theorem of Benson and Curtis, J. of Algebra, 71 (1981), 490-498.

[L5] ————. Green polynomials and signularities of unipotent classes, Adv. in Math. Vol. 42, No. 2 (1981), 169-178.

[L6] G. Lusztig. A class of irreducible representations of a Weyl group II, Indag. Math. 44 (1982), 219-226.

[L7] ———. Singularities, character formulae and a q-analog of weight multiplicities, Astérisque, Vol. 101-102 (1983), 208-229.

[L8] ———. Left cells in Weyl groups, Lecture Notes in Math. 1024 (1983), Springer.

[L9] ———. Some examples of square integrable representations of semisimple p-adic groups. Trans. Amer. Math. Soc. 277 (1983), 623-653.

[L10] ———. Characters of reductive groups over a finite field, Annals of Math. Studies 107 (1984) Princeton Univ. Press.

[L11] ———. Intersection cohomology complexes on a reductive group. Invent. Math. 75 (1984), 205-272.

[L12] ———. Cells in affine Weyl groups, to appear in Proceedings of the International Symposium on Algebraic Groups, Katata (Japan) 1983.

[L13] ———. The two-sided cells of the affine Weyl group of type $\tilde{A}_n$, Preprint.

[LV] G. Lusztig & D. Vogan. Singularities of closures of K-orbits on flag manifolds, Invent. Math. 71 (1983), 365-379.

[M] I.G. Macdonald. Symmetric functions and Hall polynomials, Oxford Univ. Press (Clarendon), Oxford 1979.

[R1] G. de B. Robinson. On the representations of the symmetric groups, Amer. J. Math. 60 (1938), 745-760; 69(1947), 286-298; 70(1948), 277-294.

[R2] ———. Representation theory of the symmetric groups, Univ. of Toronto Press, Toronto, 1961.

[Sch] C. Schensted. Longest increasing and decreasing subsequences, Canad. J. Math. 13 (1961), 179-191.

[St] R. Steinberg. Representations of algebraic groups, Nagoya Math. J. 22 (1963), 33-56.

[Verd] J.L. Verdier. A duality theorem in the étale cohomology of schemes, Proceedings of a Conference on local fields. Driebergen. Berlin-Heidelberg-New York, Springer 1967.

[Verm] D.N. Verma. The role of affine Weyl groups in the representation theory of algebraic Chevalley groups and their Lie algebras, in I.M. Gelfand (ed.), Lie groups and their representations, London 1975, 653-722.

[Vie1] G. Viennot. Quelques algorithmes de permutations, in Journées algorithmiques, Astérisque no. 38/39, Soc. Math. de France, 1976, 275-293.

[Vie2] ———. Chain and antichain families, grids and Young tableaux, Univ. de Bordeaux I. Analyse Appliquée et Informatique. no. 8307 (1983).

[Vo1] D. Vogan. A generalized τ-invariant for the primitive spectrum of a semisimple Lie algebra, Math. Ann. 242 (1979), 209-224.

[Vo2] ———. Ordering in the primitive spectrum of a semisimple Lie algebra, Math. App. 248 (1980), 195-203.

INDEX OF NOTATION

Section of definition	Symbol
§2.4	$\mathcal{V}_\lambda$
	$V_{\dot\lambda}$
	$w.\lambda$
	ChE
	C_1
§2.5	J_λ
	$\mathfrak{X}_\lambda$
	$\mathfrak{X}$
	σ_w
§2.6	$V(\lambda)$
	$L(\lambda)$
	$[V(\lambda):L(\mu)]$
	$[L(\mu):V(\lambda)]$
	A'_w
§2.7	w_λ
	$d_\mu(V_\lambda)$
	$\bar{\Lambda}_n$
	$\lambda > \mu (\lambda,\mu \in \bar{\Lambda}_n)$
	$K_{\lambda,\mu}(t)$
Ch. 3	$H^i_{\alpha_y}(\bar{B}_w)$
	$H^i_{z_\mu}(\bar{Z}_\lambda)$
Ch. 4	
§4.1	A_n
	$\underline{n}$
	Δ
§4.3	$\mathcal{D}_L(s_t)$
	$\mathcal{D}_R(s_t)$
§4.4	$r(\ ,\ )$
	$c(\ ,\ )$
	Λ_n
	$c_k(P)$
	$a_k(P)$
	$(A_\ell, A_{\ell-1}, \dots, A_1)$

Section of definition	Symbol
	C
	$\hat{C}$
	$\check{C}$
Ch. 5	$\underline{\Delta}$
	$\Pi:\underline{\Delta} \to \Lambda_n$
(6.3)	$\sigma:A_n \to \Lambda_n$
	W_J
	w^J_o
	$\delta:A_n \to \Lambda_n$
Ch. 6	
§6.1	$H_{ij,k}$
	$H^m_{ij;k}$
	$\mathbb{A}$
	$\bar{w}(w \in \mathbb{A}_n)$
	k^w_{ij}
§6.2	r^w_{ij}
Ch. 7	
§7.1	$\bar{S}$
	S
	$\zeta:\mathbb{A} \to S$
	$\hat{\sigma}:S \to \Lambda_n$
	$R(X)\ (X \in S)$
§7.2	$C(S)$
	$X \underset{L}{\sim} Y (X,Y \in S)$
	$\tau:S \to C(S)$
	$\bar{\sigma}:C(S) \to \Lambda_n$
Ch. 8	
§8.1	$x \xleftarrow{*(i+1,r)} y$
	$x \xleftarrow{*(i+1,r,m)} y$
	$x \xrightarrow{*(i+1,r)} y$
	$x \xrightarrow{*(i+1,r,m)} y$

Section of definition	Symbol
§8.3	$\rho^{A_1}_{A_2}(w)\ (w \in A_n)$
	$\theta^{A_2}_{A_1}(w)\ (w \in A_n)$
	$(a_1,\dots,a_\ell)^{(om)}$
	$(a_1 < \dots < a_\ell)^{(om)}$
§8.4	$\rho^{A_j,\dots,A_1}_{A_{j+1},\dots,A_\ell}(w)(w \in A_n)$
	$\theta^{A_{j+1},\dots,A_\ell}_{A_j,\dots,A_1}(w)(w \in A_n)$
	$\xi(w,\rho^{A_j,\dots,A_1}_{A_{j+1},\dots,A_\ell})$
Ch. 9	
§9.2	F
	F_i
	$\phi_i : F_i \to \Lambda_n$
(§11.3,§19.4)	$\Phi : A_n \to A_n$
§9.3	H_λ
Ch. 10	N_λ
	$\bar{N}_\lambda$
Ch. 11	
§11.1	$\bar{A}_n$
§11.2	$\eta : A_n \to \tilde{A}_n$
§11.3	$\tilde{\sigma} : \tilde{A}_n \to \Lambda_n$
§11.4	$\ell(\tilde{w})\ (\tilde{w} \in \tilde{A}_n)$
	$\mathcal{L}(\tilde{w})\ (\tilde{w} \in \tilde{A}_n)$
	$\mathcal{R}(\tilde{w})\ (\tilde{w} \in \tilde{A}_n)$
§11.5	$\rho^{A_j,\dots,A_1}_{A_{j+1},\dots,A_\ell}(\tilde{w})(\tilde{w} \in \tilde{A}_n)$
	$\theta^{A_{j+1},\dots,A_\ell}_{A_j,\dots,A_1}(\tilde{w})(\tilde{w} \in \tilde{A}_n)$
§11.6	$\{\mathcal{R}, \langle\ ,\ \rangle_{\mathcal{R}}\}$
	$(\xi,\zeta) \sim (\xi',\zeta')$
Ch. 12	
§12.2	$H^m_{\lambda,r}$
	$\xi(w,r)\ (w \in H_\lambda)$
§12.3	$\tilde{F}(\lambda,m)$
	$d(\lambda,m) : \tilde{F}(\lambda,m) \to \tilde{A}_{n-rm}$
§12.4	$\tilde{N}_\lambda$
	$\tilde{H}_{\lambda,k}$
§12.5	$\xi(\tilde{w},k)\ (\tilde{w} \in \tilde{N}_\lambda)$
§12.6	$\xi(w,k)\ (w \in N_\lambda)$
§12.7	$\bar{e}((x),E_u)$
§12.8	$D^{x,A(x)}_{y,B(x)}(E)$
Ch. 13	
§13.2	$L_E(w)\ (w \in A_n)$
	$L^E(w)\ (w \in A_n)$
Ch. 14	
§14.1	$\check{\mathcal{C}}_\lambda$
	$T : N_\lambda \to \check{\mathcal{C}}_\lambda$
	$\bar{T} : \bar{N}_\lambda \to \check{\mathcal{C}}_\lambda$
§14.2	$A(X,\alpha)$
§14.3	$X^\vee$
	X_λ
§14.4	$N_\lambda(a)$

Section of definition	Symbol
Ch. 15	H_λ^{-1}
Ch. 16	P_n
Ch. 19	N
	$T:A_n \to \check{C}$
	$\tilde{X}$ $(X \in C)$
(§19.4)	$\hat{T}:A_n \to \hat{C}$
§19.1	Ω
	$\xi:\Omega \to C$
	$d(w)$ $(w \in \zeta^{-1}(C))$
§19.2	$\tilde{\Lambda}_n$
	$f:C \to \tilde{\Lambda}_n$
	$\phi:\tilde{\Lambda}_n \to \Lambda_n$
§19.3	$\bar{D}_R(s_t)$
	$T^*(T \in \bar{D}_R(s_t))$
§19.4	$N(a)$
§19.5	S^{min}
§19.6	$\hat{H}_\lambda$, $\hat{H}_\lambda^{-1}$
	$H_{\lambda,\tau,k}$, $H_{\lambda,\tau,k}^{-1}$
	$\hat{H}$, $\hat{H}^{-1}$
	S^{max}
	C'
Ch. 20	
§20.1	$H_\gamma(X)$
	$L_\gamma(X)$
Ch. 21	
§21.1	$x \xrightarrow[*(i+1,\alpha)]{} y$
	$x \xrightarrow[*(i+1,\alpha,m)]{} y$
§21.3	$x \xrightarrow{(i+1,k,h)} y$

INDEX

Section of definition